TEUBNER-TEXTE zur Physik Band 33

M. Bonitz

Quantum Kinetic Theory

TEUBNER-TEXTE zur Physik

This regular series includes the presentation of recent research developments of strong interest as well as comprehensive treatments of important selected topics of physics. One of the aims is to make new results of research available to graduate students and younger scientists, and moreover to all people who like to widen their scope and inform themselves about new developments and trends.

A larger part of physics and applications of physics and also its application in neighbouring sciences such as chemistry, biology and technology is covered. Examples for typical topics are: Statistical physics, physics of condensed matter, interaction of light with matter, mesoscopic physics, physics of surfaces and interfaces, laser physics, nonlinear processes and selforganization, ultrafast dynamics, chemical and biological physics, quantum measuring devices with ultimately high resolution and sensitivity, and finally applications of physics in interdisciplinary fields.

Quantum Kinetic Theory

Von Dr. Michael Bonitz
Universität Rostock

B. G. Teubner Stuttgart · Leipzig 1998

Dr. rer. nat. habil. Michael Bonitz

Born in 1960 in Leningrad. Studied physics at the Lomonosov University in Moscow. Masters degree in 1987 under Yu. L. Klimontovich with a work on structure formation far from equilibrium. Since 1987 at Rostock University. PhD in 1991 under D. Kremp with a thesis devoted to quantum kinetic foundation of reaction-diffusion equations for dense plasmas. In 1992/93 postdoc at the University of Arizona under S. W. Koch, research on plasmons and optical properties of semiconductor structures.

Research interests: nonequilibrium processes in dense plasmas and solids, laser-matter interaction, many-body theory.

micha@elde.mpg.uni-rostock.de
http://elde.mpg.uni-rostock.de/contrib/micha

Gedruckt auf chlorfrei gebleichtem Papier.

Die Deutsche Bibliothek – CIP-Einheitsaufnahme

Bonitz, Michael:
Quantum kinetic theory / von Michael Bonitz. –
Stuttgart ; Leipzig : Teubner, 1998
(Teubner-Texte zur Physik ; Bd. 33)
ISBN 3-519-00238-8

Printed in Germany
Satz und Druck: Druckhaus „Thomas Müntzer“ GmbH, Bad Langensalza
Umschlaggestaltung: E. Kretschmer, Leipzig

Preface

We are at the beginning of a new revolution in science and technology, which is fueled by the tremendous progress in short–pulse laser technology over the last 10–15 years. Fundamental physical principles which have been known before from abstract theory only, are suddenly becoming accessible to direct experimental observation. These includes the quantum states of single atoms, Heisenberg's uncertainty principle, low temperature phenomena such as Bose condensation and Wigner crystallization, as well as ultra–intense electromagnetic fields and the possiblity to create, in small university facilities, relativistic electrons or hot dense plasmas which eventually will lead to thermonuclear fusion. In view of the technological applications promised by these developments, many countries have established special research projects, including, the National Ignition Facility in the U.S. or the new Schwerpunkt "Laserfelder" of the Deutsche Forschungsgemeinschaft in Germany.

These developments are a major challenge for Theoretical Physics – to understand and predict the interaction of matter with electromagnetic fields ranging from milliwatt to petawatt (10^{-3}W-10^{15}W) powers and lasting from seconds to femtoseconds (10^{-15}s). Within this very complex frame, the current monograph is devoted to *quantum many-particle systems* under *extreme nonequilibrium* conditions. It attempts to answer the question how these systems behave immediately after and also during their creation, thereby focusing on a general approach.

The statistical description of many–particle systems in nonequilibrium began with L. Boltzmann's famous kinetic equation [Bol72]. Since then, numerous theoretical methods have been developed to describe nonequilibrium processes in various fields, including fluids, dense plasmas, solids and nuclear matter, which are often very efficient but, at the same time, so specialized that they are understood only by a few immediate specialists. And this, despite the fact that the underlying physics is often very similar if it is uncovered from the shell of (certainly needed) field–specific jargon and notation as well as system–inherent details and parameters. Moreover, many of these specialized theories have common roots, and a return to them, from time to time, can be extremely fruitful, even for an experienced specialist, and lead to new ideas. On the other hand, for students or researchers new in the field, the best choice is to start from the roots.

This is particularly true for the problems related to ultrafast relaxation. Here exist three main branches of the theory: the two *statistical methods* based on density operators and nonequilibrium Green's functions, respectively, and the *mechanical approach* of Molecular Dynamics. All three of them are discussed and compared in this book. The simplest and most intuitive one is the *density operator theory* which, for this reason, is chosen as the basic concept. A very general and comprehensive introduction to this approach is given in Chapters 2-3. First applications to many-particle effects are given in Chs. 4–6, to collective phenomena (plasmons and instabilities) and to correlations. Students with *basic knowledge in quantum mechanics and statistical physics* will have no problems in following this treatment.[1]

More advanced applications follow in Chs. 7-11. It is demonstrated that the density operator approach is very efficient in deriving generalized non-Markovian *quantum kinetic equations* with *memory* and *initial correlations*, and that it further allows to incorporate selfenergy, dynamical screening, strong coupling effects, as well as the interaction with electromagnetic fields.

An introduction to the method of *nonequilibrium Green's functions* is given in Ch. 12, starting from a fully relativistic formulation. Here, the main results are coupled *Kadanoff-Baym equations for carriers and photons* for which various approximations are considered, including the non-relativistic limit and the extension to multi-band systems. Moreover, *recent numerical results* are presented which will be of interest also for experienced readers. Finally, Ch. 13 is devoted to the *Molecular dynamics* approach, where also modern developments in the field quantum MD are discussed. Furthermore, a detailed comparison of the concepts of quantum statistics and Molecular dynamics is performed, and the idea of numerical comparisons of the two is developed.

For illustration of the theoretical methods, a variety of recent numerical results on quantum kinetic equations, non-Markovian effects and correlation phenomena have been included. *Numerical analysis* is an important part of modern kinetic theory. Therefore, throughout this book, the theoretical results are cast in a form best suitable for numerical evaluation. Furthermore, to enable the reader to do numerical work by himself, an extensive introduction

[1] Sections which are not necessary for the understanding of the basic concepts and which may be skipped on first reading are marked with an asterisk. Readers interested in derivations and details beyond the basic material, can find the main ideas outlined in footnotes and Appendices. Extensive references are provided to the specialized literature of various fields and to classical works as well.

to the concepts of numerical solution of quantum kinetic equations is supplied in Appendix F.

Thus, I hope the reader will *gain broad fundamental knowledge* in quantum kinetic theory in general, and on the theoretical description of ultrafast relaxation in particular, which should allow him to creatively adapt these concepts to any field of many-particle physics.

Physics would be not even half as exciting and rewarding without continuous discussions and sharing of ideas with colleagues - so I am grateful to Martin Axt, Lazi Banyai, Karim ElSayed, Hartmut Haug, Klaus Henneberger, Frank Jahnke, Tilman Kuhn, Günter Manzke, Klaus Morawetz, Ronald Redmer, Hartmut Ruhl, Wilfried Schäfer, Chris Stanton, Heinrich Stolz and Günter Zwicknagel. I very much enjoyed working together with Rolf Binder, Thomas Bornath, Jim Dufty, Andreas Förster, Dirk Gericke, Yuri L'vovich Klimontovich, Stephan Koch, Sigurd Köhler, Sylvio Kosse, Wolf Kraeft, Dietrich Kremp, Nai Kwong, Thomas Ohde, Manfred Schlanges, Don Scott and Dirk Semkat. It is my great pleasure to thank my remarkable teachers Yuri L'vovich Klimontovich, Dietrich Kremp and Stephan Koch, who guided me through various fields of physics and life, and Werner Ebeling and Wilfried Schäfer for continuous encouragement of my work.

The results which are the basis for this book would not have been possible without the generous support from the Deutscher Akademischer Austauschdienst, grants from the Deutsche Forschungsgemeinschaft and grants for CPU–time at the HLRZ Jülich and the CCIT of the University of Arizona. This book greatly benefited from comments of Nai Kwong and Dirk Semkat who also assisted me in preparing several figures for Ch. 12, as did Renate Nareyka and my wife.

Finally, I thank my father for raising my interest in physics, Christine for her tremendous patience and encouragement and Sebastian and Martin for their willingness to give up our PC for this book.

Rostock, June 1998 Michael Bonitz

Contents

Used symbols and definitions

Symbol	Meaning	Definition
$\mathbf{A}$	Vector potential	
$d^{\gtrless}$	Nonequilibrium photon Green's (correlation) function	Ch. 12
$\mathcal{E}$	Electric field	
$\epsilon^{\pm}$	Retarded/advanced dielectric function	Eq. (4.21)
$\epsilon_{\alpha\beta}$	Dielectric tensor	Eq. (4.36)
$f_{1\ldots s}$	$s-$particle Wigner distribution	Eq. (2.51)
$F_{1\ldots s}$	Reduced $s-$particle density operator	Eq. (2.13)
$F_a^{\gtrless}$	One-particle/hole operator	Eq. (3.18)
ϕ	Scalar potential	
$g_{1\ldots s}$	$s-$particle correlation operator	Eq. (2.98)
$g^{\gtrless}$	Nonequilibrium particle Green's (correlation) function	Ch. 12
$H_{1\ldots s}$	$s-$particle Hamilton operator	Eq. (2.3)
H_i	One-particle Hamilton operator of particle i	Eq. (2.4)
$\bar{H}_i$	Effective one-particle Hamiltonian (contains U^H)	Eq. (2.100)
H_i^{HF}	Hartree-Fock potential	Eq. (3.9)
$\langle H \rangle$	Total energy	Eq. (2.34)
$\Lambda_{1\ldots s}^{\pm}$	$s-$particle (anti-)symmetrization operator	Eq. (3.6)
μ	Chemical potential	
n	Number density	
$\Omega^{\pm}$	Retarded/advanced Møller operator	Eq. (8.30)
$P_{1\ldots s}$	$s-$particle permutation operator	Chapter 3
$\mathcal{P}$	Principal value	App. A
$\Pi^{\pm}, \Pi^{R/A}$	Retarded/advanced polarization function	Eq. (4.18)
$\rho_{1\ldots N}$	N-particle density operator	Eq. (2.8)
$\Sigma_i^{\pm}$	Retarded/advanced one-particle selfenergy	Eq. (7.83)
$\sigma^{\gtrless}$	Particle selfenergy	Ch. 12

$T^{\pm}$	Retarded/advanced T-operator	Eq. (8.33)
$\langle T \rangle$	Mean kinetic energy	Eq. (2.25)
$\mathcal{U}_i$	External potential acting on particle i	Eq. (2.4)
$U^{\pm}$	Retarded/advanced propagator	Appendix D
$U^{0\pm}$	Retarded/advanced quasi-particle propagator	Appendix D
U_i^H	Hartree field	Eq. (2.103)
$\mathcal{V}$	System volume	
V_{ij}	Binary interaction potential	Eq. (2.5)
$V_{ij}^{\pm}$	Potential with exchange contribution	Eq. (3.9)
$\langle V \rangle$	Mean potential energy	Eq. (2.26)

Chapter 1

Introduction

Subject of this book. This book is devoted to quantum systems of *many particles* in *nonequilibrium.* More precisely, we will be interested in *many-body effects* (collective and correlation effects) and, in particular, how these effects show up on *very short time scales.* What means "short" depends on the actual system, but also on the observer. For us, "short" and "ultrafast" will refer to the initial stage of relaxation, to *times shorter than the correlation time* of the system, $t < \tau_{cor}$, where the conventional statistical description, the traditional kinetic theory fails.

Why are these initial or transient processes of interest? The reason is the recent remarkable progress in short pulse lasers, which allow to excite and to probe many-particle systems during an extremely short time which is often comparable or even shorter than the correlation time. This yields deep insight into the behavior of matter under conditions very far from equilibrium which have not been accessible for systematic quantitative analysis before. For example, it is becoming possible to follow in detail the formation of a plasma, including the buildup of the screening cloud and of the correlations between the charge carriers.

Theoretical studies of ultrafast processes are only in their beginning, and only recently systematic numerical investigations became possible. So this book discusses first results and, more importantly, outlines the different theoretical approaches to ultrafast relaxation phenomena: the density operator formalism, nonequilibrium Green's functions and Molecular Dynamics techniques. Our main focus will be on the first, which will be used to derive generalized quantum kinetic equations, which are applicable to correlated many-particle systems in general and to the initial stage of relaxation in particular.

Interestingly, related problems have been discussed already as early as in the 1950ies by a number of authors in kinetic theory, plasma physics, fluid and condensed matter theory or nuclear matter. While these had to be purely theoretical studies, many contain brilliant concepts which are worth to be reconsidered today, including possible extension beyond their original field of application.

1.1 Correlated many-particle systems. Many-body effects

Many-body effects. Before considering ultrafast relaxation phenomena in correlated systems, we briefly discuss the many-body effects which govern their properties in equilibrium as well as in nonequilibrium. These are effects resulting from the mutual interaction of a large number ($\gg 1$) of particles in the system. Their strength is naturally measured by the **nonideality parameter** Γ_a - the ratio of mean potential to mean kinetic energy of particle species "a",

$$\boxed{\Gamma_a = \frac{|\langle V_a \rangle|}{\langle T_a \rangle}; \quad \langle T_a \rangle_{cl}^{EQ} = \frac{i_a}{2} k_B T; \quad \langle T_a \rangle_{q}^{EQ} = \frac{i_a}{2} k_B T \frac{I_{3/2}(\mu_a^{id}/k_B T)}{I_{1/2}(\mu_a^{id}/k_B T)}} \tag{1.1}$$

where $\langle T_a \rangle_{cl}^{EQ}$ and $\langle T_a \rangle_{q}^{EQ}$ denote the classical and quantum kinetic energy in equilibrium, i_a is the number of degrees of freedom (which equals three for free elementary particles), I_ν is the Fermi integral (see Appendix A), and μ_a^{id} the ideal chemical potential.[1] For $\Gamma \to 0$, the system is ideal, while for $\Gamma < 1 (> 1)$ it is weakly (strongly) nonideal. The line $\Gamma = 1$ gives a qualitative boundary to the region in the density-temperature plane, where nonideality effects are important, see Fig. 1.1. Obviously, many-particle systems are ideal at sufficiently high temperatures, but also at very low densities, because there the mean interparticle distance is large and $\langle V_a \rangle \to 0$. On the other hand, ideal behavior is restored at very high densities (independently of temperature), because in that case, the (quantum) kinetic energy increases more rapidly than the potential energy. For example, for systems with Coulomb interaction[2] $V_{ab} = e_a e_b / \epsilon_b r$, ($\epsilon_b$ is the dielectric constant of the medium surrounding the charges e_a and e_b), and the upper part of the curve $\Gamma = 1$ can be written as the following condition on the **Brueckner parameter** r_s which is the ratio of the mean interparticle distance to the Bohr radius [KKER86]

$$\boxed{r_s \equiv \frac{d}{a_B} \approx 0.7; \quad \frac{4\pi}{3} d^3 = \frac{1}{n}; \quad a_B^H = \frac{\epsilon_b \hbar^2}{e_a e_b m_{ab}}} \tag{1.2}$$

[1] The ideal chemical potential is given by $\chi_a = I_{1/2}(\mu_a^{id}/k_B T)$, where χ_a is the degeneracy parameter (1.3).

[2] Throughout this book, we use *Gaussian units*.

where a_B^H corresponds to hydrogen-like bound states with the reduced mass $m_{ab} = m_a m_b/(m_a + m_b)$. For high densities such that $r_s < 1$, free particles come closer than the Bohr radius, what leads to the break up of bound states.

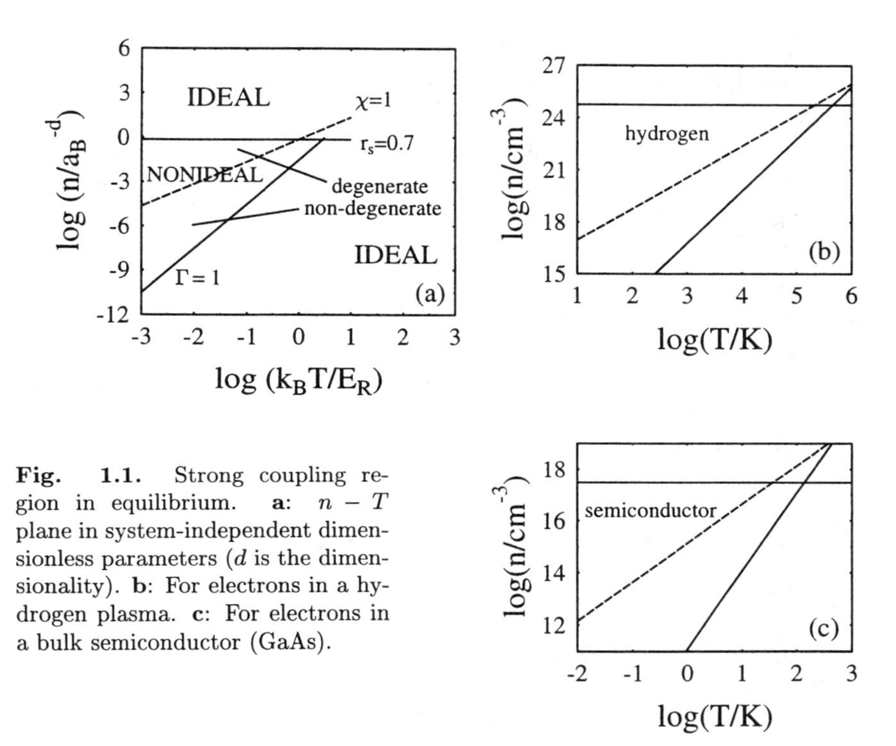

Fig. 1.1. Strong coupling region in equilibrium. **a**: $n - T$ plane in system-independent dimensionless parameters (d is the dimensionality). **b**: For electrons in a hydrogen plasma. **c**: For electrons in a bulk semiconductor (GaAs).

It is this "*corner of correlations*" where the structure of matter differs from the trivial state of nearly independent elementary particles. This region contains all cooperative phenomena from simple bound states such as atoms, excitons or nuclei, to living organisms. There are two main kinds of many-particle effects: (1) collective effects, which correspond to "synchronized" excitation of many particles, such as plasma oscillations, waves and instabilities and (2), correlation effects which usually involve particles which are located at distances smaller than the range r^{int} of the binary interaction potential. Examples for the latter are polarization and screening, correlation induced renormalization of single particle energies and two-particle states, including pressure ionization (Mott effect) and so on. Correspondingly, the potential energy splits into two parts, $\langle V_a \rangle = \langle V_a^{\rm coll} \rangle + \langle V_a^{\rm corr} \rangle$.

Another important quantity for the characterization of matter is the **de-**

generacy parameter χ

$$\boxed{\chi_a = \frac{n_a \Lambda_a^3}{2s_a+1}; \quad \Lambda_a^2 = \frac{h^2}{2\pi m_a k_B T}} \tag{1.3}$$

where Λ_a is the De Broglie wave length of particle "a". For $\chi > 1$, (above the line $\chi = 1$ in Fig. 1.1), the interparticle distance is smaller than Λ, i.e. particles feel their wave nature, and the system is essentially quantum or degenerate. Notice that these parameters are different for different particle species. The degeneracy parameter sensitively depends on the particle mass, heavier particles become degenerate only at higher densities (lower temperatures) than light particles. On the other hand, the coupling parameter Γ increases rapidly for multiply charged particles (e.g. ions). Furthermore, Γ is reduced for particles embedded into a medium, due to the dielectric properties of the latter ($\epsilon > 1$), which is the case for ions in a liquid (electrolyte) or electrons in a solid. This leads to a shift of the whole correlation region as well as the degeneracy line in the $n - T$ plane for different systems. A more or less invariant picture[3] is obtained in terms of dimensionless parameters, cf. Fig. 1.1.

1.2 Equilibrium properties of correlated systems

Although equilibrium properties are not our subject, it is important to understand the effect of correlations first for this simplest case. In a correlated system, the ground state as well as the thermodynamic equilibrium state may be strongly modified compared to an ideal gas. In particular, the thermodynamic functions, such as free energy, internal energy, chemical potential, pressure and so on contain additional interaction contributions, e.g. $F = F^{id} + F^{corr}$, $\mu = \mu^{id} + \mu^{corr}$. Fig. 1.2. shows, as an example, the chemical potential of electrons and protons in a partially ionized hydrogen plasma. At low temperatures, correlation effects lead to a lowering of the chemical potential and eventually to monotonically decreasing (as a function of carrier density) regions which may be related to a phase transition (plasma phase transition) [NS68], for recent discussions, see e.g. [KKER86, SBT95].

Furthermore, correlation effects may have a drastic impact on the equation of state $p = p(n,T) = p^{id} + p^{cor}$, as well as on the chemical composition of partially ionized or partially dissociated systems, where the chemical composition follows from the condition of equal chemical potentials of the reactants.

[3]The curve $\chi = 1$ depends, in addition to E_R and a_B, on the material specific ratio of particle mass to reduced mass.

For example, the detailed balance in the ionization/recombination of atomic hydrogen as well as molecular dissociation is given by $\mu_e + \mu_p = \mu_H$, and $2\mu_H = \mu_{H_2}$ where all chemical potentials contain correlation contributions. This leads to the mass action law (Saha equation) for nonideal systems which is readily solved numerically for the density of free electrons (degree of ionization) [SBT95]. Results for hydrogen are shown in Fig. 1.2. The formation of atoms and molecules at intermediate densities is an obvious correlation effect. Bound states vanish again at high densities due to screening induced lowering of the binding energy (Mott effect). Similar effects are found in partially ionized electron-hole-exciton plasmas in semiconductors, e.g. [EKK76, Zim87], and in nuclear matter.

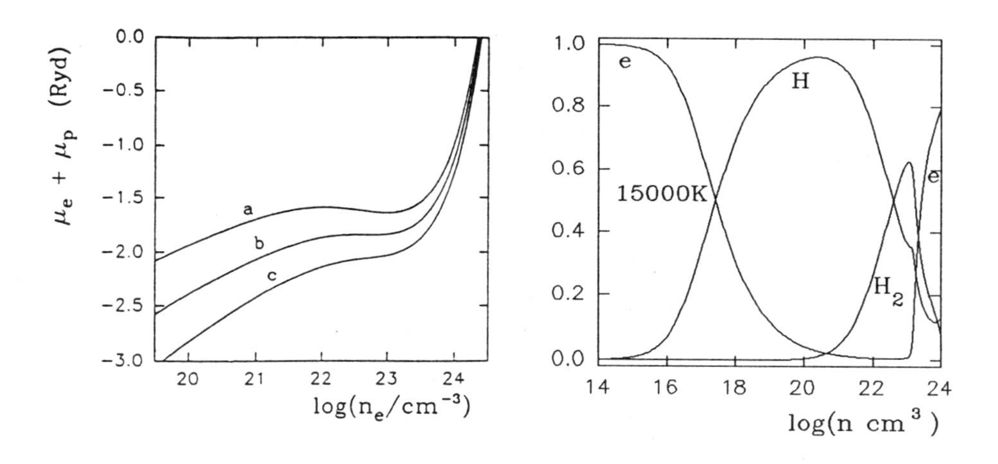

Fig. 1.2. Chemical composition of partially ionized hydrogen. **Left Fig.**: Correlated equilibrium chemical potentials of electrons and protons vs. free electron density for $T = 14000K$ (a), $17000K$ (b) and $20000K$ (c). The minimum is due to correlations, ($1Ryd = E_H = 13.6eV$). **Right Fig.**: Fractions of electrons which are free or bound in atoms or molecules vs. total electron density. The region of bound states indicates the influence of correlations. At high densities the *effective binding energy* I^{eff} of atoms and molecules decreases due to screening, and bound states are no longer stable, from Ref. [SBT95].

Theoretical approaches to equilibrium properties. Here, we list important theoretical methods which were developed to describe the equilibrium properties of correlated many-particle systems.

1. Quantum statistics, field-theoretical concepts, equilibrium Green's functions theory, diagrammatic expansions. For textbook discussions of these methods, see [AGD62, KB89, EKK76, KKER86];

2. Density operators, equilibrium correlation functions [Bog46], see Sec. 2.7;

3. Variational approaches, in particular Density functional theory of Kohn and co-workers [KL57] which is very successful in many fields of statistical physics, including very complex systems;

4. Integral equation techniques - these techniques have been developed in the theory of classical fluids (electrolytes) and have proved to be very efficient also in partially ionized dense plasmas;

5. Path integral methods, see the classical monograph of Feynman and Hibbs [FH65];

6. Stochastic modeling: in particular, Monte Carlo simulations which follow the classical algorithm of Metropolis et al. [MRR+53] or similar concepts, for an overview see [Bin79];

7. Dynamic (mechanic) modeling: Molecular Dynamics techniques, for details see Ch. 13.

These methods are often highly specialized for the investigation of equilibrium properties of various many-particle systems. For more details and further references, see e.g. [KKER86]. But we will see below that not all of these concepts are applicable to situations far from equilibrium.

1.3 Ultrafast nonequilibrium phenomena

Let us now assume that our equilibrium N-particle system at some moment $t = t_0$ is influenced by an *external excitation.* This can be the compression of the system, a heating process, the penetration of a particle beam into the system or the applications of an external field. As a result, the system will respond to the excitation - by reaching a new pressure, temperature or chemical composition, or it will adapt by assuming a new charge distribution - until it, eventually, comes to a new equilibrium state. It is the main subject of *nonequilibrium theories*, to understand this relaxation process and to predict how the final state will look like. Based on this knowledge, one may suggest a specific form of excitation which allows to reach a well-defined desired state.

Thus, nonequilibrium theories have to solve two problems: (1) What are the properties of various external excitations, what are their time scales, how do they interact with the particle system and how much energy in what spectral composition do they allow to "feed" into the system? And (2), what are the dominant *relaxation mechanisms* in a given many-particle system, how can they be activated, how much momentum and energy do they allow to transform? Obviously, both questions are closely related and require a detailed knowledge of the microscopic properties of many-particle systems and of the character of the interaction of the particles with the excitation under nonequilibrium conditions.

1.3.1 Electromagnetic field–matter interaction

Femtosecond lasers. To be more specific, we will mainly be interested in the excitation by *electromagnetic fields*, including longitudinal electric fields and the radiation field of a laser. Especially lasers have, due to recent developments,[4] become a quite unique excitation source, supplying energy in an extraordinarily wide range of time duration and power (intensity). Fig. 1.3. sketches the range of currently available pulses of sub-picosecond ($< 10^{-12}s$) duration which reach peak powers from watts to terawatts ($10^0 - 10^{12}W$) and, eventually petawatts ($10^{15}W$), dotted lines in Fig. 1.3. This allows to deliver electromagnetic field energies on the target which exceed the binding energies of solids and even heavy atoms and which allow to ionize Uranium within the duration of the pulse. (For comparison, typical binding energies of different systems are shown in the right part of Fig. 1.3.) This has an exceptional potential for applications, ranging from technology, basic research, medicine to inertial confinement fusion.

But not only the extreme peak power is of interest. Equally important are pulses with peak powers from watts to kilowatts, where the main interest is in the prospects of unprecedented temporal resolution in the range of tens of or even a few femtoseconds. Such pulses, mostly in the visible part of the spectrum, are now one of the basic tools for high accuracy investigation of solids, in particular semiconductors [Sha89, DR96], where they yield detailed information on the microscopic properties of bulk materials and low-dimensional nanostructures. Moreover, femtosecond pulses are also becoming a powerful instrument for high precision characterization of very dense plasmas [THWS96]. The reason is that these pulses have a duration which is comparable to typical response and relaxation times in these materials. This makes short-pulse lasers a unique tool both, to excite matter into a strong nonequilibrium state in a very well defined way[5], and to probe its response in its time-dependence. We

[4] such as "chirped" pulse amplification, where the frequency of the field varies in time. One succeeds in generating very short and intense pulses without damage to the optical system by dispersively stretching the pulse, and compressing it again only after amplification [SM85].

[5] The photon energy, laser intensity and pulse duration can be chosen accurately in a

illustrate this more in detail in the next Section, where we consider important microscopic relaxation mechanisms.

Field–matter interaction processes. This interaction is readily understood within the photon picture: Field energy can be absorbed by matter in portions of $\hbar\omega$. The photon energy determines what kind of absorption process is possible, while the field intensity determines the average number of photons which can interact with the material simultaneously. Obviously, multiphoton processes become relevant only at sufficiently high intensities. On the other hand, there is the inverse mechanism possible, where particles emit radiation. There is a large variety of interaction mechanisms, among them are

i) *free charge acceleration* in the oscillating electric field of the wave: The average oscillation energy of an electron ("quiver" or ponderomotive energy) in the field of amplitude E and frequency ω follows from Newton's equation: $U_{\text{pond}} = e^2E^2/(4m_e\omega)$. With modern high intensity lasers electrons are easily accelerated up to relativistic energies[6];

ii) *carrier-photon scattering*: In scattering processes with photons electrons may gain (lose) energy, re-emitting a photon of lower (higher) energy (Thomson scattering). At relativistic energies this process is called "Compton scattering", which may also involve multiple photons;

iii) *excitation of collective plasma oscillations* of the charge carriers: This is an efficient absorption mechanism for photons with an energy in resonance with the plasma (Langmuir) frequency $\omega_{pl}^2 = (4\pi ne^2/m)$ or similarly for other plasmon modes. At high laser intensities the plasma wave will have very high field amplitudes and may itself act as an accelerator for electrons ("wake field" accelerator);

iv) *emission of radiation by moving charges*: Freely moving charges emit Cherenkov radiation[7] while charges which are slowed down, e.g. in the field of an ion, emit "Bremsstrahlung" [8];

v) *excitation processes* in atoms or molecules or *interband transitions* in solids: This is the most important mechanism at low field intensities, which is

---

broad parameter range.

[6]For a laser wavelength $\lambda = 1\mu m$ and an intensity of $10^{13}W/cm^2$, $U_{\text{pond}} \approx 1eV$, the electron rest mass (about $0.5MeV$) is reached at an intensity of $1.3710^{18}W/cm^2$, see e.g. [vdL94]. The relativistic average oscillation energy is $E_{\text{osc}} = m_ec^2\left([1+U_{\text{pond}}/m_ec^2]^{1/2}-1\right)$ [PM94].

[7]Charges moving with the velocity v emit radiation of frequency ω and wave vector k on a cone around $\vec{v}$, $\omega = \vec{v}\cdot\vec{k}$.

[8]For sufficiently high velocities, the frequencies easily reach the range of x-rays.

widely used in spectroscopy of atomic or condensed matter systems. The excited electrons gain a kinetic energy of $E_{kin} = \hbar\omega - \Delta E$, where ΔE is the energy difference of the final and initial energy level;

vi) the inverse processes of v): *de-excitation* of bound particles or interband transitions to lower lying bands which is associated with the emission of a photon. These processes may occur spontaneously or coherently by many electrons, where it leads to lasing;

vii) *Ionization of atoms or molecules*: If in v) electrons are excited into the continuum, bound states become ionized. The kinetic energy of electrons and ions is given by $E_{kin} = \hbar\omega - I^{\text{eff}}$, where the ionization potential I may be modified by medium effects (screening, selfenergy etc. [SBK88]);

viii) *Multi-photon ionization*: At high intensities, multiple photons may be absorbed simultaneously by the atom to bridge the ionization gap. On the other hand, ionization is possible also off-resonance [Kel64b]: for photon energies below the gap ("tunnel ionization") and also far above the gap ("above threshold ionization"). The kinetic energy of the electrons is $E_{kin} = n\hbar\omega - I^{\text{eff}} - U_{\text{pond}}$;

ix) *Relativistic photon–particle transitions*: Photons with energies above $2m_ec^2 \approx 1MeV$ (γ quants) may generate electron-positron pairs. On the other hand, one also expects pair creation from lasers of ultra-high intensity in multi-photon processes[9], e.g. [vdL94].

Obviously, at relativistic energies a clear separation of some processes is no longer possible. There, a unified treatment of charge carriers and electromagnetic radiation is necessary, which is given by relativistic quantum electrodynamics (see Ch. 12).

Modifications on short times. Traditional concepts have, furthermore, to be revised if ultrafast processes are considered. The most striking effect results from *Heisenberg's uncertainty principle*: The picture of electromagnetic radiation consisting of portions with a sharp energy $\hbar\omega$ breaks down on short times. Laser pulses with a duration of only a few femtoseconds which, in the optical range, corresponds to only a few oscillation periods, consist of photons which are "smeared out" energetically. [The corresponding relation between energy and time scales is shown in Fig. 1.3. by the line $Et = h$: For processes with a given time duration t, the minimal energy uncertainty is given by the crossing point with this line, and real processes are confined to

[9] Pair creation in a two-photon process was already discussed by Breit and Wheeler [BW34].

the range above this line.] Instead of a photon with a fixed energy, an electron or atom will interact with radiation in a broad spectral range around $\hbar\omega$. Obviously, distinctions between resonant and off-resonant processes, above or below threshold etc. become meaningless.

These effects will be an important issue in all short-time investigations, including the present book. To gain first insight, our analysis will concentrate on the simplest types of field-matter interaction, mainly on processes iii)–vi) in the nonrelativistic limit.

1.3.2 Overview of relaxation processes

Excitation and relaxation. Let us now return to the investigation of many-particle systems being brought out of equilibrium by some external excitation (e.g. a laser pulse) and consider the relaxation into a new stationary state. The character of this process depends strongly on the relevant time scales τ_p, t_{rel} and τ_{cor}- the duration of the excitation (pulse duration), the relaxation and the correlation time[10] of the particle system. If $\tau_p \ll \tau_{cor}$, the excitation is "instantaneous", and the relaxation starts from some initial state created by the excitation. In the opposite limit, $\tau_p \gg t_{rel}$, the excitation is quasistationary, and the system is effectively in equilibrium at all times, where the equilibrium state changes slowly with the excitation. Inbetween both limits, excitation and relaxation cannot be separated. This is the most interesting, but, at the same time, the most difficult situation.

Time scales of relaxation processes. While τ_p is determined by the technical characteristics of the exciting laser, the relaxation time varies from one system to another and also with the parameters, such as density, chemical composition (e.g. degree of ionization) etc. Fig. 1.3. shows typical values of the relaxation time for different systems (see bottom). For example, electron-hole plasmas in semiconductors have a lifetime[11] in the range of $10^{-9}s \dots 10^{-6}s$. Typical relaxation times are in the range from $100fs$ (at high densities of the order of $10a_B^{-d}$, where d is the dimensionality of the structure) to several pico-

[10] There may be various relaxation times, each related to another relaxation mechanism. Here, we have in mind the relaxation time of the momentum distribution. Typically $\tau_{cor} < t_{rel}$. We will discuss these time scales more in detail in Ch. 5.

[11] the time until electrons recombine from the conduction band to the valence band

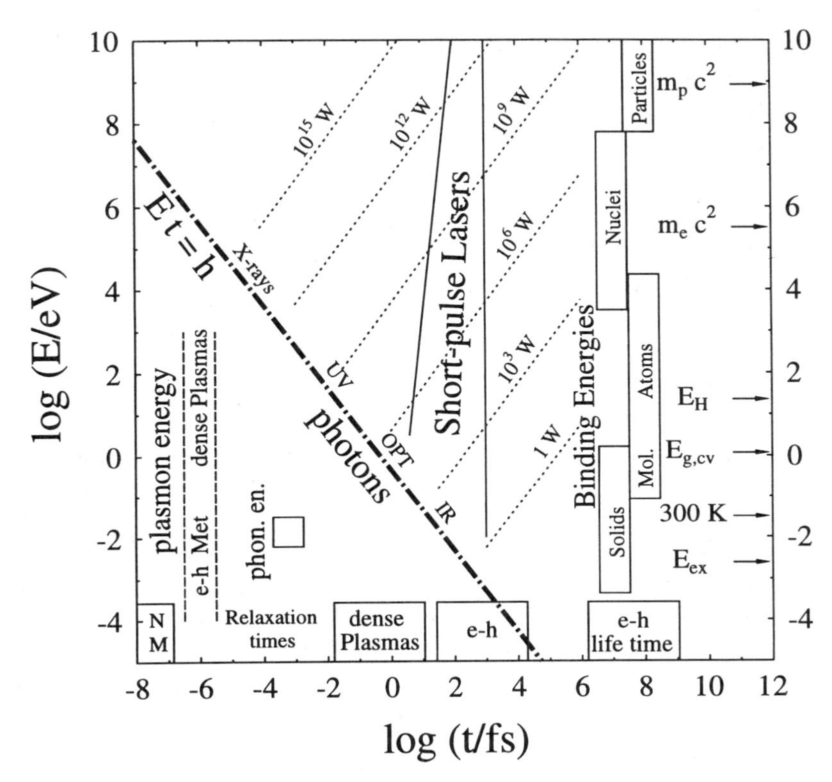

Fig. 1.3. Typical energy and time scales of relaxation in various systems. *Along right border*: range of binding energies of solids, atoms, molecules, nuclei and elementary particles, arrows correspond to excitons (ex), hydrogen (H), electron and proton rest mass and typical band gap in semiconductors. *Along left border*: typical energies of photons and plasmons in semiconductors (e-h), metals and dense plasmas. *Along lower border*: life time of e-h plasmas and typical relaxation time in semiconductors, dense plasmas and nuclear matter (for different densities, increasing inside each box from right to left). *Dash-dotted line* corresponds to Heisenberg's principle: for a given observation time, the energy uncertainty is above (or on) this line. At the same time, this line relates energies and oscillation periods for photons (frequency ranges are indicated). Dotted lines indicate constant power (in Watts). The *laser* area indicates energy vs. pulse duration (below one picosecond) of modern lasers.

seconds (at low densities, $n < 0.1a_B^{-d}$). On the other hand, dense plasmas have a very broad range of relaxation times: for example, for fully ionized hydrogen $t_{rel} \sim 10fs$ at a density of about $10^{20}cm^{-3}$, and it decreases continuously with increasing density. The shortest relaxation times are found in nuclear matter, they are around $10^{-23} \dots 10^{-22}s$. Thus, interesting overlap of relaxation (and correlation) times with the pulse duration of modern lasers is observed in plasmas of moderate density and especially in condensed matter systems.

Relaxation mechanisms. In general, a laser pulse activates several relaxation mechanisms ("eigen frequencies") of the many-particle system at once. This depends mainly on the kinetic energy gained by the electrons and other particles during the excitation. For many processes, a threshold energy is required. This includes impact ionization of bound states (atoms or molecules - in gases or plasmas; excitons, impurities and so on - in solids) or excitation of collective modes, such as vibrations of the crystal lattice (phonons). Here, the typical energy exchanged between the electrons and the scattering partner is the effective binding energy in the medium $I^{\rm eff}$ or the phonon energy $\hbar\omega_{ph}$, respectively.

The most important and general mechanism is carrier–carrier scattering, e.g. electron-electron, electron-ion(or hole) and hole-hole (ion-ion) scattering. Here, the typical scattering energies are proportional to the scattering cross section σ, e.g. for Coulomb interaction in second Born approximation[12], $\sigma \sim V_{ab}^2(q) = [4\pi e_a e_b/\epsilon q]^2$. From this expression it is clear that in systems with large background dielectric constant (e.g. fluids or solids) the energy exchange is strongly reduced and, therefore, the relaxation towards equilibrium takes longer. Another carrier-carrier scattering mechanism is the excitation of collective modes of the plasma (plasma oscillations, instabilities and so on), where the energy transfer is of the order of the plasmon energy $\hbar\omega_{pl}$. Typical values for the plasmon and phonon energies are indicated in Fig. 1.3. (lower left part). In this book, we will concentrate on carrier–carrier scattering and the underlying correlations. For a discussion of other scattering mechanisms in semiconductors and plasmas, see, respectively [HJ96] and [KKER86].

Finally we mention that usually there exists a *hierarchy* of relaxation processes with respect to the typical time, length or energy scales, which greatly simplifies the theoretical treatment of the relaxation, for details see Ch. 5.

1.3.3 Theoretical concepts. Outline of this book

The theoretical analysis of nonequilbrium behavior of many-particle systems is essentially more involved than the description of its equilibrium properties. Therefore, many of the methods discussed in Sec. 1.2 are not applicable to time dependent phenomena. Depending on the dominating relaxation mechanism, the theoretical treatment of the evolution towards equilibrium varies greatly. For situations which are close to equilibrium, one may use rate equations or hydrodynamic equations. If the momentum distributions of the particles deviates significantly from the equilibrium distribution, the appropriate concept

---

[12]An improved treatment leads to cross sections in T-matrix approximation, see Ch. 8.

are kinetic equations, such as the famous Boltzmann equation. However, as was mentioned in the beginning, conventional kinetic equations have two major deficiencies:

A. They assume implicitly that all correlations have already reached their equilibrium form (in particular, they assume that initial correlations are completely weakened), and thus they are not applicable to times shorter than the correlation time, and

B. They conserve only kinetic (single-particle) energy and are, therefore, not applicable to correlated many-particle systems. In particular, they yield the equilibrium distribution of an ideal gas, completely neglecting correlation corrections.

In fact, we will see that both points are very closely related. Therefore, the description of ultrafast processes ($t < \tau_{cor}$) as well as the relaxation of strongly coupled systems require generalized kinetic equations. Derivation, investigation and numerical solution of these equations is the main topic of this book. But before we give an outline, it is appropriate to summarize some of the important early results in this field.

Historical remarks. A straightforward extension of equilibrium theories to nonequilibrium is the *Linear response* approach developed by Kubo, Mori, Zubarev and others, e.g. [Kub57, Zub71]. This approach uses a generalized equilibrium statistical operator which depends on additional observables. This method has been very successful for the description of transport processes in correlated systems close to equilibrium, see also [Kli75, Eea84, KKER86, Röp87, ZMR96].

A very general nonequilibrium method is based on the hierarchy of equations for the reduced *density operators, the BBGKY-hierarchy*, which was developed by Bogolyubov, Born, Green, Kirkwood, Yvon and others, e.g. [Bog46, Bog61] and which was lateron generalized to quantum systems. It is well suited for the derivation of generalized quantum kinetic equations. Furthermore, there have been proposed other concepts to derive *Generalized (non-Markovian) kinetic equations*, by Prigogine [Pri63], Resibois [Res65], Zwanzig [Zwa63], Balescu [Bal63], Klimontovich and Silin [KS60] and others.[13]

[13] In particular, the kinetic equations derived by Prigogine contain an initial correlation term and a non-Markovian scattering term $dF(t)/dt = I_{IC}(t-t_0) + \int_{t_0}^{t} d\tau K(t-\tau)F(\tau)$. It could be shown in very general form that the initial correlation term is damped. We will recover this general form from the BBGKY-hierarchy in Ch. 7.

Klimontovich developed the *method of microscopic phase space densities* [Kli57] which proved to be very general[14] and allowed him to derive a great variety of generalized kinetic equations for nonideal gases and plasmas [Kli75]. Bärwinkel and Grossmann were the first to show that total energy conservation in kinetic equations is closely related to the time structure (non-Markovian or retardation effects) of the collision integral [BG67, Bär69].

A powerful approach to generalized quantum kinetic equations which derived from field theory is the method of *Nonequilibrium Green's functions.* Here, major contributions are due to Martin and Schwinger [MS59, Sch61], Kadanoff and Baym [BK61, KB89], Keldysh [Kel64a]. The incorporation of electromagnetic fields was studied by Korenman [Kor66] and Dubois [DuB67]. Extensions to the relativistic case have been developed by Akhiezer and Berestezki [AB59] as well as Dubois and Bezzerides [DuB67, BD72].

Finally, nonequilibrium processes are successfully treated within purely mechanical concepts, i.e. by *Molecular Dynamics simulations.* These methods are very straightforward in application to classical particles where they yield very high accuracy results. The extension of these techniques to quantum systems is currently actively discussed, for an overview and references, see Ch. 13.

Naturally, we can mention only a very small part of the tremendous literature. Further references will be given in the Chapters below (for an overview on the more recent literature on correlation effects in strongly coupled plasmas, solids and nuclear matter, see, respectively, [KKER86, Mah81, HJ96, RT93]. Summarizing these developments, we conclude that today there exist three main approaches which are applicable to ultrafast phenomena:[15]

I. Density operator techniques. BBGKY-hierarchy;

II. Nonequilibrium Green's functions theory;

III. Molecular Dynamics simulations.

---

[14] The fluctuating phase space density is defined in the $6N-$dimensional phase space $[x = (\mathbf{r}, \mathbf{p})]$ according to $N(x,t) = \sum_{i=1}^{N} \delta[x - x_i(t)]$, where $x_i(t)$ is the exact trajectory of particle "i". This method straightforwardly incorporates density and field fluctuations and is, in fact, the classical analogue to the second quantization method of field theory.

[15] For completeness, we mention that there exist excellent investigations of ultrafast phenomena which use Monte Carlo techniques, which we cannot discuss here, see e.g. [RHK94] and references therein. This approach is closesly related to points I. or II., since it is based on quantum kinetic equations too.

Outline of this book. We will consider all three methods, although we focus on the density operator approach, because it is conceptually simple. In Chs. 2 and 3, we give a detailed introduction into the method of reduced density operators, based on the BBGKY-hierarchy, discuss its properties, the treatment of correlations and important decoupling approximations. In Ch. 3 we generalize these results by the incorporation of spin statistics effects.

In Chs. 4–10, we discuss important approximations, beginning with the mean-field (Hartree/Vlasov) approximation, which describes collective phenomena (plasmons) in the absence of correlations. Correlation effects are introduced in Chs. 5 and 6 and discussed further in Ch. 7, where we focus on Non-Markovian behavior and selfenergy. This discussion is extended to more complex correlation phenomena in Chs. 8–10, including strong correlations and bound states as well as dynamical screening and screening buildup.

The consistent incorporation of electromagnetic fields into the BBGKY-hierarchy and the derivation of generalized Bloch equations is the subject of Ch. 11 and concludes our discussion of the density operator approach.

The method of nonequilibrium Green's functions is summarized in Ch. 12, where we start from a fully relativistic formulation. We derive relativistic Kadanoff-Baym equations for particles and photons and their nonrelativistic limit and compare the Green's functions results to that of the density operator theory. Finally, in Ch. 13 we leave the statistical theories and turn to the alternative mechanical approach of Molecular Dynamics simulations. We show that both approaches have their advantages and problems and complement each other almost perfectly.

Chapter 2

The Method of Reduced Density Operators

2.1 Density operator. Von Neumann equation

Definitions. We consider a macroscopic system of N particles in volume $\mathcal{V}$ in the thermodynamic limit

$$N \longrightarrow \infty, \qquad \mathcal{V} \longrightarrow \infty, \qquad n = \frac{N}{\mathcal{V}} = \text{const}, \tag{2.1}$$

which are subject to an external field $\mathcal{U}$. A unified description of this complex system can be based on a Lagrange functional, or on a Hamilton operator which contains a particle contribution $H^{particle}$, a field contribution H^{field} and an interaction part H^{int}

$$H = H^{particle} + H^{int} + H^{field} = \tilde{H}^{particle} + H^{field} \tag{2.2}$$

and can be given in terms of quantum field theory, e.g. [AB59, DuB67], and will be discussed briefly in Ch. 12. Here, our goal is different: we want to describe primarily the particle sub-system including correlation effects caused by binary interactions V_{ij}. We will also be interested in the modification of the particle behavior in the presence of a field, which, however, will be regarded as externally given (the action of the particles on the field will not be considered). Thus, we will rewrite the Hamiltonian $\tilde{H}^{particle}$ leaving out the field part,

$$H_{1\ldots N} = \sum_{i=1}^{N} H_i + \sum_{1 \le i < j \le N} V_{ij}, \tag{2.3}$$

$$H_i(t) = \frac{p_i^2}{2\, m_i} + \mathcal{U}_i(t), \tag{2.4}$$

where we will use the operator notation $\mathcal{U}$ for the interaction Hamiltonian. (More generally, one can replace the kinetic energy by the one-particle energy eigenvalue of the system $p^2/2m \to E(p)$.) We consider general binary interaction potentials between the particles. In many cases they can be assumed to depend on the inter-particle distance,

$$V_{ij} \to V_{ij}(|r_i - r_j|), \tag{2.5}$$

e.g. Coulomb interaction between point charges.[1] Examples for $\mathcal{U}$ are longitudinal or transverse electromagnetic fields, a gravitational potential and so on. More specific examples will be discussed below. This system can be in either one of M possible quantum states given by the vectors $|\psi^{(1)}_{1\dots N}\rangle$, $|\psi^{(2)}_{1\dots N}\rangle$ $\dots|\psi^{(M)}_{1\dots N}\rangle$ which form a complete orthonormal basis,

$$\langle\psi^{(k)}_{1\dots N}|\psi^{(l)}_{1\dots N}\rangle = \delta_{k,l}, \tag{2.6}$$

$$\sum_{k=1}^{M} |\psi^{(k)}_{1\dots N}\rangle\langle\psi^{(k)}_{1\dots N}| = 1. \tag{2.7}$$

Each of these micro-states will be realized with a certain probability W_k, with $W_k \geq 0$ and $\sum\limits_{k=1}^{M} W_k = 1$. The **N-particle density operator** is defined, according to J. von Neumann [vN55], as the sum of the projections onto all states weighted by the probabilities W_k,

$$\boxed{\rho_{1\dots N} = \sum_{k=1}^{M} W_k \, |\psi^{(k)}_{1\dots N}\rangle\langle\psi^{(k)}_{1\dots N}|} \tag{2.8}$$

The main properties of the density operator are

1. $\mathrm{Tr}_{1\dots N}\, \rho_{1\dots N} = 1$
2. For a pure quantum state "k_0", $W_k = \delta_{k,k_0}$, i.e. ρ is the projection operator onto this state. On the other hand, in a mixed state, ρ has a statistical (probabilistic) meaning, which is introduced by W_k.
3. $\rho_{1\dots N}$ is hermitean, $\rho^{\dagger}_{1\dots N} = \rho_{1\dots N}$
4. The density operator is invariant with respect to particle exchange, $\rho_{1\dots i\dots j\dots N} = \rho_{1\dots j\dots i\dots N}$, (identical particles)

[1] In other systems, such as large atoms or molecules, or nuclear matter, the interaction is more complicated, so we consider both cases.

5. Averages of an arbitrary quantity A are given by

$$\langle A \rangle = \mathrm{Tr}_{1\ldots N}\, A\, \rho_{1\ldots N} \tag{2.9}$$

6. $\rho_{1\ldots N}$ contains the full quantum-mechanical information on the N-particle state without any losses.

7. *Matrix representation of the density operator:* If the state of the N-particle system is given by a complete orthogonal set of normalized vectors $|\psi^{(1)}_{1\ldots N}\rangle$ in a certain representation b, $|b_{1\ldots N}\rangle = |b_1 \ldots b_N\rangle$, the density operator in this basis will be given by the density matrix $\langle b_N \ldots b_1|\rho_{1\ldots N}|b'_1 \ldots b'_N\rangle$. The i-th diagonal element of this matrix corresponds to the probability of finding the system in the state described by the $i-$th basis vector, where $\langle b_N \ldots b_1|\rho_{1\ldots N}|b_1 \ldots b_N\rangle \geq 0$, whereas the off-diagonal elements correspond to the probability of transitions between two different states. Calculating the trace (2.9) in this basis means to sum up the diagonal elements of the operator $A\rho_{1\ldots N}$.

Important examples for representations of the density operator, such as the coordinate, Wigner, momentum or Bloch representation will be discussed below.

Von Neumann equation. The dynamical properties of the N-particle system are governed by the evolution of the density operator. The corresponding evolution equation follows directly from the equation of motion of the state vectors $|\psi^{(k)}_{1\ldots N}\rangle$, which is the N-particle Schrödinger equation supplemented by an initial condition (and boundary conditions which we will not specialize here)

$$i\hbar \frac{\partial}{\partial t}|\psi^{(k)}_{1\ldots N}(t)\rangle = H_{1\ldots N}|\psi^{(k)}_{1\ldots N}(t)\rangle \tag{2.10}$$

$$|\psi^{(k)}_{1\ldots N}(t_0)\rangle = |\psi^{(0k)}_{1\ldots N}\rangle. \tag{2.11}$$

According to the definition, Eq. (2.8), this equation yields the equation of motion of the N-particle density operator which is called **von Neumann equation** (quantum Liouville equation)[2]

$$\boxed{i\hbar \frac{\partial}{\partial t}\rho_{1\ldots N} - [H_{1\ldots N}, \rho_{1\ldots N}] = 0} \tag{2.12}$$

with the initial condition $\rho_{1\ldots N}(t_0) = \rho^{(0)}_{1\ldots N} = \sum_{k=1}^{M} W_k\, |\psi^{(0k)}_{1\ldots N}\rangle\langle\psi^{(0k)}_{1\ldots N}|$. Eq. (2.12) describes the dynamics of the N-particle system, provided its initial

[2] $[A, B]$ denotes the commutator, cf. App. A.

state (the complete set of possible quantum states) is known. This equation has the form of Heisenberg's equation of motion. However, the time dependence of the density operator arises from the time dependence of the N−particle state vectors, which corresponds to the Schrödinger picture of quantum mechanics. From Eq. (2.12) we conclude two further dynamical properties of the density operator:

8. In the course of the time evolution of $\rho_{1\ldots N}$, total energy is conserved,
$$E(t) = \langle H_{1\ldots N}\rangle = \mathrm{Tr}_{1\ldots N} H_{1\ldots N}\, \rho_{1\ldots N}(t) = \mathrm{Tr}_{1\ldots N} H_{1\ldots N}\, \rho^{(0)}_{1\ldots N} = E(t_0).$$

9. The evolution of $\rho_{1\ldots N}$ is time reversible (see Sec. 2.2.1).

2.2 BBGKY-hierarchy

2.2.1 Reduced density operators. Equations of motion

Instead of the full N-particle density operator, it is often convenient to consider simpler quantities, the **reduced density operators** F_1, $F_{12}\ldots F_{1\ldots s}$, which are defined by the partial trace over the remaining particles

$$\boxed{F_{1\ldots s} = \mathcal{V}^s\, \mathrm{Tr}_{s+1\ldots N}\, \rho_{1\ldots N}, \qquad \frac{1}{\mathcal{V}^s}\mathrm{Tr}_{1\ldots s} F_{1\ldots s} = 1} \tag{2.13}$$

where the normalization condition follows from the normalization condition for $\rho_{1\ldots N}$. Using the reduced density operators, we can calculate averages of one, two or s-particle quantities by

$$\langle A_{1\ldots s}\rangle = \frac{n^s}{s!}\mathrm{Tr}_{1\ldots s} A_{1\ldots s} F_{1\ldots s} \tag{2.14}$$

The properties of the reduced density operators can follow directly from the properties of $\rho_{1\ldots N}$. We mention only an additional consistency requirement connecting operators of different orders which follows from the definition (2.13),

$$F_{1\ldots s} = \mathcal{V}^{-k}\, \mathrm{Tr}_{s+1\ldots s+k}\, F_{1\ldots s+k} \tag{2.15}$$

In order to make the following considerations more transparent, we will consider first a quantum system of spinless particles.[3] The effects of the spin

[3] In principle, the equations derived below, are applicable to Bose or Fermi particles also. We would only have to define the trace operations to be performed respectively in the symmetric or antisymmetric subspace of the Hilbert space.

statistics will be included explicitly into the equations of motion below, in Ch. 3.

Using the above definitions and the von Neumann equation (2.12), we can write down the hierarchy of equations of motion for the reduced density operators, the **BBGKY-hierarchy**, [Bog46, Bog61, BD79, KKK87]

$$\boxed{i\hbar\frac{\partial}{\partial t}F_{1\ldots s} - [H_{1\ldots s}, F_{1\ldots s}] = n\mathrm{Tr}_{s+1}\sum_{i=1}^{s}[V_{i,s+1}, F_{1\ldots s+1}]} \tag{2.16}$$

Here, $H_{1\ldots s}$ is the s-particle Hamilton operator which follows from $H_{1\ldots N}$ by substituting $N \to s$. The equations of the hierarchy differ from the von Neumann equation due to the terms on the r.h.s, which contain the coupling of the s particles to the remainder of the system via all possible binary interactions.[4] The complete hierarchy is, obviously equivalent to the von Neumann equation.

Properties of the hierarchy. Formal solution. Due to linearity, each of the hierarchy equations can be solved formally [KKK87]:

$$F_{1\ldots s}(t) = U_{1\ldots s}(t-t_0)\; F^{(0)}_{1\ldots s}\; U^{\dagger}_{1\ldots s}(t-t_0)$$
$$+\frac{n}{i\hbar}\int_0^{t-t_0} d\tau U_{1\ldots s}(\tau)\; \mathrm{Tr}_{s+1}\sum_{i=1}^{s}[V_{is+1}, F_{1\ldots s+1}(t-\tau)]\; U^{\dagger}_{1\ldots s}(\tau). \tag{2.17}$$

The first term corresponds to the independent evolution of the $s-$particle complex, which starts from its initial state given by $F^{(0)}_{1\ldots s}$. The second term contains the influence of the remaining part of the system, including both mean field and correlation effects (these two effects are not separated in $F_{1\ldots s}$). In Eq. (2.17), the propagator $U_{1\ldots s}$ is the solution of the homogeneous equation

$$\left\{i\hbar\frac{\partial}{\partial t} - H_{1\ldots s}\right\} U_{1\ldots s} = 0, \quad U_{1\ldots s}(t) = T\; \exp[-\frac{i}{\hbar}\int_{t_0}^{t} d\tau\, H_{1\ldots s}(\tau)], \tag{2.18}$$

where T is the time ordering operator which accounts for the fact that the Hamiltonian may be time dependent via the external potential. One readily proofs that $U_{1\ldots s}$ in fact generates the solution to Eq. (2.16) with the initial condition (2.16). For a more detailed discussion of the propagators, see Appendix D.

The fruitfulness of the concept of the hierarchy is due to the fact, that the vast majority of physical phenomena and observables of the N-particle system, can be described by a quite limited number of reduced density operators,

[4]Strictly speaking, the prefactor of the trace in Eq. (2.16) is $(N-s)/\mathcal{V}$ instead of n. However, for sufficiently large particle numbers, cf. Eq. (2.1), we may use $s \ll N$. For the modifications of the hierarchy in few-body (e.g. nuclear) systems, see e.g. [WC85, RT94].

typically not exceeding the order s=4. Therefore, it is reasonable to decouple the hierarchy at some appropriate level, thereby drastically reducing the complexity of the problem. The influence of the higher order contributions is neglected or treated approximately. Fortunately, the vast majority of physical phenomena in many-particle systems can be well described using only the first three equations:

$$i\hbar\frac{\partial}{\partial t}F_1 - [H_1, F_1] = n\mathrm{Tr}_2[V_{12}, F_{12}], \tag{2.19}$$

$$i\hbar\frac{\partial}{\partial t}F_{12} - [H_{12}, F_{12}] = n\mathrm{Tr}_3[V_{13} + V_{23}, F_{123}], \tag{2.20}$$

$$i\hbar\frac{\partial}{\partial t}F_{123} - [H_{123}, F_{123}] = n\mathrm{Tr}_4[V_{14} + V_{24} + V_{34}, F_{1234}]. \tag{2.21}$$

An important consistency requirement for any approximation is that higher order equations must yield all lower order equations after calculation of the corresponding partial traces, cf. Eq. (2.15). While the full hierarchy, obviously, has the same properties as the von Neumann equation, the properties of a reduced set of equations may depend on the actual decoupling approximation and may have to be investigated for each approximation separately. Nevertheless, several important properties, such as time reversibility and energy conservation can be discussed in a rather general way.

Time reversibility of the BBGKY hierarchy. To investigate the reversibility of approximations to the hierarchy, consider first the transformation of relevant quantities and expressions on inversion of the time arrow $t \longrightarrow -t$. One easily checks the following transformation properties of the coordinate, the momentum, the wave function, the density operator and the Hamilton operator:

$$\begin{aligned} \mathbf{r} &\longrightarrow \mathbf{r}, \\ \mathbf{p} &\longrightarrow -\mathbf{p}, \\ \psi &\longrightarrow \psi^*, \\ \rho_{1...N} &\longrightarrow \rho_{1...N}, \\ H_{1...N} &\longrightarrow H_{1...N}. \end{aligned} \tag{2.22}$$

Furthermore, the commutator of ρ and H transforms according to

$$[H_{1...N}, \rho_{1...N}] \longrightarrow -[H_{1...N}, \rho_{1...N}].$$

The transformation properties of the commutators are most easily seen from their classical limit, i.e. the Poisson brackets, cf. Sec. 2.3.2. These transformation symmetries immediately yield the time reversibility of the von Neumann equation, and therefore, of the full BBGKY hierarchy also.

Regarding the reversibility of reduced sets of equations, we have to investigate the properties of the reduced density operators. Since the trace operation does not affect the time symmetry, we find the transformations

$$\begin{aligned} F_{1\ldots s} &\longrightarrow F_{1\ldots s}, \\ H_{1\ldots s} &\longrightarrow H_{1\ldots s}, \\ [H_{1\ldots s}, F_{1\ldots s}] &\longrightarrow -[H_{1\ldots s}, F_{1\ldots s}], \\ [V_{ij}, F_{1\ldots s}] &\longrightarrow -[V_{ij}, F_{1\ldots s}]. \end{aligned} \tag{2.23}$$

We may conclude that the original symmetry of the full hierarchy is retained by any closure relation which

i) consists in the neglect of terms of commutator structure only and

ii) which is still exact with respect to the time dependence of all quantities.

This means, that the equations are solved as an initial value problem, including the initial values of all density operators up to the corresponding order exactly. In contrast, irreversibility is introduced if approximations to the time structure are made (e.g. retardation or gradient expansions, Markov limit, local approximations and so on, see Chs. 7,8), if the initial values are given only approximately (or if there is some sort of averaging over the initial values) or if the initial time is shifted to minus (plus) infinity. The latter limiting procedure transforms the initial value problem into a boundary value one (on the time-arrow), allowing only evolution in one direction, that of increasing (decreasing) time.

2.2.2 Conservation of density and total energy

Density balance. Particle number N is a single-particle observable, so density conservation can be investigated based on the first hierarchy equation. To include the possibility of particle number changes, we have to permit a time-dependent normalization of F_1, Eq. (2.13), and we obtain $N(t) = n\mathcal{V}(t) = n\mathrm{Tr}_1\, F_1(t)$. Calculating the trace over "1" of the first hierarchy equation, we have

$$\frac{d}{dt}N = \frac{n}{i\hbar}\mathrm{Tr}_1\,[\mathcal{U}_1, F_1(t)] + \frac{n^2}{i\hbar}\mathrm{Tr}_{12}\,[V_{12}, F_{12}(t)], \tag{2.24}$$

where we took into account that the commutator $[p_1^2, F_1]$ vanishes under the trace. We see that the particle number may change (1) as a result of the action of the external field and (2) due to interaction with the other particles. The first effect is observed e.g. in laser-matter interaction, where the electromagnetic field may generate carriers (excitation of electron-hole pairs in solids, photoionization etc.). These effects will be studied in Ch. 11. On the other hand, the number of particles of a given species may change in inelastic scattering processes (e.g. impact ionization/recombination processes), see Sec. 2.6.1.

Total energy conservation.[5] Like the von Neumann equation, the complete BBGKY-hierarchy conserves total energy. On the contrary, Markovian kinetic equations, such as the conventional Boltzmann equation, Landau equation or Balescu equation, are known to conserve kinetic energy (or quasi-particle energy) only. The reasons will be discussed below. It is therefore, important to have criteria that allow to single out those decoupling approximations that retain the conservation of total energy[6]. To do this, we consider the time evolution of kinetic and potential energy, the general nonequilibrium definitions of which are

$$\langle T\rangle = n\mathrm{Tr}_1\frac{p_1^2}{2m}F_1 \tag{2.25}$$

$$\langle V\rangle = \frac{n^2}{2}\mathrm{Tr}_{12}V_{12}F_{12}, \tag{2.26}$$

and which require only knowledge of the one and two-particle density operators. Therefore, criteria for energy conservation have to be given primarily on the closure approximation for the three-particle density operator. We consider first the time derivative of the kinetic energy $\langle T\rangle$, which is calculated with the solution $F_1(t)$ of the first hierarchy equation:

$$\begin{aligned} i\hbar\frac{d}{dt}\langle T\rangle &= i\hbar\, n\,\mathrm{Tr}_1\,\frac{p_1^2}{2m}\frac{\partial}{\partial t}F_1 \\ &= n\mathrm{Tr}_1\frac{p_1^2}{2m}[H_1,F_1] + n^2\mathrm{Tr}_{12}\,\frac{p_1^2}{2m}[V_{12},F_{12}]. \end{aligned} \tag{2.27}$$

It is convenient to perform a simple transformation of the first term on the r.h.s. of Eq. (2.27)

$$\begin{aligned} n\mathrm{Tr}_1\frac{p_1^2}{2m}[H_1,F_1] &= n\mathrm{Tr}_1\,\{H_1-\mathcal{U}_1(t)\}\,[H_1,F_1] \\ &= -n\mathrm{Tr}_1\mathcal{U}_1(t)[H_1,F_1], \end{aligned} \tag{2.28}$$

---

[5] Part of the results of these Section have been obtained in collaboration and discussion with Th. Bornath, J.W. Dufty, D. Kremp and N.H. Kwong.

[6] The corresponding analysis for the Martin–Schwinger hierarchy of nonequilibrium Green's functions has been given in [BK61], see also [KB89].

where use was made of $n\mathrm{Tr}_1 H_1 [H_1, F_1] = 0$. We now transform the last term of Eq. (2.27),

$$\begin{aligned} n^2 \,\mathrm{Tr}_{12} \frac{p_1^2}{2m}[V_{12}, F_{12}] &= \frac{n^2}{2}\mathrm{Tr}_{12} \frac{p_1^2 + p_2^2}{2m}[V_{12}, F_{12}] \\ &= \frac{n^2}{2}\mathrm{Tr}_{12}\Big\{H_{12} - \mathcal{U}_1(t) - \mathcal{U}_2(t)\Big\}[V_{12}, F_{12}]. \quad (2.29) \end{aligned}$$

In the last equation, we took into account that, due to the invariance of the trace, $\mathrm{Tr}_{12}(V_{12}[V_{12}, F_{12}]) = 0$. On the other hand, the time derivative of the potential energy is

$$\begin{aligned} i\hbar \frac{d}{dt}\langle V \rangle &= i\hbar \frac{n^2}{2}\mathrm{Tr}_{12}\, V_{12}\frac{\partial}{\partial t}F_{12} \qquad (2.30) \\ &= \frac{n^2}{2}\mathrm{Tr}_{12}\, V_{12}[H_{12}, F_{12}] + \frac{n^3}{2}\mathrm{Tr}_{123}\, V_{12}[V_{13} + V_{23}, F_{123}] \end{aligned}$$

Collecting together these results, we can now calculate the time derivative of the total energy,

$$i\hbar \frac{d}{dt}\langle T + V \rangle = \frac{n^3}{2}\mathrm{Tr}_{123}(V_{12}[V_{13} + V_{23}, F_{123}]) + i\hbar P_{\mathcal{U}} \qquad (2.31)$$

The last term in Eq. (2.31) contains all contributions which are explicitly related to the external potential $\mathcal{U}(t)$ (recall that $\mathcal{U}$ is just the particle-field interaction Hamiltonian):

$$i\hbar P_{\mathcal{U}} = -n\mathrm{Tr}_1\mathcal{U}_1(t)[H_1, F_1] - \frac{n^2}{2}\mathrm{Tr}_{12}\left\{\mathcal{U}_1(t) + \mathcal{U}_2(t)\right\}[V_{12}, F_{12}]. \qquad (2.32)$$

What is left now is to recognize that Eq. (2.32) is directly related to the time derivative of the mean value of the external potential

$$-P_{\mathcal{U}} = \frac{d}{dt}\langle \mathcal{U}_1(t)\rangle - \langle\frac{\partial \mathcal{U}_1(t)}{\partial t}\rangle \qquad (2.33)$$

where the last term is the average power of the external field given by $\langle\frac{\partial \mathcal{U}_1(t)}{\partial t}\rangle = n\mathrm{Tr}_1 \frac{\partial \mathcal{U}_1(t)}{\partial t} F_1$. So, our final result for the **time derivative of total energy** of the particles is [Bon]

$$\boxed{\frac{d}{dt}\langle T + V \rangle = \frac{n^3}{2\, i\hbar}\mathrm{Tr}_{123}\, V_{12}[V_{13} + V_{23}, F_{123}] + \left\langle \frac{\partial \mathcal{U}(t)}{\partial t}\right\rangle - \frac{d}{dt}\langle \mathcal{U}(t)\rangle} \qquad (2.34)$$

We see that the energy of the particles is influenced by the three-particle density operator and also by the particle-field interaction. On the other hand we recall that the total system includes the field, cf. Eq. (2.2), hence we expect $\langle H \rangle = \langle H^{particle} + H^{int} + H^{field} \rangle$ to be conserved. Therefore, in a full theory, which includes the field selfconsistently, a change of the particle energy caused by the field (last two terms in Eq. (2.34)) must be compensated by the corresponding change of the field energy. In our case, where the field dynamics is excluded from the consideration, the conserved quantity can be found by integrating Eq. (2.34), we obtain the **general result for the total energy balance**

$$\boxed{\begin{aligned} \langle T+V\rangle(t) \quad &- \quad \langle T+V\rangle(t_0) + \int_{t_0}^{t} d\bar{t}\, n \mathrm{Tr}_1\, \mathcal{U}_1(\bar{t})\, \frac{\partial F_1(\bar{t})}{\partial \bar{t}} \\ &= \quad \frac{n^3}{2\, i\hbar} \int_{t_0}^{t} d\bar{t}\, \mathrm{Tr}_{123}\, V_{12}[V_{13}+V_{23}, F_{123}(\bar{t})] \end{aligned}} \tag{2.35}$$

We briefly summarize the main consequences of Eq. (2.35):

i) If there is no external potential, total energy is the sum of kinetic + potential energy, which is conserved if the term with the trace over the three-particle density operator is zero.

ii) In the presence of a time-independent external potential, total energy includes the mean value of the potential $T+V \longrightarrow T+V+\mathcal{U}$ and is conserved under the same condition.

iii) In case of a time-dependent external potential $\mathcal{U} = \mathcal{U}(t)$, the conserved quantity is given by the l.h.s. of Eq. (2.35), provided, the term with F_{123} is zero. In this equation, the energy gained (lost) by the particles due to the action of the field is canceled by the integral term.

iv) We thus found a very general energy conservation criterion which is the same with or without a time-dependent external potential: the vanishing of the trace term on the r.h.s. of Eq. (2.34). One readily proofs that a sufficient condition for this to be fulfilled is

$$P_{123}F_{123} = F_{123}, \tag{2.36}$$

where P_{123} denotes an arbitrary permutation of three particles [DB89].

v) Condition (2.36) is, of course, fulfilled for the complete BBGKY-Hierarchy. On the other hand, this condition is a very important criterion to check if a closure relation to the hierarchy does conserve total energy. On the other hand, many closure relations cannot be expressed in terms of

F_{123}, but are formulated in terms of correlation operators. In that case, condition (2.36) does not apply, and the question of energy conservation has to be investigated separately. We will return to this problem in Ch. 2.5 when we study important closure relations to the hierarchy.

vi) For purely distance dependent interaction (2.5), the trace term in Eq. (2.34) vanishes exactly (this will be shown using the coordinate representation in Sec. 2.3.1, cf. Eq. (2.43)). In this case, any closure $F_{123} \to F_{123}^{cl}$ which retains the commutator structure under the trace in Eq. (2.34) is conserving.

vii) The derivation of the expression (2.34) did assume that $F_1(t), F_{12}(t)$ and $F_{123}(t)$ are exact solutions to the actual system of equations of motion (BBGKY-hierarchy + closure relation). This means, that no approximation with respect to the time structure of the density operators has been made (this has also been pointed out in Ref. [BM90]).

viii) Our derivation was made in abstract operator notation and is, therefore, not limited to a particular representation or system. In particular, it can be directly extended to multi-component systems. Furthermore, systems with spin statistics (Bose or Fermi systems) are covered by our analysis too. In this case, the trace operation should be regarded as running over the corresponding subspace of the full Hilbert space of states (for more details see Ch. 3). All the results remain valid.

ix) Moreover, our analysis also includes solid state systems (multiband systems). In this case, one may use the Bloch representation (or some similar representation) of the hierarchy. Again, this will only affect the explicit computation of the trace in Eq. (2.34). The questions related to the Bloch representation will be discussed in Sec. 2.4.1.

These are the main results that can be derived from a very general analysis of the conservation properties of the BBGKY-hierarchy. These results will be used in the further Sections, where they will be applied to particular closure relations to the hierarchy.

Concluding this Section, we give some comments on our results. The analysis of energy conservation is very simple within the density operator formalism, this is one of the advantages of this method. On the other hand, it is instructive to compare the criterion (2.36) with the corresponding energy conservation criterion found in the Green's functions theory. There, a necessary condition for energy conserving approximations is again the symmetry with respect to the permutation of the particle indices. But instead of a condition for three-particle quantities, there it is the permutation invariance of the two-particle correlation function $g_{12}^{<}$ [KB89]. The reason is, of course the dependence of $g_{12}^{<}$ on two times. (For the same reason, the potential energy is calculated from the one-particle correlation function $g_1^{<}(t,t')$ instead of the two-particle density operator in our approach.) A brief comparison with the results of the nonequilibrium Green's functions approach is given in Ch. 12.

2.3 Basic representations of the hierarchy

In this section we consider the most important representations of the BBGKY-hierarchy for the reduced density operators, such as the coordinate, momentum and Wigner representation. First, we need to transform the basis vectors $|\psi\rangle$ into a chosen representation. Then, any $s-$particle operator will be represented by an $s \times s-$matrix.

2.3.1 Coordinate representation

We will denote the basis vectors in the coordinate representation by

$$|\psi_{1\dots N}\rangle \longrightarrow |x_1 \dots x_N\rangle, \qquad x_i = \mathbf{r}_i\, s_i, \tag{2.37}$$

where s_i is the spin projection of particle i. For the matrix elements of an $s-$particle quantity, such as $F_{1\dots s}$, we will use the notation

$$\langle x_1 \dots x_s | F_{1\dots s} | x_s' \dots x_1' \rangle = F_{1\dots s}(x_1, \dots; x_s, x_1', \dots, x_s'), \tag{2.38}$$

We give two elementary examples. First, consider the matrix element of the product of two–particle operators $A_{12}B_{12}$, which is obtained after introduction of the unity–operator

$$\langle x_1 x_2 | A_{12} B_{12} | x_2' x_1' \rangle = \sum_{\bar{s}_1 \bar{s}_2} \int d\bar{r}_1 d\bar{r}_2 \, \langle x_1 x_2 | A_{12} | \bar{x}_2 \bar{x}_1 \rangle \langle \bar{x}_1 \bar{x}_2 | B_{12} | x_2' x_1' \rangle. \tag{2.39}$$

Secondly, we rewrite the normalization condition of the $s-$particle density operator in coordinate representation, calculating the sum of the diagonal elements of the matrix (i.e. $\mathbf{r}_1' = \mathbf{r}_1, s_1' = s_1, \dots \mathbf{r}_s' = \mathbf{r}_s, s_s' = s_s$),

$$\mathcal{V}^s \mathrm{Tr}_{1\dots s} F_{1\dots s} = \sum_{s_1 \dots s_s} \int dr_1 \dots dr_s \, F(x_1, x_2, \dots; x_s, x_1, \dots, x_s).$$

For the derivation of the hierarchy equations in coordinate representation we need the matrix elements of the Hamilton operator (2.3):

$$\langle x_i|H_i|x_i'\rangle = \left\{-\frac{\hbar^2}{2m_i}\nabla_i^2 + \mathcal{U}(x_i)\right\}\delta(x_i - x_i'), \tag{2.40}$$

$$\langle x_j x_i|V_{ij}|x_j' x_i'\rangle = V(\mathbf{r}_i - \mathbf{r}_j)\,\delta(x_i - x_i')\,\delta(x_j - x_j'), \tag{2.41}$$

where $\delta(x_m - x_m') = \delta(\mathbf{r}_m - \mathbf{r}_m')\,\delta_{s_m,s_m'}$. We see that the advantage of the coordinate representation is that not only the matrices of the one-particle operators are diagonal, but also the matrix of the binary interaction potential[7]. Using Eqs. (2.40), (2.41), we obtain the **coordinate representation of the BBGKY-hierarchy** [8]

$$\left\{i\hbar\frac{\partial}{\partial t} - H_{1\dots k}(x_1,\dots,x_k) + H_{1\dots k}(x_1',\dots,x_k')\right\} F_{1\dots k}(x_1\dots x_k; x_1'\dots x_k', t)$$
$$= n\sum_{i=1}^{k}\sum_{s_{k+1}}\int d\mathbf{r}_{k+1}\left\{V(\mathbf{r}_i - \mathbf{r}_{k+1}) - V(\mathbf{r}_i' - \mathbf{r}_{k+1})\right\}$$
$$\times F_{k+1}(x_1,\dots,x_{k+1}; x_1',\dots,x_k', x_{k+1}, t) \tag{2.44}$$

For example, we obtain the coordinate representation of the first hierarchy equation

$$\left\{i\hbar\frac{\partial}{\partial t} + \frac{\hbar^2}{2m_1}(\nabla_1^2 - \nabla_{1'}^2) - \mathcal{U}(x_1,t) + \mathcal{U}(x_1',t)\right\} F_1(x_1; x_1', t) =$$

---

[7]For example, we can calculate commutators involving the potential according to Eq. (2.39)

$$[V_{ij}, A_{ij}] \quad\rightarrow\quad \langle x_i x_j|V_{ij}A_{ij}|x_j' x_i'\rangle = \sum_{\bar{s}_i\bar{s}_j}\int d\bar{r}_i d\bar{r}_j\ \times$$
$$\left\{\langle x_i x_j|V|\bar{x}_j\bar{x}_i\rangle\,\langle\bar{x}_i\bar{x}_j|A|x_j' x_i'\rangle - \langle x_i x_j|A|\bar{x}_j\bar{x}_i\rangle\,\langle\bar{x}_i\bar{x}_j|V|x_j' x_i'\rangle\right\}, \tag{2.42}$$

which simplifies for distance dependent (and spin independent) potentials (2.41) to

$$\int d\bar{r}_i d\bar{r}_j\ \{V(\mathbf{r}_i - \mathbf{r}_j) - V(\mathbf{r}_i' - \mathbf{r}_j')\}\ \langle\bar{x}_i\bar{x}_j|A|x_j' x_i'\rangle\,,$$

Obviously, this expression vanishes for the diagonal elements, $\mathbf{r}_i = \mathbf{r}_i'$ and $\mathbf{r}_j = \mathbf{r}_j'$, which leads to the conclusion

$$\mathrm{Tr}_{ij}\,[V_{ij}, A_{ij}] = 0, \qquad \text{for distance-dependent potentials.} \tag{2.43}$$

Similar conclusions hold for more complicated expressions involving commutators with binary potentials, if the trace runs over all indices of the potentials. In particular, we immediately conclude that the trace term in Eq. (2.34) vanishes for distance-dependent potentials.

[8]For binary potentials which depend not only on the distance, the r.h.s. of Eq. (2.44) contains additional integrations, cf. Eq. (2.42).

$$n\sum_{s_2}\int d\mathbf{r}_2\,\{V(\mathbf{r}_1-\mathbf{r}_2)-V(\mathbf{r}_1'-\mathbf{r}_2)\}\,F_{12}(x_1,x_2;x_1',x_2,t). \tag{2.45}$$

Analogously, we find for the second hierarchy equation

$$\left\{i\hbar\frac{\partial}{\partial t}-H_{12}(x_1,x_2,t)+H_{12}(x_1',x_2',t)\right\}F_{12}(x_1,x_2;x_1',x_2',t)= \tag{2.46}$$

$$n\sum_{i=1}^{2}\sum_{s_3}\int d\mathbf{r}_3\,\{V(\mathbf{r}_i-\mathbf{r}_3)-V(\mathbf{r}_i'-\mathbf{r}_3)\}\,F_{123}(x_1,x_2,x_3;x_1',x_2',x_3,t),$$

$$\text{with}\quad H_{12}(x_1,x_2,t)=\sum_{i=1}^{2}\left\{-\frac{\hbar^2}{2m_i}\nabla_i^2+\mathcal{U}(x_i,t)\right\}+V(\mathbf{r}_1-\mathbf{r}_2) \tag{2.47}$$

In the following, we will not write the spin indices, i.e. we will write $x_i \to r_i$, where r_i is understood as vector $\mathbf{r}_i$.

2.3.2 Wigner representation

Another important representation, due to its close analogy to classical kinetic equations, is the Wigner representation [Wig32]. To derive it, we first introduce center of mass and relative coordinates, R_i and r_i for each particle, according to

$$r_i' = R_i+\frac{r_i}{2},\quad r_i'' = R_i-\frac{r_i}{2},\quad \text{or, vice versa,} \tag{2.48}$$

$$R_i = \frac{r_i'+r_i''}{2},\quad r_i = r_i'-r_i''. \tag{2.49}$$

So, for example $F_1(r_1';r_1'',t) \longrightarrow F_1(R_1+\frac{r_1}{2};R_1-\frac{r_1}{2},t)$. Introducing the new variables (2.48) into the coordinate representation of the hierarchy (2.44), we immediately obtain the representation in terms of center of mass and relative variables. We will use the following notation:

$$F_{1\ldots s}(R_1+\frac{r_1}{2},\ldots,R_s+\frac{r_s}{2};R_1'-\frac{r_1}{2},\ldots,R_s'-\frac{r_s'}{2},t)=\tilde{F}_{1\ldots s}(R_1,r_1,\ldots,R_s,r_s,t),$$

and further take into account that

$$\nabla^2_{r_1'}-\nabla^2_{r_1''}=2\,\nabla_{R_1}\nabla_{r_1}.$$

We now introduce the Wigner transform of the $s-$particle density matrix $\tilde{F}_{1\ldots s}$, which yields the quasiclassical $s-$particle distribution function which depends on the coordinates and the momenta of all particles,

$$F_{1\ldots s}(R_1,p_1,\ldots,R_s,p_s,t) = \int\frac{dr_1}{(2\pi\hbar)^3}\cdots\frac{dr_s}{(2\pi\hbar)^3}\exp\{-i\,(p_1r_1+\ldots p_sr_s)\,/\hbar\}$$
$$\times\;\tilde{F}_{1\ldots s}(R_1,r_1,\ldots,R_s,r_s,t). \tag{2.50}$$

However, more common is the distribution

$$f_{1\ldots s}(R_1, p_1, \ldots, R_s, p_s, t) = n^s \, F_{1\ldots s}(R_1, p_1, \ldots, R_s, p_s, t) \tag{2.51}$$

which is normalized to the particle number according to

$$\int \frac{dr_1 dp_1}{(2\pi\hbar)^3} \cdots \frac{dr_s dp_s}{(2\pi\hbar)^3} f_{1\ldots s}(R_1, p_1, \ldots, R_s, p_s, t) = N^s(t). \tag{2.52}$$

For particles with spin, one has additionally on the l.h.s. the spin summations $\sum_{s_1} \ldots \sum_{s_s}$. If the distribution $f_{1\ldots s}$ does not depend on the spin explicitly, the sums yield the factors $(s_1 + 1) \cdot \ldots \cdot (s_s + 1)$.

The BBGKY-hierarchy in Wigner representation is then given by (for details cf. App. B)

$$\begin{aligned}
&\left\{ \frac{\partial}{\partial t} + \sum_{i=1}^{k} \frac{p_i}{m_i} \nabla_{R_i} \right\} f(R_1 p_1, \ldots, R_k p_k, t) && (2.53)\\
&- \frac{1}{i\hbar} \sum_{1 \le i < j \le k} V_k^{(ij)} - \frac{1}{i\hbar} \sum_i U_k^{(i)} = \frac{1}{i\hbar} \sum_{i=1}^{k} F_{k+1}^{(i)}, \qquad \text{with}
\end{aligned}$$

$$\begin{aligned}
V_k^{(ij)} &= \int \frac{dr_{ij}}{(2\pi\hbar)^3} d\bar{p}_i \, \exp\{-i\,(p_i - \bar{p}_i)\, r_{ij}/\hbar\} && (2.54)\\
&\times \left(V(R_i - R_j + \frac{r_{ij}}{2}) - V(R_i - R_j - \frac{r_{ij}}{2}) \right)\\
&\times f_{1\ldots k}(R_1, p_1, \ldots, R_i, \bar{p}_i, \ldots, R_j, p_i - \bar{p}_i + p_j, \ldots, R_k, p_k, t),\\
U_k^{(i)} &= \int \frac{dr_i}{(2\pi\hbar)^3} d\bar{p}_i \, \exp\{-i\,(p_i - \bar{p}_i)\, r_i/\hbar\} && (2.55)\\
&\times \left(\mathcal{U}(R_i + \frac{r_i}{2}) - \mathcal{U}(R_i - \frac{r_i}{2}) \right)\\
&\times f_{1\ldots k}(R_1, p_1, \ldots, R_{i-1}, p_{i-1} R_i, \bar{p}_i, R_{i+1}, p_{i+1}, \ldots, R_k, p_k, t),\\
F_{k+1}^{(i)} &= \int \frac{dr_i}{(2\pi\hbar)^3} d\bar{p}_i dR_{k+1} \, dp_{k+1} \, \exp\{-i\,(p_i - \bar{p}_i)\, r_i/\hbar\} && (2.56)\\
&\times \left(V(R_i - R_{k+1} + \frac{r_i}{2}) - V(R_i - R_{k+1} - \frac{r_i}{2}) \right)\\
&\times f_{1\ldots k+1}(R_1, p_1, \ldots, R_i, \bar{p}_i, \ldots, R_k, p_k, R_{k+1}, p_{k+1}, t).
\end{aligned}$$

As an example, we give the **first hierarchy equation in Wigner representation**:

$$\begin{aligned}
&\left\{\frac{\partial}{\partial t}+\frac{p_1}{m_1}\nabla_{R_1}\right\} f(R_1,p_1,t) \\
-&\ \frac{1}{i\hbar}\int \frac{dr_1}{(2\pi\hbar)^3}\,d\bar{p}_1\,\exp\{-i\,(p_1-\bar{p}_1)\,r_1/\hbar\} \\
\times&\ \left(\mathcal{U}(R_1+\frac{r_1}{2})-\mathcal{U}(R_1-\frac{r_1}{2})\right) f(R_1,\bar{p}_1,t) \\
=&\ \frac{1}{i\hbar}\int \frac{dr_1}{(2\pi\hbar)^3}\,d\bar{p}_1 dR_2\,dp_2\,\exp\{-i\,(p_1-\bar{p}_1)\,r_1/\hbar\} \\
\times&\ \left(V(R_1-R_2+\frac{r_1}{2})-V(R_1-R_2-\frac{r_1}{2})\right) f_{12}(R_1,\bar{p}_1,R_2,p_2,t)
\end{aligned} \tag{2.57}$$

The advantage of this representation is that all functions depend on the same phase space variables as in the classical case. Nevertheless, one has to recall that the Wigner distributions $f(R,p,t)$, $f_{12}(R_1,p_1,R_2,p_2,t),\ldots$ are not identical to classical distribution functions. In particular, they are not confined to values between 0 and 1. Most importantly, however, is that averages of observables which are calculated with the Wigner functions, are well defined and do not show these problems.

Classical limit. Quantum corrections. The classical limit of the BBGKY-hierarchy is readily calculated from the Wigner representation (2.53). To this end, the potentials V_{ij} and U_i are expanded into a Taylor series with respect to the relative coordinates r_{ij} and r_i, respectively. After straightforward calculations we obtain from the two lowest expansion terms (details of the calculations, including the complete expansion are given in the Appendix B):

$$\left\{\frac{\partial}{\partial t}+\sum_{i=1}^{k}\frac{p_i}{m_i}\nabla_{R_i}\right\} f(R_1p_1,\ldots,R_kp_k,t) \tag{2.58}$$

$$-\ \frac{1}{i\hbar}\sum_{1\le i<j\le k} V^{(ij)}_{cl,k}-\frac{1}{i\hbar}\sum_i U^{(i)}_{cl,k}=\frac{1}{i\hbar}\sum_{i=1}^{k}F^{(i)}_{cl,k+1}+O(\hbar^4),\qquad \text{with}$$

$$\begin{aligned}
\frac{1}{i\hbar}V^{(ij)}_{cl,k} &= \left\{\nabla_{R_i}V(R_i-R_j)\,\nabla_{p_i}+\frac{(i\hbar)^2}{24}\nabla^3_{R_i}V(R_i-R_j)\,\nabla^3_{p_i}\right\} \\
&\times\ f_{1\ldots k}(R_1,p_1,\ldots,R_k,p_k,t),
\end{aligned} \tag{2.59}$$

$$\begin{aligned}
\frac{1}{i\hbar}U^{(i)}_{cl,k} &= \left\{\nabla_{R_i}\mathcal{U}(R_i)\,\nabla_{p_i}+\frac{(i\hbar)^2}{24}\nabla^3_{R_i}\,\mathcal{U}(R_i)\,\nabla^3_{p_i}\right\} \\
&\times\ f_{1\ldots k}(R_1,p_1,\ldots,R_k,p_k,t),
\end{aligned} \tag{2.60}$$

$$\frac{1}{i\hbar}F^{(i)}_{cl,k+1} = \int dR_{k+1}\,dp_{k+1}$$

$$\times \quad \left\{ \nabla_{R_i} V(R_i - R_j) \, \nabla_{p_i} + \frac{(i\hbar)^2}{24} \nabla^3_{R_i} V(R_i - R_j) \, \nabla^3_{p_i} \right\}$$
$$\times \quad f_{1\ldots k+1}(R_1, p_1, \ldots, R_{k+1}, p_{k+1}, t), \tag{2.61}$$

where the differentiations with respect to R_i act only on the potentials. In the parenthesis in $V^{(ij)}_{cl,k}$, $U^{(i)}_{cl,k}$, and $F^{(i)}_{cl,k+1}$, the first terms correspond to the classical limit, whereas the second ones which are of the order of $\hbar^2$ give the first quantum corrections. Again, we give as an example, the **first equation of the classical hierarchy**, which follows by neglecting the quantum corrections,

$$\begin{aligned} &\left\{ \frac{\partial}{\partial t} + \frac{p_1}{m_1} \nabla_{R_1} - \nabla_{R_1} \mathcal{U}(R_1) \, \nabla_{p_1} \right\} f(R_1, p_1, t) \\ = &\int dR_2 \, dp_2 \, \nabla_{R_1} V(R_1 - R_2) \, \nabla_{p_1} f_{12}(R_1, p_1, R_2, p_2, t) \end{aligned} \tag{2.62}$$

At the end of this Section, we mention another useful notation of the classical form of the hierarchy (2.58) which is given by

$$\frac{\partial}{\partial t} f_{1\ldots s} - \{H_{1\ldots s}, f_{1\ldots s}\} = \mathrm{Tr}_{s+1} \sum_{i=1}^{s} \{V_{i,s+1}, f_{1\ldots s+1}\}, \qquad \text{with} \tag{2.63}$$

$$\{A_{1\ldots s}, B_{1\ldots s}\} = \sum_{i=1}^{s} \left(\nabla_{p_i} A_{1\ldots s} \nabla_{R_i} B_{1\ldots s} - \nabla_{p_i} B_{1\ldots s} \nabla_{R_i} A_{1\ldots s} \right), \tag{2.64}$$

where we introduced the Poisson brackets, and $A_{1\ldots s}$ and $B_{1\ldots s}$ are arbitrary functions of the phase space variables. We, thus obtain the well-known result that the classical limit of the quantum-mechanical equations of motion can be obtained by replacing the commutators by the Poisson brackets according to $[A_{1\ldots s}, B_{1\ldots s}] \longrightarrow \frac{1}{i\hbar} \{A_{1\ldots s}, B_{1\ldots s}\}$ and the trace over $s+1$ by the phase space integral $\int dR_{s+1} dp_{s+1}$.

2.3.3 Spatially homogeneous systems. Momentum representation

We now return to the quantum form of the equations of motion for the reduced density operators $F_{1\ldots s}$. Similarly as for the coordinate representation, we now have to transform the basis vectors into the momentum representation.

$$|\psi_{1\ldots N}\rangle \longrightarrow |x_1 \ldots x_N\rangle, \qquad x_i = \mathbf{p}_i \, s_i. \tag{2.65}$$

The matrix element of an $s-$particle quantity, such as $F_{1\ldots s}$ is then again given by Eq. (2.38) where only x_i contains the momentum. There is no need to repeat the transformations of Sec. 2.3.1. Instead, we will consider here the very important case of spatially homogeneous systems. In this case, according to Noether's theorem, the total momentum is conserved, and therefore, the momentum representation is the most convenient one. Moreover, the results of the momentum representation can be applied to weakly inhomogeneous systems too, where the space dependence is reduced to a parametric dependence on the center of mass coordinates R_i (zeroth order gradient expansion). For this reason, most of the following considerations will be carried out in the momentum representation, and we, therefore, provide some details of the derivation. The case of strong inhomogeneities, where the appropriate representation is e.g. the Wigner representation, is discussed in Ch. 4 in connection with the quantum Vlasov equation.

From elementary quantum mechanics it is known that any function of momentum (i.e. of the momentum operator) $R(\mathbf{p})$ will in the momentum representation be given by a diagonal matrix, i.e. $\langle \mathbf{p}|R(\mathbf{p})|\mathbf{p}'\rangle = R(\mathbf{p})\delta(\mathbf{p}-\mathbf{p}')$. So, for example, the operator of the momentum itself or of kinetic energy will be diagonal. On the other hand, functions of the coordinate $Q(\mathbf{r})$ are more complicated. Recall that the momentum representation of $\mathbf{r}$ is [LL62] $\mathbf{r} \to i\hbar\frac{\partial}{\partial \mathbf{p}}$, therefore,

$$\langle \mathbf{p}|Q(\mathbf{r})|\mathbf{p}'\rangle = Q\left(i\hbar\frac{\partial}{\partial \mathbf{p}}\right)\delta(\mathbf{p}-\mathbf{p}'). \tag{2.66}$$

Using this result, e.g. the momentum representation of space or distance dependent external potentials are readily computed. Let us now consider the density operators (to simplify the notation, we again drop the spin variables). In the homogeneous case, one-particle quantities depend only on the difference coordinate, e.g. we have for F_1,

$$\langle \mathbf{r}_1|F_1|\mathbf{r}_1'\rangle = F_1(\mathbf{r}_1 - \mathbf{r}_1'), \tag{2.67}$$

and for the matrix elements we obtain using, for example a basis of free-particle states $\langle \mathbf{p}|\mathbf{r}\rangle = (2\pi\hbar)^{-3/2}e^{-\frac{i}{\hbar}\mathbf{pr}}$,

$$\begin{aligned}\langle \mathbf{p}|F_1|\mathbf{p}'\rangle = \int \frac{d\mathbf{r}d\mathbf{r}'}{(2\pi\hbar)^3}e^{-\frac{i}{\hbar}(\mathbf{pr}-\mathbf{p}'\mathbf{r}')}\,F_1(\mathbf{r}-\mathbf{r}') &= \\ \delta(\mathbf{p}-\mathbf{p}')\int d(\mathbf{r}-\mathbf{r}')\,e^{-\frac{i}{\hbar}\mathbf{p}(\mathbf{r}-\mathbf{r}')}\,F_1(\mathbf{r}-\mathbf{r}') &= F(\mathbf{p})\,\delta(\mathbf{p}-\mathbf{p}'). \end{aligned} \tag{2.68}$$

Eq. (2.68) yields the normalization condition

$$\begin{aligned}1 = \frac{2s+1}{\mathcal{V}}\int d\mathbf{p}_1 F_1(\mathbf{p}_1,\mathbf{p}_1) &= \frac{2s+1}{\mathcal{V}}\int d\mathbf{p}_1 F(\mathbf{p}_1)\,\delta(0) \\ &= (2s+1)\int \frac{d\mathbf{p}_1}{(2\pi\hbar)^3}F(\mathbf{p}_1), \end{aligned} \tag{2.69}$$

where use had been made of $\delta(0) = \frac{\mathcal{V}}{(2\pi\hbar)^3}$. Similarly, we find for $f = nF_1$

$$n(t) = (2s+1) \int \frac{d\mathbf{p}_1}{(2\pi\hbar)^3} f(\mathbf{p}_1, t). \tag{2.70}$$

For the binary interaction potential V_{12}, we obtain from Eq. (2.41)

$$\begin{aligned} \langle \mathbf{p}_1\mathbf{p}_2|V_{12}|\mathbf{p}_2'\mathbf{p}_1'\rangle &= \int d\mathbf{r}_1 d\mathbf{r}_2 \, \langle \mathbf{p}_1\mathbf{p}_2|\mathbf{r}_2\mathbf{r}_1\rangle \, \langle \mathbf{r}_1\mathbf{r}_2|\mathbf{p}_2'\mathbf{p}_1'\rangle \, V_{12}(\mathbf{r}_1 - \mathbf{r}_2) & (2.71) \\ &= \frac{1}{(2\pi\hbar)^{12/2}} \int d\mathbf{r}_1 d\mathbf{r} \, e^{-\frac{i}{\hbar}(\mathbf{p}_1 - \mathbf{p}_1' + \mathbf{p}_2 - \mathbf{p}_2')\mathbf{r}_1} \, e^{-\frac{i}{\hbar}(\mathbf{p}_2 - \mathbf{p}_2')\mathbf{r}} \, V(\mathbf{r}) \\ &= \frac{1}{(2\pi\hbar)^3} V(\mathbf{p}_1 - \mathbf{p}_1') \, \delta(\mathbf{p}_1 + \mathbf{p}_2 - \mathbf{p}_1' - \mathbf{p}_2'), & (2.72) \end{aligned}$$

where we introduced $\mathbf{r} = \mathbf{r}_1 - \mathbf{r}_2$, and the Fourier transform of the potential $V(\mathbf{p}_1 - \mathbf{p}_1') = \int d\mathbf{r} e^{-\frac{i}{\hbar}(\mathbf{p}_1 - \mathbf{p}_1')\mathbf{r}} V(\mathbf{r})$ and used the property of the δ-function $\delta(\mathbf{p}) = \frac{1}{(2\pi\hbar)^3} \int d\mathbf{r} e^{-\frac{i}{\hbar}\mathbf{p}\mathbf{r}}$. For an arbitrary two-particle quantity, we have, in the homogeneous situation, taking the density operator as an example

$$\begin{aligned} \langle \mathbf{r}_1\mathbf{r}_2|F_{12}|\mathbf{r}_2'\mathbf{r}_1'\rangle &= F_{12}(\mathbf{r}_1 - \mathbf{r}_1', \mathbf{r}_2 - \mathbf{r}_1', \mathbf{r}_2' - \mathbf{r}_1'), & (2.73) \\ \langle \mathbf{p}_1\mathbf{p}_2|F_{12}|\mathbf{p}_2'\mathbf{p}_1'\rangle &= \frac{1}{(2\pi\hbar)^3} F_{12}(\mathbf{p}_1, \mathbf{p}_2; \mathbf{p}_2') \, \delta(\mathbf{p}_1 + \mathbf{p}_2 - \mathbf{p}_1' - \mathbf{p}_2'), \\ F_{12}(\mathbf{p}_1, \mathbf{p}_2; \mathbf{p}_2') &= \int d\mathbf{r} d\bar{\mathbf{r}} d\bar{\bar{\mathbf{r}}} \, e^{\frac{i}{\hbar}(\mathbf{p}_1\mathbf{r} + \mathbf{p}_2\bar{\mathbf{r}} - \mathbf{p}_2'\bar{\bar{\mathbf{r}}})} \, F_{12}(\mathbf{r}, \bar{\mathbf{r}}, \bar{\bar{\mathbf{r}}}). \end{aligned}$$

The normalization condition of F_{12} is given by

$$\begin{aligned} 1 &= \frac{(2s+1)^2}{\mathcal{V}^2} \int d\mathbf{p}_1 d\mathbf{p}_2 \langle \mathbf{p}_1\mathbf{p}_2|F_{12}|\mathbf{p}_2\mathbf{p}_1\rangle \\ &= \frac{(2s+1)^2}{\mathcal{V}} \int \frac{d\mathbf{p}_1}{(2\pi\hbar)^3} \frac{d\mathbf{p}_2}{(2\pi\hbar)^3} F_{12}(\mathbf{p}_1, \mathbf{p}_2; \mathbf{p}_2). & (2.74) \end{aligned}$$

For the function f_{12} we obtain

$$\begin{aligned} f_{12}(\mathbf{p}_1, \mathbf{p}_2; \mathbf{p}_2', t) &= n^2 F_{12}(\mathbf{p}_1, \mathbf{p}_2; \mathbf{p}_2', t) \\ Nn &= (2s+1)^2 \int \frac{d\mathbf{p}_1}{(2\pi\hbar)^3} \frac{d\mathbf{p}_2}{(2\pi\hbar)^3} f_{12}(\mathbf{p}_1, \mathbf{p}_2; \mathbf{p}_2). & (2.75) \end{aligned}$$

Analogously, we have for an arbitrary three-particle quantity, in the homogeneous situation,

$$\langle \mathbf{r}_1\mathbf{r}_2\mathbf{r}_3|F_{123}|\mathbf{r}_3'\mathbf{r}_2'\mathbf{r}_1'\rangle = F_{123}(\mathbf{r}_1 - \mathbf{r}_1', \mathbf{r}_2 - \mathbf{r}_1', \mathbf{r}_3 - \mathbf{r}_1', \mathbf{r}_2' - \mathbf{r}_1', \mathbf{r}_3' - \mathbf{r}_1'),$$

$$\begin{aligned}
\langle \mathbf{p}_1\mathbf{p}_2\mathbf{p}_3|F_{123}|\mathbf{p}_3'\mathbf{p}_2'\mathbf{p}_1'\rangle &= \frac{1}{(2\pi\hbar)^6}F_{123}(\mathbf{p}_1,\mathbf{p}_2,\mathbf{p}_3;\mathbf{p}_2',\mathbf{p}_3')\times \\
&\quad \delta(\mathbf{p}_1+\mathbf{p}_2+\mathbf{p}_3-\mathbf{p}_1'-\mathbf{p}_2'-\mathbf{p}_3'), \qquad (2.76)\\
F_{123}(\mathbf{p}_1,\mathbf{p}_2,\mathbf{p}_3;\mathbf{p}_2',\mathbf{p}_3') &= \int d\mathbf{r}_1 d\mathbf{r}_2 d\mathbf{r}_3 d\bar{\mathbf{r}}_2 d\bar{\mathbf{r}}_3 e^{\frac{i}{\hbar}(\mathbf{p}_1\mathbf{r}_1+\mathbf{p}_2\mathbf{r}_2+\mathbf{p}_3\mathbf{r}_3-\mathbf{p}_2'\bar{\mathbf{r}}_2-\mathbf{p}_3'\bar{\mathbf{r}}_3)}\\
&\quad \times F_{123}(\mathbf{r}_1,\mathbf{r}_2,\mathbf{r}_3,\bar{\mathbf{r}}_2,\bar{\mathbf{r}}_3). \qquad (2.77)
\end{aligned}$$

For F_{123} we obtain the normalization condition

$$1=\frac{(2s+1)^3}{\mathcal{V}^2}\int\frac{d\mathbf{p}_1}{(2\pi\hbar)^3}\frac{d\mathbf{p}_2}{(2\pi\hbar)^3}\frac{d\mathbf{p}_3}{(2\pi\hbar)^3}F_{123}(\mathbf{p}_1,\mathbf{p}_2,\mathbf{p}_3;\mathbf{p}_2,\mathbf{p}_3), \qquad (2.78)$$

and for the function f_{123}, analogously,

$$\begin{aligned}
&f_{123}(\mathbf{p}_1,\mathbf{p}_2,\mathbf{p}_3;\mathbf{p}_2',\mathbf{p}_3',t)=n^3F_{123}(\mathbf{p}_1,\mathbf{p}_2,\mathbf{p}_3;\mathbf{p}_2',\mathbf{p}_3',t),\\
&Nn^2=(2s+1)^3\int\frac{d\mathbf{p}_1}{(2\pi\hbar)^3}\frac{d\mathbf{p}_2}{(2\pi\hbar)^3}\frac{d\mathbf{p}_3}{(2\pi\hbar)^3}f_{123}(\mathbf{p}_1,\mathbf{p}_2,\mathbf{p}_3;\mathbf{p}_2,\mathbf{p}_3,t). \qquad (2.79)
\end{aligned}$$

Due to the fact that the matrix elements $\langle\mathbf{r}_1|A_1|\mathbf{r}_1'\rangle$, $\langle\mathbf{r}_1\mathbf{r}_2|A_{12}|\mathbf{r}_2'\mathbf{r}_1'\rangle$ and $\langle\mathbf{r}_1\mathbf{r}_2\mathbf{r}_3|A_{123}|\mathbf{r}_3'\mathbf{r}_2'\mathbf{r}_1'\rangle$ etc. are real quantities, we have for the matrix elements in momentum space

$$\begin{aligned}
\langle\mathbf{p}_1'|A_1|\mathbf{p}_1\rangle &= \langle\mathbf{p}_1|A_1|\mathbf{p}_1'\rangle^*,\\
\langle\mathbf{p}_1'\mathbf{p}_2'|A_{12}|\mathbf{p}_2\mathbf{p}_1\rangle &= \langle\mathbf{p}_1\mathbf{p}_2|A_{12}|\mathbf{p}_2'\mathbf{p}_1'\rangle^*,\\
\langle\mathbf{p}_1'\mathbf{p}_2'\mathbf{p}_3'|A_{123}|\mathbf{p}_3\mathbf{p}_2\mathbf{p}_1\rangle &= \langle\mathbf{p}_1\mathbf{p}_2\mathbf{p}_3|A_{123}|\mathbf{p}_3'\mathbf{p}_2'\mathbf{p}_1'\rangle^*. \qquad (2.80)
\end{aligned}$$

With these results, we can derive the momentum representation of the BBGKY-hierarchy. Due to the homogeneity, the momentum-dependent contributions (kinetic energy, momentum-dependent external potentials) in the Hamilton operators vanish, and we obtain

$$\begin{aligned}
&\frac{\partial}{\partial t}f_{1\ldots k}(p_1,\ldots,p_k;p_1',\ldots,p_k',t)\\
&-\frac{1}{i\hbar}\sum_{1\le i<j\le k}V_{p,k}^{(ij)}-\frac{1}{i\hbar}\sum_{i=1}^{k}U_{p,k}^{(i)}=\frac{1}{i\hbar}\sum_{i=1}^{k}F_{p,k+1}^{(i)}, \qquad (2.81)\\
V_{p,k}^{(ij)} &= (2s+1)\int\frac{dq}{(2\pi\hbar)^3}V(q)\\
&\times\Big\{f_{1\ldots k}(p_1,\ldots,p_i+q,p_j-q,\ldots,p_k;p_1',\ldots,p_k',t)\\
&-f_{1\ldots k}(p_1,\ldots,p_k;p_1',\ldots,p_i+q,p_j-q,\ldots,p_k',t)\Big\}, \qquad (2.82)\\
U_{p,k}^{(i)} &= \Big\{\mathcal{U}_{(x)}(p_i)+\mathcal{U}_{(x)}(p_i')\Big\}f_{1\ldots k}(p_1,\ldots,p_k;p_1',\ldots,p_k',t), \qquad (2.83)
\end{aligned}$$

$$\begin{aligned} F^{(i)}_{p,k+1} &= (2s+1)\int \frac{dq}{(2\pi\hbar)^3}\frac{dp_{k+1}}{(2\pi\hbar)^3}V(q) \\ &\times \Big\{ f_{1\dots k+1}(p_1,\dots,p_i+q,\dots,p_{k+1}-q;p_1',\dots,p_{k+1}',t) \\ &- f_{1\dots k+1}(p_1,\dots,p_{k+1};p_1',\dots,p_i+q,\dots,p_{k+1}'-q,t)\Big\}, \end{aligned} \tag{2.84}$$

where, due to momentum conservation, the whole equation has to be considered for $p_1+\ldots+p_k = p_1'+\ldots+p_k'$. $\mathcal{U}_{(x)}$ is the matrix element of the coordinate-dependent part of the external potential, $\langle p_1|\mathcal{U}_{(x)1}|p_1'\rangle = \delta(p_1-p_1')\mathcal{U}_{(x)}(p_1)$. For example, the potential of a longitudinal electric field $\mathcal{E}$ is $\mathcal{U}_1 = -e_1\mathcal{E}\mathbf{r}_1$, and the corresponding matrix element follows, using Eq. (2.66), $\mathcal{U}_{(x)}(\mathbf{p}_1) = -i\hbar e_1\mathcal{E}\partial/\partial\mathbf{p}_1$. The **momentum representation of the first hierarchy equation** looks particularly simple:

$$\boxed{\begin{gathered}\left\{\frac{\partial}{\partial t} - \frac{1}{i\hbar}\mathcal{U}_{(x)}(p_1)\right\} f(p_1,t) = \\ \frac{2s+1}{\hbar}\int \frac{dq}{(2\pi\hbar)^3}\frac{dp_2}{(2\pi\hbar)^3}V(q)\,\mathrm{Im}\, f_{12}(p_1+q,p_2-q;p_1,p_2,t)\end{gathered}} \tag{2.85}$$

2.4 Multi-component and multi-band systems

The results obtained so far can be straightforwardly generalized to systems consisting of several species. We now have to generalize the state vectors $|b_N\ldots b_1\rangle$ to $b_i = \lambda_i x_i$. For clarity, we will write the component index λ_i explicitly. This new index comprises all quantum numbers which are related to the species (such as the quantum numbers of atomic levels, the band index in the case of a solid, subband index of quantum confined structures and so on). We first consider inhomogeneous systems, where we use the coordinate representation.

Inhomogeneous systems. The matrix element of an $s-$particle quantity, such as $f_{1\ldots s} = n^s F_{1\ldots s}$ is now[9]

$$\langle \lambda_1 x_1\ldots\lambda_s x_s|f_{1\ldots s}|\lambda_s' x_s'\ldots\lambda_1' x_1\rangle = f_{1\ldots s}(\lambda_1,x_1,\ldots,\lambda_s,x_s;\lambda_1',x_1',\ldots,\lambda_s',x_s').$$

The matrix elements of the Hamilton operator (2.3) contain terms of the fol-

[9]Notice that the normalization, e.g. of f_1, reads now $\int dx_1 f(\lambda_1,x_1;\lambda_1,x_1) = N_{\lambda_1}$, where N_{λ_1} is the number of particles on level (or of species) "λ_1".

lowing type (for example) [10]:

$$\begin{aligned}
\langle \lambda_i x_i | H_i | \lambda_i' x_i' \rangle &= \left\{ -\frac{\hbar^2}{2\, m_i} \nabla_i^2 \, \delta_{\lambda_i, \lambda_i'} + \mathcal{U}^{\lambda_i \lambda_i'}(x_i) \right\} \delta(x_i - x_i'), \\
\langle \lambda_i x_i \lambda_j x_j | V_{ij} | \lambda_j' x_j' \lambda_i' x_i' \rangle &= V^{\lambda_i \lambda_j}(\mathbf{r}_i - \mathbf{r}_j) \, \delta_{\lambda_i, \lambda_i'} \, \delta_{\lambda_j, \lambda_j'} \, \delta(x_i - x_i') \, \delta(x_j - x_j').
\end{aligned}$$

Now we can rewrite the BBGKY-hierarchy for a multi-component system in coordinate representation:

$$\begin{aligned}
&\left\{ i\hbar \frac{\partial}{\partial t} - \sum_{i=1}^{k} H_{1\ldots k}^{(i)} - \sum_{1 \le i < j \le k} H_{1\ldots k}^{(ij)} \right\} f_{1\ldots k}(\lambda_i, x_i, \ldots; \lambda_k', x_k', t) = \\
&\sum_{i=1}^{k} \sum_{s_{k+1}} \sum_{\lambda_{k+1}} \int d\mathbf{r}_{k+1} \left\{ V^{\lambda_i \lambda_{k+1}}(\mathbf{r}_i - \mathbf{r}_{k+1}) - V^{\lambda_i' \lambda_{k+1}}(\mathbf{r}_i' - \mathbf{r}_{k+1}) \right\} \times \\
&f_{k+1}(\lambda_1, x_1, \ldots, \lambda_{k+1}, x_{k+1}; \lambda_1' x_1' \ldots, \lambda_{k+1}, x_{k+1}, t), \quad \text{with} \qquad (2.86) \\
H_{1\ldots k}^{(i)} &= -\frac{\hbar^2}{2\, m_i} (\nabla_i^2 - \nabla_{i'}^2) \, f_{1\ldots k}(\ldots) + \\
&\sum_{\bar{\lambda}_i} \left\{ \mathcal{U}^{\lambda_i \bar{\lambda}_i}(x_i) \, f_{1\ldots k}(\ldots, \lambda_i \to \bar{\lambda}_i) - f_{1\ldots k}(\ldots, \lambda_i' \to \bar{\lambda}_i) \mathcal{U}^{\bar{\lambda}_i \lambda_i'}(x_i) \right\}, \\
H_{1\ldots k}^{(ij)} &= \left\{ V^{\lambda_i \lambda_j}(\mathbf{r}_i - \mathbf{r}_j) - V^{\lambda_i' \lambda_j'}(\mathbf{r'}_i - \mathbf{r'}_j) \right\} f_{1\ldots k}(\ldots),
\end{aligned}$$

where only those arguments of the distribution functions are shown which are different from the ones on the l.h.s., and the trace on the r.h.s. adds now a summation over the band index.

Homogeneous systems. We now consider the momentum representation, i.e. use $b_i = \lambda_1 p_i s_i$, where the spin index will be suppressed. We will use the following notations for the matrix elements of one-particle and two-particle quantities, where we explicitly account for the homogeneity

$$\langle \lambda_1 p_1 | A_1 | \lambda_1' p_1' \rangle = A^{\lambda_1 \lambda_1'}(p_1) \, \delta(p_1 - p_1'), \tag{2.87}$$

$$\langle \lambda_1 p_1 \lambda_2 p_2 | B_{12} | \lambda_2' p_2' \lambda_1' p_1' \rangle = B_{\lambda_2 \lambda_2'}^{\lambda_1 \lambda_1'}(p_1 p_2; p_2') \, \frac{\delta(p_1 + p_2 - p_1' - p_2')}{(2\pi\hbar)^3}, \tag{2.88}$$

To derive the hierarchy equations for a multi-component system we have to calculate the following matrix elements

$$\begin{aligned}
\langle \lambda p | H^0 | \lambda' p' \rangle &= E_\lambda(p) \, \delta_{\lambda, \lambda'} \, \delta(p - p'), \\
\langle \lambda p | \mathcal{U}(t) | \lambda' p' \rangle &= \mathcal{U}^{\lambda \lambda'}(p, t) \, \delta(p - p'), \\
\langle \lambda_1 p_1 \lambda_2 p_2 | V_{12} | \lambda_2' p_2' \lambda_1' p_1' \rangle &= \frac{V^{\lambda_1 \lambda_2}(p_1 - p_1')}{(2\pi\hbar)^3} \, \delta_{\lambda_1, \lambda_1'} \, \delta_{\lambda_2, \lambda_2'} \, \delta(p_1 + p_2 - p_1' - p_2'),
\end{aligned}$$

[10]This example describes binary interaction potentials in a multicomponent system without inelastic processes. Other situations are treated analogously.

where H_i^0 is the one-particle energy operator with the eigenvalue E_i. Also, we recall that $\mathcal{U}(p,t)$ may be an operator expression in case of coordinate dependent potentials, as it was discussed in Sec. 2.3.3. Furthermore, we take into account that the trace over an arbitrary operator A is now given by $\mathrm{Tr}_1 A_1 = \sum_{\lambda_1} \int dp_1 \, \langle \lambda_1 p_1 | A | \lambda_1 p_1 \rangle$. Using these results[11], we obtain the BBGKY-hierarchy of a spatially homogeneous multi-component system

$$
\begin{aligned}
&\frac{\partial}{\partial t} f_{1\ldots k}(\lambda_1 p_1, \ldots, \lambda_k p_k, \lambda_1' p_1', \ldots, \lambda_k' p_k', t) \\
-\;& \frac{1}{i\hbar} \sum_{i=1}^{k} \left(E_k^{(i)} + U_k^{(i)} \right) - \frac{1}{i\hbar} \sum_{1 \le i < j \le k} V_{p,k}^{(ij)} = \frac{1}{i\hbar} \sum_{i=1}^{k} F_{p,k+1}^{(i)}, \qquad (2.92)\\
E_k^{(i)} \;=\;& \left\{ E_{\lambda_i}(p_i) - E_{\lambda_i'}(p_i) \right\} f_{1\ldots k}(\ldots), \\
U_k^{(i)} \;=\;& \sum_{\bar\lambda_i} \left\{ \mathcal{U}^{\lambda_i \bar\lambda_i}(p_i) \, f_{1\ldots k}(\ldots, \lambda_i \to \bar\lambda_i) - f_{1\ldots k}(\ldots, \lambda_i' \to \bar\lambda_i) \, \mathcal{U}^{\bar\lambda_i \lambda_i'}(p_i) \right\}, \\
V_{p,k}^{(ij)} \;=\;& \int \frac{dq}{(2\pi\hbar)^3} V(q) \Big\{ f_{1\ldots k}(\ldots, p_i \to p_i + q, p_j \to p_j - q) \\
-\;& f_{1\ldots k}(\ldots, p_i' \to p_i + q, p_j' \to p_j - q) \Big\}, \\
F_{p,k+1}^{(i)} \;=\;& \sum_{\lambda_{k+1}} \int \frac{dq}{(2\pi\hbar)^3} \frac{dp_{k+1}}{(2\pi\hbar)^3} V(q) \times \Big\{
\end{aligned}
$$

[11] We also need to calculate the commutators $[H_i^0, F_i]$, $[\mathcal{U}_i, F_i]$ and $[V_{ij}, F_{ij}]$. For the first commutator we find

$$
\langle \lambda_i p_i | [H_i^0, F_i] | \lambda_i' p_i' \rangle = \left\{ E_{\lambda_i}(p_i) - E_{\lambda_i'}(p_i) \right\} f^{\lambda_i \lambda_i'}(p) \, \delta(p_i - p_i'). \qquad (2.89)
$$

This term obviously vanishes for diagonal elements $\lambda_i = \lambda_i'$ (homogeneous case). Only for off-diagonal elements, i.e. for transitions from component i to i', the commutators contribute being proportional to the energy difference of both states. Next, consider the commutator with the external potential. For example, the term appearing in the first hierarchy equation is given by

$$
\langle \lambda_i p_i | [\mathcal{U}_i(t), F_i] | \lambda_i' p_i' \rangle = \delta(p_i - p_i') \sum_{\bar\lambda_i} \left\{ \mathcal{U}^{\lambda_i \bar\lambda_i'}(p_i) \, f^{\bar\lambda_i \lambda_i'}(p_i) - f^{\lambda_i \bar\lambda_i'}(p_i) \, \mathcal{U}^{\bar\lambda_i \lambda_i'}(p_i) \right\}. \qquad (2.90)
$$

For the commutators involving the binary interaction potential, we find expressions of the form (here given for the two-particle density operator)

$$
\begin{aligned}
\langle \lambda_i p_i \lambda_j p_j | [V_{ij}, F_{ij}] | \lambda_j' p_j' \lambda_i' p_i' \rangle \;&=\; \delta(p_i + p_j - p_i' - p_j') \int \frac{dq}{(2\pi\hbar)^3} V^{\lambda_i \lambda_j}(q) \\
&\times\; 2 \, \mathrm{Im} \, f_{\lambda_j \lambda_j'}^{\lambda_i \lambda_i'}(p_i + q, p_j - q; p_i', p_j'). \qquad (2.91)
\end{aligned}
$$

$$f_{1\dots k+1}(\dots, p_i \to p_i + q, p_{k+1} \to p_{k+1} - q, \lambda'_{k+1} \to \lambda_{k+1}, p'_{k+1} \to p_{k+1})$$
$$- \quad f_{1\dots k+1}(\dots, p'_i \to p_i + q, p'_{k+1} \to p_{k+1} - q, \lambda'_{k+1} \to \lambda_{k+1})\Big\},$$

where, only those arguments of the distribution functions are shown which are different from the ones on the l.h.s, and, due to momentum conservation, the whole equation has to be considered for $p_1 + \dots + p_k = p'_1 + \dots + p'_k$. Eq. (2.92) is a very general result. It contains the equations of motion for all matrix elements of the reduced $s-$particle density operator, with $s = 1, \dots N-1$. The diagonal matrix elements correspond to the probability to find the population of certain states of the $s-$particle complex. In many situations, only the diagonal elements are nonzero. This is the case if the multi–component system consists of different physical or chemical species. We discuss the role of the off-diagonal elements separately in the next section.

2.4.1 Bloch representation of the hierarchy

An important special case of multi-component systems is the one where each component (given by the index λ_i) denotes not a different species, but a different state of a complex system. This can be energy levels of bound complexes, such as atoms, molecules, clusters, nuclei etc., as well as bands or subbands in a solid or a low-dimensional semiconductor structure, respectively. We will assume that the particles "sitting" on these levels are all of the same kind. This is the case in solids or plasmas, where these particles are electrons distributed over the energy bands or bound state levels, respectively. We will see in the following, that our approach which is based on the BBGKY-hierarchy in abstract operator form, is very well suited for the investigation of the dynamics of these systems. Its advantage is, that most of the questions which are related to the decoupling of the hierarchy, to physical approximations etc., can be solved on the compact operator level. Only thereafter, one can derive the equations in the Bloch representation by a simple projection onto the Bloch basis [DKBB97, BDK98]. The details of this approach will be discussed below in Chapter 11.

Using the Bloch basis, the matrix $f_{1\dots s}$ describes the joint probability of given momenta (momentum representation) of s particles together with their "affiliation" to one of the s levels. The time evolution of this function is, in the inhomogeneous case, given by Eq. (2.86), and, for homogeneous systems, by Eq. (2.92), respectively. The diagonal matrix elements give the joint probability of one particular configuration of momenta and levels for all particles, whereas the off-diagonal elements yield the probability of transitions between two different configurations. This description is well-known from atomic optics, and the resulting system of equations for the matrix elements of the single-particle density operator are usually called Bloch equations. The hierarchy is the natural generalization of these equations to joint two-particle, three-particle etc. probability densities. In solids, additional effects related to the interaction between the particles, have to be taken into account, which leads to generalizations of the Bloch equations by the inclusion of mean-field and correlation (scattering, screening etc.) effects.

For the important case of optical excitation, the external potential $\mathcal{U}$ is determined by the characteristics of the electro-magnetic field. The specific form of the field–matter interaction part of the Hamiltonian, H^{int}, strongly depends on the actual situation, both on the material (selection rules etc.) and on the characteristics of the light (spectral properties, intensity and so on). The most important example is the dipole interaction.

Example: Electro-magnetic field in dipole approximation. To illustrate our results, we consider now the important example of charged particles in an electromagnetic field. We will consider the simplest case - that of dipole interaction, where the external potential is related to a nonrelativistic electric field by

$$\mathcal{U}_i = -\mathbf{d_i} \cdot \mathcal{E}. \tag{2.93}$$

$\mathcal{E}$ is the operator of the electric field and $\mathbf{d}_i = -e_i\mathbf{r}$ is the operator of the field-induced dipole momentum of particle i (for electrons, $e_i < 0$, and d is positive). A derivation of this expression along with a more general treatment of the electromagnetic field - matter interaction will be given later in Ch. 11. The matrix elements of the potential $\mathcal{U}_i$ are now (we consider the spatially homogeneous case)

$$\langle \lambda_i p_i | \mathcal{U}_i | \lambda_i' p_i' \rangle = \mathbf{d}^{\lambda_i \lambda_i'}(p_i)\, \delta(p_i' - p_i)\, \mathcal{E}(t), \tag{2.94}$$

where the dipole matrix elements have the properties

$$\begin{aligned} \mathbf{d}^{\lambda\lambda}(p) &= 0, \\ \mathbf{d}^{\lambda\lambda'}(p) &= [\mathbf{d}^{\lambda'\lambda}(p)]^*, \quad \lambda' \neq \lambda, \\ \text{or, compactly,}\quad \mathbf{d}^{\lambda\lambda'}(p) &= \mathbf{d}^{\lambda\lambda'}(p)(1 - \delta_{\lambda\lambda'}). \end{aligned} \tag{2.95}$$

The BBGKY-hierarchy for a homogeneous system with dipole interaction is then given by Eq. (2.92), where the only modification is in the term $U_k^{(i)}$:

$$U_k^{(i)} = -\mathcal{E} \sum_{\bar{\lambda}_i} \left\{ \mathbf{d}^{\lambda_i \bar{\lambda}_i}(p_i)\, f_{1\ldots k}(\ldots, \lambda_i \to \bar{\lambda}_i) - f_{1\ldots k}(\ldots, \lambda_i' \to \bar{\lambda}_i)\, \mathbf{d}^{\bar{\lambda}_i \lambda_i'}(p_i) \right\}$$

For example, the first hierarchy equation in Bloch representation for the case of dipole interaction, i.e. the **optical Bloch equations** read:

$$\boxed{\begin{aligned}&\frac{\partial}{\partial t} f^{\lambda\lambda'}(p,t) - \{E_\lambda(p) - E_{\lambda'}(p)\} f^{\lambda\lambda'}(p,t)\\ &+\mathcal{E}(t) \sum_{\bar\lambda} \left\{\mathbf{d}^{\lambda\bar\lambda}(p)\, f^{\bar\lambda\lambda'}(p,t) - f^{\lambda\bar\lambda}(p,t)\, \mathbf{d}^{\bar\lambda\lambda'}(p)\right\}\\ &= \sum_{\lambda_2} \int \frac{dq}{(2\pi\hbar)^3} \frac{dp_2}{(2\pi\hbar)^3} V(q) \times \\ &\left\{ f^{\lambda\lambda'}_{\lambda_2\lambda_2}(p+q, p_2-q, p, p_2, t) - f^{\lambda\lambda'}_{\lambda_2\lambda_2}(p, p_2, p+q, p_2-q, t)\right\}\end{aligned}} \tag{2.96}$$

The physical meaning of this equation is obvious: The single–particle density operator of an arbitrary electron is now represented by a $N \times N$ matrix (N is the number of levels). The N diagonal elements correspond to the probability to find an electron in a particular band, while the off–diagonal terms $f^{\lambda\lambda'}$ correspond to the probability of transitions from band λ to band λ'. One clearly sees that the transition probabilities have a tendency to oscillate in time with difference of the band energies. Also, the influence of the field on the level population and the interband transition is clear: It leads to a coupling of off–diagonal elements to diagonal ones and vice versa. This means, in the presence of the field, the band populations are driven by the transitions and vice versa. The terms on the r.h.s. describe the coupling of the chosen electron to the other electrons in the system: it contains mean field effects and also correlations. These effects are fully included in the matrix elements of the two–particle density operator. The equation of motion for this quantity, Eq. (2.97) couples to the three–particle matrix and so on.

For completeness, the second hierarchy equation is given below (due to momentum conservation, $p_1 + p_2 = p_1' + p_2'$)

$$\begin{aligned}&\frac{\partial}{\partial t} f^{\lambda_1\lambda_1'}_{\lambda_2\lambda_2'}(p_1,p_2,p_1',p_2',t) - \{E_{\lambda_1}(p_1) + E_{\lambda_2}(p_2) - E_{\lambda_1'}(p_1') - E_{\lambda_2'}(p_2')\} f^{\lambda_1\lambda_1'}_{\lambda_2\lambda_2'}(\ldots)\\ &\quad +\mathcal{E}(t) \sum_{\bar\lambda} \left\{\mathbf{d}^{\lambda_1\bar\lambda}(p_1)\, f^{\bar\lambda\lambda_1'}_{\lambda_2\lambda_2'}(\ldots) + f^{\lambda_1\bar\lambda_1}_{\lambda_2\lambda_2'}(\ldots)\, \mathbf{d}^{\bar\lambda\lambda_1'}(p_1')\right\}\\ &\quad -\mathcal{E}(t) \sum_{\bar\lambda} \left\{\mathbf{d}^{\lambda_2\bar\lambda}(p_2)\, f^{\lambda_1\lambda_1'}_{\bar\lambda\lambda_2'}(\ldots) - f^{\lambda_1\lambda_1'}_{\lambda_2\bar\lambda_2}(\ldots)\, \mathbf{d}^{\bar\lambda\lambda_2'}(p_2')\right\}\\ &= \sum_{\lambda_3} \int \frac{dq}{(2\pi\hbar)^3} \frac{dp_3}{(2\pi\hbar)^3} V(q) \times\\ &\qquad \Big\{ f_{123}(\lambda_1, p_1+q, \lambda_2, p_2, \lambda_3, p_3-q, \lambda_1', p_1', \lambda_2', p_2', \lambda_3, p_3, t)\\ &\qquad - f_{123}(\lambda_1, p_1, \lambda_2, p_2, \lambda_3, p_3, \lambda_1', p_1+q, \lambda_2', p_2', \lambda_3, p_3-q, t)\\ &\qquad + f_{123}(\lambda_1, p_1, \lambda_2, p_2+q, \lambda_3, p_3-q, \lambda_1', p_1', \lambda_2', p_2', \lambda_3, p_3, t)\end{aligned}$$

$$-f_{123}(\lambda_1, p_1, \lambda_2, p_2, \lambda_3, p_3, \lambda_1', p_1, \lambda_2' + q, p_2', \lambda_3, p_3 - q, t)\Big\}. \tag{2.97}$$

The hierarchy equations derived in this section are generally valid for quantum systems with dipole interaction. They contain no limitation on the number of atomic levels (energy bands) in the system, i.e. the indices λ_i or all density matrices and in the sums run over all possible levels. In practice, however, the relevant number of levels and, correspondingly, the number of dipole transitions are limited: Only those levels will be of importance which have a spacing close to the energy of the incoming photon. This may strongly depend on the state of the system, in particular on the density, which affects the level spacing. Moreover, the exciting field may strongly influence the "selection" of the active transitions, and this selection may evolve in time, because the spectral width of the field and also the system's response change. Thus, one observes a very complex dynamic interplay of matter with the electro–magnetic field which, in general, requires a selfconsistent quantum–kinetic treatment. We will study these problems more in detail in Ch. 11.

2.4.2 Remarks on general properties of the BBGKY-hierarchy

Before moving on to special approximations of the hierarchy, we conclude this section with some remarks on the results obtained so far.

i) We considered a variety of representations for the BBGKY-hierarchy. These results include the corresponding representation of the von Neumann equation of the full density operator $\rho_{1...N}$ (substitute $s \to N$ and neglect the coupling terms on the r.h.s. of the hierarchy equations).

ii) We considered only the most general representations of the BBGKY-hierarchy. This is sufficient to understand the principal points. The extension to more complex situations, in particular to specific other systems of basis vectors, is straightforward.

iii) So far we considered only the reduced density operators $F_{1...s}$ and discussed their equations of motion. In correlated many-particle systems, it is often more convenient to consider instead of $F_{12}, \dots, F_{1...s}$ the correlation operators $g_{12}, \dots, g_{1...s}$ and their dynamics. This will be done in Sec. 2.5.1.

iv) Up to now, we did not consider effects of the spin statistics of Bose or Fermi particles which give rise to Pauli blocking and exchange. While the

BBGKY-hierarchy in the operator form (2.16) is correct in this case too (the spin statistics can be taken into account in calculating the trace), the various representations of the hierarchy discussed above are not complete. Nevertheless, they contain the basic physical effects, so their investigation is very instructive. The inclusion of effects of Fermi or Bose statistics will be done in Ch. 3.

2.5 Correlations in many-particle systems

There exists an enormous literature on statistical theory in general or on its application to different fields, many remarkable books have been devoted to this subject. All these investigations are centered around the problem of correlations in interacting many-particle systems. And the major differences in all these treatments is the way correlations are accounted for or, in other words, the choices for the decoupling of the hierarchy - both physically appropriate and practically feasible. We, therefore, will discuss the decoupling problem in detail in Sec. 2.6 where we compare the most important approaches.

Of special relevance for the derivation of quantum kinetic equations applicable to short times is a recently developed approach which allows to include selfenergy effects into the density operator concept [BK96, KBKS97]. We, therefore, will focus on this method in applications in Chapters 7-10. This approach will be developed step by step as we proceed in the analysis of correlation effects, beginning with Sec. 2.6.1.

2.5.1 BBGKY-hierarchy for correlation operators

It is advantageous to rewrite the equations of the BBGKY-hierarchy, Eq. (2.16) in terms of correlation operators g_{12}, g_{123} etc. These are given e.g. by the well-known **cluster expansion for the density operators**[12]

$$\boxed{\begin{aligned} F_{12} &= F_1F_2 + g_{12} \\ F_{123} &= F_1F_2F_3 + g_{23}F_1 + \ldots + g_{123} \\ F_{1234} &= F_1F_2F_3F_4 + g_{34}F_1F_2 + \ldots + g_{12}g_{34} + \ldots + g_{234}F_1 + \ldots + g_{1234} \\ &\ldots \end{aligned}} \tag{2.98}$$

The dots denote contributions arising from permutations of the particle indices in the previous term. As one can see, the density operators F_{12}, F_{123} etc.

[12] A generalization of this expansion which includes the spin statistics explicitly is discussed in Ch. 3.

contain one-particle and higher order contributions. Products of one-particle density operators $F_1 F_2 \dots F_s$ correspond to the uncorrelated superposition of s particles, whereas g_{12}, $g_{123}, \dots$ describe correlations of two, three or more particles which are caused by their interaction. We now rewrite the first three **hierarchy equations in terms of correlation operators**, (details of the derivation can be found in Appendix C):

$$
\begin{gathered}
i\hbar \frac{\partial}{\partial t} F_1 - [\bar{H}_1, F_1] = n\mathrm{Tr}_2[V_{12}, g_{12}] \\
i\hbar \frac{\partial}{\partial t} g_{12} - [\bar{H}_{12}, g_{12}] = [V_{12}, F_1 F_2] + \\
n\mathrm{Tr}_3\Big\{[V_{13}, F_1 g_{23}] + [V_{23}, F_2 g_{13}] + [V_{13} + V_{23}, g_{123}]\Big\} \\
i\hbar \frac{\partial}{\partial t} g_{123} - [\bar{H}_{123}, g_{123}] = [V_{12} + V_{13} + V_{23}, F_1 F_2 F_3] + \\
[V_{13} + V_{23}, F_3 g_{12}] + [V_{12} + V_{23}, F_2 g_{13}] + [V_{12} + V_{13}, F_1 g_{23}] + \\
n\mathrm{Tr}_4[V_{14} + V_{24}, g_{12} g_{34}] + n\mathrm{Tr}_4[V_{14} + V_{34}, g_{13} g_{24}] + n\mathrm{Tr}_4[V_{24} + V_{34}, g_{14} g_{23}] \\
+ n\mathrm{Tr}_4[V_{14}, F_1 g_{234}] + n\mathrm{Tr}_4[V_{24}, F_2 g_{134}] + n\mathrm{Tr}_4[V_{34}, F_3 g_{124}] \\
+ n\mathrm{Tr}_4[V_{14} + V_{24} + V_{34}, g_{1234}]
\end{gathered}
\tag{2.99}
$$

where we introduced the effective Hamiltonians which contain an effective potential (Hartree potential or mean field) U^H:

$$
\begin{aligned}
\bar{H}_1 &= H_1 + U_1^H, & (2.100)\\
\bar{H}_{12} &= \bar{H}_1 + \bar{H}_2 + V_{12}, & (2.101)\\
\bar{H}_{123} &= \bar{H}_1 + \bar{H}_2 + \bar{H}_3 + V_{12} + V_{13} + V_{23} & (2.102)\\
U_1^H &= n\mathrm{Tr}_2\, V_{12}\, F_2. & (2.103)
\end{aligned}
$$

We will also need the effective free Hamilton operators which do not contain the interaction potential, e.g.

$$
\begin{aligned}
\bar{H}_1^0 &= H_1 + U_1^H, & (2.104)\\
\bar{H}_{12}^0 &= \bar{H}_1 + \bar{H}_2, & (2.105)
\end{aligned}
$$

and so on. The external field $\mathcal{U}$ is contained in H_1, cf. Eq. (2.4), and will not be written explicitly.

Equations (2.99) are still exact. They are coupled to the higher order equations via g_{1234}. The generalization to the higher order equations is cumbersome but straightforward. The general structure of these equations

$$i\hbar\frac{\partial}{\partial t}g_{1\ldots s} - [\bar{H}_{1\ldots s}, g_{1\ldots s}] = I_{1\ldots s} + \Pi_{1\ldots s} + C_{s+1}. \tag{2.106}$$

Here, $\bar{H}_{1\ldots s} = \sum_{i=1}^{s} \bar{H}_i + \sum_{i<j} V_{ij}$. The terms coupling to $g_{1\ldots s+1}$ are $C_{s+1} = n\mathrm{Tr}_{s+1}\sum_{i=1}^{s}[V_{i,s+1}, g_{1\ldots s+1}]$. $I_{1\ldots s}$ comprises all inhomogeneity contributions which involve only operators of lower order than s, and $\Pi_{1\ldots s}$ is the generalized polarization term which contains $g_{1\ldots s}$ with index combinations different from the one on the l.h.s. (s-particle polarization terms) and lower order polarization contributions involving $g_{12}\ldots g_{1\ldots s-1}$ under the trace over $s+1$.

2.5.2 Energy conservation condition in terms of correlation operators

Using the hierarchy in terms of correlation operators, one can systematically derive approximations. The most important consistency criterion for any decoupling approximation to the BBGKY-hierarchy is that the conservation properties of the exact hierarchy are retained. A general necessary and sufficient criterion was found above, cf. Eq. (2.34). However, this result was given in terms of the three-particle density operator F_{123}. We will now express condition (2.34) in terms of correlation operators. Consider the following term which enters expression (2.34),

$$\begin{aligned} V_{13}F_{123} &= V_{13}F_1F_2F_3 + V_{13}F_2g_{13} \\ &+ V_{13}F_3g_{12} \\ &+ V_{13}F_1g_{23} \\ &+ V_{13}\,g_{123}, \end{aligned} \tag{2.107}$$

and analogously for $1 \leftrightarrow 2$. The terms in the first line are required for deriving the second equation (2.99), (they are canceled exactly by the equation for F_1F_2). The term in the second line gives the Hartree term $U_1^H g_{12}$, and the third line gives rise to the polarization contributions (see Appendix C). We now consider the properties of each term in Eq. (2.107). The term $F_1F_2F_3$ is trivially symmetric with respect to index permutations and is thus, due of the sufficient condition (2.36), conserving. Consider now one of the polarization terms (full commutator):

$$\mathrm{Tr}_{123}V_{12}\left\{V_{13}F_1g_{23} - F_1g_{23}V_{13}\right\} = \mathrm{Tr}_{123}\left\{V_{12}V_{13}F_1g_{23} - V_{13}F_1g_{23}V_{12}\right\} = 0,$$

where we interchanged the indices $2 \leftrightarrow 3$ in the second term. Using he invariance of the trace, one verifies that both terms cancel. Consider next the contribution of the Hartree terms to Eq. (2.34),

$$\begin{aligned} &\mathrm{Tr}_{123} V_{12} \left\{ V_{13} F_3 g_{12} - F_3 g_{12} V_{13} \right\} = \\ &\mathrm{Tr}_{123} \left\{ V_{12} V_{13} - V_{13} V_{12} \right\} F_3 g_{12} = \frac{d}{dt} \langle H_1^H \rangle . \end{aligned} \tag{2.108}$$

This expression is, in general, nonzero, but vanishes, for example, in the spatially homogeneous situation. It also vanishes if the potentials depend only on the interparticle distance (to show this, use the coordinate representation, where these potentials are diagonal, cf. Eq. (2.41)). Finally, consider the second term on the r.h.s. in (Eq. 2.107),

$$\begin{aligned} &\mathrm{Tr}_{123} V_{12} \left\{ V_{13} F_2 g_{13} - F_2 g_{13} V_{13} \right\} = \mathrm{Tr}_{123} \left\{ V_{12} V_{13} - V_{13} V_{12} \right\} F_2 g_{13} = \\ &\mathrm{Tr}_{123} \left\{ V_{13} V_{12} - V_{12} V_{13} \right\} F_3 g_{12} = -\frac{d}{dt} \langle H_1^H \rangle . \end{aligned} \tag{2.109}$$

The last line was obtained by interchanging $2 \leftrightarrow 3$ and yields just minus the Hartree contribution. We summarize the main conclusions:

1. Inclusion/neglect of the polarization terms does not influence the conservation properties (though it affects the absolute value of the energy),

2. Terms on the l.h.s. of the second equation (2.99), e.g. the ladder term, have no influence on the conservation properties,

3. For *distance-dependent potentials*, the contribution of the Hartree term and of the second term on the r.h.s. of (Eq. 2.107) vanish separately. This means, neglect of the Hartree term does not violate energy conservation.

4. For general potentials, the Hartree term must be included in inhomogeneous systems in order to compensate the second term on the r.h.s. of Eq. (2.107),

5. The above statements hold, strictly speaking, only if $g_{123} = 0$. Otherwise, ternary correlations may alter the behavior, and the conservation properties of the closure have to be investigated separately. The above results remain valid, if the contribution of g_{123} is separately conserving, for example, if g_{123} is symmetric, $P_{123}\, g_{123} = g_{123}$.

2.6 Decoupling of the BBGKY-hierarchy

The concept of the hierarchy of equations for the reduced density operators or the correlation operators is only useful, if the behavior of the many-particle system can be sufficiently well described with a small number of density operators up to a certain order s. This means, that the dynamics of the higher order operators may be regarded as not relevant for the current problem. For example, certain correlations in the system are negligibly small, so $g_{1\ldots s+1} \approx 0$, $g_{1\ldots s+2} \approx 0$ and so on. Or, more generally, all operators of the orders higher than s are known functionals of the lower order operators $g_{1\ldots s+1} = g_{1\ldots s+1}[F_1, g_{12}, \ldots, g_{1\ldots s}]$.[13] These relations which express higher order correlations by lower order ones are called decoupling (closure) approximations. In this section, we give a brief overview on the most important hierarchy closure schemes.

There exist at least three basic schemes for deriving approximations for many-body systems which are applied to the BBGKY-hierarchy:

A. Perturbation theory with respect to internal system parameters, such as strength of the interaction or particle density: The first gives rise to an expansion in powers of the interaction potential (Born series), where all terms up to a certain power of V_{ij} are included. This expansion converges well for weak potentials of short range. However, this does not apply to plasmas, due to the long range of the Coulomb interaction. Here, expansions have to be based on the screened interaction. Furthermore, for strong interaction (compared to the characteristic kinetic energy), there are situations, where the expansion does not converge at all. An example are attractive potentials which give rise to bound states which cannot be treated by perturbation theory.

   On the other hand, density expansions (virial expansions or cluster expansion etc.) use a certain power of the density as the closure criterion. One example is the limitation to two-particle, three-particle or higher collisions. It is obvious that in a dilute system the probability for the encounter of three particles is much lower than that of two particles.[14] However, this expansion, has its limitations too. For example, it is known that in 3D, the classical four-particle collision integral diverges (as func-

[13] Fortunately, this is almost always the case, provided the system consists of a macroscopically large number of particles. On the other hand, for small particle numbers we expect large influence of statistical fluctuations (around the averages), and, eventually a statistical description is not suitable at all.

[14] Analogously, in a quantum treatment, this is expressed by the overlap of the wave functions of three (two) particles.

tion of time), whereas in (strictly) 2D systems this occurs already with the three-particle integral [GF67, DC67]. In general, both expansions are closely related. This can be seen clearly from the hierarchy for the correlation operators: The coupling of one equation to the next always adds terms of the next order in both the density and the interaction potential.

B. Perturbative expansions in external parameters: here, the primary example is the magnitude of the externally controlled potential $\mathcal{U}$, which is assumed to be weak. Since the potential $\mathcal{U}$ is contained in each Hamiltonian $H_{1...s}$, each hierarchy equation for $F_{1...s}$ contains on the l.h.s. terms of the order $\mathcal{U}^0$ and $\mathcal{U}^1$. The order of the terms on the r.h.s. (coupling to $F_{1...s+1}$) depends on the closure approximation. If, e.g. F_{12} is expressed in terms of F_1, one can by iteration determine which powers of $\mathcal{U}$ are contained in F_{12}. This analysis is even more complex in the Bloch picture, where the diagonal and off-diagonal elements of the density operators depend on $\mathcal{U}$ in different ways. A well-known example for this kind of closure scheme is the $\chi^{(n)}$–approach in semiconductor optics which was developed recently by Axt and Stahl [AS94a, AS94b], see also [LHBK94, BK95]. This approach is of particular relevance in the weak field limit where the perturbation expansion converges sufficiently fast. Obviously, this scheme leads to completely different approximations compared to A, and a comparison of both schemes is complicated. We will not consider this approach further. For a recent review, see [AM98].

C. Non-perturbative schemes: As we have seen, perturbation expansions of type A have serious limitations. Furthermore, there is a variety of physical phenomena which cannot be described at all on the basis of a perturbation theory, but require the summation of the whole infinite series. The most important examples are screening phenomena in plasmas and bound states. Interestingly, this does not mean that one needs an infinite number of hierarchy equations. As we will see below, these effects are already well described on the level of the second hierarchy equation. The crucial point is, that these phenomena require different closures where the selection of the relevant contributions is based on topological criteria (e.g. diagrammatic expansions), while other terms of different structure are neglected despite being of the same order in density or interaction. This gives rise to selfconsistent approximation schemes. Such procedures are well elaborated e.g. in fluid theory (integral equations techniques) or in Green's functions theory (based on the concept of selfenergy, e.g. [KB89, KKER86].

In the following, we will consider important closure relations of both, type A and C. These relations are best classified in terms of orders of correlations which are taken into account:

I. *No correlations* $g_{12} = g_{123} = \ldots = 0$: this leads to mean field approximations, such as the Hartree-Fock approximation. The most important example of a mean-field kinetic equation is the Vlasov equation [Har28, Vla38].

II. *Zero three-particle correlations*, non-vanishing binary correlations, $g_{123} = g_{1234} = \ldots = 0$. With approximations of this kind it is possible to derive the most important kinetic equations, such as the Landau, Boltzmann [Bol72] and Balescu-Lenard [Bal60, Len60] equation. Depending on the approximations with respect to the time arguments of the density operators, one can derive either Markovian or generalized (non-Markovian) kinetic equations.

III. *Zero four-particle and higher correlations*, non-vanishing three-particle correlations. With approximations of this type one can describe a large variety of physical effects. This includes the scattering of free particles on atoms or molecules, the formation and destruction of bound states by impact (inelastic) collisions and so on. Also, selfenergy effects can be introduced on this level.

IV. *Higher order correlations.* Certain phenomena, in particular kinetic processes in bound complexes, inelastic atom–atom scattering etc., involve four-particle correlations. Besides, in numerous approximation schemes, higher order correlations are included partially[15] (see below).

V. *Correlations due to spin statistics.* In systems of quantum particles with nonzero spin, there exists a different kind of correlations, which is due to the spin statistics theorem. Even if no correlations, in the classical sense do exist (cf. I.), the particles "feel" the existence of others because certain quantum states are no longer available (Pauli blocking). We will consider the modifications in the hierarchy due to spin statistics separately, in Chapter 3.

In the following section, we give a brief overview on approximations I.-IV.

---

[15] This applies in particular to selfconsistent schemes, which are equivalent to summation to all orders of special classes of diagrams.

2.6.1 Correlation effects

Mean field approximation. If binary and higher correlations are neglected, the BBGKY-hierarchy reduces to the first equation

$$i\hbar\frac{\partial}{\partial t}F_1 - [H_1 + U_1^H, F_1] = 0, \tag{2.110}$$

where the one-particle Hamiltonian contains the kinetic part, the external field $\mathcal{U}_1$ and the mean field (2.103). This approximation was introduced by Hartree [Har28], see also [Har48]. For particles with spin, the complete mean-field approximation contains additionally an exchange contribution (Fock term) which we will discuss in Ch. 3. The most important example for a mean field kinetic equation is the Vlasov (or Hartree) equation, where the mean field is just the electrostatic field generated by all particles [Vla38, Vla45]. This field is self-consistently related to the one-particle density operator and obeys Poisson's equation.[16]

Neglecting correlations completely, this closure approximation cannot describe scattering processes and irreversible relaxation. The neglected term is a full commutator, $n\text{Tr}_2[V_{12}, g_{12}]$, which on time inversion changes its sign. Therefore, Eq. (2.110) remains time reversible too, cf. Eq. (2.23). Obviously, Eq. (2.110) conserves energy. With $F_{123} = 0$, the symmetry condition (2.36) is of course fulfilled. Due to the neglect of correlations, the potential energy, Eq. (2.26), contains only a mean field contribution, i.e.

$$\langle V\rangle = \langle U_1^H\rangle = \frac{n^2}{2}\text{Tr}_{12}V_{12}F_1F_2. \tag{2.111}$$

If $\mathcal{U}$ is time-independent, the sum of particle energy + the particle-field interaction energy $\langle H\rangle = \langle T + U^H + \mathcal{U}_1\rangle$ is constant, in the general case, the conserved quantity is given by Eq. (2.35). This approximation will be studied in Ch. 4, generalizations to the relativistic case are discussed in Ch. 12.

Neglect of three-particle correlations. If three-particle correlations are neglected, the hierarchy reduces to the first and second equations of (2.99), which can be rewritten in the form

$$\begin{aligned} i\hbar\frac{\partial}{\partial t}F_1 - [\bar{H}_1, F_1] &= n\text{Tr}_2[V_{12}, g_{12}], \\ i\hbar\frac{\partial}{\partial t}g_{12} - [\bar{H}_{12}^0, g_{12}] &= [V_{12}, F_1F_2] + L_{12} + \Pi_{12}, \quad \text{with} \qquad (2.112) \\ L_{12} = [V_{12}, g_{12}]; \quad \Pi_{12} &= n\text{Tr}_3\Big\{[V_{13}, F_1g_{23}] + [V_{23}, F_2g_{13}]\Big\}, \qquad (2.113) \end{aligned}$$

[16]In case of an electromagnetic field, one has to solve Maxwell's equations.

where, L_{12} and Π_{12} denote the two-particle ladder and polarization terms, respectively. The first term on the r.h.s. of Eq. (2.112) is an inhomogeneity which depends only on the one-particle density operator. As we will see, this term is the origin of the scattering contributions in all kinetic equations with two-particle collision integrals. Depending on whether L_{12} and Π_{12} are taken into account, there are four basic approximations possible:

1. *Second Born approximation*, $L_{12} = \Pi_{12} = 0$: this is the simplest and most widely used approximation for the treatment of correlations (the collision integrals are quadratic in the matrix element of the static binary potential V_{ij}). This is, at the same time, the static limit of the polarization approximation (Balescu-Lenard equation, 3. below), and the weak-coupling limit of the ladder approximation (Boltzmann equation, 2. below), and it is equivalent to the second Born approximation for the selfenergy in Green's functions theory [KB89]. In case of plasmas, from this approximation one can derive the Landau kinetic equation [BK96, BBK97], what will be discussed in Ch. 6.

2. *Ladder approximation*, $L_{12} \neq 0$, $\Pi_{12} = 0$: this approximation allows to describe interactions of arbitrary strength (strong coupling) including, in the case of different particle species (e.g. oppositely charged particles), bound states. It leads to the non-Markovian Boltzmann equation [KBKS97] which contains in the scattering kernel the T-matrix [BG58, BG57], and is equivalent to the T-matrix approximation for the selfenergy in Green's functions theory [Puf61, KB89, KKER86]. This approximation is considered in Ch. 8, see also Fig. 2.1.

3. *Polarization approximation, RPA*, $L_{12} = 0$, $\Pi_{12} \neq 0$: this approximation corresponds to the weak-coupling case also, but in addition to the Born approximation, it takes dynamical polarization effects fully into account. This is crucially for polarizable media with long range interaction, such as plasmas. This ansatz yields the Balescu-Lenard equation [Bal60, Len60] and is equivalent to the random phase approximation for the selfenergy [BP53, KB89, KKER86]. We will derive the non-Markovian generalization of this equation in Ch. 9, [BKDK], see also Fig. 2.1.

4. *Screened ladder approximation*, $L_{12} \neq 0$ and $\Pi_{12} \neq 0$: This approximation includes both ladder and polarization terms and, therefore, is able to describe strongly coupled polarizable media (screened ladder approximation [KKER86]). We will discuss this ansatz in Ch. 10.

While the Born approximation results from perturbation expansion (scheme A above) with respect to density or the interaction potential, the ladder approximation 2. and the polarization approximation 3. correspond to selfconsistent decoupling schemes (C above).[17] Fig. 2.1 shows, that the ladder approximation corresponds to an infinite ladder summation (summation of the complete Born series). On the other hand, the polarization approximation is equivalent to an infinite "bubble" summation over all polarization contributions.

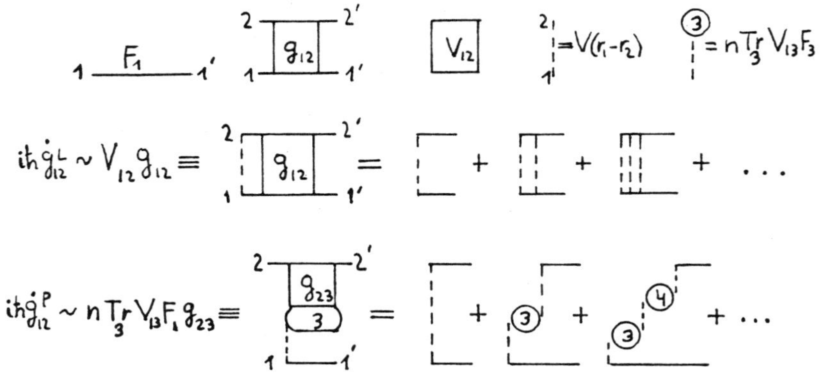

Fig. 2.1.: Graphical representation of F_1, g_{12}, V_{12}, $V(r_1 - r_2)$ (distance dependent potential) and the trace (**1st line**). **2nd/3rd line**: the "ladder"/"polarization" term of the 2nd Eq. (2.99). Both series start with the first Born approximation, $\sim V_{12}F_1F_2$ (only first term of commutator shown), each successive term follows by iteratively inserting the previous one into the ladder/polarization term.

Consider now the reversibility of approximations 1.-4. The neglect of ternary correlations corresponds to the neglect of a complete commutator $n\mathrm{Tr}_3[V_{13} + V_{23}, g_{123}]$, and thus does not affect the reversibility properties. The same is true for the ladder and polarization terms. Finally, consider the question of energy conservation of approximations 1.-4. Here, we may use the results of Sec. 2.5.2, where it was shown that polarization and ladder terms have no influence on energy conservation. Thus, it follows trivially that, if the Hartree terms are included, each of the approximations 1.-4. conserves total energy. As we will see, the complete neglect of ternary correlations corresponds to the neglect of selfenergy effects.

Partial account of three-particle correlations. We now consider more complex closure relations which allow to describe a large number of additional phenomena. Here, we neglect four-particle correlations completely, $g_{1234} = 0$, and in the third equation (2.99), we neglect on the r.h.s. the terms containing

[17] Formally, this selfconsistency arises from the fact that in these approximations the correlation operator appears not only on the l.h.s. of Eq. (2.112), but also in the ladder and polarization term. Therefore, with the ladder or/and polarization term included, the solution for g_{12} of Eq. (2.112) is essentially more complicated as in the case of the Born approximation, cf. Chs. 8 and 9.

products of two-particle correlation operators.

$$
\begin{aligned}
i\hbar\frac{\partial}{\partial t}g_{123} - [\bar{H}^0_{123}, g_{123}] &= I_{123} + L_{123} + \Pi_{123}, & (2.114)\\
I_{123} &= [V_{12} + V_{13} + V_{23}, F_1F_2F_3] + [V_{13} + V_{23}, F_3 g_{12}], \\
&\quad +[V_{12} + V_{23}, F_2 g_{13}] + [V_{12} + V_{13}, F_1 g_{23}] & (2.115)\\
L_{123} &= [V_{12} + V_{13} + V_{23}, g_{123}], & (2.116)\\
\Pi_{123} &= n\mathrm{Tr}_4[V_{14}, F_1 g_{234}] + n\mathrm{Tr}_4[V_{24}, F_2 g_{134}] \\
&+ n\mathrm{Tr}_4[V_{34}, F_3 g_{124}], & (2.117)
\end{aligned}
$$

where I_{123} is the inhomogeneity, L_{123} is the three-particle ladder term and Π_{123} contains the three-particle polarization terms. First we notice that I_{123} contains various three–particle interaction processes: scattering of three particles as well as scattering of free particles on correlated pairs. This is important for dense correlated systems, but will not be considered here.

2.6.2 *Selfenergy effects

The decoupling approximations 1.-4. have been extremely successful in the statistical description of many-particle systems in a large variety of fields. Nevertheless, they have a number of serious deficiencies. First, the dynamics on short time scales is not described correctly, initial correlations do not vanish. Moreover, the equilibrium solution of the corresponding kinetic equations turns out to be that of an ideal quantum gas, it does not describe the effect of correlations (for more details, see Ch. 6). Different theories offer different approaches to solve this problem. For example, Green's function theory successfully uses the concept of selfenergy. It has been a long-standing problem, if and how this idea can be introduced into the density operator theory. Only recently, solutions have been proposed - for electron-phonon scattering in Refs. [SKM94, SKM95] and for carrier–carrier scattering in Refs. [BK96, BKS$^+$]. The crucial point is that selfenergy terms arise from particular contributions to three-particle correlations. Here, we briefly list the approximations which allow to "upgrade" approximations 1.-4. above by the inclusion of selfenergy (a detailed discussion will be given in Ch. 7).

The approximation for g_{123}, which has to be inserted into the equation for g_{12}, is based on the following inhomogeneity [BKS$^+$, BK96]

$$
I_{123} \longrightarrow [V_{13} + V_{23}, F_3 g_{12}], \qquad (2.118)
$$

where, compared to Eq. (2.115), we neglected the term containing the product of three one-particle operators, and also the permutations of the expression

(2.118). This inhomogeneity couples the particle pair $1-2$ to third particles which allows us to include medium effects on the pair properties. By solving for g_{123} and inserting the result into the equation for g_{12}, one is able to renormalize the two-particle Hamiltonian, including damping (finite lifetime) effects. As we will see in Ch. 7, this is essential for the correct treatment of the correlation dynamics, including the correlation build-up and the decay of initial correlations. We mention that the approximation (2.118) accounts only for the simple case of spinless particles. The generalization to bosons and fermions will be performed in Sec. 3.4.1 after the discussion of the (anti-)symmetrization of the BBGKY-hierarchy in Ch. 3. The following approximations which correspond to the cases 1.-4. above, are of particular importance:

1. *Generalized second Born approximation*, $L_{12} = L_{123} = \Pi_{12} = \Pi_{123} = 0$, this leads to the the non-Markovian Landau equation with selfenergy corrections in second Born approximation [BKS$^+$, BK96, BBK97], which will be derived in Ch. 7.

2. *Generalized binary collision approximation*, $L_{12} \neq 0, L_{123} \neq 0$, $\Pi_{12} = \Pi_{123} = 0$, this approximation corresponds to the ladder or T-matrix approximation, with selfenergy included on the T-matrix level [KBKS97]. Using this approximation, we will derive the non-Markovian Boltzmann equation in Ch. 8.

3. *Generalized polarization approximation*, $L_{12} = L_{123} = 0$, $\Pi_{12} \neq 0, \Pi_{123} \neq 0$, this approximation yields the selfconsistent random phase approximation (RPA) of Green's functions theory. Using this approximation, we will derive the non-Markovian Balescu-Lenard equation with RPA selfenergy included [BKDK] in Ch. 9.

4. *Generalized screened ladder approximation*, $L_{12} \neq 0, L_{123} \neq 0$, $\Pi_{12} \neq 0, \Pi_{123} \neq 0$, this approximation gives rise to kinetic equations with selfenergy on the level of the screened ladder approximation.

The incorporation of selfenergy effects into the BBGKY-hierarchy and the various non-Markovian kinetic equations will be an important point in the Chapters below.[18] We will discuss these questions in detail for the Born ap-

[18] Interestingly, we will see below, that these non-Markovian kinetic equations for the Wigner function are exactly the same which are derived from Green's functions theory by using the generalized Kadanoff-Baym ansatz (GKBA) [LvV86]. On the other hand, with the generalized hierarchy closure approximations 1.-4. one is able to derive the GKBA for the respective approximation. Furthermore, it is interesting to underline that these closures have no restrictions for the free propagators which appear in the GKBA, they yield the full equations of motion of the Dyson-type for the retarded and advanced propagators.

proximation in Ch. 7. Generalizations to the remaining cases are then straightforward and will be given in Chs. 8-10.

Let us briefly discuss the properties of the generalized hierarchy closures 1.-4. They are obviously time reversible. The reason is again, that only terms of commutator form have been neglected. Consider now the conservation properties. Now the analysis of the symmetry of F_{123} (Sec. 2.2.2) leads to an analysis of g_{123}. The remaining contributions to F_{123} are fully included and are, therefore, symmetric with respect to index permutations. The symmetry of g_{123} depends on the symmetry properties of the r.h.s. of Eq. (2.114). An inspection shows that each of the contributions L_{123} and Π_{123} consists of three terms which are cyclic permutations of each other and are thus conserving. However, I_{123} of Eq. (2.118) is not symmetric. Therefore, the question of energy conservation with selfenergy effects included has to be studied for each case separately.[19]

Further problems. Higher order correlations. With the inclusion of binary and ternary correlations two-particle and three-particle scattering processes can be described. This includes also scattering of free particles on two-particle bound complexes. Of particular interest in dense systems is the description of inelastic collisions which may lead to excitation/deexcitation of particles with internal degrees of freedom or to impact ionization/recombination of bound states (complex atoms, molecules, excitons, clusters and so on). These questions are beyond the scope of this book, and the reader is referred to the density operator results of Ebeling, Klimontovich, Kremp and coworkers [Ebe76, KK81, KKK87] and McLennan [McL89a]. Green's functions results and further references can be found in [AP77, KKER86].

We mention that the classification given above in terms of correlation operators cannot include all existing approaches. In particular, there have been proposed many selfconsistent approximation schemes for binary or ternary density or correlation operators which, in fact, involve higher order contributions too. As an example, we mention Kirkwood's superposition approximation[20] or fluid integral equations[21] which are very successful in the description of neutral and ionic fluids (electrolytes) as well as dense partially ionized plasmas, for further reference, see [KEKS83, Fal71].

---

[19] For the special case of distance dependent interaction, it is sufficient to show that the trace term in Eq. (2.34) involving g_{123} retains the commutator structure. We will see that this is indeed fulfilled.

[20] It consists in the ansatz (for a homogenous medium) $F_{123} = F_{12}F_{13}F_{23}$, which describes many properties of fluids or dense gases qualitatively correct, see e.g. [Gre53].

[21] They are based on the Ornstein–Zernicke equation $g_{12} = c_{12} + n\mathrm{Tr}_3 c_{13} g_{32}$ which defines the direct correlation function c_{12} and which leads to powerful theoretical schemes, such as the Percus-Yewick or the hypernetted chain (HNC) approximation.

2.7 Relation to equilibrium correlation functions

Before proceeding with the analysis of nonequilibrium phenomena on the basis of the BBGKY–hierarchy, we briefly discuss, the relation of the latter to equilibrium theories. In fact, the concept of a hierarchy of equations for distribution functions (density operators) has been first introduced for systems in thermodynamic equilibrium [Bog46].

Let us consider the equilibrium state of the N-particle system governed by the Hamilton operator (2.3). Using the canonical ensemble (i.e. volume $\mathcal{V}$, particle number N, temperature T and external potential $\mathcal{U}$ are fixed), the N–particle density operator is given by the canonical distribution

$$\rho^{eq}_{1\dots N} = e^{[F-H_{1\dots N}]/kT}; \quad \mathrm{Tr}_{1\dots N}\rho^{eq}_{1\dots N} = 1. \tag{2.119}$$

The normalization condition (second condition) yields the free energy $F = F(N, \mathcal{V}, T, \mathcal{U})$ or, equivalently, the partition function Z

$$Z = \mathrm{Tr}_{1\dots N} e^{-H_{1\dots N}/kT}; \quad F = -kT \ln Z, \tag{2.120}$$

which is completely defined if the eigenvalue spectrum of $H_{1\dots N}$ would be known. One major problem of equilibrium theories is to find approximate results for this spectrum. To obtain the equations for the equilibrium distribution functions in configuration space, we use the Wigner representation for the density operator and to perform an average over the momenta,

$$\rho^{eq}(\mathbf{r}_1, \dots \mathbf{r}_N) = \int d\mathbf{p}_1, \dots d\mathbf{p}_N \, \rho^{eq}(\mathbf{r}_1, \dots \mathbf{r}_N; \mathbf{p}_1, \dots \mathbf{p}_N). \tag{2.121}$$

Further progress can be made with the quasi–classical approximation, i.e. approximating the Hamilton operator by a continuous phase space function $H_{1\dots N} \to H(\mathbf{r}_1, \dots \mathbf{r}_N; \mathbf{p}_1, \dots \mathbf{p}_N) = \sum_i p_i^2/2m + \Phi(\mathbf{r}_1, \dots \mathbf{r}_N)$, where $\Phi = \sum_i \mathcal{U}(\mathbf{r}_i) + \frac{1}{2}\sum_{ij} V(|\mathbf{r}_i - \mathbf{r}_j|)$. Then, obviously, all momentum integrations in (2.121) can be carried out leading to

$$\rho^{eq}_{cl}(\mathbf{r}_1, \dots \mathbf{r}_N) = \frac{1}{\mathcal{V}^N Z_{cl}} \, e^{-\Phi/kT}; \quad Z_{cl} = \frac{1}{\mathcal{V}^N} \int d\mathbf{r}_1, \dots d\mathbf{r}_N \, e^{-\Phi/kT}, \tag{2.122}$$

with the Z_{cl} being the classical configuration integral. As for the BBGKY–hierarchy [cf. Eq. (2.13)], one defines classical reduced *equilibrium* s-particle distributions by

$$\begin{aligned} f^{eq}_{cl}(\mathbf{r}_1, \dots \mathbf{r}_s) &= \mathcal{V}^s \int d\mathbf{r}_{s+1}, \dots d\mathbf{r}_N \, \rho^{eq}_{cl}(\mathbf{r}_1, \dots \mathbf{r}_N), \\ 1 &= \frac{1}{\mathcal{V}^s} \int d\mathbf{r}_1, \dots d\mathbf{r}_s \, f^{eq}_{cl}(\mathbf{r}_1, \dots \mathbf{r}_s), \end{aligned} \tag{2.123}$$

which obey the following classical hierarchy of equations [Bog46], (in compact notation)

$$\nabla_{\mathbf{r}_1} f_1 + \frac{1}{kT}\nabla_{\mathbf{r}_1}\mathcal{U} = -\frac{n}{kT}\int d\mathbf{r}_2\, f_{12}\nabla_{\mathbf{r}_1}V_{12}, \quad (2.124)$$

$$\nabla_{\mathbf{r}_1} f_{12} + \frac{1}{kT} f_{12}\nabla_{\mathbf{r}_1}\mathcal{U} + f_{12}\nabla_{\mathbf{r}_1}V_{12} = -\frac{n}{kT}\int d\mathbf{r}_3\, f_{123}\nabla_{\mathbf{r}_1}V_{13}, \quad (2.125)$$

$$\dots \quad \dots \quad \dots$$

To illuminate the correspondence between the nonequilibrium and equilibrium distribution functions, let us consider the case $\mathcal{U} = const$. Then, the system is homogeneous, and $f_1(\mathbf{r}_1) \equiv 1$, $f_{12}(\mathbf{r}_1, \mathbf{r}_2) = f_{12}(|\mathbf{r}_1 - \mathbf{r}_2|)$ and so on. The cluster expansion [compare with Eq. (2.98)] now takes the form

$$f_{12}(|\mathbf{r}_1 - \mathbf{r}_2|) = 1 + g_{12}(|\mathbf{r}_1 - \mathbf{r}_2|), \quad (2.126)$$

$$\begin{aligned} f_{123}(|\mathbf{r}_1 - \mathbf{r}_2|, |\mathbf{r}_1 - \mathbf{r}_3|) &= 1 + g_{12}(|\mathbf{r}_1 - \mathbf{r}_2|) + g_{13}(|\mathbf{r}_1 - \mathbf{r}_3|) \\ &+ g_{23}(|\mathbf{r}_2 - \mathbf{r}_3|) + g_{123}(|\mathbf{r}_1 - \mathbf{r}_2|, |\mathbf{r}_1 - \mathbf{r}_3|), \end{aligned} \quad (2.127)$$

etc. For example, in binary collision approximation ($g_{123} = 0$), the r.h.s. of Eq. (2.125) vanishes, and we obtain $f_{12}(|\mathbf{r}_1 - \mathbf{r}_2|) = e^{-V(|\mathbf{r}_1-\mathbf{r}_2|)/kT}$, or, for the pair correlation function,

$$g_{12}(|\mathbf{r}_1 - \mathbf{r}_2|) = e^{-V(|\mathbf{r}_1-\mathbf{r}_2|)/kT} - 1. \quad (2.128)$$

The full correlation function which appears in the BBGKY–hierarchy follows from Eq. (2.128) by multiplication with the product of two equilibrium momentum distribution functions $f^{eq}(p_1)f^{eq}(p_2)$, see also Sec. 7.3.4. With the result for f_{12} or g_{12} all thermodynamic quantities of the *nonideal* equilibrium system can be calculated.[22]

One readily confirms that the hierarchy (2.124, 2.125) is a special case[23] of the classical limit of the BBGKY-hierarchy, obtained in Sec. 2.3.2. This agreement is important, as it allows one to make use of many results of equilibrium theory. In particular, standard decoupling approximations of the hierarchy (2.124, 2.125) can be applied to the nonequilibrium hierarchy as well.[24] Moreover, the equilibrium case is essentially simpler, so many approximations for nonequilibrium many-particle systems can be easily tested first on the equilibrium situation.

[22]For more details, see e.g. Ch.2 of Ref. [Kli75].

[23]Take the stationary limit $\partial/\partial t = 0$ and use equilibrium momentum distributions.

[24]In fact, many of the approximations discussed in the previous sections, have their equilibrium counterparts.

Chapter 3

*Correlations due to the Spin Statistics

There exists a specific type of inter-particle correlations in the case of Bose or Fermi particles - correlations arising from the spin statistics theorem[1]. This is obvious if one for example recalls the Pauli exclusion principle which reduces the space of quantum states available for each particle. Even if there is no (or negligibly weak) interaction between particles, each of them will eventually "feel" the existence of the others if it approaches a quantum state that is already occupied ("blocked").

There are different approaches to account for the spin statistics in kinetic theory. One concept is to use as the starting point quantities which are intrinsically (anti-)symmetrized because they obey the respective (anti-)commutation relations, such as the field operators Ψ (or creation/annihilation operators $a^\dagger, a$) and to construct the theory based on the equations of motion for these quantities. This concept is used in field theory or Green's functions theory [KB89, KKER86], (see Ch. 12), or projection operator techniques and has the advantage that all results "automatically" contain spin statistics effects completely.

On the contrary, our density operator approach is based on the equations for spinless particles which have the advantage to be very simple and physically transparent. However, in order to be able to apply the results to fermions or bosons, we need a method which allows to perform the (anti-)symmetrization of the equations of motion for the reduced density operators and correlation operators. This problem will be solved in this Chapter. In principle, exchange and phase space filling effects can be very easily incorporated into the equations derived above. The only place where symmetry properties of the system enter

[1]This chapter (as all sections marked with "*") may be skipped on first reading.

the BBGKY-hierarchy is in the calculation of the trace. So it is possible to retain the same operator form of the hierarchy for spinless particles and for bosons or fermions as well, but to take their specific statistical properties into account via a different prescription for the calculation of the traces. In particular, we have to recall only that instead of the full Hilbert space $\mathcal{H}_N$ of N-particle states, in the case of bosons (fermions) we have to consider the (anti-)symmetric subspace $\mathcal{H}_N^+$ ($\mathcal{H}_N^-$). The simplest way of doing this is, to define all trace operations which appear in the hierarchy equations in such a way that they are carried out in the corresponding subspace of N-particle states. The projection onto this subspace is achieved by acting with an (anti-)symmetrization operator $\Lambda^\pm$ on the state vector with the result

$$|\Psi_{1...N}>^\pm = \frac{1}{N!}\Lambda^\pm_{1...N}|\Psi_{1...N}> \tag{3.1}$$

For example, the following trace of an $s-$particle operator would be calculated according to

$$\mathrm{Tr}_{1...s}A_{1...s} = \sum_{x_1...x_s} \langle x_1 \dots x_s|A_{1...s}\Lambda^\pm_{1...s}|x_s \dots x_1\rangle, \tag{3.2}$$

where x denotes an arbitrary representation of the basis vectors (cf. Sec. 2.3). In Eq. (3.2) we used the property of projection operators $[\Lambda^\pm_{1...s}]^2 = s!\Lambda^\pm_{1...s}$, so that we need to perform an (anti-)symmetrization only of one vector. The explicit form of the operator $\Lambda^\pm_{1...s}$ will be given below.

This procedure has the advantage that the hierarchy equations in the operator form retain the same simple spin-invariant structure. This allows to investigate properties (such as energy conservation) or to derive specific approximations to the hierarchy first in the relatively simple "spinless" form. The spin properties of the actual system can be introduced lateron, when the transition from the density operators to a certain representation is made.

3.1 (Anti-)Symmetrization of the density operators

The alternative approach is to incorporate the (anti-)symmetrization operators not in the basis vectors, but in the density operators.[2] This approach allows one to make effects of the spin statistics, such as Pauli blocking and

[2]This duality is analogous to the Heisenberg and Schrödinger pictures of quantum mechanics

exchange in the equations of motion for the reduced density operators explicit, while the state vectors are unchanged belonging to the full Hilbert space. This procedure was developed by Dufty and Boercker [BD79], see also [BD81, DKBB97], and will be used in the following. According to Ref. [BD79], the reduced density operators are replaced by modified operators that contain proper (anti)symmetrization factors

$$F_{1\ldots s} \longrightarrow F_{1\ldots s}\Lambda^{\pm}_{1\ldots s}, \tag{3.3}$$

where the (anti-)symmetrization operator $\Lambda^{\pm}$ acts on the states to the right. It contains all possible permutations of s particles $P_{1\ldots s}$

$$\Lambda^{\pm}_{1\ldots s} = \sum_{P^{(s)}} \epsilon^{N(P^{(s)})}\, P^{(s)}, \tag{3.4}$$

with $\epsilon = 1$ for bosons and -1 for fermions (and zero for spinless particles). For fermions, the sign of each contribution depends on whether $P^{(s)}$ is an even or odd permutation, i.e. if $P^{(s)}$ can be decomposed into an even or odd number $N(P^{(s)})$ of binary permutations (see below). For bosons, all contributions have the same prefactor $+1$. Notice that the sum contains the trivial case of zero binary permutations $N(P^{(s)}) = 0$, which contributes $+1$ to $\Lambda^{\pm}$. Thus, the case of spinless particles is trivially included.

For example, $\Lambda^{\pm}_{12} = 1 + \epsilon P_{12}$, where the action of the binary permutation operator P_{12} on a two-particle state in an arbitrarily chosen representation x leads to $P_{12}|x_2 x_1\rangle = |x_1 x_2\rangle$. As a result, the matrix elements of a two-particle operator times the antisymmetrization operator contain a direct and an exchange term:

$$\begin{aligned}\langle x_1 x_2|A_{12}\Lambda^{\pm}_{12}|x_2' x_1'\rangle &= \langle x_1 x_2|A_{12}|x_2' x_1'\rangle + \epsilon\langle x_1 x_2|A_{12}|x_1' x_2'\rangle \\ &= A(x_1, x_2, x_1', x_2') + \epsilon A(x_1, x_2, x_2', x_1'). \end{aligned} \tag{3.5}$$

Since any permutation of s particles $P^{(s)}$ can be decomposed into successive binary permutations P_{ij}, the (anti-)symmetrization operators $\Lambda^{\pm}$ may be expressed in terms of two-particle permutations too. One readily derives the following properties:

$$\begin{aligned}\Lambda^{\pm}_{12} &= 1 + \epsilon P_{12}, \\ \Lambda^{\pm}_{123} &= \Lambda^{\pm}_{12}\,(1 + \epsilon P_{13} + \epsilon P_{23}), \\ \Lambda^{\pm}_{1234} &= \Lambda^{\pm}_{123}\,(1 + \epsilon P_{14} + \epsilon P_{24} + \epsilon P_{34}), \\ &\ldots \end{aligned} \tag{3.6}$$

Further important properties of $\Lambda^{\pm}$ and P_{ij} are

$$\begin{aligned}
\Lambda^{\pm\,2}_{1\ldots s} &= s!\,\Lambda^{\pm}_{1\ldots s} \\
P^2_{ij} &= 1, \\
\mathrm{Tr}_j P_{ij} &= 1, \\
P_{ij} A_{ij} P_{ij} &= A_{ji}, \\
P_{ij} B_{ij} &= B_{ij} P_{ij}, \quad \text{if} \quad B_{ij} = B_{ji}, \\
(1 - \epsilon P_{ij})\Lambda^{\pm}_{ij} &= 0,
\end{aligned} \tag{3.7}$$

where A_{ij} is an arbitrary operator which acts on particles i and j, and B_{ij} is a symmetric operator.

Now we can write in Eq. (3.3) for the reduced density operators explicitly

$$\begin{aligned}
F_1 &\longrightarrow F_1, \\
F_{12} &\longrightarrow F_{12}(1 + \epsilon P_{12}), \\
F_{123} &\longrightarrow F_{123}(1 + \epsilon P_{12})(1 + \epsilon P_{13} + \epsilon P_{23}), \\
F_{1234} &\longrightarrow F_{1234}(1 + \epsilon P_{12})(1 + \epsilon P_{13} + \epsilon P_{23})(1 + \epsilon P_{14} + P_{24} + \epsilon P_{34}), \\
&\ldots
\end{aligned} \tag{3.8}$$

The (anti-)symmetrized correlation operators have the same properties. We want to stress, that we keep the previous notation where $F_{12}, F_{123}, g_{12}, g_{123}$ etc. are spin-independent. The influence of the spin statistics will be made explicit only via the (anti-)symmetrization factors. Then we will retain the same cluster expansion for the reduced density operators, given by Eqs. (2.98). The full (anti-)symmetrized cluster expansion then follows simply by means of the substitutions (3.8).

3.2 Exchange and phase space filling effects

In this section we demonstrate how important physical effects which arise from the spin statistics, such as quantum–mechanical exchange and Pauli blocking, are introduced into the BBGKY-hierarchy explicitly. To this end, we derive a number of useful relations which will allow us to simplify terms involving the operators $\Lambda^{\pm}$ in the hierarchy equations.

i) *Quantum-mechanical exchange:*

Consider the trace over the following commutator which arises from the trace over F_{12}:

$$\begin{aligned}
n\mathrm{Tr}_2[V_{12}, F_1 F_2 \Lambda^{\pm}_{12}] &= [H_1^{HF}, F_1], \\
\text{with} \quad H_1^{HF} &= n\mathrm{Tr}_2 V^{\pm}_{12} F_2 = n\mathrm{Tr}_2 V_{12} F_2 \Lambda^{\pm}_{12}; \quad V^{\pm}_{12} = V_{12}\Lambda^{\pm}_{12},
\end{aligned} \tag{3.9}$$

being the Hartree-Fock Hamiltonian. The expression for H^{HF} generalizes the previously introduced mean-field operator (Hartree field) U^H, Eq. (2.103). So everywhere we may replace $U^H \to H^{HF}$. We mention that in the language of Green's functions, this term gives the Hartree-Fock selfenergy Σ^{HF} [KB89, KKER86].

ii) *Phase space filling (Pauli blocking):*
Next, consider the expression $Tr_3 V_{13} F_1 F_2 F_3 P_{23}$ which appears in the calculation of the trace over F_{123}. This term describes the interaction of particles one and three with each other, where additionally, due to the exchange operator, also particle two gets involved into the scattering process. We will see below that these terms, give rise to phase space occupation (Pauli blocking) effects. In matrix representation $\langle x_1 x_2 | \dots | x_2' x_1' \rangle$, we have

$$\begin{aligned} \int dx_3 \, \langle x_2 | F_2 | x_3 \rangle \, \langle x_1 x_3 | V_{13} F_1 F_3 | x_2' x_1' \rangle &= \\ \int dx_3 d\bar{x}_1 \, \langle x_1 x_2 | F_2 | x_3 \bar{x}_1 \rangle \, \langle \bar{x}_1 x_3 | V_{13} F_1 F_3 | x_2' x_1' \rangle &= \\ \langle x_1 x_2 | F_2 V_{12} F_1 F_2 | x_2' x_1' \rangle. & \end{aligned} \tag{3.10}$$

As a result, we have a modified binary interaction between particles one and two (notice the index change compared to the original pair), with an effective potential $V_{12} \to F_2 \, V_{12}$. Therefore, in operator notation we can write the following expression and its permutations

$$\begin{aligned} \mathrm{Tr}_3 V_{13} F_1 F_2 F_3 P_{23} &= F_2 V_{12} F_1 F_2, \\ \mathrm{Tr}_3 V_{23} F_1 F_2 F_3 P_{13} &= F_1 V_{12} F_1 F_2. \end{aligned} \tag{3.11}$$

Similarly, we have for operator products with the permutation operator acting from the left,

$$\begin{aligned} \mathrm{Tr}_3 P_{23} F_1 F_2 F_3 V_{13} &= F_1 F_2 V_{12} F_2, \\ \mathrm{Tr}_3 P_{13} F_1 F_2 F_3 V_{23} &= F_1 F_2 V_{12} F_1. \end{aligned} \tag{3.12}$$

In the trace over F_{123}, we have combinations of these expressions of the following form

$$n \mathrm{Tr}_3 V_{13} F_1 F_2 F_3 (\epsilon P_{23} + \epsilon P_{13}) = n\epsilon (F_1 + F_2) \, V_{12} \, F_1 F_2. \tag{3.13}$$

Examination of the hierarchy equations (see below) yields that the previous expressions of the type $V_{12} \, F_1 F_2$ in the case of particles with spin can always be combined with terms of the form (3.13) to

$$V_{12} \, F_1 F_2 + n \mathrm{Tr}_3 V_{13} F_1 F_2 F_3 (\epsilon P_{23} + \epsilon P_{13}) = \hat{V}_{12} \, F_1 F_2, \tag{3.14}$$

where we introduced the operator of the shielded potential [BD79]

$$\hat{V}_{12} = (1 + n\epsilon F_1 + n\epsilon F_2)V_{12}. \tag{3.15}$$

In the same manner, substituting for products of one-particle operators correlation operators, we can derive the expressions

$$\begin{aligned}
\mathrm{Tr}_3 V_{13} F_2 g_{13} P_{23} &= F_2 V_{12} g_{12}, \\
\mathrm{Tr}_3 V_{23} F_1 g_{23} P_{13} &= F_1 V_{12} g_{12}, \\
\mathrm{Tr}_3 P_{23} F_2 g_{13} V_{13} &= g_{12} V_{12} F_2, \\
\mathrm{Tr}_3 P_{13} F_1 g_{23} V_{23} &= g_{12} V_{12} F_1.
\end{aligned} \tag{3.16}$$

Notice that the exchange corrections destroy the commutator form of the collision term $[V_{12}, F_{12}]$. Instead, now the commutator is replaced by the difference of (3.14) minus its hermitean conjugate,[3] which sometimes is conveniently rewritten as

$$\begin{aligned}
\hat{V}_{12}\, F_1 F_2 - F_1 F_2 \hat{V}_{12}^{\dagger} &= (1 + n\epsilon F_1)(1 + n\epsilon F_2)\, V_{12} F_1 F_2 \\
&- F_1 F_2 V_{12}\, (1 + n\epsilon F_1)(1 + n\epsilon F_2) = I_{12}^{>} - I_{12}^{<},
\end{aligned} \tag{3.17}$$

where the symmetry of this expression suggests to define

$$\begin{aligned}
I_{ab}^{\gtrless} &= F_a^{\gtrless} F_b^{\gtrless}\, V_{ab}\, F_a^{\lessgtr} F_b^{\lessgtr}, \\
F_a^{<} &= F_a, \\
F_a^{>} &= 1 + n\epsilon F_a.
\end{aligned} \tag{3.18}$$

While $F_a^{<}$ is just the operator of the one-particle density, $F_a^{>}$ is related to the complementary probability and is sometimes called "hole" operator[4]. $I^{\gtrless}$ turn out to be the kernels of the two-particle scattering integrals, describing the scattering "into" (<) or "out" (>) of a certain state, cf. Eq. (3.22).

---

[3] The destruction of the commutator form has nontrivial consequences for the analysis of the time reversibility and the conservation properties, which we discuss in Sec. 3.4.2.

[4] from nuclear matter or solid state terminology where excitation of a particle from a certain energy level or band leaves behind a "hole" which behaves like a particle itself. We mention that the appearance of the density factor comes from the fact that F_a is normalized to the volume. It cancels if the operators are transformed into a given representation.

3.3 (Anti-)Symmetrization of the first and second hierarchy equations

The generalization of the first hierarchy equation (2.19) follows, if we use on the r.h.s. for F_{12} the (anti-)symmetrized expression (3.8)

$$i\hbar\frac{\partial}{\partial t}F_1 - [H_1, F_1] = n\mathrm{Tr}_2[V_{12}, F_1F_2 + g_{12}]\Lambda_{12}^{\pm}. \tag{3.19}$$

Using the definition of the Hartree-Fock Hamiltonian (3.9), we obtain the **(anti-)symmetrized first hierarchy equation**

$$\boxed{\begin{aligned} i\hbar\frac{\partial}{\partial t}F_1 - [\bar{H}_1, F_1] &= n\mathrm{Tr}_2[V_{12}^{\pm}, g_{12}] \\ \bar{H}_1 = H_1 + H_1^{HF} &= H_1 + n\mathrm{Tr}_2 V_{12}F_2\Lambda_{12}^{\pm} \end{aligned}} \tag{3.20}$$

For the derivation of the second hierarchy equation, we have to go back to the equation for F_{12} (2.20). It remains the same for the (anti-)symmetrized operators. However, modifications appear if F_{12} and F_{123} are expressed by lower order operators, since we have to use the (anti-)symmetrized version of the Ursell-Mayer expansion, Eq. (3.8). This yields

$$\begin{aligned} & i\hbar\frac{\partial}{\partial t}\big(F_1F_2 + g_{12}\big)\Lambda_{12}^{\pm} - [H_{12}, F_1F_2 + g_{12}]\Lambda_{12}^{\pm} \\ = & \ \Big\{n\mathrm{Tr}_3[V_{13} + V_{23}, F_1F_2F_3] + n\mathrm{Tr}_3[V_{13} + V_{23}, F_1g_{23}] \\ + & \ n\mathrm{Tr}_3[V_{13} + V_{23}, F_2g_{13}] + n\mathrm{Tr}_3[V_{13} + V_{23}, F_3g_{12}] \\ + & \ n\mathrm{Tr}_3[V_{13} + V_{23}, g_{123}]\Big\}\Lambda_{123}^{\pm}. \end{aligned} \tag{3.21}$$

Due to the factorization property of $\Lambda_{123}^{\pm}$, Eq. (3.6), an overall factor $\Lambda_{12}^{\pm}$ can be canceled. Further, eliminating the derivatives of the one-particle operators using Eq.(3.20) and taking into account relations (3.9,3.11 and 3.16), we obtain [BD79, DKBB97] (details are given in Appendix C) the **(anti-)symmetrized second hierarchy equation**

$$\boxed{\begin{aligned} i\hbar\frac{\partial}{\partial t}g_{12} - [\bar{H}_{12}^0, g_{12}] - (\hat{V}_{12}g_{12} - g_{12}\hat{V}_{12}^{\dagger}) &= (\hat{V}_{12}F_1F_2 - F_1F_2\hat{V}_{12}^{\dagger}) \\ +\Pi_{12}^{(1)} + \Pi_{12}^{(2)} &+ n\mathrm{Tr}_3[V_{13} + V_{23}, g_{123}](1 + \epsilon P_{13} + \epsilon P_{23}) \\ \bar{H}_{12}^0 = H_1 + H_2 + H_1^{HF} + H_2^{HF}, &\quad \Pi_{12}^{(1)} = n\mathrm{Tr}_3[V_{13}^{\pm}, F_1]g_{23}\Lambda_{23}^{\pm} \end{aligned}} \tag{3.22}$$

where $\Pi^{(1,2)}$ denote the polarization terms, and $\Pi_{12}^{(2)}$ follows from the substitution $1 \longleftrightarrow 2$.

As we will see in the Chapters below (starting with Ch. 6), the first term on the r.h.s. gives rise to the two-particle collision integrals including phase space occupation effects. The difference corresponds to the balance of scattering "in" minus scattering "out" of a given state and can be written as difference of the collision kernels $I^> - I^<$, cf. Eq. (3.18). This symmetric representation is a feature of the spin statistics which allowed us to introduce in addition to the one-particle density operator $F = F^<$ a symmetric counterpart, the hole operator $F^>$. Of course, this idea can be extended to two-particle and more complex operators. Notice that $F^>$ contains contributions of different orders in the density. At low densities $F^> \to 1$, only at high densities effects of the spin statistics become important.

3.4 (Anti-)Symmetrization of the third hierarchy equation

To derive the (anti-)symmetrized third hierarchy equation we consider the equation for F_{123} (2.21) using again the (anti-)symmetrized version of the Ursell-Mayer expansion, Eq. (3.8). This yields

$$\begin{aligned}
& i\hbar\frac{\partial}{\partial t}\Big(F_1F_2F_3 + g_{12}F_3 + \ldots + g_{123}\Big)\Lambda^{\pm}_{123} \\
& -[H_{123}, F_1F_2F_3 + g_{12}F_3 + \ldots + g_{123}]\Lambda^{\pm}_{123} \\
= & \Big\{n\mathrm{Tr}_4[V_{14}+V_{24}+V_{34}, F_1F_2F_3F_4] \\
+ & \; n\mathrm{Tr}_4[V_{14}+V_{24}+V_{34}, F_1F_2g_{34}] + \ldots + n\mathrm{Tr}_4[V_{14}+V_{24}+V_{34}, g_{12}g_{34}] + \ldots \\
+ & \; n\mathrm{Tr}_4[V_{14}+V_{24}+V_{34}, F_1g_{234}] + \ldots \\
+ & \; n\mathrm{Tr}_4[V_{14}+V_{24}+V_{34}, g_{1234}]\Big\}\Lambda^{\pm}_{1234},
\end{aligned} \tag{3.23}$$

where "..." denotes all permutations of the preceding term. Due to the factorization property of $\Lambda^{\pm}_{1234}$, Eq. (3.6), an overall factor $\Lambda^{\pm}_{123}$ can be canceled. Further, eliminating the derivatives of the one-particle and two-particle operators using Eqs.(3.20), (3.22) and taking into account relations (3.9,3.11) and 3.16), we obtain (details are given in Appendix C)

$$\begin{aligned}
& i\hbar\frac{\partial}{\partial t}g_{123} - [\bar{H}^0_{123}, g_{123}] - \Big\{\Big(\hat{V}_{12}+\hat{V}_{13}+\hat{V}_{23}\Big)\,g_{123} - g_{123}\,\Big(\hat{V}^\dagger_{12}+\hat{V}^\dagger_{13}+\hat{V}^\dagger_{23}\Big)\Big\} \\
= & \Big\{(\hat{V}_{12}F_1F_2F_3 + (\hat{V}_{13}+\hat{V}_{23})F_3g_{12} \\
- & \; \epsilon nF_3(F_1V_{13}+F_2V_{23})g_{12} - \epsilon n(g_{13}V_{13}+g_{23}V_{23})g_{12}
\end{aligned}$$

$$
\begin{aligned}
&+ \quad n\mathrm{Tr}_4[V_{14} + V_{24}, g_{12}g_{34}](1 + \epsilon P_{14} + \epsilon P_{24} + \epsilon P_{34}) \\
&+ \quad n\mathrm{Tr}_4[V_{14}^{\pm}, F_1 g_{234}](1 + \epsilon P_{24} + \epsilon P_{34}) \\
&+ \quad n\mathrm{Tr}_4[V_{14}, g_{1234}](1 + \epsilon P_{14} + \epsilon P_{24} + \epsilon P_{34}) + \mathcal{P}(123) - \mathrm{h.c.}\big\},
\end{aligned} \tag{3.24}
$$

where $\mathcal{P}(123)$ denotes cyclic permutations of indices $1, 2, 3$ of all terms on the r.h.s., and $\bar{H}^0_{123} = H_1 + H_2 + H_3 + H_1^{HF} + H_2^{HF} + H_3^{HF}$. The hermitean conjugation applies to the whole r.h.s. too, and we underline that it leads to a change in the order of the operators. Notice that the three-particle scattering terms can be transformed in analogy to the two-particle terms $I_{ab}^{\gtrless}$, Eq. (3.18). Indeed, we find

$$
\left(\hat{V}_{12} + \hat{V}_{13} + \hat{V}_{23}\right) F_1 F_2 F_3 - h.c. = I_{123}^{>} - I_{123}^{<}, \tag{3.25}
$$

$$
\text{with} \quad I_{abc}^{\gtrless} = F_a^{\gtrless} F_b^{\gtrless} F_c^{\gtrless} (V_{ab} + V_{ac} + V_{bc}) F_a^{\lessgtr} F_b^{\lessgtr} F_c^{\lessgtr}. \tag{3.26}
$$

Eq. (3.24) is very important, since it allows for the derivation of the vast majority of practically relevant approximations of many-particle theory, thereby fully including the effects of Bose or Fermi statistics. Let us briefly discuss this equation and the changes which appeared in comparison to the third hierarchy equation without spin statistics (2.99). The terms on the l.h.s. are trivial generalizations, where simply the Hartree potentials U^H are replaced by the Hartree-Fock potential H^{HF} and the bare interaction potential V by the shielded one $\hat{V}$. The same applies to the three-particle scattering term (first term on the r.h.s.). The polarization terms and the four-particle correlation term (3rd - 5th lines on the r.h.s.) are generalized by including all possible exchange contributions. The only terms that are qualitatively new are related to selfenergy effects, second term and second line on the r.h.s., and we, therefore, discuss them in more detail.

3.4.1 (Anti-)Symmetrization of the selfenergy terms

The selfenergy term in the case of spinless particles, cf. Eq. (2.118), $[V_{13} + V_{23}, F_3 g_{12}]$, is now replaced by

$$
\left\{(\hat{V}_{13} + \hat{V}_{23})F_3 - \epsilon n F_3(F_1 V_{13} + F_2 V_{23}) - \epsilon n(g_{13}V_{13} + g_{23}V_{23})\right\} g_{12} - \mathrm{h.c.} \tag{3.27}
$$

While the first term is a straightforward generalization of the previous expression (2.118), the remaining ones are qualitatively new and solely due to the spin statistics (they vanish with $\epsilon \to 0$). With the definition of $\hat{V}$, Eq. (3.15), the first and second term in Eq. (3.27) can be combined using the identity,

$$
\begin{aligned}
&\hat{V}_{13}F_3 - \epsilon n\, F_1\, F_3 V_{13} \\
&= (1 + n\epsilon F_1)(1 + n\epsilon F_3)V_{13}F_3 - \epsilon F_1 F_3 V_{13}(1 + n\epsilon F_3),
\end{aligned} \tag{3.28}
$$

to yield

$$\Big\{\Big((1+n\epsilon F_1)(1+n\epsilon F_3)V_{13}F_3 - \epsilon F_1 F_3 V_{13}(1+n\epsilon F_3)\Big)\, g_{12} - \text{h.c.} + (1 \leftrightarrow 2)\Big\}$$
$$= \Big(S^{>}_{13} - \epsilon S^{<}_{13} + S^{>}_{23} - \epsilon S^{<}_{23}\Big)\, g_{12} - \text{h.c.}, \tag{3.29}$$

where the symmetry of this expression suggested to define

$$S^{\gtrless}_{ab} = F^{\gtrless}_{a} F^{\gtrless}_{b} \, V_{ab} \, F^{\lessgtr}_{b}, \tag{3.30}$$

with $F^{\gtrless}$ given by Eq. (3.18). As it will turn out below, $S^{\gtrless}_{ab}$ are the kernels of the two-particle scattering rates $\Sigma^{\gtrless}_{ab}$. The structure of these terms is quite similar to that of the kernels of the collision terms $I^{\gtrless}_{ab}$ which were introduced in Eq. (3.18). These quantities are closely related:

$$I^{\gtrless}_{ab} \;=\; S^{\gtrless}_{ab} \, F^{\lessgtr}_{a}. \tag{3.31}$$

We will see in Ch. 7 that the kernels $S^{\gtrless}$ give rise to the carrier-carrier scattering related selfenergy terms in the hierarchy.

In analogy to the hierarchy equations for the density operators, one can (anti-)symmetrize the corresponding equations in a given representation (matrix equations), including the Bloch equations, cf. Sec. 2.4.1. However, it is more convenient to first perform the (anti-)symmetrization in the compact operator form and also introduce approximations (a decoupling ansatz to the hierarchy) there and, only at the end, expand the final form of the equations into a given basis. Therefore, we will consider the generalized Bloch equations which include spin statistics and correlation (scattering) effects after having discussed correlation effects, in Ch. 11.

3.4.2 Energy conservation with spin statistics

The modifications from the spin statistics in the explicit form of the hierarchy equations raise the question of energy conservation again. As was discussed in Sec. 2.2.2, we have to expect that spin statistics effects do not alter the conservation behavior of a given hierarchy closure approximation. In this Section we show explicitly that this is indeed the case. As in the case without spin (Sec. 2.5.2), we consider separately the terms appearing in Eq. (2.34):

$$\begin{aligned}
V_{13}F_{123} \;&=\; V_{13}F_1F_2F_3\,(\Lambda^{\pm}_{13} \pm P_{23}) + V_{13}F_2 g_{13}\,(\Lambda^{\pm}_{13} \pm P_{23}) \\
&+\; V_{13}\,(F_3 g_{12} \pm g_{23}F_1 P_{13}) \\
&+\; V_{13}\,\Big\{F_1 g_{23}\,\Lambda^{\pm}_{23} \pm g_{12}F_3(P_{13}+P_{23})\Big\},
\end{aligned} \tag{3.32}$$

and analogously for $1 \leftrightarrow 2$. Again, the terms in the first line are required for deriving Eq. (2.99) (they are canceled exactly by the equation for $F_1F_2\Lambda^{\pm}_{12}$). The term in the second line gives the Hartree-Fock term $H_1^{HF}g_{12}$, and the third line gives the polarization contributions including the exchange polarization.

Our argumentation now follows that of the spinless case. Again, the complete expression $F_{123}\,\Lambda^{\pm}_{123}$ with $g_{123} = 0$ fulfills condition (2.36) and is thus conserving. Also, the term containing the one-particle operators is trivially symmetric. We consider therefore, only the contributions containing binary correlations. We begin with the Pauli blocking terms, i.e. terms with P_{23} in the first line of Eq. (3.32),

$$\begin{gathered}\mathrm{Tr}_{123}V_{12}\left\{V_{13}(F_1F_3+g_{13})P_{23}-P_{23}(F_1F_3+g_{13})V_{13}\right\}\Lambda^{\pm}_{12}+1\leftrightarrow 2=\\ \mathrm{Tr}_{12}\left\{V_{12}(F_1+F_2)V_{12}(F_1F_2+g_{12})-V_{12}(F_1F_2+g_{12})V_{12}(F_1+F_2)\right\}\Lambda^{\pm}_{12}=0.\end{gathered}$$

Due to the invariance of the trace, both terms cancel. This means, Pauli blocking does not contribute to the time derivative of total energy, and we may omit P_{23} in the first line of Eq. (3.32). Consider now one of the Hartree-Fock terms (second line of Eq. (3.32)).

$$\begin{aligned}\mathrm{Tr}_3V_{13}\left\{F_3g_{12}\pm g_{23}F_1P_{13}\right\} &=\\ \mathrm{Tr}_3V_{13}\left\{F_3g_{12}\pm P_{13}g_{12}F_3\right\}=\mathrm{Tr}_3V^{\pm}_{13}\,F_3g_{12}.&\end{aligned}\tag{3.33}$$

Their contribution to the derivative of energy is

$$\mathrm{Tr}_{123}\left\{V^{\pm}_{12}V^{\pm}_{13}-V^{\pm}_{13}V^{\pm}_{12}\right\}F_3g_{12}=\frac{d}{dt}\left\langle H_1^{HF}\right\rangle \tag{3.34}$$

plus the symmetric term $(1 \leftrightarrow 2)$ and differs from the spinless case only by substitution $V \to V^{\pm}$. Similarly, we find for the term on the first line of Eq. (3.32)

$$\mathrm{Tr}_{123}\left\{V^{\pm}_{12}V^{\pm}_{13}-V^{\pm}_{13}V^{\pm}_{12}\right\}F_2g_{13}=-\frac{d}{dt}\left\langle H_1^{HF}\right\rangle, \tag{3.35}$$

where we again used the fact that under the trace we may interchange the indices $2 \leftrightarrow 3$. Thus again we find that the terms on the first and second line of Eq. (3.32) compensate. Now it remains to consider the polarization terms which can be transformed in the following way (cf. Appendix C), [DKBB97]:

$$\mathrm{Tr}_3V_{13}\left\{F_1g_{23}\,\Lambda^{\pm}_{23}\pm g_{12}F_3(P_{13}+P_{23})\right\}=\mathrm{Tr}_3[V^{\pm}_{13},F_1]g_{23}\Lambda^{\pm}_{23}, \tag{3.36}$$

and analogously for $1 \leftrightarrow 2$. We first show that the term in $\Lambda^{\pm}_{23}$ that contains P_{23}, does not contribute to the energy:

$$\begin{aligned}&\mathrm{Tr}_{123}V^{\pm}_{12}\left\{V^{\pm}_{13}\,F_1\,g_{23}\,P_{23}-F_1\,g_{23}\,P_{23}\,V^{\pm}_{13}\right\}=\\ &\qquad\mathrm{Tr}_{123}\left\{V^{\pm}_{12}\,V^{\pm}_{13}-V^{\pm}_{13}\,V^{\pm}_{12}\right\}F_1\,g_{23}\,P_{23}=0,\end{aligned}\tag{3.37}$$

which follows again from interchanging $2 \leftrightarrow 3$ in the second term. Thus, the polarization contributions to (3.32) are

$$\mathrm{Tr}_{123} V_{12}^{\pm}[V_{13}^{\pm}, F_1] g_{23} + 1 \leftrightarrow 2 = 0, \tag{3.38}$$

which vanishes exactly as in the spinless case (we only have to substitute $V \rightarrow V^{\pm}$), so the polarization plus exchange polarization terms are energy conserving.

Summarizing the properties of the terms in Eq. (3.32), we conclude that the terms which contribute to the time derivative of total energy can be written in the compact form

$$i\hbar \frac{d}{dt}\langle T + V \rangle = \frac{n^3}{2} \mathrm{Tr}_{123} \left(V_{12}^{\pm}[V_{13}^{\pm} + V_{23}^{\pm}, F_{123}] \right), \tag{3.39}$$

where all the effects of the spin statistics have been transferred onto the binary interaction potentials. Furthermore, we have succeeded in restoring the commutator form in the trace term in Eq. (3.39). We therefore may conclude that all the approximations considered for spinless particles above, remain conserving for bosons and fermions too.[5] Notice that in the presence of an external potential $\mathcal{U}$ the conservation law of Eq. (3.39) has again to be generalized, as it was discussed in Sec. 2.2.2.

As a conclusion, we mention that extensive investigations of the (anti-)symmetrization procedure of the hierarchy have been performed in nuclear matter theory, see e.g. [WC85, SRT90]. For a recent overview and further references, see [RT93].

[5] On the other hand, an approximation that neglects all exchange terms consistently (in the collision term of the first hierarchy equation, in the Hartree-Fock terms in $\bar{H}_{12}$, in the polarization terms and so on), but includes the Pauli blocking terms, will still be conserving, since the latter have no influence on the conservation properties. This is most easily verified from Eq. (3.32), where the direct and exchange terms may be collected in two groups of terms which each independently has the noted properties.

Chapter 4

Mean–Field Approximation. Quantum Vlasov Equation. Collective Effects

The first closure of the BBGKY–hierarchy which we are going to study is the mean–field approximation (Hartree-Fock approximation), [Har28, Har48]. As mentioned in Sec. 2.6.1, this approximation is applicable, if all correlations in the system are negligibly small, i.e. $g_{12} \approx g_{123} \approx \ldots \approx 0$. As a result, the BBGKY-hierarchy reduces to the first equation

$$i\hbar\frac{\partial}{\partial t}F_1 - [H_1 + H_1^{HF}, F_1] = 0; \qquad H_1^{HF} = n\mathrm{Tr}_2 V_{12} F_2 \Lambda_{12}^{\pm}, \tag{4.1}$$

where the Hamiltonian H_1 contains the kinetic part and the external field $\mathcal{U}_1$. Eq. (4.1) describes the evolution of the one-particle properties under the influence of the external potential $\mathcal{U}$ and of an average field created by all particles. This approximation is the simplest way to include the action of an external field into a kinetic description, and it has been extensively used in many fields of physics, in particular, in plasma physics, but also in solid state theory and nuclear matter theory[1] . External fields give rise to a large variety of effects in many–particle systems. In this Chapter, we focus on one particular phenomenon: *collective particle response to a longitudinal time-independent inhomogeneous external field.* A more general discussion of field effects together with the effect of correlations will be given in Chs. 11 and 12.

[1] In the latter case, this approximation is usually called TDHF (time-dependent Hartree-Fock approximation). For examples of its application to nuclear systems see [Köh85] and references therein.

To treat the effect of inhomogeneous fields, it is convenient to transform Eq. (4.1) to either the coordinate or the Wigner representation, cf. Secs. 2.3.1 and 2.3.2, respectively. We provide both forms. In the coordinate representation, we obtain from Eq. (2.45), with $g_{12} = 0$ and exchange contributions included, and denoting the matrix elements of $\mathcal{U}$ by $U(r,t)$,

$$\left\{ i\hbar\frac{\partial}{\partial t} + \frac{\hbar^2}{2m}(\nabla_{r'}^2 - \nabla_{r''}^2) - U(r',t) + U(r'',t) \right\} f(r',r'',t) = \sum_{\bar{s}} \int d\bar{r} \times$$
$$\{V(r'-\bar{r}) - V(r''-\bar{r})\} \left[f(r',r'',t) f(\bar{r},\bar{r},t) \mp f(r',\bar{r},t) f(\bar{r},r'',t) \right], \qquad (4.2)$$

where the spin summation yields the factor (2s+1) if the interaction potential is spin independent, what we shall assume. Also, to simplify the analysis, in the following we will neglect the exchange term. Then the external potential and the Hartree term can be combined to an effective potential which modifies the external field

$$U^{\text{eff}}(r,t) = U(r,t) + (2s+1) \int d\bar{r}\, V(r-\bar{r})\, f(\bar{r},\bar{r},t). \qquad (4.3)$$

This means, a particle in point r feels an effective field which is composed by the external field plus the field created by all particles at all positions $\bar{r}$ at the same moment t (i.e. the field is instantaneous). After introduction of center of mass and difference coordinates $R = (r' + r'')/2$ and $r = r' - r''$, we obtain

$$\left\{ i\hbar\frac{\partial}{\partial t} + \frac{\hbar^2}{m}\nabla_R \nabla_r - U^{\text{eff}}(R + \frac{r}{2}, t) + U^{\text{eff}}(R - \frac{r}{2}, t) \right\} \tilde{f}(R,r,t) = 0 \qquad (4.4)$$

where now $U^{\text{eff}}(R,t) = U(R,t) + (2s+1) \int d\bar{R}\, V(R-\bar{R}) \tilde{f}(\bar{R},0,t)$.

Consider now the Wigner representation. Then Eq. (4.1) transforms into Eq. (2.57), where on the r.h.s. $g_{12} \to 0$, and we perform the obvious generalization to the multi-component case, cf. Sec. 2.4,

$$\left\{ \frac{\partial}{\partial t} + \frac{p}{m_a}\nabla_R \right\} f_a(R,p,t) - \frac{1}{i\hbar} \int \frac{dr}{(2\pi\hbar)^3}\, d\bar{p}\, \exp\{-i\,(p-\bar{p})\, r/\hbar\}$$
$$\times \left(U_a^{\text{eff}}(R + \frac{r}{2}, t) - U_a^{\text{eff}}(R - \frac{r}{2}, t) \right) f_a(R,\bar{p},t) = 0 \qquad (4.5)$$
$$\text{with} \quad U_a^{\text{eff}}(R,t) = U_a(R,t) + \sum_b \int d\bar{R}\, V_{ab}(R-\bar{R})\, n_b(\bar{R},t), \qquad (4.6)$$

and n_b is the density, $n_b(R,t) = (2s+1) \int dp f_b(R,p,t)/(2\pi\hbar)^3$. The most important example for a mean–field kinetic equation is the Vlasov (or Hartree) equation, where U_a^{eff} is just the electrostatic potential ϕ, generated by all particles, times the charge e_a [Vla38, Vla45]. Then, Eq. (4.6) is just the solution of Poisson's equation $\Delta\phi = -4\pi \left(\sum_b e_b n_b + \rho^{ext}\right)$. [2]

Due to the Hartree term in U^{eff}, Eq. (4.5) is nonlinear in the Wigner distribution, which inhibits an analytical solution. In cases, where the external excitation is weak, it is reasonable to use linearization procedures which allow for analytical solutions, cf. Sec. 4.1. However, if the system is subject to a high-amplitude external field, linearizations break down, and one has to take the nonlinear terms seriously. There exist several schemes which allow to include nonlinearity effects approximately, such as quasilinear theories, which will be discussed in Sec. 4.5. Numerical solutions to the full quantum Vlasov equation will be discussed in Sec. 4.6.

4.1 Linearization of the quantum Vlasov equation. Dielectric function

We consider the system sufficiently long before the external potential $U(t)$ has been turned on and denote by f_{0a} the solution of the field-free equation (4.5),

$$\left\{\frac{\partial}{\partial t} + \frac{p}{m_a}\nabla_R\right\} f_{0a}(R,p,t) - \frac{1}{i\hbar}\int \frac{dr}{(2\pi\hbar)^3}\, d\bar{p}\, \exp\{-i\,(p-\bar{p})\, r/\hbar\}$$
$$\times \left(U_a^H(R+\frac{r}{2},t) - U_a^H(R-\frac{r}{2},t)\right) f_{0a}(R,\bar{p},t) = 0, \tag{4.7}$$

where only the Hartree-field is left from U^{eff}. If the system would be in equilibrium, f_{0a} would be independent of R and t. But in general, we do not need to assume this, and f_{0a} could also be the solution of an equation with an time-independent or slowly varying[3] external field. Let us now turn on the external potential U, assuming that it is weak, where the criterion is, that the solution of the full nonlinear equation (4.5) is close to that of Eq. (4.7),

$$\begin{aligned} f_a(R,p,t) &= f_{0a}(R,p,t) + f_{1a}(R,p,t), \quad \text{with} \\ |f_{1a}(R,p,t)| &\ll f_{0a}(R,p,t). \end{aligned} \tag{4.8}$$

In fact, as we will see below, there are more restrictive conditions on f_{1a}. Using Eqs. (4.8) and (4.7), we may linearize Eq. (4.5) and obtain

$$\left\{\frac{\partial}{\partial t} + \frac{p}{m_a}\nabla_R\right\} f_{1a}(R,p,t) - \frac{1}{i\hbar}\int \frac{dr}{(2\pi\hbar)^3}\, d\bar{p}\, \exp\{-i\,(p-\bar{p})\, r/\hbar\}$$

[2] The more general case, where U^{eff} contains also the influence of transverse (electromagnetic) fields, will be discussed in Chs. 11 and 12.

[3] much slower than the disturbing potential $U(t)$

$$\times \left(U_{1a}^{\text{eff}}(R + \frac{r}{2}, t) - U_{1a}^{\text{eff}}(R - \frac{r}{2}, t) \right) f_{0a}(R, \bar{p}, t) = 0, \qquad (4.9)$$

$$\text{with} \quad U_{1a}^{\text{eff}}(R,t) = U_a(R,t) + \sum_b \int d\bar{R}\, V_{ab}(R - \bar{R})\, n_{1b}(\bar{R}, t),$$

and $n_{1a}(R,t) = (2s_a + 1) \int dp f_{1a}(R,p,t)/(2\pi\hbar)^3$. The major achievement is that in the integral term of Eq. (4.9) appears only f_{0a}, rendering an equation which is linear in the disturbance f_{1a}. Obviously, the neglect of $f_{1a}(R,p,t)$ in the integral term of Eq. (4.9) for all values of the radius vector, momentum and for all times, is a further strong restriction on the validity of the linearization. For example, in the classical limit (or for the the long-wavelength limit of quantum systems), this leads to the condition $|\partial f_{1a}/\partial p| \ll |\partial f_{0a}/\partial p|$. Therefore, approximation (4.8) may break down also for weak external potentials if the evolution leads to growth of f_{1a} in time. Obviously, if the field-free solution f_{0a} has extrema (except for the trivial case $p = 0$), this inequality will be violated and we expect deviations from the linear behavior. We will see below that, this is indeed the case. In particular, if f_{0a} has a minimum, fluctuations may grow spontaneously giving rise to plasma instabilities, cf. Sec. 4.3.

Linear response function and RPA polarization. Due to linearity, Eq. (4.9) can be solved by Fourier-Laplace transform, where we take into account the initial condition $\lim_{t \to -\infty} U_a(t) = 0$. The external potential may be expanded in terms of monochromatic plane waves

$$U_a(\vec{R}, t) = \int_C \frac{d\omega}{2\pi} \int \frac{d\vec{k}}{(2\pi)^3}\, U_a(\vec{k}, \omega)\, e^{i\vec{k}\vec{R} - i\omega t}, \qquad (4.10)$$

where $\mathrm{Im}\,\omega > 0$ is required to assure that the initial condition is fulfilled. The contour "C" will be specified more precisely below Similarly we expand f_{1a} and U_{1a}^{eff} with the result for the spectral components of the effective potential

$$U_{1a}^{\text{eff}}(\vec{k}, \omega) = U_a(\vec{k}, \omega) + \sum_b V_{ab}(\vec{k})\, n_b(\vec{k}, \omega). \qquad (4.11)$$

Inserting the expansions of f_{1a} and U_{1a}^{eff} into Eq. (4.9) yields, in the integral term, two delta functions $\delta(\bar{p} - p \mp \hbar k/2)$, which result in the solution

$$f_{1a}(\vec{p}, \vec{k}, \omega) = -U_{1a}^{\text{eff}}(\vec{k}, \omega) \frac{f_{0a}(\vec{p} + \frac{\hbar\vec{k}}{2}) - f_{0a}(\vec{p} - \frac{\hbar\vec{k}}{2})}{\omega - \frac{\vec{k}\vec{p}}{m_a}}. \qquad (4.12)$$

Integrating Eq. (4.12) over the momentum $\vec{p}$ and summing over the spin, we obtain the spectral component of the density disturbance

$$n_{1a}(\vec{k}, \omega) = U_{1a}^{\text{eff}}(\vec{k}, \omega)\, \Pi_a^R(\vec{k}, \omega), \qquad (4.13)$$

where we introduced the retarded density response function (polarization function in case of polarizable media)

$$\Pi_a^R(\vec{k},\omega) = (2s_a+1)\hbar \int \frac{d\vec{p}}{(2\pi\hbar)^3} \frac{f_a(\vec{p}) - f_a(\vec{p}+\hbar\vec{k})}{\hbar\omega + \frac{p^2}{2m_a} - \frac{(\vec{p}+\hbar\vec{k})^2}{2m_a} + i\delta}. \tag{4.14}$$

Here, we have dropped the index "0" of the distributions to simplify the notation in the subsequent formulas which will not lead to confusion. Also, in Eq. (4.14), the arguments in the distributions have been shifted by means of the substitution $\vec{p} \to \vec{p} + \hbar\vec{k}/2$, and we replaced the complex frequency by $\omega \to \omega + i\delta$, with $\delta > 0$, so the frequency argument ω of Π^R is real.[4] Eq. (4.14) is the well-known Lindhard polarization (RPA polarization), which was obtained by many authors [KS52a, KS52b, BP53, Lin54, EC59].

Classical limit. The classical limit of the RPA polarization is readily obtained by taking the long wavelength limit, $k \to 0$, of Eq. (4.14). Keeping in the numerator and denominator only terms up to first order in k, and introducing the velocity $\vec{v} = \vec{p}/m_a$, we obtain the Vlasov polarization function,

$$\Pi_a^R(\vec{k},\omega) = -(2s_a+1)\hbar \int \frac{d\vec{v}}{(2\pi\hbar)^3} \frac{\vec{k}\frac{\partial f_a(\vec{v})}{\partial \vec{v}}}{\omega - \vec{k}\vec{v} + i\delta}. \tag{4.15}$$

Notice the scalar products in the numerator and denominator which mean, that the derivative is to be taken along the propagation direction $\vec{k}$ of the external wave.[5] From Eq. (4.16) and relation (4.21), immediately follows the classical dielectric function, which was obtained by Vlasov [Vla38].

General linear response functions. It is instructive to rewrite the quantum expression (4.14) in a more general form, by replacing the free kinetic energy by a generalized single-particle dispersion $p^2/2m_a \to E_p^a$, which

---

[4] The infinitesimal positive imaginary part of the complex frequency assures the existence of the Laplace transform (4.10). The case of negative imaginary parts is more difficult and requires analytical continuation of Π^R what will be discussed below.

[5] We underline that the above long wavelength limit is not uniquely defined if we take the vector character of $\vec{k}$ serious. The result will depend on the path in momentum space (except for one-dimensional systems or isotropic 2D or 3D systems). Another way to look at this problem is to introduce the Cartesian components of $\vec{k}$ and $\vec{v}$, $(\alpha,\beta) = x,y,z$, which leads to a tensor expression for the polarization

$$\Pi_{\alpha\beta}^{aR}(\vec{k},\omega) = -(2s_a+1)\hbar \int \frac{d\vec{v}}{(2\pi\hbar)^3} \frac{k_\alpha \frac{\partial f_a(\vec{v})}{\partial v_\beta}}{\omega - \vec{k}\vec{v} + i\delta}. \tag{4.16}$$

The previous result (4.15) is just the trace of this tensor, i.e. $\Pi_a^R = \Pi_{xx}^{aR} + \Pi_{yy}^{aR} + \Pi_{zz}^{aR}$.

Of course, the full quantum expression (4.14) has the same tensor structure. We will return to this question in Sec. 4.2, where we consider the electrodynamic definition of the dielectric function.

is relevant, e.g. for electrons in a solid.[6] If we further introduce generalized distribution functions $f(p) \to f(E_p)$, we may rewrite Eq. (4.14),

$$\Pi_a^R(\vec{k},\omega) = (2s_a+1)\hbar \int \frac{d\vec{p}}{(2\pi\hbar)^3} \frac{f_a(E^a_{\vec{p}}) - f_a(E^a_{\vec{p}+\hbar\vec{k}})}{\hbar\omega + E^a_{\vec{p}} - E^a_{\vec{p}+\hbar\vec{k}} + i\delta}. \tag{4.17}$$

Π is the density response to the external field in linear approximation, cf. Eq. (4.13), which is naturally defined as [KB89]

$$\Pi_a^R(\vec{k},\omega) = \left(\frac{\delta\, n_{1a}}{\delta\, U_a}\right)|_{U_a\to 0}(\vec{k},\omega+i\delta). \tag{4.18}$$

Furthermore, we may insert the result for the density, Eq. (4.13) into Eq. (4.11) and obtain

$$U_a(\vec{k},\omega) = \sum_b U_{1b}^{\rm eff}(\vec{k},\omega)\left\{\delta_{ab} - V_{ab}(\vec{k})\Pi_b^R(\vec{k},\omega)\right\}. \tag{4.19}$$

We can rewrite this equation in compact matrix form, introducing the vectors (in the space of components) U, $U^{\rm eff}$, and obtain $U(\vec{k},\omega) = U^{\rm eff}(\vec{k},\omega)\epsilon_R(\vec{k},\omega)$, or equivalently, $U^{\rm eff}(\vec{k},\omega) = \epsilon_R^{-1}(\vec{k},\omega)U(\vec{k},\omega)$, with ϵ_R^{-1} being the retarded linear response function, defined as

$$\epsilon_R^{-1}(\vec{k},\omega) = \left(\frac{\delta\, U_{1a}^{\rm eff}}{\delta\, U_a}\right)|_{U_a\to 0}(\vec{k},\omega+i\delta). \tag{4.20}$$

ϵ_R is related to the density response function Π^R by

$$\epsilon_R(\vec{k},\omega) = 1 - \sum_{ab} V_{ab}(\vec{k},\omega)\Pi_b^R(\vec{k},\omega). \tag{4.21}$$

In the case of polarizable media, such as charged particle systems (plasmas), ϵ_R turns out to be the dielectric function of the plasma.

We see that in the low excitation regime, the spectral components of the density response and the effective potential are linear functions of the external field. Each spectral component of the external field generates a response of the system with the same spatial and temporal harmonics. This is, of course, a consequence of the linearization of the Vlasov equation. We will see below, that for stronger excitation, the response of the system is more complex, including nonlinear effects, such as generation of higher harmonics and mode coupling.

[6] There, in fact, the kinetic equation contains on the l.h.s. the single-particle energy $E(p)$, which is due to the effect of the lattice on the electrons, as will be shown in Ch. 11.

Finally, we may transform the linear relations back to real space, where we obtain, for example for the density and the effective potential,

$$n_a(\vec{R},t) = \int_{-\infty}^{t} d\bar{t} \int d\vec{\bar{R}}\, \Pi_a(\vec{R}-\vec{\bar{R}}, t-\bar{t})\, U_a(\vec{\bar{R}},\bar{t}), \tag{4.22}$$

$$U_{1a}^{\rm eff}(\vec{R},t) = \int_{-\infty}^{t} d\bar{t} \int d\vec{\bar{R}}\, \epsilon_R^{-1}(\vec{R}-\vec{\bar{R}}, t-\bar{t})\, U_a(\vec{\bar{R}},\bar{t}), \tag{4.23}$$

which clearly show the causal (retarded) relationship between exciting field and system response. The condition for the existence of the inverse function ϵ_R^{-1} as well as for the existence of solutions to Eq. (4.19) is the vanishing of the determinant

$$||\epsilon_R(\vec{k},\omega)|| = 0, \tag{4.24}$$

which constitutes the dispersion relation for the linear inhomogeneous equation (4.11) for $U_1^{\rm eff}$. The solutions of Eq. (4.24) are the complex functions $\Omega_s(\vec{k})$, which are the eigen–frequencies (modes) of the medium, which are labelled by the discrete index "s". We will investigate the mode (plasmon) spectrum in some detail below. However, before doing this, we have to be aware that in order to find the complex solutions Ω_s, $\epsilon(\vec{k},\omega)$ has to be defined in the whole complex frequency plane. On the other hand, the Laplace transform for $U(t)$, $U^{\rm eff}$ and f_1 exists only in the upper frequency half plane. To find solutions on the real axis and in the lower half plane, we need to continue these functions analytically, what we will do next.

Analytic continuation of the dielectric function. Since the analytic properties of the dielectric function are completely defined by the polarization functions of all components, cf. Eq. (4.21), it is sufficient to consider the analytic continuation of the latter. The analytic continuation of the polarization into the complex z-plane can be written in the form of a Cauchy integral (we omit the component index) [Sto74, KKER86]

$$\Pi(k,z) = \int \frac{d\omega}{2\pi} \frac{\hat{\Pi}(k,\omega)}{z-\omega}, \tag{4.25}$$

with $\hat{\Pi}$ being the spectral function of the polarization given by Eq. (4.27). As usual, this integral defines two functions: the retarded function Π^R and the advanced function Π^A, which are analytic in the upper and the lower half-plane, respectively.[7] The analytic continuation of the retarded function into the lower half plane (and, respectively, the advanced function into the upper

[7] $\Pi(k,z)$ has a branch cut along the real axis. With the Plemlj formula we have ($\mathcal{P}$

half plane) is then given by

$$\begin{aligned} \Pi^R(k,z) &= \Pi^A(k,z) + \hat{\Pi}(k,z), \qquad \mathrm{Im}\, z < 0, \\ \Pi^A(k,z) &= \Pi^R(k,z) - \hat{\Pi}(k,z), \qquad \mathrm{Im}\, z > 0, \end{aligned} \tag{4.28}$$

where $\hat{\Pi}(k,z)$ is the analytic continuation of the spectral function. The analytic continuations of Π^R and Π^A can have singularities, which are those of $\hat{\Pi}(k,z)$. Furthermore, the analytic continuation of the spectral function ensures that all derivatives with respect to $\mathrm{Im}\, z$ are continuous at $\mathrm{Im}\, z = 0$ too.

We thus have obtained one function which is analytic in the whole complex frequency plane and, therefore can be used for the determination of the complete collective excitation spectrum. We can write down the result for the continuation of the retarded polarization function which we will denote by $\tilde{\Pi}$, $(z = \omega - i\gamma)$

$$\tilde{\Pi}^R(k,\omega,\gamma) = \begin{cases} \Pi^R(k,\omega,\gamma) & , \quad \gamma < 0 \\ \mathcal{P}\displaystyle\int \frac{d\bar{\omega}}{2\pi} \frac{\hat{\Pi}(k,\bar{\omega})}{\omega - \bar{\omega}} - \pi i\, \hat{\Pi}(k,\omega) & , \quad \gamma = 0 \\ \Pi^A(k,\omega,\gamma) - 2\pi i\, \hat{\Pi}(k,\omega,\gamma) & , \quad \gamma > 0. \end{cases} \tag{4.29}$$

The analytic continuation of the dielectric function was first proposed by Landau [Lan46][8] for the case of the Vlasov dielectric function. Notice that this procedure is applicable to arbitrary approximations of the dielectric function, and is not restricted to the result of Eq. (4.17). In particular, this result is applicable to plasmas of arbitrary symmetry, in particular to 1D, 2D and 3D plasmas [BBS+93].

---

denotes the principal value, cf. App. A)

$$\Pi(k,\omega \pm i\delta) = \mp\pi i \hat{\Pi}(k,\omega) + \mathcal{P}\int \frac{d\bar{\omega}}{2\pi} \frac{\hat{\Pi}(k,\bar{\omega})}{\omega - \bar{\omega}}, \tag{4.26}$$

where the discontinuity at the cut is defined by the spectral function

$$\hat{\Pi}(k,\omega) = \frac{1}{i}\left\{\Pi^R(k,\omega + i\delta) - \Pi^A(k,\omega - i\delta)\right\}. \tag{4.27}$$

[8]We mention that this approach which gives rise to ("Landau") damping of collective excitations in a collisionless theory, was heavily debated in the 1950ies. Van Kampen demonstrated that the Vlasov equation has solutions ("Van Kampen modes") which are not damped [Kam55, Kam57], giving rise e.g. to the Bernstein waves [BGK57]. Later it was shown that both concepts agree if the superposition of all modes is considered [Cas59].

Random phase approximation. We now provide the results for the analytic continuation of the polarization function in the linear regime (RPA), Eq. (4.17). Calculating the difference of the retarded and advanced functions according to Eq. (4.27), we obtain the spectral function in RPA on the real frequency axis

$$i\hat{\Pi}(\vec{k},\omega) = (2s+1)\int \frac{d\vec{p}}{(2\pi\hbar)^3}\left\{f(E_{\vec{p}}) - f(E_{\vec{p}+\hbar\vec{k}})\right\}\delta[\hbar\omega + E_{\vec{p}} - E_{\vec{p}+\hbar\vec{k}}]. \quad (4.30)$$

Using the result (4.30) and the spectral representation (4.25), we obtain the analytic continuation of the RPA polarization which is valid on the whole complex frequency plane,

$$\Pi(\vec{k},z) = (2s+1)\int \frac{d\vec{p}}{(2\pi\hbar)^3}\frac{f(E_{\vec{p}}) - f(E_{\vec{p}+\hbar\vec{k}})}{\hbar z + E_{\vec{p}} - E_{\vec{p}+\hbar\vec{k}}}. \quad (4.31)$$

Again, Π has to be understood as retarded function Π^R for $\mathrm{Im}\, z > 0$ and, respectively, as advanced function Π^A for $\mathrm{Im}\, z < 0$. Additionally, we have the result on the real frequency axis

$$\mathrm{Re}\Pi(\vec{k},\omega) = (2s+1)\,\mathcal{P}\int \frac{d\vec{p}}{(2\pi\hbar)^3}\frac{f(E_{\vec{p}}) - f(E_{\vec{p}+\hbar\vec{k}})}{\hbar\omega - E_{\vec{p}} + E_{\vec{p}+\hbar\vec{k}}}, \quad (4.32)$$

$$\mathrm{Im}\Pi(\vec{k},\omega) = \frac{i}{2}\,\hat{\Pi}(\vec{k},\omega). \quad (4.33)$$

This is a general result which does not depend on the symmetry of the plasma. Notice that in most treatments, only the limiting case $\gamma = 0$ (i.e. the polarization on the real frequency axis) is considered. This is justified for weakly damped collective excitations only, e.g. at zero temperature. However, at elevated temperatures, the plasmons may be strongly damped. Furthermore, in nonequilibrium, there may be unstable modes which have a finite growth rate ($\gamma < 0$), where this approximation fails. In this case, it is essential to use the retarded polarization given on the whole complex frequency plane, Eq. (4.34), what assures the correct results for the plasmon frequency and damping/growth rate.[9]

4.2 Collective plasma excitations (plasmons)

Electrodynamic definition of the dielectric tensor. It is instructive to consider the dielectric properties of plasmas also from the view point of

[9] Inserting Eqs. (4.31,4.32) into Eq. (4.29), we can write down the result for the retarded

electrodynamics. There, the dielectric function ϵ is introduced as the response of the medium to an external electric field $(\alpha, \beta = x, y, z)$, cf. e.g. [ABR84],

$$D_\alpha(\mathbf{r},t) = E_\alpha(\mathbf{r},t) + \sum_\beta \int_{-\infty}^{t} dt' d\mathbf{r}' \epsilon_{\alpha\beta}(\mathbf{r}-\mathbf{r}', t-t') E_\beta(\mathbf{r}', t') + O(E^2), \quad (4.34)$$

where $\mathbf{D}$ is the dielectric displacement vector, and we neglected all nonlinear response terms. Assuming homogeneous systems $(\bar{\mathbf{r}} = \mathbf{r} - \mathbf{r}', \bar{t} = t - t')$ we can take the Fourier/Laplace transform of (4.34) according to

$$\epsilon_{\alpha\beta}(\mathbf{q}, \omega) = \int_0^\infty d\bar{t} \int d\bar{\mathbf{r}} \epsilon_{\alpha\beta}(\bar{\mathbf{r}}, \bar{t}) e^{i\omega\bar{t} - i\mathbf{q}\bar{\mathbf{r}}}, \quad (4.35)$$

which yields

$$D_\alpha(q, \omega) = \sum_\beta \epsilon_{\alpha\beta}(\mathbf{q}, \omega) E_\beta(\mathbf{q}, \omega). \quad (4.36)$$

The tensor $\epsilon_{\alpha\beta}(\mathbf{q}, \omega)$ contains the information about the temporal and spatial dispersion of the plasma (in linear response) and is connected, e.g. with the

---

RPA polarization function on the whole complex frequency plane

$$\frac{\tilde{\Pi}^R(\vec{k}, \omega, \gamma)}{2s+1} = \begin{cases} \int \frac{d\vec{p}}{(2\pi\hbar)^3} \frac{f(E_{\vec{p}}) - f(E_{\vec{p}+\hbar\vec{k}})}{\hbar\omega - i\hbar\gamma - E_{\vec{p}} + E_{\vec{p}+\hbar\vec{k}}} & , \quad \gamma < 0 \\ \mathcal{P} \int \frac{d\vec{p}}{(2\pi\hbar)^3} \frac{f(E_{\vec{p}}) - f(E_{\vec{p}+\hbar\vec{k}})}{\hbar\omega - E_{\vec{p}} + E_{\vec{p}+\hbar\vec{k}}} - & \\ \frac{i}{2} \int \frac{d\vec{p}}{(2\pi\hbar)^2} \{f(E_{\vec{p}}) - f(E_{\vec{p}} + \hbar\omega)\} & \\ \quad \times \delta[\hbar\omega + E_{\vec{p}} - E_{\vec{p}+\hbar\vec{k}}] & , \quad \gamma = 0 \\ \int \frac{d\vec{p}}{(2\pi\hbar)^3} \frac{f(E_{\vec{p}}) - f(E_{\vec{p}+\hbar\vec{k}})}{\hbar\omega - i\hbar\gamma - E_{\vec{p}} + E_{\vec{p}+\hbar\vec{k}}} - & \\ i \int_{AC} \frac{d\vec{p}}{(2\pi\hbar)^2} \{f(E_{\vec{p}}) - f(E_{\vec{p}} + \hbar\omega)\} & \\ \quad \times \delta[\hbar\omega + E_{\vec{p}} - E_{\vec{p}+\hbar\vec{k}}] & , \quad \gamma > 0. \end{cases}$$

In the last line the symbol "AC" denotes that after integration with ω being real, the result has to be analytically continued into the lower frequency half plane. Usually, this reduces to the substitution of $\omega \to \omega - i\gamma$ in the argument of the distribution functions. We mention that the complex frequency leads to a complex (momentum) argument in the distribution function. This sometimes gives rise to oscillations in the dielectric function vs. $\mathrm{Im}\,\omega$, in particular, if f contains exponentials, as is the case in equilibrium (Maxwell or Fermi/Bose distributions), see for example Fig. 4.2.a.

conductivity tensor via

$$\epsilon_{\alpha\beta}(\mathbf{q},\omega) = \delta_{\alpha\beta} + i\frac{4\pi}{\omega}\sigma_{\alpha\beta}(\mathbf{q},\omega). \tag{4.37}$$

While $\epsilon(\mathbf{r},t)$ is real, the Fourier/Laplace transformed tensor is complex with the properties

$$\epsilon_{\alpha\beta}(\mathbf{q},\omega) = \epsilon^*_{\alpha\beta}(-\mathbf{q},-\omega). \tag{4.38}$$

It is sometimes useful to decompose the dielectric tensor into a Hermitean and anti-Hermitean part,

$$\begin{aligned} \epsilon_{\alpha\beta} &= \epsilon^h_{\alpha\beta} + i\,\epsilon^a_{\alpha\beta}, \\ \epsilon^h_{\alpha\beta} &= \frac{1}{2}\left\{\epsilon_{\alpha\beta} + \epsilon^*_{\beta\alpha}\right\}; \qquad \epsilon^a_{\alpha\beta} = \frac{1}{2}\left\{\epsilon_{\alpha\beta} - \epsilon^*_{\beta\alpha}\right\}. \end{aligned} \tag{4.39}$$

Dispersion relation. The dispersion relation of collective plasma excitations follows from Maxwell's equations (11.2)which, due to linearity in the electromagnetic field, can be reduced to algebraic equations by Fourier/Laplace transform, cf. Eq.(4.35),

$$\left\{q^2\delta_{\alpha\beta} - q_\alpha q_\beta - \frac{\omega^2}{c^2}\epsilon_{\alpha\beta}(\mathbf{q},\omega)\right\} E_\beta = 0. \tag{4.40}$$

This is a system of three homogeneous equations for the electrical field components which has nontrivial solutions if the determinant vanishes, i.e.

$$\left|q^2\delta_{\alpha\beta} - q_\alpha q_\beta - \frac{\omega^2}{c^2}\epsilon_{\alpha\beta}(\mathbf{q},\omega)\right| = 0, \tag{4.41}$$

which constitutes the general form of the plasmon dispersion relation, e.g. [Che87, ABR84, Dav89].[10] Furthermore, one has to keep in mind, that $\epsilon_{\alpha\beta}(\mathbf{q},\omega)$

[10] Depending on the system symmetry and the application of external electric or magnetic fields, the solution of Eq. (4.41) can be very complicated. In special cases, simplifications are possible. These include

i) Dispersion of longitudinal oscillations: if there exists a potential ϕ with $\mathbf{E}(\mathbf{q},\omega) = -i\mathbf{q}\phi(\mathbf{q},\omega)$, the dispersion relation is given by $\sum_{\alpha\beta}\frac{q_\alpha q_\beta}{q^2}\epsilon_{\alpha\beta}(\mathbf{q},\omega) = 0$.

ii) Isotropic plasmas, see below,

iii) Two-dimensional plasmas: the exact dispersion relation reads $\frac{\omega^4}{c^4}[\epsilon_{11}\epsilon_{22} - \epsilon_{12}\epsilon_{21}] - \frac{\omega^2}{c^2}q_1q_2[\epsilon_{12}+\epsilon_{21}] - q_1^2q_2^2 = 0$

iv) One-dimensional plasmas: the dielectric function is a scalar, and Eq. (4.41) reduces to $\epsilon(\mathbf{q},\omega) = 0$.

is a complex tensor that depends on a complex frequency variable $\hat{\omega} = \omega - i\gamma$. Therefore, Eq.(4.41) has to be fulfilled simultaneously for its real and imaginary part. The solutions are the complex frequencies of the excited field modes (s) as a function of the wavenumber

$$\hat{\Omega}_s(\mathbf{q}) = \Omega_s(\mathbf{q}) - i\Gamma_s(\mathbf{q}), \tag{4.42}$$

where Ω_s and Γ_s define, respectively, the oscillation frequency and the damping rate of the mode s. If $\Gamma < 0$, the field amplitude corresponding to this particular wave number and frequency, grows in time,[11] which corresponds to a plasma instability. In special cases, there are approximate solutions of the dispersion relation (4.41) possible, (see below).

It is interesting to investigate the energy exchange between the oscillations and the plasma. To this end, we calculate the time derivative of the electromagnetic field energy which follows from Maxwell's equations (11.2),

$$\frac{dW}{dt} = \frac{1}{4\pi} \int_{\mathcal{V}} d^3r \left(\mathbf{B}\frac{\partial \mathbf{B}}{\partial t} + \mathbf{E}\frac{\partial \mathbf{D}}{\partial t} \right), \tag{4.43}$$

and which yields, with Eq. (4.36), after Fourier-Laplace transform, and averaging over time, for each spectral component

$$\frac{1}{\mathcal{V}} \frac{d\overline{W}(\mathbf{q},\omega)}{dt} = -\frac{i\,\omega}{4\pi} \sum_{\alpha\beta} \left[\epsilon^*_{\alpha\beta}(\mathbf{q},\omega) - \epsilon_{\beta\alpha}(\mathbf{q},\omega) \right] E_\alpha(\mathbf{q},\omega) E^*_\beta(\mathbf{q},\omega). \tag{4.44}$$

From Eq. (4.44), we immediately conclude that a plane monochromatic wave of frequency ω and wave number $\mathbf{q}$ will propagate through the plasma without damping if the dielectric tensor of the medium is purely Hermitean, $\epsilon_{\alpha\beta} = \epsilon^h_{\alpha\beta}$.[12] Since arbitrary external fields can be expanded into monochromatic plane waves, this result is generally valid.

Isotropic media. The explicit results for the dielectric function simplify essentially for special symmetries. In particular, in isotropic systems, where $\epsilon_{\alpha\beta} = \epsilon_{\alpha\beta}(|\mathbf{q}|,\omega)$, the dielectric tensor has only two independent components, a longitudinal "l" and a transverse "tr" one:

$$\epsilon_{\alpha\beta}(q,\omega) = \left(\delta_{\alpha\beta} - \frac{q_\alpha q_\beta}{q^2} \right) \epsilon^{tr}(q,\omega) + \frac{q_\alpha q_\beta}{q^2}\, \epsilon^l(q,\omega), \tag{4.45}$$

and the dispersion relation Eq. (4.41) decouples into two separate equations

$$\epsilon^l(q,\hat{\omega}) = 0, \qquad q^2 - \frac{\hat{\omega}^2}{c^2}\, \epsilon^{tr}(q,\hat{\omega}) = 0. \tag{4.46}$$

[11] Alternatively, one can solve Eq. (4.42) for complex $\mathbf{q}$ as a function of real ω, which yields the spatial behavior of the modes.

[12] The term in brackets is just $-\epsilon^a$, cf. Eq. (4.40.)

The second relation corresponds to electro-magnetic waves with the field vectors being perpendicular to the wave vector, and the first one to longitudinal electrostatic (potential) oscillations with $\mathbf{E}$ being parallel to $\mathbf{q}$. Finally, the energy loss of a given field component (4.44) transforms in an isotropic medium into (q and ω are real):

$$\frac{1}{\mathcal{V}}\frac{d\overline{W}(\mathbf{q},\omega)}{dt} = -2\,\frac{\omega}{q^2}\{\operatorname{Im}\epsilon^l(q,\omega)|\mathbf{q}\,\mathbf{E}|^2 + \operatorname{Im}\epsilon^{tr}(q,\omega)|\mathbf{q}\times\mathbf{E}|^2\}. \tag{4.47}$$

Since in equilibrium collective excitations cannot grow, the electromagnetic field energy of plasma oscillations (which are excited e.g. by thermal fluctuations) can only decrease via energy loss to the particles. This requires the following conditions to hold simultaneously in equilibrium ($\omega > 0$)

$$\operatorname{Im}\epsilon^l(q,\omega) \geq 0, \qquad \operatorname{Im}\epsilon^{tr}(q,\omega) \geq 0. \tag{4.48}$$

Or, vice versa, growth of a field mode with given $\mathbf{q}$ and ω, i.e. a plasma instability, requires that the imaginary part of the longitudinal and/or transverse dielectric function becomes negative for these parameters ("negative damping"), which is possible in special nonequilibrium situations. However, this is only a necessary condition for an instability. In addition, there has to exist an eigenfrequency of the plasma for these values of $\mathbf{q}$ and ω (a solution of Eqs. (4.46)), which can take up the energy from the charged particles. A more detailed analysis of the conditions for plasma instabilities will be carried out in Sec. 4.3.

Small damping approximations. If the plasma excitations are only weakly damped (or weakly unstable), one can solve the dispersion relations (4.41) or (4.46) approximately. Assuming $|\operatorname{Im}\omega|/\operatorname{Re}\omega \sim \eta \ll 1$ and $|\epsilon^a_{ij}|/|\epsilon^h_{ij}| \sim \eta \ll 1$, we expand (4.41) up to first order in η,

$$|R_{\alpha\beta} + iI_{\alpha\beta}| \approx 0, \tag{4.49}$$

where the zeroth and first order terms are given by

$$R_{\alpha\beta} = q^2\delta_{\alpha\beta} - q_\alpha q_\beta - \frac{\Omega^2}{c^2}\epsilon^h_{\alpha\beta}, \quad I_{\alpha\beta} = \frac{1}{c^2}\left\{\Gamma\frac{\partial}{\partial\omega}[\omega^2\epsilon^h_{\alpha\beta}] - \Omega^2\epsilon^a_{\alpha\beta}\right\}, \tag{4.50}$$

and the derivative has to be taken at $\omega = \Omega_s(q), s = 1, 2, \ldots$. In this approximation, the calculation of the dispersion and the damping of the modes decouples: $\Omega_s(\mathbf{q})$ follows from the zeroth order term in Eq. (4.49), $|R_{\alpha\beta}| = 0$, whereas $\Gamma_s = -\operatorname{Im}\hat{\Omega}_s$ is given by the first order contributions, e.g. in one-dimensional systems, we have[13]

$$\Gamma_s^{1D}(q) = \frac{\epsilon^a\left[\Omega_s(q)\right]}{\frac{\partial}{\partial\omega}\epsilon^h\left[\Omega_s(q)\right]} = \frac{\mathrm{Im}\,\epsilon\left[\Omega_s(q)\right]}{\frac{\partial}{\partial\omega}\mathrm{Re}\,\epsilon\left[\Omega_s(q)\right]}, \tag{4.53}$$

Eq. (4.53) is commonly used. Nevertheless, one has to keep in mind that all these relations provide only an estimate for the plasmon dispersion and, in some cases, may be far from the correct result (especially the damping). This can be seen clearly from Fig. 4.1. below. Therefore, the solution of the complex dispersion relation is always preferable.

4.3 Plasma instabilities

General instability criteria In this point we list some necessary and sufficient conditions for plasma instabilities which are readily verified based on the general properties of the dielectric function:

i) Existence of at least one complex eigenfrequency $\hat{\Omega}_s(\mathbf{q})$ with a positive imaginary part $\mathrm{Im}\,\hat{\Omega}_s(\mathbf{q}) = -\Gamma_s(\mathbf{q}) > 0$ is a necessary and sufficient condition for an instability, for the given wavenumber.

ii) A necessary condition for an instability in anisotropic systems is that the anti-Hermitean part of the dielectric tensor has at least one negative component (cf. Eq. (4.44)) and, furthermore, $\sum_{\alpha\beta} E_\alpha E_\beta\, \epsilon^a_{\alpha\beta}(\mathbf{q},\omega) < 0$, which depends, in particular, on the angle between the field and the wave vector.

iii) If ϵ is scalar (e.g. in 1D), or the dispersion of several modes decouples, (e.g. in isotropic systems), the necessary condition is $\mathrm{Im}\,\epsilon(\mathbf{q},\omega) < 0$

---

[13]Here we took into account that, in this case, the Hermitean and anti-Hermitean part of ϵ coincide with the real and imaginary part, respectively. This result applies also to longitudinal oscillations in an isotropic plasma (with $\epsilon \to \epsilon^l$). In 2D, one has to solve $R_{11}I_{22} + R_{22}I_{11} - R_{12}I_{21} - R_{21}I_{12} = 0$ with the result

$$\Gamma_s^{2D}(\mathbf{q}) = \Omega_s^2 \frac{R_{11}\epsilon^a_{22} + R_{22}\epsilon^a_{11} R_{21}\epsilon^a_{12} + R_{12}\epsilon^a_{21}}{R_{11}[\omega^2\epsilon^h_{22}]' + R_{22}[\omega^2\epsilon^h_{11}]' - R_{21}[\omega^2\epsilon^h_{12}]' + R_{12}[\omega^2\epsilon^h_{21}]'}, \tag{4.51}$$

and, in 3D, $I_{11}R_{22}R_{33} + \ldots - I_{33}R_{21}R_{12} = 0$, with the solution

$$\Gamma_s^{3D}(\mathbf{q}) = \Omega_s^2(\mathbf{q}) \frac{R_{22}R_{33}\epsilon^a_{11} + \ldots - R_{21}R_{12}\epsilon^a_{33}}{R_{22}R_{33}[\omega^2\epsilon^h_{11}]' + \ldots - R_{21}R_{12}[\omega^2\epsilon^h_{33}]'}, \tag{4.52}$$

where "$\,'\,$" denotes the derivative with respect to ω at the eigenfrequency $\Omega_s(\mathbf{q})$.

Nyquist theorem. This is a powerful and easy to use method for identifying the number of unstable modes at a given wavenumber [Nyq32], see also [Dav89, Che87]. The scope of this statement is significantly broader than usually assumed. Therefore, we give a short discussion which is as general as possible. Consider a function Λ of wavenumber q and complex frequency $\hat{\omega} = \omega - i\gamma$, which is only required to be analytic in the upper frequency half-plane (e.g. the retarded dielectric function). The idea is to find a simple criterion for the existence of complex zeroes of Λ in this half-plane. We introduce a function G (omitting the q variable)

$$G(\hat{\omega}) = \frac{1}{\Lambda(\hat{\omega})} \frac{\partial \Lambda}{\partial \hat{\omega}}, \tag{4.54}$$

which has poles at $\hat{\omega}_1 \dots \hat{\omega}_m$, the location of the zeroes of Λ. The contour integral over G transforms, with the help of Cauchy's theorem, into a sum of the residua of G at the zeroes of Λ,

$$I = \frac{1}{2\pi i} \int_C G(\hat{\omega}) d\hat{\omega} = \sum_{i=1}^{m} \text{Res}\, G(\hat{\omega} = \hat{\omega}_i), \tag{4.55}$$

where the contour C extends from $\hat{\omega} = -\infty + i\delta$ to $\hat{\omega} = \infty + i\delta$ ($\delta \to +0$) and closes on an infinite semicircle in the upper half-plane. Near $\hat{\omega} = \hat{\omega}_i$ we can write $\Lambda(\hat{\omega}) = A(\hat{\omega} - \hat{\omega}_i)^{s_i} + B(\hat{\omega} - \hat{\omega}_i)^{s_i+1} + \dots, \quad s_i = 1, 2, \dots$ with s_i being the order of the pole at $\hat{\omega}_i$. Therefore, close to $\hat{\omega}_i$

$$G(\hat{\omega}) = \frac{s_i}{\hat{\omega} - \hat{\omega}_i}, \tag{4.56}$$

and the integral yields $I = \sum_{i=1}^{m} s_i = M$. Consequently, if M is a positive integer, then Λ has zeroes in the upper frequency half-plane. On the other hand, we can calculate the integral I by expressing G via Λ:

$$I = \frac{1}{2\pi i} \int\limits_C \frac{1}{\Lambda(\hat{\omega})} \frac{\partial \Lambda}{\partial \hat{\omega}} d\hat{\omega} = \frac{1}{2\pi i} \ln \left[\frac{\Lambda(\hat{\omega}_s)}{\Lambda(\hat{\omega}_e)} \right], \tag{4.57}$$

where $\hat{\omega}_s$ and $\hat{\omega}_e$ are, respectively, the initial and the end points on the contour C which differ only by a phase factor $e^{i2\pi}$. Since we found before that this integral equals M, we conclude that $\Lambda(\hat{\omega}_e) = \Lambda(\hat{\omega}_s) e^{i2\pi M}$. Therefore, the stability analysis is transformed into a graphical problem: The number of unstable modes is equal to the number of times $\Lambda(\hat{\omega})$ encircles the origin $\Lambda = 0$, when mapped along the contour C. That means, one has to plot ImΛ vs. ReΛ for $\hat{\omega}$ going from $-\infty + i0$ to $\infty + i0$ (Nyquist diagram) and to check if/how many times this curve encircles the origin in the complex Λ plane. Since Λ is an arbitrary function, this is a very general method. In particular, using for Λ the determinant (4.41), it can be used for the analysis of the existence and the number of unstable plasmon modes. The Nyquist theorem can be efficiently applied to classical and quantum plasmas as well [SBBK94].

The microscopic instability criteria of Newcomb and Penrose. In addition to the instability criteria on the general dielectric tensor, there exist criteria based on the microscopic properties of the plasma, in particular, on the momentum distribution. Here we confine ourselves to longitudinal excitations. *Newcomb's theorem* connects the stability of a plasma with the monotonic properties of the distribution functions: If the distribution function of each component of a 1D plasma has only one maximum, the system is stable against small amplitude longitudinal fluctuations (Nyquist theorem). This can be shown also for quantum systems within the RPA (cf. sec. 69 of [Bal63]).[14] A 3D plasma with isotropic momentum distribution $f_a = f_a(p^2)$ is always stable, regardless the form of the distribution functions. This statement holds for classical plasmas [Eck72] and for quantum plasmas, as well [Bon94, Bon95], see section 4.4.1 below. In other words, it is necessary for an instability to occur that the distribution function of at least one carrier species f_a has at least two maxima. Only under this condition, the imaginary part of the dielectric function can become negative. But even if $\mathrm{Im}\epsilon^l < 0$ for a certain frequency-wavevector range, it still remains to obtain sufficient conditions for an instability. These conditions depend on the excitation mechanism, the number of components, the dimensionality etc. The most prominent example is the *Penrose criterion* for a one-dimensional distribution function F with two maxima (one-component classical plasma) [Pen60]. From the Nyquist theorem one can derive the following necessary and sufficient condition for an instability

$$\int_{-\infty}^{\infty} dv \frac{F(v) - F(v_0)}{(v - v_0)^2} > 0, \tag{4.58}$$

where v_0 denotes the location of the minimum. Eq. (4.58) shows, that for an instability occurs only, if the minimum of the distribution is sufficiently deep.

The question of instabilities in classical plasmas has been studied in great detail, and more results can be found e.g. in [Kad68, Mik75, ABR84, Dav89]. Much less work has been done for quantum plasmas. We, therefore, briefly summarize the main peculiarities of instabilities in quantum systems.

Instabilities in quantum plasmas. Historically, the analysis of collective excitations in charged particle systems was first done for ionized gases which are usually non-degenerate [Lan28, TL29, Vla38]. Only with the increasing

[14] This statement holds also for homogeneous 3D and 2D systems, if their 1D distribution (integrated over the transverse directions) has only one maximum.

interest in solid state theory and overdense ionized gases, quantum plasmas became the subject of investigation, e.g. [Gol47, Tom50, Sil52, KS52a, PS61]. Since the classical case is recovered from the more general quantum case simply by taking the long wavelength limit, cf. Eq. 4.15, we expect differences between both for finite wave numbers. Indeed, the main difference in Eqs. (4.14) and (4.15) is the appearance of the difference of the distributions $f(p) - f(p + \hbar k)$ in the first, and the derivative $\partial f(p)/\partial p$ in the second. As we have seen above, the necessary condition for an instability is a positive sign of $\mathrm{Im}\Pi$, which is possible for non-monotonic distributions. Then $\partial f(p_0)/\partial p > 0$ in the vicinity of a certain p_0, independently of the wave number q of the plasma oscillation. This is a local condition. In contrast, in the quantum case, we have a "global" condition: $\mathrm{Im}\Pi > 0$ requires $f(p) - f(p + \hbar q) > 0$, which is fulfilled only up to a maximum wave number q_{cr}. This means, only plasma waves with wave numbers in the range $0 \leq q \leq q_{cr}$, may be amplified[15].

We mention that there are numerous predictions of plasma instabilities in quantum plasmas in solids, e.g. [PS61, Har62, BN63, Has65, RS67, SR69], of surface wave instabilities [GP64, BB66], current instabilities, related to negative differential conductivity, e.g. [Poz81], see also the reviews [Kog62, Pus69, KY75]. There is active research on instabilities in 1D and 2D semiconductor structures which we briefly discuss in Sec. 4.4.1. Here we mention a rather general obstacle in these systems: due to the low carrier mobility, one can achieve only rather low drift velocities, which makes it very difficult to create suitable nonequilibrium carrier distributions.

4.4 Examples: Plasmons in quantum systems

In this Section, we wish to illustrate the general results obtained so far by applying them to various simple systems[16]. We discuss collective plasma excitations in one-dimensional quantum plasmas more in detail and give some remarks on 2D and 3D systems.

[15] Basically, the wavenumber must "fit" in the minimum in order for the wave to gain energy from the particles. We will explicitly confirm this for one-dimensional plasmas below. Interestingly, it turns out that this is not an artifact of the linear approximation. This sensitivity to the wave number is confirmed also in solutions of the full nonlinear kinetic equation, cf. Sec. 4.6.

[16] This is an exceptionally broad field where much work has been done, both experimentally and theoretically. For illustration purposes, we limit ourselves to simple examples of longitudinal plasma oscillations in isotropic systems. For the discussion of transverse (electromagnetic) modes, surface plasmons, interband excitations or magnetic field effects, we will refer to the more specialized literature.

4.4.1 One-dimensional quantum plasmas

The case of one-dimensional plasmas is particularly simple to treat. We, therefore, present some derivations for this case. The obtained results can be straightforwardly generalized to 2D and 3D systems.

There are two types of one-dimensional plasmas: three-dimensional plasmas which are homogeneous in one plane and "true" 1D systems. Examples for the latter are metallic wires (though their width is usually still large compared to the interparticle distance) and plasmas in 1D quantum confined systems (quantum wires). In the last case, the lateral confining potential lets carriers move freely only in one dimension. As a result, the matrix element of the Coulomb potential[17], screening properties and so on are different from the 3D case. This gives rise to different properties of the collective excitations in the two cases, as can be seen in Fig. 4.3. below.

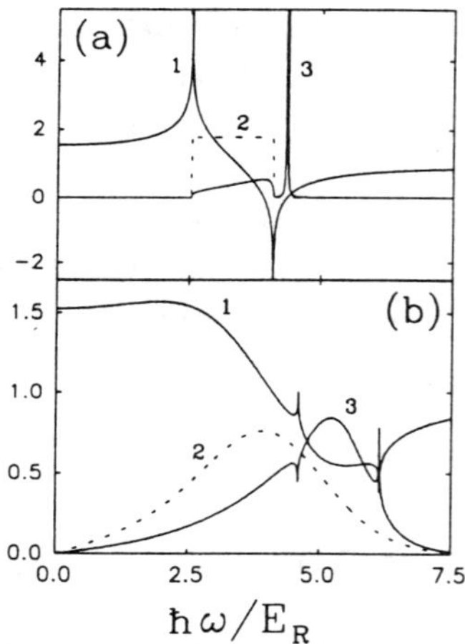

Fig. 4.1. Real (1) and imaginary (2) part of the dielectric function, and spectral function (3) for a quantum wire of thickness $2/3a_B$ (GaAs, $\epsilon_b = 12.7$, $a_B = 135$ Å, $E_R = 4.2$ meV, $n = 10^6 cm^{-1}$, $k = 1/a_B$). All functions are shown for *zero damping*, i.e. $\gamma = 0$, for $T = 0K$ **(a)** and $100K$ **(b)**. At $0K$, Reϵ has two zeroes, while Im ϵ is positive inside the pair continuum (Fig. 4.2.b) and zero elsewhere. At $100K$, the zeroes of Reϵ have vanished. Nevertheless, *complex zeroes* of ϵ do still exist, however, *at nonzero* γ, see Fig. 4.2.

The spectral function of the polarization, Eq. (4.30), reduces in 1D to

$$\hat{\Pi}_a(q,\omega) = \Pi_a^>(q,\omega) - \Pi_a^<(q,\omega) = \frac{m_a}{q}\left\{f_a(p_a^+) - f_a(p_a^-)\right\}, \tag{4.59}$$

with $p_a^\pm = \frac{m_a}{q}[\hbar\omega \pm E_q^a]$ and $E_a(q) = \frac{\hbar^2 q^2}{2m_a}$. Here the normalization condition is $2\int \frac{dk}{2\pi} f_a(k) = n_a$, where n_a is the average 1D density of carriers of component "a". The retarded/advanced RPA polarization function for 1D follows from

[17] In contrast to 3D systems, where the Coulomb potential is $V(q) = 4\pi e^2/(\epsilon_b q^2)$, in a quantum confined system (e.g. quantum wire), $V(q)$ is better approximated by $V(q) = 2\,e^2 K_0(qd)/\epsilon_b$, where K_0 is the modified Bessel function, and ϵ_b and d are the background dielectric constant and the wire diameter, respectively.

Eq. (4.31):

$$\Pi_a^{R/A}(\omega,\gamma,q) = \frac{m_a}{q}\int_0^\infty \frac{dk}{\pi} f(k) \left\{ \frac{1}{\tilde{p}_a^- - k} - \frac{1}{\tilde{p}_a^+ - k} + \frac{1}{\tilde{p}_a^- + k} - \frac{1}{\tilde{p}_a^+ + k} \right\}, \tag{4.60}$$

where $\tilde{p}_a^\pm = p_a^\pm - i\frac{m_a}{q}\hbar\gamma$. $\Pi^{R/A}$ are analytic, respectively, in the upper ($\gamma < 0$) and in the lower ($\gamma > 0$) frequency half plane. Using Eq. (4.29) together with Eqs. (4.59,4.60), one can explicitly calculate the real and imaginary part of the 1D RPA polarization, on the whole complex frequency plane [BBS+94]. Here, we limit ourselves to the simplest case of zero temperature.

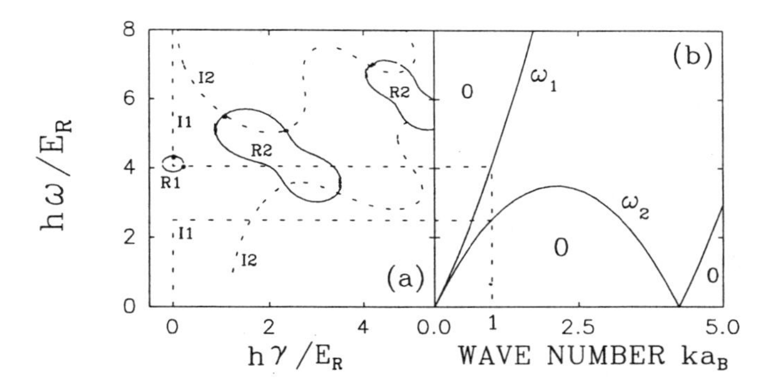

Fig. 4.2. **(a)**: Zeroes of the dielectric function in the complex frequency plane. Along the full lines (closed lines R), $\mathrm{Re}\,\epsilon = 0$, while along the dotted lines (I), $\mathrm{Im}\,\epsilon = 0$. Complex zeroes of ϵ correspond to crossings of these lines (thick dots). Results are shown for $T = 0K$ (R1, I1) and $T = 100K$ (R2, I2). Notice the deformation of these lines and their periodic replicas at $100K$, which arise from the analytic continuation of the Fermi function. **(b)**: Pair continuum and undamping regions on the real axis ($\gamma = 0$). Between the lines labelled ω_1 and ω_2 (see definition below), $\mathrm{Im}\,\epsilon > 0$, elsewhere it is zero. Parameters are the same as in Fig. 4.1.

Equilibrium plasmons at $T = 0K$. With the distribution function $f(k) = \Theta(k_F - |k|)$, where k_F is the Fermi momentum, the integration in Eq. (4.60) can be carried out analytically, (see also footnote 9).[18] The exact result

[18] For the analytic continuation of the step function, we use the continuous representation

$$f(k) = \lim_{\Delta\to 0} F_\Delta(k), \qquad F_\Delta(k) = \frac{1}{\pi}\left[\arctan\frac{k+k_F}{\Delta} - \arctan\frac{k-k_F}{\Delta}\right]. \tag{4.61}$$

For complex k, the *arctan* is a complex function, but the imaginary part vanishes if Δ goes to zero, and we can drop it.

for the real and imaginary part of the polarization function is [BBS+94]

$$\mathrm{Re}\,\Pi(\omega,\gamma,q) = \frac{m}{\pi q}\ln\frac{\left[\omega^2-\omega_2(q)^2\right]^2+2\gamma^2\left[\omega^2+\omega_2(q)^2\right]+\gamma^4}{\left[\omega^2-\omega_1(q)^2\right]^2+2\gamma^2\left[\omega^2+\omega_1(q)^2\right]+\gamma^4}, \tag{4.62}$$

$$\mathrm{Im}\,\Pi(\omega,\gamma,q) = \frac{m}{q}\times \begin{cases} F_\gamma(p^+) - F_\gamma(p^-) & , \ \gamma < 0, \\ F_\Delta(p^+) - F_\Delta(p^-) & , \ \gamma = 0, \\ F_\gamma(p^+) - F_\gamma(p^-) + \\ 2\mathrm{Re}[F_\Delta(p^+ - i\delta) - F_\Delta(p^- - i\delta)] & , \ \gamma > 0, \end{cases} \tag{4.63}$$

where $\hbar\omega_{1,2}(q) = \hbar^2 q k_F/m \pm E(q)$ and $\delta = \gamma m/q$, and the limit $\Delta \to 0$ has to be taken.[19]

Using Eq. (4.21) and the results (4.62), (4.63), one can simultaneously solve the equations $\mathrm{Re}\,\epsilon = 0$ and $\mathrm{Im}\,\epsilon = 0$ in the complex frequency plane for the plasmon dispersion $\Omega(q)$ and damping $\Gamma(q)$ [BBK93]. The results are shown in Figs. 4.1 and 4.2.: First, we find that $Re(\epsilon)$ has singularities on the real frequency axis at the values $\omega_s(q) = |\omega_{1,2}(q)|$, which are the boundaries of the pair continuum (see Fig. 4.2.b) [20], whereas for nonzero γ, all singularities disappear. Collective plasma excitations are readily found graphically: they are given by the crossing points of the lines $\mathrm{Re}\,\epsilon = 0$ and $\mathrm{Im}\,\epsilon = 0$, cf. Fig. 4.2.a. In a one-component quantum plasma, there exist two crossings. While the high frequency one corresponds to the optical plasmon which is undamped at $T = 0$, the other one is usually attributed to single particle excitations[21]. The momentum dependence of the two zeroes is shown in Fig. 4.3. For a quantum wire, both start at zero frequency, whereas in a 1D plasma without quantum confinement (3D Coulomb potential), the optical plasmon starts at the plasma frequency, cf. e.g. [Tom50, DK68]. Notice that due to zero damping, the dispersion of the optical plasmon at $T = 0K$ can be found analytically

[19] The $\Delta \to 0$ terms arise from the residua of the polarization function's denominator (4.60). This guarantees that ImΠ will be continuous when it crosses the real frequency axis (notice $F_{-\alpha} = -F_\alpha$).

[20] These lines include, for example, the $\omega = 0$ divergencies at $q = 0$ and $q = 2k_F$, which are related to the Peierls instability [Pei55].

[21] This mode follows the upper edge of the pair continuum $\Omega_{ac}(q) \approx \omega_1(q)$ and is always strongly damped. Notice also that there exists an undamping region (where Im$\Pi = 0$) which is enclosed by the line $\omega_-(q)$ and the momentum axis [WB74]. This is a peculiarity of 1D systems, which occurs in 2D or 3D only in non-equilibrium.

from Eq. (4.62)[22]

$$\Omega^2(q) \approx q^2 d^2 \omega_0^2 \ln(qd) + O(q^2), \qquad \text{with} \quad \omega_0^2 = \frac{2\,n\,e^2}{\epsilon_b\,m\,d^2}. \tag{4.66}$$

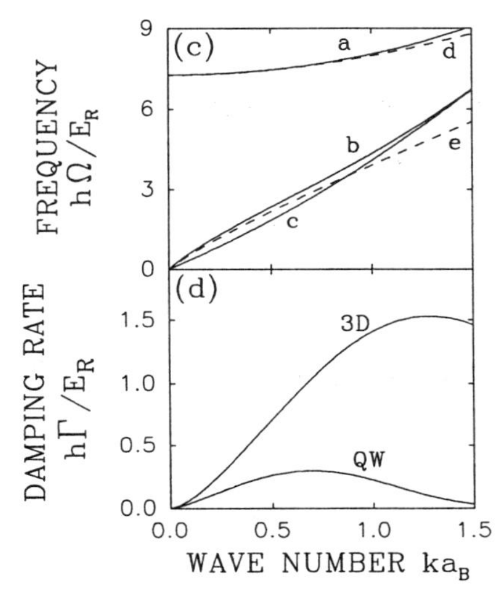

Fig. 4.3. Plasmon dispersion Ω and damping Γ calculated from the complex zeroes of ϵ at $T = 0K$. Shown are the optical plasmon of a quantum wire (b) and a 1D electron gas with a 3D Coulomb potential matrix element $V(q) \sim 1/q^2$ (a). Line (c) is the second zero of the dielectric function (cf. Figs. 4.1., 4.2.a), its damping is shown for the two cases (3D and quantum wire "QW") in the lower figure. The dashed lines are the optical modes calculated not with the RPA dielectric function but with its classical limit, the Vlasov dielectric function.

Two-component plasmas. The analysis is readily extended to multi-component plasmas by replacing $\Pi \to \sum_a \Pi_a$. The result for an equilibrium electron-hole plasma is shown in Fig. 4.4.

Spectral function and plasmon pole approximations. A very important quantity, due to its direct relation to experimentally measurable quantities, is the spectral function -Imϵ^{-1}, which is readily found from $\hat{\Pi}$, Eq. (4.59). It has a delta shaped peak at the frequency of the optical plasmon and a broader peak in the pair continuum with the maximum at the low frequency zero of

[22]Since the mode is undamped, one has to solve only $\mathrm{Re}\,\epsilon = 0$ at $\gamma = 0$, and we recover from Eqs. (4.62,4.63) the result of Ref. [WB74]

$$\mathrm{Re}\Pi(\omega, \gamma = 0, q) = \frac{m}{\pi q} \ln \frac{|\omega^2 - \omega_2(q)^2|}{|\omega^2 - \omega_1(q)^2|}, \tag{4.64}$$

and the imaginary part is $\mathrm{Im}\Pi(\omega, \gamma = 0, q) = \frac{m}{\pi q}$ for $\omega_- \leq \omega \leq \omega_1$ and zero otherwise. The solution of $\mathrm{Re}\,\epsilon(\omega, \gamma = 0, q) = 1 - V(q)\mathrm{Re}\,\Pi(\omega, \gamma = 0, q) = 0$ is [LS91]

$$\Omega^2(q) = \frac{A(q)\omega_1^2 - \omega_2^2}{A(q) - 1}, \quad \text{with} \quad A(q) = \exp[\frac{q\pi}{mV(q)}]. \tag{4.65}$$

The long-wavelength ($q \to 0$) limit for the real part of the polarization is $\mathrm{Re}\Pi(\omega, \gamma = 0, q) = \frac{2k_F}{\pi m}\frac{q^2}{\omega^2} + O(q^4)$, where the Fermi momentum is connected with the 1D density via $k_F = n\pi/2$. Taking into account $\lim_{q\to 0} K_0(qd) \approx -\ln(qd)$, Eq.(4.65), gives Eq. (4.66).

Re(ϵ), cf. Figs. 4.1. and 4.4. This quantity can be measured in inelastic polarized light scattering (Raman scattering) experiments. In particular, the measurements of Goñi et al. on quantum wires, [Ge91], agree well with our results.

Since the plasmon spectrum is dominated by peaks at the mode positions, one may try to derive analytical approximations for ϵ, which reproduce at least these peaks sufficiently well and which are called "plasmon pole approximations". This is a very useful and general concept, which, in particular, is well applicable to 1D systems[23], and one finds [BBS+94]

$$\epsilon(q,\omega) = 1 - \frac{q^2 V(q) n/m}{\omega^2 - \Omega(q)^2 + q^2 V(q) n/m}, \tag{4.67}$$

where $\Omega(q)$ is the plasmon dispersion, given e.g. by Eq. (4.66), and $V(q)$ is the Coulomb potential of the system of interest.

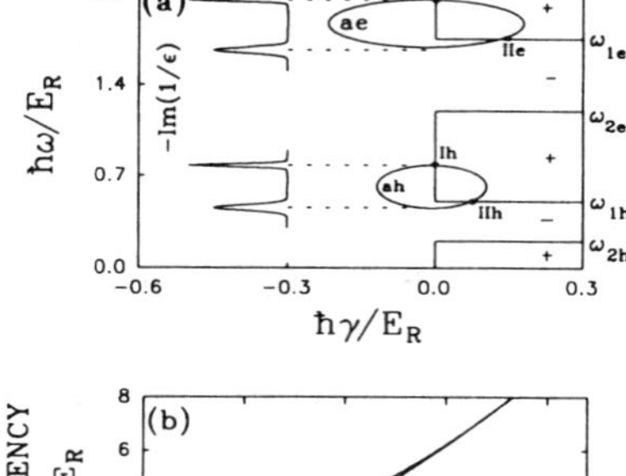

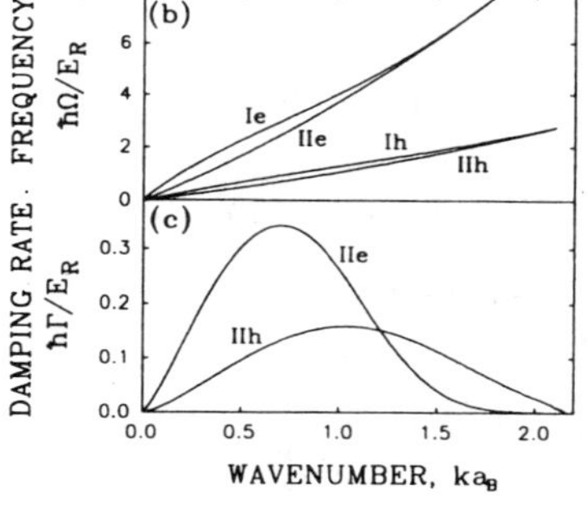

Fig. 4.4. Zero temperature plasmons in a two-component electron-hole plasma in a quantum wire. **(a)**: Along the elliptic lines (ae and ah), Re ϵ = 0, along the rectangular lines, Im ϵ = Im Π = 0, giving rise to *four crossings*. Areas of positive/negative signs of Im Π are labelled "+/-". The spectral function is sketched on the left, it has peaks at the zeroes of the dielectric function. **(b)**: Dispersion Ω of the four zeroes of ϵ. Modes 1e and 1h are undamped. **(c)**: Damping Γ of the other two zeroes. Parameters as in Fig. 4.1, except wire thickness $d = 100$Å.

Plasmons at finite temperature. Let us now consider equilibrium quantum 1D plasmas at elevated temperature with the Fermi distribution $f(k) = 1/[\exp(\beta(k^2/2m - \mu)) + 1]$, where β and $\mu = \mu(n, \beta)$ are the inverse temperature and the chemical potential, respectively. The smooth edge of the Fermi

[23]Formulas of this type are most easily derived from the classical limit of the inverse dielectric function, i.e. the Vlasov dielectric function (4.15). They can be improved phenomenologically to reproduce also the pair continuum. Furthermore, these formulas are straightforwardly generalized to multi-component systems.

function causes a number of changes compared to the zero temperature case: The pair continuum has no longer sharp boundaries, and, therefore, all modes, including the optical plasmon, become damped. This plasmon damping which appears here in a completely "collisionless" theory has been found by Landau (Landau damping) and has the following simple qualitative explanation: While charged particles and plasma waves exchange energy in both directions, in equilibrium, there are always more slow carriers (which gain energy from the wave) than faster ones (which lose energy) due to the monotonically decreasing shape of the distribution. This results in a total energy loss (damping) of the plasma waves in equilibrium, which is consistent with our general considerations before, cf. e.g. Eq. (4.47).

Fig. 4.2.a shows the zeroes of the real and imaginary part of the dielectric function at $T = 100K$ for a one-component electron plasma, which are essentially more complicated than at $T = 0K$. Interestingly, there appear additional crossing points of the two curves. However, it can be verified that only the crossings which are closest to the real axis are of physical relevance[24]. The effect of temperature on the plasmon dispersion is obvious: The complex zeroes of the dielectric function shift steadily toward higher damping values. At the same time, the frequencies of all plasmons shift upwards, see Fig. 4.2.a. This is due to the fact that plasma oscillations are now excited in a medium of faster moving particles what leads to an increase of the screening length r_{sc}[25]. Notice that there exists a critical momentum $q_{qr}(T)$, beyond which $\mathrm{Re}(\epsilon)$ has no longer zeroes at $\gamma = 0$, cf. Fig. 4.1.a. Here, the small damping approximations (see above) fail, while our approach, still yields complex zeroes of the dielectric function which are strongly damped and which are resolved in the spectral function as broad peaks, Fig. 4.1.a, [BBS+94].

Nonequilibrium plasmas. Plasma instabilities. We now consider nonequilibrium situations, where in addition to thermal carriers, the system contains a portion of nonequilibrium (fast) particles. If the distribution of at least one carrier component has more than one maximum, we expect qualitative changes in the mode spectrum, in particular, the possibility of unstable, growing modes, see Sec. 4.3. Then the above argument for monotonically

---

[24] The reason for the more complicated shape of the curves $\mathrm{Im}\epsilon = 0$ and $\mathrm{Re}\epsilon = 0$ and their additional crossing points is the complicated pole structure of the analytic continuation of the Fermi function. These poles are located at the Matsubara frequencies and give rise to a periodic fractal-like pattern. Only the "original" points carry physical information, and their replicas should be excluded from the plasmon analysis. We mention that this question has been extensively discussed for classical (Maxwell) plasmas, e.g. [Hay63, Den65, Sae65]

[25] The same effect occurs in 2D and 3D. For example, for small q, one finds for the frequency of the optical plasmon $\Omega^2(q) \approx \omega_p^2[1 + \alpha(q, n, T)]$, with $\alpha^{3D} = q^2 r_{3D}^2$ and $\alpha^{2D} = q r_{2D}$, e.g. [HK93].

decaying distributions does not apply, and in the vicinity of the minima, there may be more fast carriers than slow ones, so a wave may gain energy.

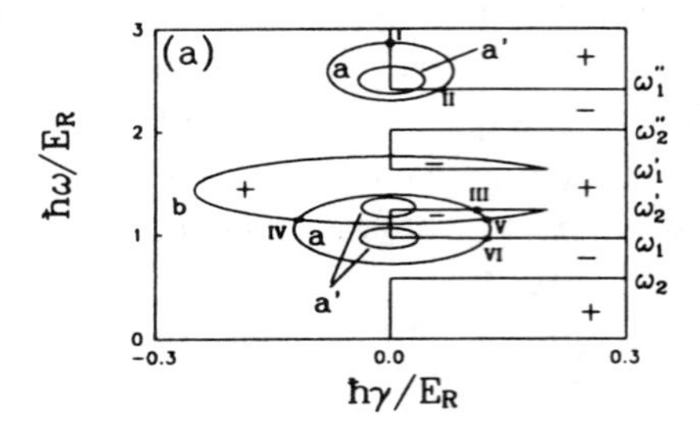

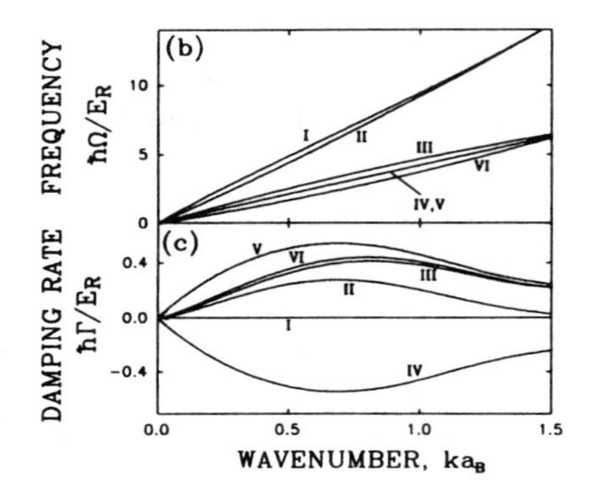

Fig. 4.5. Plasmons in a nonequilibrium electron plasma in a quantum wire with the distribution $f_{\rm NEQ}$ of footnote 26. **(a)**: Zeroes of $\mathrm{Re}\,\epsilon$ and $\mathrm{Im}\,\epsilon$, for explanation see Fig. 4.4. The nonequilibrium carriers give rise to an inversion region between $\omega'_{1,2} = qk_3/m \pm E(q)$, where the line $\mathrm{Im}\,\epsilon = 0$ extends into the left half plane ($\gamma < 0$). The crossing point "IV" corresponds to an unstable mode. The lines $\mathrm{Re}\,\epsilon = 0$ "a" and "a'" correspond to two different wires of $d = 100$Åand $d = 200$Å, respectively. Only in the first case, there exists an unstable mode **(b)**: Dispersion Ω of the six zeroes of ϵ. **(c)**: Damping Γ of all zeroes. The optical plasmon "I" is undamped, the unstable mode "IV" has negative "damping".

The determination of the complex zeroes of the dielectric function and the analysis of the mode spectrum is performed in complete analogy to the equilibrium case. However, only for distributions of special shape analytical results can be obtained [BBS+94], but they are too complicated to reproduce them here. In Fig. 4.5. the zeroes of the real and imaginary part of the dielectric function for the case of a nonequilibrium distribution which contains carriers at $T = 0K$ and a sharp second maximum at $k = k_3$ are shown[26]. One clearly sees that there appears now one crossing point located at "negative damping", which corresponds to an instability. Interestingly, the unstable mode has an almost linear dispersion (q-dependence),

$$\Omega_{inst}(q) \approx (k_F + k_3)q/m, \tag{4.68}$$

i.e., it corresponds to an acoustic instability [BBK93, BCK90]. Above a critical value q_{cr}, the growth rate becomes negative, and the instability disappears. This confirms our above discussion, Sec. 4.3, that a mode can be unstable

[26]The following model distribution was used, $f_{\rm NEQ}(k) = \Theta(k_F - k)\Theta(k) + \Theta(k_4 - k)\Theta(k - k_3)$, $k_F < k_3 < k_4; k \geq 0; k_F = 1.9/a_B, k_3 = 1.84k_F, k_4 = k_3 + k_F$.

only as long as its wave number "fits" into the minimum of the momentum distribution.

Changing the system parameters, such as density and shape of the nonequilibrium distribution or the geometry (wire width, cf. Fig. 4.5.), one can try to optimize the conditions for unstable modes [BBS+94]. As was mentioned in Sec. 4.3, unfortunately, the observation of plasma instabilities in quantum systems is difficult due to the low drift velocities achievable in solids and semiconductors in particular. Therefore, structures with special minibands or density modulation have been proposed which should allow to decrease the threshold drift velocity [BCK90, KBX93].

4.4.2 Plasmons in 2D and 3D quantum systems

2D quantum plasmas. Two-dimensional quantum plasmas (2D electron gas) have been investigated long ago [Rit57, Fer58, Ste67, Fet73, Fet74]. Examples were thin metallic films, but the first experimental observations of quasi-2D behavior were reported for an electron gas on liquid helium [GA76]. Very active research was carried out in the field of semiconductors already in the sixties and seventies [Cha71, SATL77, TKS78, KC78], and also the question of instabilities was studied intensively [GP64, BB66, SR69]. With the progress in semiconductor technology, two-dimensional quantum confined structures (quantum wells, superlattices etc.) became available which allowed for detailed experimental studies of 2D plasmons, e.g. [Tea80, Gea88, Don92, Wea93, RJPE93]. As in the 1D case, in the "true" 2D situation which is most closely reproduced in quantum wells, super lattices and so on, the Coulomb matrix element differs from the 3D form: in the limit of zero layer thickness, one obtains $V(q) = \frac{2\pi e^2}{\epsilon_b q}$, (cf. e.g. [BK95]), which again alters the plasmon spectrum.

The RPA polarization for isotropic 2D plasmas on the real frequency axis has been derived by Stern [Ste67]. We again give the result on the whole complex plain by performing the analytic continuation. The spectral function of the polarization is given by Eq.(4.30). The 2D result is [BBS+93]

$$\hat{\Pi}_a(q,\omega) = 2i\frac{2m_a}{q}\int_{p_a^-}^{\infty} dp\, p \frac{f_a(\frac{p^2}{2m_a}) - f_a(\frac{p^2}{2m_a} + \hbar\omega)}{\sqrt{p^2 - {p_a^-}^2}}, \tag{4.69}$$

where $p_a^{\pm} = \frac{m_a}{q}[\hbar\omega \pm E_q^a]$ and $E_q^a = \frac{\hbar^2 q^2}{2m_a}$. As in the preceding section, we calculate the retarded/advanced RPA polarization function for 2D from Eq.(4.31) to [BBS+93]

$$\Pi_a^{R/A}(\omega,\gamma,q) = \frac{2m_a}{q}\int_0^\infty \frac{dk}{\pi} k\, f(k) \left\{ \frac{\mathrm{sgn}[\hbar\omega - E(k)]}{\sqrt{\tilde{p}_a^{-\,2} - k^2}} - \frac{\mathrm{sgn}[\hbar\omega + E(k)]}{\sqrt{\tilde{p}_a^{+\,2} - k^2}} \right\}, \tag{4.70}$$

where again the retarded and advanced function are analytic, in the upper ($\gamma < 0$) and lower ($\gamma > 0$) frequency half-plane, respectively. $\tilde{p}_a^\pm = p_a^\pm - i\dfrac{m_a}{q}\hbar\gamma$, $\gamma = -\mathrm{Im}\,\hat{\omega}$. This result is valid on the whole complex frequency plane, for arbitrary distributions. The plasmon spectra in equilibrium 2D systems are similar to those in 1D. Again, the optical plasmon starts at zero frequency with $q \to 0$, although the functional form of the dispersion is different [Ste67]. Plasma instabilities in quantum wells were predicted in [HQ86, CKB88, Bak88]. Until now, no experimental evidence has been reported, what is mainly due to the rather low carrier mobilities in these systems. Various semiconductor structures have been proposed to achieve the necessary electron drift velocities [KBCX91, XKB92, KBX93].

3D quantum plasmas. In an isotropic 3D plasma, the spectral function of the polarization $\hat{\Pi}_a(q,\omega)$, Eq. (4.30), is given by

$$\hat{\Pi}_a(q,\omega) = \Pi_a^>(q,\omega) - \Pi_a^<(q,\omega) = 2i\frac{m_a}{2q}\int_{p_a^-}^\infty dp\, p \left\{ f_a(\frac{p^2}{2m_a}) - f_a(\frac{p^2}{2m_a} + \hbar\omega) \right\}, \tag{4.71}$$

with $p_a^\pm = \frac{m_a}{q}[\hbar\omega \pm E_a(q)]$ and $E_a(q) = \frac{\hbar^2 q^2}{2m_a}$. The RPA dielectric function and the corresponding dispersion relation on the real frequency axis have been derived by Lindhard and, even before him, by Klimontovich and Silin [KS52a, KS52b] and by Bohm and Pines [BP53]. We readily derive the result for the entire complex frequency plane using the general formula (4.29). Then, after analytic continuation, the retarded/advanced polarization $\tilde{\Pi}^R(p,\omega,\gamma)$ of an isotropic 3D plasma for arbitrary γ is given by

$$\Pi_a^{R/A}(\omega,\gamma,q) = \frac{2m_a}{q}\int_0^\infty \frac{dk}{2\pi^2} k\, f_a(k) \left\{ \ln\frac{\tilde{p}_a^- + k}{\tilde{p}_a^- - k} - \ln\frac{\tilde{p}_a^+ + k}{\tilde{p}_a^+ - k} \right\}, \tag{4.72}$$

where $\tilde{p}_a^\pm = p_a^\pm - i\dfrac{m_a}{q}\hbar\gamma$. $\Pi^{R/A}$ are analytic, respectively, in the upper ($\gamma < 0$) and lower ($\gamma > 0$) frequency half plane.

The 3D RPA DF at zero temperature has been calculated by numerous authors. The first quantum result (long wavelength limit, $k \ll k_F$) has been obtained already by Goldman in 1947 [Gol47], and the full analytical result was, apparently, first given by Silin in 1952 [Sil52]. The imaginary part of the RPA dielectric function for Fermi distributions is known for arbitrary temperature [Kog62, Glu71],

$$\operatorname{Im}\Pi_a(q,\omega) = \frac{(2s_a+1)k_B T m_a^2}{4\pi\hbar^4 q}\ln\left|\frac{1+\exp[\beta\mu_a - \frac{\beta}{E_q^a}(\hbar\omega+E_q^a)^2]}{1+\exp[\beta\mu_a - \frac{\beta}{E_q^a}(\hbar\omega-E_q^a)^2]}\right|, \tag{4.73}$$

where μ and β are the chemical potential and the inverse temperature, respectively, whereas, the real part is known only for limiting cases. Analytical results and numerically useful approximations for special cases can be found in numerous papers, see e.g. [KKER86, KK74] and references therein.

Consider now the case of nonequilibrium. An interesting and rather counterintuitive result is that even for distributions with several maxima, there are no plasma instabilities possible (provided the distribution is isotropic). This fact is well-known for classical plasmas [Eck72], and can be proved also for quantum plasmas within the RPA [Bon94]. To show this it has to be proved that the complex dispersion relation $\operatorname{Re}\epsilon = \operatorname{Im}\epsilon = 0$ has no complex zeroes in the upper frequency half-plane. For this, it is sufficient to show that (at least) one of $\operatorname{Re}\epsilon$ or $\operatorname{Im}\epsilon$ have always the same sign in this half-plane. It can be shown, that in fact, always $\operatorname{Im}\epsilon > 0$, for $\operatorname{Im}\hat{\omega} = -\gamma > 0$ for arbitrary isotropic distribution functions as in equilibrium.[27] Thus, one can only achieve an undamping of nonequilibrium plasmons[28]. Thus, to excite strong plasma instabilities, the distribution has to be deformed e.g. by means of an external field.

[27]Due to Eq. (4.21), it is sufficient to prove that $\operatorname{Im}\Pi_a(\omega,\gamma,q) < 0$ for any plasma component. The complex polarization function for isotropic 3D systems can be evaluated by introducing spherical coordinates. Defining $z = cos\theta, y = k\,z, u = \frac{m_a}{q}\omega$ ($\omega \geq 0$) and $\delta = \frac{m_a}{q}\gamma$, one angle integration can be carried out

$$\frac{\Pi_a(\omega,\gamma,q)}{(2\,s_a+1)} = \frac{m_a}{q}\int_0^\infty \frac{dk}{(2\pi)^2}k\,f_a(k)\int_{-k}^{k} dy\left\{\frac{1}{y-\frac{q}{2}-(u-i\delta)} - \frac{1}{y+\frac{q}{2}-(u-i\delta)}\right\}.$$

The imaginary part is easily separated:

$$\operatorname{Im}\Pi_a(\omega,\gamma,q) = -(2\,s_a+1)\frac{m_a}{q}\int_0^\infty \frac{dk}{(2\pi)^2}k\,f_a(k)\{A_+ - A_-\},$$
$$\text{with}\quad A_\pm = \arctan\frac{k-(\pm u-\frac{q}{2})}{|\delta|} - \arctan\frac{k-(\pm u+\frac{q}{2})}{|\delta|}.$$

It is readily verified that $A_+ - A_- \sim \operatorname{sign}(k)$, so it does not change its sign for non-negative k, and, therefore $\operatorname{Im}\Pi_a \leq 0$.

[28]Of course, in reality, the distribution is never strictly isotropic, already due to fluctuations.

4.5 *Quasilinear theory for classical and quantum systems

The deficiency of the linearization procedure (RPA) is that, while the field response (excitation of plasmons) is calculated, the plasma (i.e. its main distribution f_0) is considered not affected by the field. This becomes a problem in the case of an instability, which, according to linear theory, will grow without limit. On the other hand, we expect that the plasma will respond to an instability and, eventually, stabilize it.

Classical quasilinear theory. The simplest approximation which goes beyond the linear one is the so-called quasilinear theory [DP62, VVS62] which has successfully been applied to many problems in classical plasmas, see e.g. [ABR84]. The idea is to calculate the perturbation f_1 of the distribution function ($f = f_0 + f_1$, with $|f_1|/f_0 = \eta \ll 1$), Eq. (4.8), in linear approximation, and then to use this result to compute the change of f_0 (which will be of the order η^2). This expansion was discussed above in Sec. 4.1. It is possible, to define the function f_0 more generally by an averaging procedure, $f_0 = \langle f \rangle$, thus field fluctuations can be included.[29] With this ansatz introduced into the Vlasov equation (classical limit of Eq. (4.9)), and including terms up to order η^2, we obtain

$$\frac{\partial f_0}{\partial t} + e\left\langle \mathbf{E}\frac{\partial f_1}{\partial \mathbf{p}} \right\rangle = 0, \qquad \frac{\partial f_1}{\partial t} + \mathbf{v}\frac{\partial f_1}{\partial \mathbf{r}} + e\mathbf{E}\frac{\partial f_0}{\partial \mathbf{p}} = 0. \tag{4.74}$$

Fourier transforming the second equation and solving it for f_{1k} in analogy to (4.12), we obtain a closed equation for f_0

$$\frac{\partial f_0}{\partial t} = \sum_{ij} \frac{\partial}{\partial p_i} D_{ij} \frac{\partial f_0}{\partial p_j} \qquad D_{ij} = \frac{e^2}{2} \sum_k \frac{k_i k_j}{k^2} \frac{|E_{\mathbf{k}}|^2(-\Gamma_{\mathbf{k}})}{(\Omega - \mathbf{kv})^2 + \Gamma_{\mathbf{k}}^2}. \tag{4.75}$$

This equation describes the collisionless relaxation of the distribution function due to "diffusion" in velocity space.[30] The relaxation is governed by the diffusion coefficient D_{ij}, which, in turn, is determined by the plasmon spectrum. In particular, the sign of the diffusion coefficient may be positive or negative, depending on the sign of the damping increment $\Gamma_{\mathbf{k}}$. The latter and the intensity $E_{\mathbf{k}}^2$ of the spectral components of the field follow from the same equations

[29] The average can be over a time much larger than the relevant oscillation period, or it can be an average over the spectrum of field fluctuations (plasma turbulence).

[30] Due to the averaging procedure the original reversible Vlasov equation transformed into an irreversible which describes the evolution toward a stationary nonequilibrium state which is accompanied by an entropy increase [Bon91].

as in linear theory,

$$\frac{\partial |E_{\mathbf{k}}|^2}{\partial t} = -2\Gamma_{\mathbf{k}} |E_{\mathbf{k}}|^2, \qquad \Gamma_{\mathbf{k}} = -\frac{\pi}{2}\frac{\omega_{pl}^3}{k^2}\frac{\partial f_0(v = \Omega/k)}{\partial v}. \tag{4.76}$$

Eqs. (4.75, 4.76) constitute a closed system of equations. In case of an instability, i.e. if f_0 is a nonequilibrium distribution with a minimum, $\Gamma_{\mathbf{k}} < 0$, and the diffusion coefficient is positive. As a result, the curvature of the distribution function f_0 in the velocity range corresponding to the phase velocities of the excited modes, decreases. Since this is just the region of the minimum, the relaxation tends to remove the minimum and, with it, the instability condition. Obviously, the system evolves towards a quasi-stationary state, where f_0 is flat[31] in this velocity interval. Due to carrier number conservation, the height of the plateau and its boundaries are easily found from the original distribution by means of a Maxwell construction.

This simple theory gives a very intuitive and qualitatively correct picture of the *nonlinear* response of the plasma to the excitation of unstable collective excitations, and it correctly describes the internal (still collisionless) stabilization mechanism. This theory is straightforwardly generalized to many-component systems as well as to those under the influence of a magnetic field and can be applied to long-wavelength oscillations in quantum systems also. We mention that the quasilinear equations (4.75, 4.76) follow directly from the more general kinetic theory for plasmons and electrons of Klimontovich [Kli59] and Pines and Schrieffer [PS62], cf. Sec. 4.7.

Quasilinear theory for quantum systems. Consider now weakly nonlinear oscillations in a quantum plasma. Obviously, we expect that the quantum character of the collective mode behavior will be essential, most of all, for short-wavelength oscillations. Furthermore, the deviations from the classical properties will be particularly strong in the case of low-dimensional systems of small size, i.e. when the ratio between wavelength and system size is not vanishingly small. Then one has situations where only a small number of modes is excited, and no loss of phase memory will occur, in contrast to excitation of a quasi-continuum of modes in classical system.

We consider first the simplest case, where only a single mode with wavenumber k_0 is excited in a one-dimensional system. After Fourier transformation with respect to the coordinate, the collisionless Boltzmann equation (4.9) has the form (we omit the time argument of f)

$$\left\{\frac{\partial}{\partial t} + i\frac{p\,k}{m}\right\} f(p,k) = i\sum_q [f(p-\frac{q}{2}, k-q) - f(p+\frac{q}{2}, k-q)][U_q + 2\,V_q \sum_{p'} f(p', q)], \tag{4.77}$$

[31] More precisely, it has a plateau along the direction of the excited wavenumber.

where $f(p,q), U_q$ and V_q are the Fourier transforms of the Wigner function, the external potential and the Coulomb potential, respectively. The external potential is of the form $U_q = U\delta_{k_0,q} + U^*\delta_{k_0,-q}$ and excites plasma oscillations with the wavenumber k_0. We expect that, due to nonlinear effects, there will also appear higher harmonics. Therefore, we represent the Wigner function by a complete sum of spatial harmonics

$$f(p,k) = f_0(p) + \sum_{i=1}^{\infty}[f_i(p)\delta_{k,ik_0} + f_i^*(p)\delta_{k,-ik_0}], \tag{4.78}$$

where each spatial harmonic gives rise to the respective density component $n_s = 2\int \frac{dp}{2\pi} f_s(p)$, where $n_s = n_s^*$. With the abbreviations

$$\Delta f(p \mp \frac{q}{2}) = f(p - \frac{q}{2}) - f(p + \frac{q}{2}), \tag{4.79}$$

$U_1 = U + 2V_{k_0}n_{k_0}$ and $U_i = 2V_{ik_0}n_{ik_0}$, $(i > 1)$, the property of the Coulomb potential $V_{-s} = V_s$, and introducing the ansatz (4.78) into Eq. (4.77), we obtain an infinite system of equations for the spatial harmonics of the Wigner function. The equation for f_0 reads [BSBK94][32]

$$\frac{\partial}{\partial t} f_0(p) = 2\sum_{i=1}^{\infty} \mathrm{Im}[\Delta f_i(p \mp \frac{ik_0}{2})U_i^*]. \tag{4.81}$$

This system of equations is equivalent to the original kinetic equation and as complicated. However, Eq.(4.81) already reveals the basic nonlinear mechanism in collisionless quantum plasmas: The homogeneous part of the distribution is changed (slowly, on times much longer than the oscillation period) due to interaction of carriers with all harmonics. These equations are especially useful if the number of excited harmonics is small. In particular, if f_0 is influenced only by f_1, one recovers the level of classical quasilinear theory which was discussed before.

A very interesting result follows, if for f_1 and U_1 in Eq.(4.81) a single-pole approximation is used, i.e. $f_1(pt) = f_1(p)\exp[-i\hat{\Omega}_1 t]$ and $U_1(t) = U_1\exp[-i\hat{\Omega}_1 t]$,

---

[32] The general form of the equation for the n-th harmonic $(n > 0)$ is

$$\left[\frac{\partial}{\partial t} + i\, n\, k_0 \frac{p}{m}\right] f_n(p) = i\Delta f_0(p \mp \frac{nk_0}{2})U_{nk_0}$$
$$+i\sum_{l=1}^{\infty}\left[\Delta f_l(p \mp \frac{-(l-n)k_0}{2})U^*_{(l-n)k_0} + \Delta f_l^*(p \mp \frac{(l+n)k_0}{2})U_{(l+n)k_0}\right], \tag{4.80}$$

which has to be supplemented by the adjoint equation for f_n^*.

where $\hat{\Omega}_1(k_0) = \Omega_1(k_0) - i\Gamma_1(k_0)$, and $\Omega_1(k_0)$ and $\Gamma_1(k_0)$ are the linear dispersion and growth rate of an unstable mode ($\Gamma_1(k_0) < 0$), cf. Eq. (4.12),

$$f_1(p) = \frac{\Delta f_0(p \mp \frac{k_0}{2})}{pk_0/m - \hat{\Omega}_1} U_1. \tag{4.82}$$

With this expression inserted into Eq.(4.81) we obtain

$$\frac{\partial}{\partial t} f_0(p) = 2|U_1|^2(-\Gamma_1)\exp[-2\Gamma_1 t] \times$$

$$\left[\frac{f_0(p-k_0) - f_0(p)}{[(p-\frac{k_0}{2})\frac{k_0}{m} - \Omega_1]^2 + \Gamma_1^2} + \frac{f_0(p+k_0) - f_0(p)}{[(p+\frac{k_0}{2})\frac{k_0}{m} - \Omega_1]^2 + \Gamma_1^2}\right]. \tag{4.83}$$

Together with f_0, also the dispersion and growth rate become slowly time dependent. From Eq.(4.83), one can see that the excitation of an unstable mode causes a deformation of the homogeneous distribution in such a way that the instability is weakened. This becomes obvious if one considers $\partial f(p_{max})/\partial t$ where p_{max} is the position of the original nonequilibrium peak. This derivative is negative, what causes a lowering of the maximum. Moreover, new maxima start growing at $p_{max} \pm k_0$, what in turn, reduces the growth of the original mode. The analysis shows that with removal of the growth condition, all modes decay, which is accompanied by a return of the distribution to its original form. As a result the original unstable mode reappears, and the process continues periodically.[33] We mention that the cycle of this evolution is mainly defined by the linear growth rate, where stronger linear growth leads to a shorter cycle period. Furthermore, the number of the generated harmonics can be controlled by the ratio between the wavenumber k_0 and the peak position p_{max}. This behavior is fully confirmed by numerical solutions of the full nonlinear quantum Vlasov equations, see Fig. 4.6 below.

4.6 Numerical solutions of the nonlinear quantum Vlasov equation

The analytical results for the linear and quasilinear case have revealed rich information on the qualitative behavior of plasmas in the collisionless regime. To verify them quantitatively, numerical solutions of the full nonlinear quantum

[33] In contrast to the classical quasilinear equations, the system (4.80) is reversible and does not lead to a stationary state.

Vlasov equation (4.5) are of high interest. Moreover, the analytical approaches are, as a rule, limited to low amplitude excitations U, while it is an interesting problem to understand also the behavior in cases where the external field is strong or has a more complex time or space dependence.

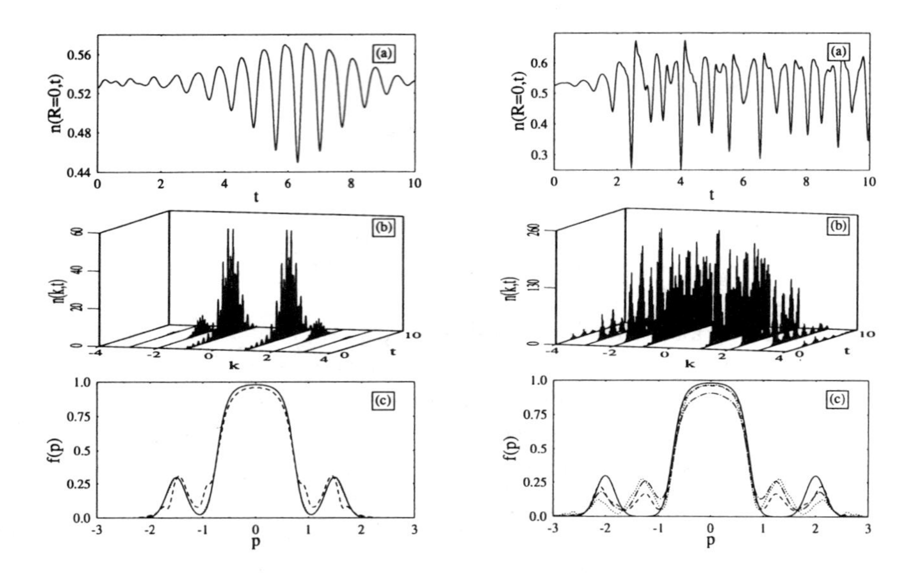

Fig. 4.6. Nonlinear evolution of a linearly unstable plasma from numerical solution of Eq. (4.84) for $k_0 a_B = 0.7$, $U = 0.05$ and $d = 0.25 a_B$. The initial 1D distributions are homogeneous along the quantum wire and given in Figs. (c). They differ only in the position or the nonequilibrium peak: $p_{max} a_B = 1.5$ in the left and $p_{max} a_B = 2.0$ in the right figures. **(a)**: Time evolution of electron density n at $R = 0$. **(b)**: Time evolution of the spatial Fourier transform of n. $k = 1$ corresponds to the wave number of the external excitation, also higher spatial harmonics are excited. **(c)**: Wigner function for different times. Left: $t = 0$ (solid line), $6.83ps$ (dashed line) and $10ps$ (dotted line, on top of solid line), right: $2.24ps$ (dash-dotted line), $3.6ps$ (dashed line) and $4.27ps$ (dotted line).

It turns out that the quantum Vlasov equation (4.5) is conveniently solved after Fourier transform with respect to the coordinate. The result is straightforward, so it is sufficient to discuss the one-dimensional case,

$$
\begin{aligned}
i\hbar \frac{\partial f(p,k)}{\partial t} &= \frac{\hbar^2 pk}{m} f(p,k) \\
&+ \int \frac{dq}{2\pi} U^{\text{eff}}(q) \left[f(p-q/2, k-q) - f(p+q/2, k-q) \right] \quad (4.84) \\
U^{\text{eff}}(q) &= U(q) + V(q) n(q); \qquad n(q) = 2 \int \frac{dp'}{2\pi} f(p', q). \quad (4.85)
\end{aligned}
$$

This equation has been solved numerically for situations for which the RPA predicts damped and unstable modes, respectively [Sco93, BSBK94]. In the limit of weak external field we expect the result to agree with the linear (RPA) behavior. This is indeed the case. Starting with an equilibrium distribution that is homogeneous in space, we observe clearly an exponentially damped (optical) plasmon. More interesting is the case of an homogeneous *initially linear unstable momentum distribution* with two extrema, where the RPA predicts a growing acoustic plasmon, see Sec. 4.4.1. For small amplitude excitation, after a short transient time which is due to the decay of the stable optical plasmon, indeed, an exponential growth is observed [BBS$^+$94]. However, with increasing amplitude of the oscillations, nonlinear effects become important for the long-time behavior. As expected, the growth of the mode slows down and eventually the plasmon becomes damped. For weak excitation, the corresponding nonlinear damping of unstable plasmons agrees well with the predictions from quantum quasi-linear theory, Sec. 4.5. However, if the amplitude of the external field (or of the density fluctuation) is not small, i.e. $|\delta n| \sim n$, the long–time behavior of the plasma becomes irregular. This is clearly seen in Fig. 4.6, cf. [BSBK94, Sco93].

4.7 *Kinetic equations for carrier–plasmon interaction

In the previous sections, we discussed the collective excitations of quantum charged particle systems using various approximations of increasing complexity: The linear approximation, Sec. 4.1, describes only the effect of the carriers on the plasmons in first order, the quasilinear approximation, Secs. 4.5, took into account additionally the feedback of the plasma oscillations on the particles to lowest order. Of course, the final goal is a fully selfconsistent treatment of carriers and plasmons. This will be considered below, in Chs. 9 and 12, but this requires the inclusion of the Coulomb correlations between the carriers (scattering effects)[34], which we start considering in the next chapters.

Here, we discuss another approach to carrier–plasmon interaction which is simplified, but very intuitive, and resembles the carrier–phonon problem. It has been derived long ago independently by Klimontovich and Pines and Schrieffer [Kli59, PS62]. Their idea was to transform the Hamiltonian of the particle-field complex into a representation in terms of two types of quasipar-

[34] A selfconsistent approach which treats particles and electromagnetic fields (including transverse fields and plasmons) fully selfconsistently, is possible only within quantum electrodynamics and will be considered in Ch. 12

ticles - charged particles and plasmons, which interact with each other, and to construct a kinetic theory for carrier-plasmon interaction from it.

Hamiltonian for carriers and plasmons. In second quantization, the electron Hamilton operator is given by [BP53]

$$H = \sum_{\mathbf{k}} E_k a_{\mathbf{k}}^\dagger a_{\mathbf{k}} + \frac{1}{2} \sum_{\mathbf{k},\mathbf{k}',\mathbf{q}\neq 0} V_q a_{\mathbf{k}-\mathbf{q}}^\dagger a_{\mathbf{k}'+\mathbf{q}}^\dagger a_{\mathbf{k}'} a_{\mathbf{k}}, \tag{4.86}$$

with $E_k = \hbar^2 k^2/2m$ and the fermionic creation and annihilation operators $a_{\mathbf{k}}^\dagger$ and $a_{\mathbf{k}}$ which obey the anti-commutation rules (we drop the spin here)

$$[a_{\mathbf{k}}, a_{\mathbf{k}'}^\dagger]_+ = \delta_{\mathbf{k},\mathbf{k}'}, \qquad [a_{\mathbf{k}}, a_{\mathbf{k}'}]_+ = [a_{\mathbf{k}}^\dagger, a_{\mathbf{k}'}^\dagger]_+ = 0, \qquad a_{\mathbf{k}}^\dagger a_{\mathbf{k}} = n_{\mathbf{k}}, \tag{4.87}$$

and n is the particle number operator. The basic idea is that the Coulomb potential V between the electrons contains short-range and long-range interactions. The long-range part involves correlations of many electrons giving rise to collective excitations, i.e. plasmons, whereas the short-range part contains the effects of collisions and strong correlations[35]. For a subdivision, one may use the fact that plasma oscillations exist as a well-defined excitation up to a critical wave number k_c (except for 1D), which thus provides a natural boundary.

Then one can transform only the long-range part ($k < k_c$) of the Coulomb interaction into a field form introducing the Bose operators $b_{\mathbf{k}}^\dagger$ and $b_{\mathbf{k}}$ with the commutator relations

$$[b_{\mathbf{q}}, b_{\mathbf{q}'}^\dagger] = \delta_{\mathbf{q},\mathbf{q}'} \qquad [b_{\mathbf{q}}, b_{\mathbf{q}'}] = [b_{\mathbf{q}}^\dagger, b_{\mathbf{q}'}^\dagger] = 0, \qquad b_{\mathbf{q}}^\dagger b_{\mathbf{q}} = N_{\mathbf{q}}, \tag{4.88}$$

where N is the plasmon number operator.[36] As a result, the Hamiltonian (4.86) is transformed into

$$H = H_e + H_{pl} + H_{e-pl}^I + H_{e-pl}^{II}, \tag{4.89}$$

where the electron part contains the kinetic energy and the short-range interactions, and the plasmon contribution reads

$$H_{pl} = \sum_{0<k<k_c} \left\{ \hbar\Omega(\mathbf{k}) \left(b_{\mathbf{k}}^\dagger b_{\mathbf{k}} + \frac{1}{2} \right) - n \frac{2\pi e^2}{k^2} \right\}, \tag{4.90}$$

[35] Of course, strictly speaking, such a subdivision is not possible. There is no unique prescription for it. Moreover, short-range and long-range interactions do overlap and influence each other. However, in the case of dilute systems, where short-range collisions are very rare, this approach may be expected to allow for a qualitative analysis.

[36] For the connection of $b_{\mathbf{q}}^\dagger$ and $b_{\mathbf{q}}$ with the field variables and subsidiary conditions on the latter which guarantee that Maxwell's equations (11.2) are satisfied, cf. [GRH91].

where the last contribution subtracts the selfinteraction. The first electron-plasmon coupling term describes absorption or emission of a plasmon by an electron,

$$H^I_{e-pl} = \sum_{\mathbf{p}} \sum_{0<k<k_c} \sqrt{\frac{V_k}{\mathcal{V}\hbar\Omega(\mathbf{k})}} \left(\frac{\hbar \mathbf{kp}}{m} - \frac{\hbar^2 k^2}{2m} \right) \times \left(b_{\mathbf{k}} - b^{\dagger}_{-\mathbf{k}} \right) a^{\dagger}_{\mathbf{k}+\mathbf{p}} a_{\mathbf{p}}, \tag{4.91}$$

whereas H^{II}_{e-pl} contains higher order processes, such as two-plasmon absorption/emission [PS62] and is usually much smaller than H^I_{e-pl} and will, therefore, be neglected.[37]

RPA kinetic equations for electrons and plasmons. Keeping in the Hamiltonian (4.89) only resonant transitions (carrier-plasmon scattering) Pines and Schrieffer derived coupled kinetic equations for the electron and plasmon distribution functions $f(\mathbf{k})$ and N_q which describe the relaxation of electrons and plasmons towards a Fermi and Bose distribution, respectively. For the spatially homogeneous situation, the plasmon kinetic equation is

$$\begin{aligned} \frac{\partial N_q}{\partial t} &= \frac{\pi}{\mathcal{V}} \sum_{\mathbf{k}} V_q \Omega(\mathbf{q}) N_q \left[f(\mathbf{k}+\mathbf{q}) - f(\mathbf{k}) \right] \delta\left(\hbar\Omega(\mathbf{q}) - E_{\mathbf{k},\mathbf{q}} \right) \\ &+ \frac{\pi}{\mathcal{V}} \sum_{\mathbf{k}} V_q \Omega(\mathbf{q})\, f(\mathbf{k}+\mathbf{q}) \left[1 - f(\mathbf{k}) \right] \delta\left(\hbar\Omega(\mathbf{q}) - E_{\mathbf{k},\mathbf{q}} \right), \end{aligned} \tag{4.92}$$

with $E_{\mathbf{k},\mathbf{q}} = E(\mathbf{k}+\mathbf{q}) - E(\mathbf{k})$. This equation has a very transparent form. The second sum corresponds to spontaneous plasmon emission by fast electrons associated with the Cherenkov effect. This contribution is always positive leading to gain of plasmons with wavenumber $\mathbf{q}$. On the other hand, the first sum arises from induced emission and absorption of plasmons by electrons and is negative in equilibrium, where it corresponds to resonant absorption of plasmons (Landau damping). This contribution is directly connected with the spectral function of the polarization, cf. Eq. (4.30). If the damping of the plasmons is small, i.e. $|\Gamma(\mathbf{q})| \ll \Omega(\mathbf{q})$, this sum can be written as $-2\Gamma(\mathbf{q})\, N_q$. For nonequilibrium electron distributions with a second maximum, $\Gamma(\mathbf{q})$ may become negative in a certain interval of wavenumbers $0 \le q \le q_{max}$. This is the case of a plasma instability, and one observes net gain of plasmons. All these

[37] Of course, there are several problems: The critical momentum k_c is not defined selfconsistently. For quantum plasmas, it is of the order of $k_c \sim \omega_p / v_F$, and for classical plasmas of the order of the inverse Debye radius, $k_c \sim k_D = \omega_p/(kT/m)^{1/2}$ [BP53]. Furthermore, the plasmon dispersion $\Omega(q)$ which enters the Hamiltonian, is difficult to calculate selfconsistently too.

processes contribute in a symmetric way to the electron kinetic equation[38]. A generalization to two-component plasmas including kinetic equations for optical and acoustic plasmons is straightforward [PS62]. Klimontovich derived classical kinetic equations for electrons and plasmons starting from a generalized BBGKY [Kli59]. His equations, however, go far beyond the RPA level, including two-particle and higher order correlations as well as cross correlations between electrons and plasmons. For generalizations of this theory, including equations for atoms and plasmons, cf. [Kli75]. We return to the questions of carrier-plasmon interaction in Ch. 9 where we derive the non-Markovian Balescu-Lenard equation and in Ch. 12, where a much more general treatment of carrier-photon coupling (including plasmons) is discussed.

Concluding remarks. We conclude this chapter with some remarks. Due to the neglect of correlations, the mean-field approximation does not contain any characteristic relaxation times. Thus, (if it is justified), it has no limitations with respect to the times and is, in particular, applicable to ultra-short times too.

On the other hand, in cases where correlation effects are not negligible, the inclusion of collision terms in the kinetic equation becomes important. The simplest approach is to perform perturbative expansions on the basis of retardation time approximations, e.g. [BG68, KF69, Mer70, Kli75]. This yields the correct qualitative behavior: collective excitations obtain and additional collisional damping (beyond the Landau damping), plasmon life time and growth rates of instabilities are reduced. Moreover, correlations modify the spectrum of the excitations too, e.g. [GK92, KRG94, KG94].

For a systematic study of correlation effects, we now go back to the BBGKY-hierarchy and use decoupling approximations that go beyond the mean-field approximation.

---

[38] The electron equation has the form

$$
\begin{aligned}
\frac{\partial f(\mathbf{k})}{\partial t} &= \frac{\pi}{\mathcal{V}} \sum_{q<q_c} V_q \Omega(\mathbf{q}) \Big\{ \\
& N_q \ \{[f(\mathbf{k}+\mathbf{q}) - f(\mathbf{k})]\, \delta\left(\hbar\Omega(\mathbf{q}) - E_{\mathbf{k},\mathbf{q}}\right) + [f(\mathbf{k}-\mathbf{q}) - f(\mathbf{k})]\, \delta\left(\hbar\Omega(\mathbf{q}) - E_{\mathbf{k},-\mathbf{q}}\right)\} \\
+ & \ \{f(\mathbf{k}+\mathbf{q})\,[1 - f(\mathbf{k})]\, \delta\left(\hbar\Omega(\mathbf{q}) - E_{\mathbf{k},\mathbf{q}}\right) + f(\mathbf{k})\,[1 - f(\mathbf{k}-\mathbf{q})]\, \delta\left(\hbar\Omega(\mathbf{q}) - E_{\mathbf{k},-\mathbf{q}}\right)\} \Big\}.
\end{aligned}
$$

The terms proportional to N_q are gain and loss of electrons, stimulated by resonant absorption of a plasmon with wavenumber $\mathbf{q}$ or $-\mathbf{q}$. In equilibrium, the first contribution is negative, whereas the second is always positive. The terms in the last line describe again spontaneous Cherenkov radiation of electrons which scatter into the corresponding states with lower momentum. Via to the Pauli principle, the scattering rates depend on the population of the final state.

Chapter 5

Correlations and their Dynamics

As we discussed in the Introduction, in real (i.e. nonideal) systems, the interaction forces between the particles give rise to a variety of mutual influences. On one hand, there appear large scale phenomena, which involve a big number of particles. In the previous Chapter, we have studied a typical example - collective plasma excitations which are the result of the long–range Coulomb interaction. There, each particle interacts, at the same time, with many others,[1] while any two particles exhibit only weak influence on one another. On the other hand, there exist other phenomena, where small scale interaction, close encounters predominate. Examples are particle–particle scattering or the formation of bound states. The approach of the BBGKY-hierarchy allows for a clear mathematical distinction of both types of phenomena, cf. first equation (2.99): Collective effects are the result of an average (mean) field U^H created by all particles, while binary correlations are absent, $g_{12} \approx 0$, whereas correlation phenomena arise from non-vanishing g_{12}, g_{123} etc. These processes become essential if the coupling parameter Γ, the ratio of the mean potential and kinetic energy, (cf. Ch. 1), is not negligible. Therefore, correlation effects are very important for nuclear systems, fluids, solids and dense gases or plasmas.

The influence of correlations on the nonequilibrium properties, on the relaxation, is the main underlying subject of the next chapters. In this chapter, we begin with a general analysis of correlations and their dynamics.

---

[1]We have seen in Secs. 4.5, 4.7 that this can be described as a diffusion process in momentum space, or particle–plasmon scattering.

5.1 Hierarchy of relaxation processes. Time scale separation

We consider a general many–particle system which is excited by some external potential $\mathcal{U}(t)$ with a characteristic duration $\tau_{\mathcal{U}}$. During this time, the system is brought into a nonequilibrium state, from which the system eventually relaxes towards a stationary state.[2] Typically, the relaxation process goes through several stages, depending which degrees of freedom have been activated by the excitation. Table 5.1 illustrates this hierarchy on a typical example, listing processes by increasing distance from the stationary state. Shortly after the initial time t_0, the relaxation is very complicated and possibly "violent", characterized by different types of correlations, and it is influenced by initial correlations. During this *initial stage*, initial correlations are expected to decay, and, as a result of scattering processes, correlations are being built up (cf. Sec. 5.2). After the correlation time τ_{cor}, the relaxation simplifies and is governed only by equations of motion for the one-particle distributions f_a (kinetic equations). This is the *kinetic stage*, during which the nonequilibrium distribution functions undergo momentum and energy relaxation, due to scattering of particles which each other or interaction with collective excitations (phonons, plasmons) etc., and, after the relaxation time t_{rel}, reach their equilibrium form[3]. Notice that this equilibrium distribution, is, in general, different from a Bose or Fermi (B/F) function, and contains correlation corrections Δf.

For times larger than the relaxation time, the evolution enters the comparably calm *hydrodynamic or gas dynamic stage*, where mainly macroscopic quantities, such as density n_a, pressure p and temperature T_a relax and which is governed by large scale transport processes, such as diffusion, heat conductivity and chemical reactions, e.g. [EKKR]. After this phase, for times large compared to the characteristic macroscopic relaxation time t_{hyd}, the system eventually reaches a stationary state, which is an equilibrium state or, in case of external potentials $\mathcal{U}$, a stationary nonequilibrium state. Here, the properties are determined by the laws of thermodynamics, the equation of state, the mass action law and so on. It has to be stressed, that this state may differ strongly from familiar equilibrium behavior of ideal systems, such as an ideal gas. Correlations between the particles remain visible also in equilibrium, leading to correlation corrections to the macroscopic quantities, such as free energy, chemical potential or pressure. Interestingly, the behavior of the system at long times is influenced by processes on the very first stage, which is due

[2]Obviously, in general, there is no separation of the excitation and the relaxation phase.

[3]In case of multiple scattering processes "α", $t_{rel} = \max_\alpha\{t_{rel}^{(\alpha)}\}$.

to the buildup of correlations and their effect on the macroscopic properties.

For a correct and efficient statistical modeling, it is essential that the full complex evolution of the N-particle system can be described as a superposition of a quite limited number of comparably simple relaxation processes. Of equal importance is that individual processes differ greatly with respect to the typical time and length scales, the number of particles involved and so on. This allows one to identify for each stage the dominating processes, while neglecting the others (or treating them as small perturbations), and to develop the appropriate theoretical approaches.

The concept of time scales is among the most fruitful for theoretical modeling of physical phenomena [Bog46]. The separation of fast processes (e.g. in oscillation theory) or the identification of "master" processes (selforganization, nonequilibrium phase transitions) allows not only to simplify the model, but often yields far deeper insight in the underlying physics. This applies in particular to kinetic theory. According to Bogolyubov, there exists a hierarchy of time scales for the equilibration of one–particle, two–particle and higher order properties (correlations): $\tau_{cor}^{(N)} < \tau_{cor}^{(N-1)} < \ldots < \tau_{cor}^{(2)}$.[4] This means that for times larger than $\tau_{cor}^{(N)}$, the N-particle correlation operator has reached its equilibrium form and is no longer explicitly time–dependent. However, being coupled to lower order operators via the BBGKY-hierarchy, it remains time dependent implicitly[5]:

$$g_{1\ldots N}(t) = g_{1\ldots N}^{EQ}\Big(\{g_{1\ldots N-1}(t)\}, \ldots, \{g_{12}(t)\}, \{F_1(t)\}\Big), \quad t > \tau_{cor}^{(N)}. \tag{5.1}$$

A substantial separation of scales is usually observed for the two–particle, one–particle and macroscopic processes,

$$\tau_{cor} \ll t_{rel} \ll t_{hyd}, \tag{5.2}$$

where $\tau_{cor} = \max_N\{\tau_{cor}^{(N)}\}$, and, in analogy to Eq. (5.1),

$$F_1(t) = F_1^{EQ}\Big(\{n(t)\}, \{p(t)\}, \{T(t)\}\Big), \quad t > t_{rel}. \tag{5.3}$$

Thus, after the relaxation time, the one-particle density operator depends on time only via the macroscopic quantities density, pressure and temperature, cf. table 5.1.

Relaxation time approximation. One can take advantage of the time scale separation (5.2) by constructing simplified kinetic theories, the most

[4]Substantial differences can be expected only for small N. Possible exceptions are long–living large–scale fluctuations, such as occur in the vicinity of phase transitions, or bound state correlations, which are not considered here, see e.g. [Kli75].

[5]This is called Bogolyubov's "functional hypothesis" [Bog46], cf. also [AP77].

Hierarchy of Relaxation Processes

Time	Stage	Quantities	Theory
$t \gg t_{hyd}$	**Equilibrium** Correlated Equil. Stat. noneq. state	n_a^{EQ}, T^{EQ}, p^{EQ} $p = p(n_1, n_2 \ldots, T)$ $n_a = n_a(n_1, n_2 \ldots, T)$ $p = p^{id} + p^{cor}$ etc. $n_a(\mathcal{U}), T(\mathcal{U}), p(\mathcal{U})$	Equil. Theory Eqn. of State, Mass action law Quasi-Equil. Th.
$[t_{rel}, t_{hyd}]$	**Hydrod. Stage** Local Equilibrium Correl. corrections	$n_a(\mathbf{R}t), p_a(\mathbf{R}t), T_a(\mathbf{R}t)$ $f_a = f_a^{EQ}\big(n(\mathbf{R}t),$ $p(\mathbf{R}t), T(\mathbf{R}t)\big)$ $= f_a^{B/F} + \Delta f_a$	Hydrodynamic, Gas-dynamic or Rate Equations
$[t_{cor}, t_{rel}]$	**Kinetic Stage** Functional hypoth. Equil. correlations Kin. energy cons.	$f_a(\mathbf{p}\mathbf{R}, t)$ $f_a(t_0)$ $g_{ab} = g_{ab}^{EQ}(\{f(t)\})$ $= g_{ab,0}^M + g_{ab,1}^M + \ldots$	Kinetic theory/ Relaxation time approximation Markov limit + Correl. correct.
$[t_0, \tau_{cor}]$	**Initial Stage** Initial Correl. Correl. Buildup Total energy cons. Higher Correl.	$g_{ab}(\mathbf{p}_a\mathbf{R}_a\mathbf{p}_b\mathbf{R}_b, t)$ $g_{ab}(t_0)$ $g_{abc}, g_{abcd}, \ldots$	Generalized Kinet. Theory/ Correlation time approximation

Table 5.1: Relaxation in correlated many–particle systems. Beginning at the initial time t_0, the evolution goes (from bottom to top) through several stages. Accordingly, the relevant observables and concepts for a statistical description change (for explanation, see text).

prominent example of which is the relaxation time approximation (RTA). The idea is that close to the relaxation time (but still $t < t_{rel}$), the deviations of F_1 from its equilibrium value are small, suggesting the linear ansatz for the Wigner distribution function (the matrix element of F_1)

$$\frac{df(\mathbf{p},\mathbf{R},t)}{dt} \approx -\frac{f(\mathbf{p},\mathbf{R}) - f^{EQ}(n,T;\mathbf{p},t)}{\tau_{rel}(n,T)}, \tag{5.4}$$

where f^{EQ} is the (local) equilibrium distribution function which depends on density and temperature n, T.[6] Taking advantage of the existence of different time scales, this approximation has been very successful for incorporating the effect of collisions (fast process) into macroscopic balance equations, describing diffusion, electrical and heat conductivity etc. (slow processes) qualitatively correct. Among the well-known representatives of this approach we mention the Drude theory of dielectric and optical properties of solids and the relaxation time approximation to the semiconductor Bloch equations. The relaxation time $t_{rel,a}$ of particle species "a" is calculated from the collision rates ν of all relevant scattering processes "γ", $t_{rel,a}^{-1} = \sum_{b\gamma} \nu_{ab}^{(\gamma)}$ which are taken from experiment or are derived from kinetic theory, e.g. [BSP+92].[7]

5.2 Correlation buildup. Correlation time approximation

In similar manner, one can construct a relaxation time approximation for the binary correlation operator [Bon96], using as the basis the first inequality in Eq. (5.2). For times $t < \tau_{cor}$ close to the correlation time, the deviations of the binary correlation operator from its equilibrium form are small, allowing us to make the ansatz

$$\frac{dg(t)}{dt} \approx -\frac{g(t) - g^{EQ}(\{f(t)\})}{\tau_{cor}}, \tag{5.5}$$

where g^{EQ} is the equilibrium binary correlation operator, which still depends on time via the nonequilibrium distributions. The solution of Eq. (5.5) together with the initial condition $g(t_0) = g_0$ is readily found and yields g in

[6] n and T, and thus also t_{rel} may be space and time dependent themselves, but on a much larger scale compared to t_{rel}.

[7] Of course, there are situations, where a relaxation time approximation is not applicable, i.e. when details of the momentum dependence of the scattering rates are important, especially if the system is far from equilibrium.

"Correlation time approximation",[8]

$$\boxed{g_{CTA}(t) = (1 - e^{-t/\tau_{cor}})\, g^{EQ}\{f(t)\} + g_0\, e^{-t/\tau_{cor}}} \tag{5.6}$$

g_{CTA} evolves from g_0 at the initial moment, gradually approaching g^{EQ}, while the influence of the initial correlations decays. If for g^{EQ} the Markov limit is used, cf. Sec. 7.3.3, the first term in Eq. (5.5) gives rise to the conventional Markovian collision integral in the kinetic equation for the Wigner distribution, which is approached exponentially reflecting the gradual "build up" of correlations. Including further corrections into g^{EQ}, one can obtain improved Markovian collision integrals. Interestingly, this intuitive approximation can be derived from the exact non–Markovian correlation operator using a retardation expansion, cf. Sec. 7.3.3.

Correlation time and initial correlations. The second term in Eq. (5.6) describes the influence of initial correlations which, obviously, dominate the early stage of the relaxation.[9] The explicit form of g_0 strongly depends on the "pre-history" of the system for $t < t_0$, on a possible previous excitation, on the presence of external fields or on the level of fluctuations in the system. With a proper choice of g_0, one can include a variety of interesting many-body phenomena in the analysis of the short-time behavior. With increasing time, the initial correlations decay, vanishing approximately at $t = \tau_{cor}$.[10] Therefore, the correlation time is an important quantity in the study of short-time phenomena. For example, in a classical system in equilibrium, it is given as the ratio of the range of the interaction potential and the thermal velocity, $\tau_{cor} = r^{int}/v_{th}$, [Kli75]. In a plasma, where r^{int} is given by the screening length, this yields the plasma period $\tau_{cor} \approx 2\pi/\omega_{pl}$.

On the other hand, τ_{cor} is much more complex in nonequilibrium where the interaction range may change and no thermal velocity exists. Numerical solutions of non-Markovian kinetic equations have revealed a very interesting general way to define τ_{cor}, which is based on the time evolution of kinetic energy [BK96] and which will be discussed in Ch. 7.

---

[8] Of course, all limitations of the relaxation time approximation apply here too. Nevertheless, with a reasonable choice of g^{EQ} and τ_{cor}, very good quantitative results may be achieved [Bon96].

[9] In cases where, before action of the external excitation $\mathcal{U}$, the system contains only a very low particle density, e.g. in nonmetallic solids at low temperature, $g_0 \approx 0$.

[10] Notice that certain types of initial correlations do not decay, for example those which correspond to bound states or large scale fluctuations, which have to be treated separately [Kli75, KK81].

Chapter 6

Correlation Dynamics and Non-Markovian Effects

After the introductory discussion of correlation dynamics in Ch. 5, we now begin with a strict investigation on the basis of the BBGKY-hierarchy. It is natural to start from the simplest case - the static Born approximation in a one-component system without external fields. This will allow us to discuss the main correlation effects in a straightforward and easy to understand way and to relate them to non-Markovian behavior. From there, we will advance to more involved approximations in the following chapters.

We will use the hierarchy closing approximation that neglects three–particle correlations completely, $g_{123} = 0$. This approximation was discussed in Sec. 2.6.1. In particular, it was shown that it leads to conservation of total energy and is thus a suitable starting point for the derivation of kinetic equations for correlated (nonideal) many-particle systems. In addition to neglecting ternary correlations, we will, in this chapter, neglect also the ladder and polarization terms (see Sec. 2.6.1), which will lead us easily to the non-Markovian generalization of the familiar Landau equation (static second Born approximation). Proceeding further, in Ch. 7 we remove the condition $g_{123} = 0$ in order to include selfenergy effects in the non-Markovian Landau equation. Then, in Chs. 8 and 9, we derive the non-Markovian Boltzmann equation (binary collision approximation) and Balescu–Lenard equation, respectively.

The Landau equation plays a central role in many fields of physics, especially in plasma and condensed matter theory. It is not only a well-defined approximation of many-body physics on its own, it is, at the same time, an important limiting case of more involved theories: it is the static limit of the Balescu–Lenard equation and the weak coupling limit of the ladder (T-matrix) approximation.

6.1 Solution for g_{12} in second Born approximation

We use the (anti-)symmetrized generalization of the first two equations (2.99), i.e. Eqs. (3.20), (3.22), where the ladder and polarization terms and the three-particle correlations are neglected. As a result, we obtain the BBGKY-hierarchy on the level of the static second Born approximation,

$$
\begin{aligned}
i\hbar\frac{\partial}{\partial t}F_1 - [\bar{H}_1^0, F_1] &= n\mathrm{Tr}_2[V_{12}, g_{12}]\,\Lambda_{12}^{\pm}, &(6.1)\\
i\hbar\frac{\partial}{\partial t}g_{12} - [\bar{H}_{12}^0, g_{12}] &= \left\{\hat{V}_{12}F_1F_2 - F_1F_2\hat{V}_{12}^{\dagger}\right\}, &(6.2)\\
F_1(t_0) = F^0, \qquad & g_{12}(t_0) = g^0, &(6.3)
\end{aligned}
$$

where the (anti-)symmetrization operator $\Lambda_{12}^{\pm}$, the shielded potential[1] $\hat{V}_{12}$ and the Hartree–Fock Hamiltonian H^{HF} which is included in $\bar{H}_1$ and $\bar{H}_{12}^0$, where introduced in Ch. 3, cf. Eqs. (3.6), (3.15) and (3.9),

$$
\begin{aligned}
\Lambda_{12}^{\pm} &= 1 \pm P_{12}, \quad P_{12}|12\rangle = |21\rangle\,, \quad \hat{V}_{12} = (1 \pm nF_1 \pm nF_2)V_{12},\\
\bar{H}_1^0 &= H_1^0 + H_1^{HF}, \quad \bar{H}_{12}^0 = \bar{H}_1^0 + \bar{H}_2^0, \quad H_1^{HF} = n\mathrm{Tr}_2 V_{12}F_2\Lambda_{12}^{\pm}.
\end{aligned}
$$

Eqs. (6.1) and (6.2), together with the initial conditions (6.3), form a well-defined initial value problem for F_1 and g_{12}. They describe the dynamics of the particle pair $1-2$, given by the coupled evolution of its one-particle and two-particle properties and contain the full information on the initial state, including arbitrary initial correlations.[2]

Notice that Eqs. (6.1), (6.2) are "local" in time, there are no retardation effects, i.e. all functions contain the same current time t. They are, in principle, well suited for numerical solution.[3] However, we will proceed differently and derive one closed equation for the one-particle density operator, i.e. a *kinetic equation*. This will turn out to have a number of advantages for the physical understanding and numerical analysis as well.

[1] In Chs. 6–8, V_{12} is assumed to be a short range potential, for Coulomb systems it is understood as the (statically screened) Debye potential. It's derivation, based on dynamical screening of the bare Coulomb potential, will be given in Ch. 9

[2] There are some general limitations on g^0 which are not related to the approximations made. For example, consistency of Eqs. (6.1) and (6.2) requires $\mathrm{Tr}_2 g_{12}(t) = 0$ and also $g_{21} = g_{12}$ for all times including the initial moment, see also footnote 10 of Ch. 5.

[3] This has been done for various approximations by several authors, e.g. [TH93, GKRT93]. The main obstacle, even for a homogeneous and isotropic system, is that storage of the binary correlation operator requires excessive amounts of computer memory.

To do this, we solve Eq. (6.2) for $g_{12}(t)$, using the fact that it is linear in g_{12}, and that g_{12} appears only under the time derivative and in the commutator on the l.h.s. The result is, in standard way, given as the solution of the homogeneous equation (with the r.h.s. put equal to zero) plus a source term which accounts for the inhomogeneity:

$$\begin{aligned} g_{12}(t) &= U_{12}^{0+}(tt_0)\, g^0\, U_{12}^{0-}(t_0t) \\ &+ \frac{1}{i\hbar}\int_{t_0}^{\infty} d\bar{t}\, U_{12}^{0+}(t\bar{t})\left\{\hat{V}_{12}F_1F_2 - F_1F_2\hat{V}_{12}^{\dagger}\right\}|_{\bar{t}}\, U_{12}^{0-}(\bar{t}t) \end{aligned} \tag{6.4}$$

where $U_{12}^{0\pm}$ are retarded/advanced propagators which have the properties $U_{12}^{0\pm} = U_1^{0\pm}\, U_2^{0\pm}$, $U^{0\pm}(tt') = [U^{0\mp}(t't)]^{\dagger}$. U_1^{0+} is defined by

$$\left\{i\hbar\frac{\partial}{\partial t} - \bar{H}_1^0\right\} U_1^{0+}(tt') = i\hbar\,\delta(t-t'), \tag{6.5}$$

and U^{0-} obeys the adjoint equation. A detailed discussion of the propagators is given in Appendix D.

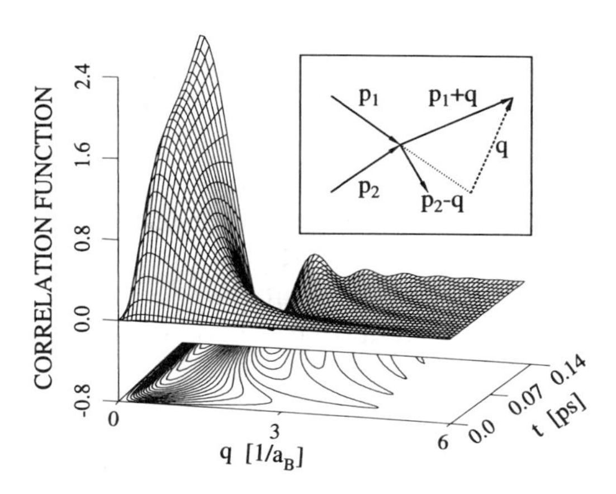

Fig. 6.1.: Relaxation of the imaginary part of g_{12} for static electron-electron scattering in a bulk semiconductor. The initial distribution is a Gaussian centered at $k = 3a_B^{-1}$, $n = 3.64 \times 10^{17} cm^{-3}$, zero initial correlations have been used. The inset explains the scattering process and the notation. In the figure, the initial momenta of the particle pair are $p_1 = p_2 = 3\hbar/a_B$ with $\mathbf{p}_1$ and $\mathbf{p}_2$ being parallel.

The solution (6.4) has a characteristic "*non-Markovian*" (non-local in time) form: the time integral contains the distribution functions not only from the current moment t, but also for all previous times, what is usually called "*memory effect*". Eq. (6.4) is fully equivalent to the local differential equation (6.1). Thus, we see clearly that the non-Markovian behavior is a (rather formal) consequence of the fact that the correlations have their own dynamics.

$U^{0\pm}$ are the Hartree–Fock propagators, which include the mean field, (in principle, we may include in $\bar{H}_1^0$ also an external potential, what we discuss in Ch. 11, where we will derive the non-Markovian generalizations of the semiconductor Bloch equations). Here we consider the simplest case, where the external field and the mean–field contributions are neglected. Then, $\bar{H}_1^0$ is replaced by the free Hamiltonian, $\bar{H}_1^0 \longrightarrow H_1^0$.

Figure 6.1. gives an example of the relaxation of the binary correlation function at short times. Starting from an uncorrelated nonequilibrium state one clearly sees the buildup of correlations. First Img is very broad what is due to the Heisenberg uncertainty principle, but it quickly develops a quasi-stationary structure which exhibits already equilibrium features: one dominating peak close to zero momentum. The periodic structure is a transient phenomenon. It is of the form sinxt/x, $(x \sim q)$ and evolves towards a delta function with increasing time. The long time limit is discussed in detail in Sec. 7.3.3.

6.2 Non-Markovian quantum Landau equation

With only the free Hamiltonian included, the system is spatially homogeneous, and the appropriate representation is the momentum representation, cf. Sec. 2.3.3. Then, the matrix element of the one–particle Hamiltonian is $E_1^0 = E^0(\mathbf{p}_1)$; for example, in the case of a parabolic energy dispersion, $E_1^0 = p_1^2/2m_1$. Due to the fact that the free Hamiltonian is time independent, the solution of Eq. (6.5) is particularly simple, with the matrix elements

$$U_1^{0+}(tt') \longrightarrow U^{0+}(\mathbf{p}_1, t-t') = \Theta(t-t')\, e^{-\frac{i}{\hbar}E_1^0(t-t')}. \tag{6.6}$$

Inserting the result for g_{12}, Eq. (6.4), with the propagators (6.6) into the r.h.s. of the first hierarchy equation (6.1), we obtain the non-Markovian Landau equation for a spatially homogeneous system,

$$\frac{d}{dt} f(\mathbf{p}_1, t) = I(\mathbf{p}_1, t) + I^{IC}(\mathbf{p}_1, t), \tag{6.7}$$

with the collision integral

$$
\begin{aligned}
I(\mathbf{p}_1,t) &= \frac{2}{\hbar^2}\int\limits_0^{t-t_0} d\tau \int \frac{d\mathbf{p}_2}{(2\pi\hbar)^3}\int \frac{d\bar{\mathbf{p}}_1}{(2\pi\hbar)^3}\int \frac{d\bar{\mathbf{p}}_2}{(2\pi\hbar)^3}(2\pi\hbar)^3\delta(\mathbf{p}_{12}-\bar{\mathbf{p}}_{12}) \\
&\times\; V(\bar{\mathbf{p}}_1-\mathbf{p}_1)\Big(V(\mathbf{p}_1-\bar{\mathbf{p}}_1)\pm V(\mathbf{p}_1-\bar{\mathbf{p}}_2)\Big)\cos\Big\{\frac{E^0_{12}-\bar{E}^0_{12}}{\hbar}\tau\Big\} \\
&\times\; \Big\{\bar{f}_1\bar{f}_2[1\pm f_1][1\pm f_2]\;-\;f_1f_2[1\pm\bar{f}_1][1\pm\bar{f}_2]\Big\}\Big|_{t-\tau},
\end{aligned}
\tag{6.8}
$$

and an additional collision integral due to the initial correlations

$$
\begin{aligned}
I^{IC}(\mathbf{p}_1,t) &= -\frac{2}{\hbar}\int \frac{d\mathbf{p}_2}{(2\pi\hbar)^3}\int \frac{d\bar{\mathbf{p}}_1}{(2\pi\hbar)^3}\int \frac{d\bar{\mathbf{p}}_2}{(2\pi\hbar)^3}(2\pi\hbar)^3\delta(\mathbf{p}_{12}-\bar{\mathbf{p}}_{12}) \\
&\times\; \Big(V(\mathbf{p}_1-\bar{\mathbf{p}}_1)\pm V(\mathbf{p}_1-\bar{\mathbf{p}}_2)\Big) \\
&\times\; \mathrm{Im}\Big\{e^{-\frac{i}{\hbar}(E^0_{12}-\bar{E}^0_{12})(t-t_0)}g_0(\mathbf{p}_1,\mathbf{p}_2,\bar{\mathbf{p}}_1,\bar{\mathbf{p}}_2)\Big\},
\end{aligned}
\tag{6.9}
$$

where we denoted $E^0_{12}=E^0_1+E^0_2$ and $\mathbf{p}_{12}=\mathbf{p}_1+\mathbf{p}_2$.

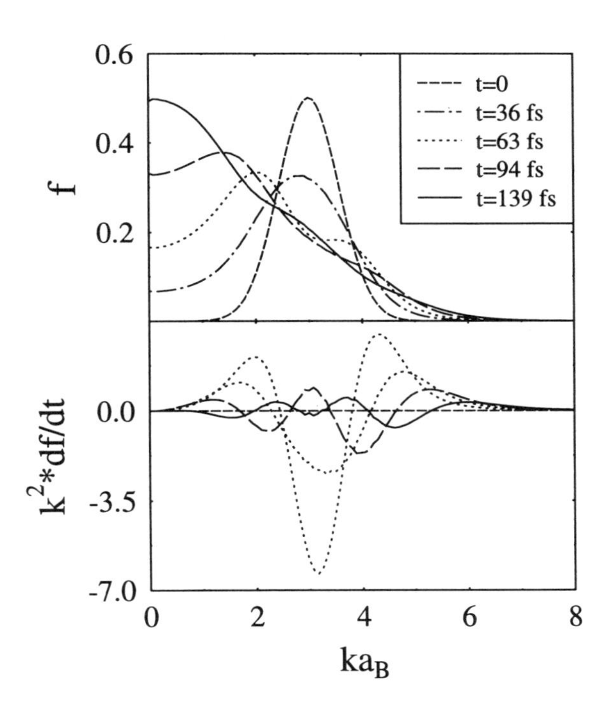

Fig. 6.2.: Relaxation of the Wigner distribution function - **upper fig.** and its time derivative - **lower fig.** for the situation of Fig. 6.1. The interaction potential has been increased artificially to magnify non-Markovian effects.

This kinetic equation is an important extension of the conventional Landau equation to times shorter than the correlation time and generalizes the non-Markovian result of Silin [Sil67] and Klimontovich and Ebeling [KE72] to systems with initial correlations and spin statistics. We will discuss its properties in Ch. 7. Fig. 6.2. shows the time evolution of $f(k)$ starting from a nonequilibrium initial distribution. Furthermore its time derivative is shown (essentially the derivative of the density $n(k)$, which is more sensitive to details of the relaxation than f). In contrast to Markovian kinetic equations which yield a very smooth and monotonic relaxation, here memory (retardation) effects appear. Consider in detail the region of the nonequilibrium peak: There is some overshooting, the peak of f is decreased too strong (the system "remembers" the peak even after it has been removed and keeps scattering out particles). Eventually, df/dt becomes positive in this region to restore the distribution, what leads to intrinsically oscillatory relaxation behavior. The reason for this is that the non-Markovian relaxation of f is equivalent to a *second order local differential equation* (corresponding to the two coupled hierarchy equations for f and g, see also Sec. 7.3). However, it has to be noted that this effect is weak (see Sec. 7.3.1, Fig. 7.1.). The reason is that the collision integral involves averaging over momenta (for different momenta, this effect appears on different momentum and time scales), so memory effects are reduced due to destructive interference.[4] We mention that oscillatory behavior may be much stronger in the case of nonzero initial correlations, which are localized in k-space.

This leads to essential problems with the collision integrals (6.8) and (6.9): These integrals describe a time–reversible dynamics of the Wigner distribution. The collision integrals are nonlocal in time with an *unlimited memory depth*, ($I(t)$ depends on the complete history of the system, where $f(t)$ is remembered completely, back to the initial state t_0), and the *initial correlations are not being weakened.* This is, of course, an unphysical long–time behavior. The reason is that the coupled equations (6.2) and (6.1) describe the isolated dynamics of the particle pair $1-2$, what corresponds to two–particle states of infinite lifetime. We will show in the following, how finite lifetime (damping) effects can be incorporated into the BBGKY–hierarchy to resolve these problems.

---

[4]Such an averaging is not present in the simpler case of carrier-phonon scattering, where memory effects are visible more clearly, e.g. [Zim92].

Chapter 7

Non-Markovian Kinetic Equations with Selfenergy. Generalized Landau Equation

As was discussed in Ch. 6, generalized non-Markovian kinetic equations allow to describe the coupled dynamics of one-particle and two-particle properties. They are, thus applicable to the initial time period $t_0 \leq t \leq \tau_{cor}$, where binary correlations relax towards their equilibrium form. On the other hand, we have seen that the isolated dynamics of two-particles which follows from $g_{123} = 0$, leads to unphysical effects. Therefore, in this chapter, we will remove this restriction and derive more realistic generalized kinetic equations. Again, we consider the simplest case, the static Born approximation.

We will take advantage of our previous analysis of three-particle correlations, cf. Secs. 2.6.1 and 3.4.1, where we already identified the terms which correspond to selfenergy effects. Here, we demonstrate explicitly that these terms, indeed lead to renormalization of the energy spectrum and to finite lifetime (damping) effects. In Sec. 7.1, we start with a pragmatic discussion to demonstrate the basic concept. The consequences for the non-Markovian Landau equation and for its properties will be investigated in Sec. 7.3.

7.1 *Selfenergy in density operator approach

To simplify the notation, we begin with the case without spin. Inspection of the equation of motion for g_{123} [third hierarchy equation, Eq. (2.99)], leads to the following approximation (cf. Sec. 2.6.1)

$$\boxed{i\hbar\frac{\partial}{\partial t}g_{123} - \left\{H_{123}^{0\text{eff}}\, g_{123} - g_{123}\, H_{123}^{0\text{eff}\,\dagger}\right\} = [V_{13}+V_{23}, F_3\, g_{12}]} \tag{7.1}$$

where the effective Hamiltonian is, in general, non-hermitean with $H_{123}^{0\text{eff}} = \bar{H}_1+\bar{H}_2+\bar{H}_3$ (the index "0" indicates the neglect of the interaction potentials)[1], and the definition of $\bar{H}_1$ will be given later.[2] Eq. (7.1) is solved for g_{123} in similar way as we solved the second equation in Sec. 6.1 (here it is convenient to use the simpler propagators $U = U^+ + U^-$, cf. Appendix D),

$$\begin{aligned} g_{123}(t) &= U_{123}^0(tt_0)\, g_{123}^0\, U_{123}^{0\dagger}(tt_0) \\ &+ \frac{1}{i\hbar}\int_{t_0}^t d\bar{t}\, U_{123}^0(t\bar{t})\, [V_{13}+V_{23}, F_3 g_{12}]|_{\bar{t}}\, U_{123}^{0\dagger}(t\bar{t}), \end{aligned} \tag{7.2}$$

where $U_{123}^0 = U_1\, U_2\, U_3$, and U_1 is defined by

$$\left\{i\hbar\frac{\partial}{\partial t} - \bar{H}_1\right\} U_1(tt') = 0, \qquad U_1(tt) = 1. \tag{7.3}$$

To shorten the notation, we will not write the term with g_{123}^0, but restore it in the final expression (7.8). Also, we will not proceed in the operator notation here (this more general treatment will be performed below in Sec. 7.4, see also [BK96]), but, for sake of transparency, use the momentum representation assuming spatial homogeneity.

Momentum representation. Introducing the matrix elements of the Hamiltonian $\bar{H}_1|p_1\rangle = \epsilon_1|p_1\rangle$, the one–particle propagator has the form

$$U_1(tt') = \exp\{-\frac{i}{\hbar}\int_{t'}^t d\tau\, \epsilon_1(\tau)\}. \tag{7.4}$$

---

[1]Notice that the r.h.s. of Eq. (7.1) is not permutation invariant in the particle indices. This is because the symmetric terms containing g_{13} and g_{23} have been dropped since they do not contribute to the renormalization of the Hamiltonian H_{12} in the equation for g_{12}. These terms are generalized polarization terms which are beyond the static Born approximation. This asymmetry indicates that there could be problems with the energy conservation of this approximation. We will return to this question in Sec. 7.3.2.

[2]The original form of the third hierarchy equation (2.99) contains the $H_{123}^0 = H_1^0 + H_2^0 + H_3^0$ (the three-particle ladder terms are beyond the Born approximation). This is one possible choice, however, a consistent treatment will require a renormalization $H_{123}^0 \to H_{123}^{0\text{eff}}$, which we derive below. We further mention that an analogous approach for the simpler case of carrier-phonon scattering has been given by Kuhn, Schilp and Mahler, [SKM94, SKM95].

The momentum representation of the solution (7.2) is given by (we denote $p_1 \to 1$ and use $1+2+3 = 1'+2'+3'$ due to homogeneity)

$$
\begin{aligned}
g_{123}(123;1'2'3',t) &= \frac{1}{i\hbar}\int_0^{t-t_0} d\tau e^{-\frac{i}{\hbar}\int\limits_{t_o}^{\tau} d\bar\tau\left(\epsilon_{123}(\bar\tau)-\epsilon^{*'}_{123}(\bar\tau)\right)} V(3-3') \qquad (7.5)\\
&\times \Big\{f_3'\Big(g(1+3-3',2;1',2')+g(1,2+3-3';1',2')\Big)\\
&- f_3\Big(g(1,2;1'+3'-3,2')+g(1,2;1',2'+3'-3)\Big)\Big\}|_{t-\tau},
\end{aligned}
$$

with $\epsilon_{123} = \epsilon_1+\epsilon_2+\epsilon_3$. Now we insert expression (7.5) for g_{123} on the r.h.s. of the second hierarchy equation (2.99). The result consists of 16 terms:

$$
\begin{aligned}
i\hbar\frac{\partial}{\partial t}g(1,2,1',2') &= \ldots + \frac{1}{i\hbar}\int_0^{t-t_0} d\tau \int \frac{d3}{(2\pi\hbar)^3}\frac{dq}{(2\pi\hbar)^3}V^2(q)\times\Bigg\{\\
& e^{-\frac{i}{\hbar}\int\limits_{t_0}^{\tau} d\bar\tau\left(\epsilon_{1-q}(\bar\tau)+\epsilon_2(\bar\tau)+\epsilon_{3+q}(\bar\tau)-\epsilon^*_{1'}(\bar\tau)-\epsilon^*_{2'}(\bar\tau)-\epsilon^*_3(\bar\tau)\right)}\\
&\times\Big[f_3\left(g(1,2;1',2')+g(1-q,2+q;1',2')\right)\\
&\quad -f_{3+q}\left(g(1-q,2;1'-q,2')+g(1-q,2;1',2'-q)\right)\Big]\\
&+e^{-\frac{i}{\hbar}\int\limits_{t_0}^{\tau} d\bar\tau\left(\epsilon_1(\bar\tau)+\epsilon_2(\bar\tau)+\epsilon_{3+q}(\bar\tau)-\epsilon^*_{1'-q}(\bar\tau)-\epsilon^*_{2'}(\bar\tau)-\epsilon^*_{3+q}(\bar\tau)\right)}\\
&\times\Big[f_3\left(g(1,2;1',2')+g(1,2;1'-q,2'+q)\right)\\
&\quad -f_{3+q}\left(g(1-q,2;1'-q,2')+g(1,2-q;1'-q,2')\right)\Big]\\
&+e^{-\frac{i}{\hbar}\int\limits_{t_0}^{\tau} d\bar\tau\left(\epsilon_1(\bar\tau)+\epsilon_{2-q}(\bar\tau)+\epsilon_{3+q}(\bar\tau)-\epsilon^*_{1'}(\bar\tau)-\epsilon^*_{2'}(\bar\tau)-\epsilon^*_3(\bar\tau)\right)}\\
&\times\Big[f_3\left(g(1,2;1',2')+g(1+q,2-q;1',2')\right)\\
&\quad -f_{3+q}\left(g(1,2-q;1'-q,2')+g(1,2-q;1',2'-q)\right)\Big]\\
&+e^{-\frac{i}{\hbar}\int\limits_{t_o}^{\tau} d\bar\tau\left(\epsilon_1(\bar\tau)+\epsilon_2(\bar\tau)+\epsilon_3(\bar\tau)-\epsilon^*_{1'}(\bar\tau)-\epsilon^*_{2'-q}(\bar\tau)-\epsilon^*_{3+q}(\bar\tau)\right)}\\
&\times\Big[f_3\left(g(1,2;1',2')+g(1,2;1'+q,2'-q)\right)\\
&\quad -f_{3+q}\left(g(1-q,2;1',2'-q)+g(1,2-q;1',2'-q)\right)\Big]\quad\Bigg\}, \qquad (7.6)
\end{aligned}
$$

where on the r.h.s., f and g are to be taken at the time $t - \tau$.

Selfenergy terms. Of these 16 terms, in four (first terms after each exponential) g has the same momentum arguments as on the l.h.s. and may, therefore, be taken out of the p_3 and q integrals, which will allow us to combine them with the Hamilton operator H_{12}^0 on the l.h.s. This gives rise to a renormalization of H_{12}^0, what is nothing but a selfenergy effect. Neglecting the remaining 12 terms[3], we introduce an operator $\tilde{\Sigma}^{(12)}$, which allows us to rewrite the four selfenergy terms as

$$\tilde{\Sigma}^{(12)}(12;1'2',t)\, g(12;1'2',t) =$$
$$\frac{1}{i\hbar}\int_0^{t-t_0} d\tau g(12;1'2',t-\tau) \int \frac{d3}{(2\pi\hbar)^3}\frac{dq}{(2\pi\hbar)^3} V^2(q) f_3(t-\tau) \times$$
$$\left\{ e^{-\frac{i}{\hbar}\int\limits_{t_0}^{\tau} d\bar\tau \left(\epsilon_{1-q}+\epsilon_2+\epsilon_{3+q}-\epsilon^*_{1'}-\epsilon^*_{2'}-\epsilon^*_3\right)\big|_{\bar\tau}} + e^{-\frac{i}{\hbar}\int\limits_{t_0}^{\tau} d\bar\tau \left(\epsilon_1+\epsilon_2+\epsilon_3-\epsilon^*_{1'-q}-\epsilon^*_{2'}-\epsilon^*_{3+q}\right)\big|_{\bar\tau}} + \right.$$
$$\left. e^{-\frac{i}{\hbar}\int\limits_{t_0}^{\tau} d\bar\tau \left(\epsilon_1+\epsilon_{2-q}+\epsilon_{3+q}-\epsilon^*_{1'}-\epsilon^*_{2'}-\epsilon^*_3\right)\big|_{\bar\tau}} + e^{-\frac{i}{\hbar}\int\limits_{t_0}^{\tau} d\bar\tau \left(\epsilon_1+\epsilon_2+\epsilon_3-\epsilon^*_{1'}-\epsilon^*_{2'-q}-\epsilon^*_{3+q}\right)\big|_{\bar\tau}} \right\}. \quad (7.7)$$

Due to symmetry in the momentum arguments, $\tilde{\Sigma}^{(12)}$ may be decomposed according to

$$\tilde{\Sigma}^{(12)}(12;1'2',t) = \tilde{\Sigma}_1(t) + \tilde{\Sigma}_2(t) - \tilde{\Sigma}_1'^*(t) - \tilde{\Sigma}_2'^*(t) \qquad \text{with}$$
$$\tilde{\Sigma}_1(t) g(12;1'2',t) = \tilde{\Sigma}_1^{IC}(t) g(12;1'2',t) + \frac{1}{i\hbar}\int_0^{t-t_0} d\tau g(12;1'2',t-\tau)$$
$$\times \int \frac{d3}{(2\pi\hbar)^3}\frac{dq}{(2\pi\hbar)^3} V^2(q) f_3(t-\tau) e^{\frac{i}{\hbar}\int\limits_{t_0}^{\tau} d\bar\tau \left(\epsilon_{1-q}+\epsilon_2+\epsilon_{3+q}-\epsilon^*_{1'}-\epsilon^*_{2'}-\epsilon^*_3\right)\big|_{\bar\tau}}, \quad (7.8)$$

where the complex conjugation in $\tilde{\Sigma}_1'$ and $\tilde{\Sigma}_2'$ is understood as not to affect g under the integral. By adding the term $\tilde{\Sigma}_1^{IC}$, we restored the initial three–particle correlations from Eq. 7.2. Consistent with approximation (7.1), they are of the form $g_{123}^0 \sim [V_{13}+V_{23}, F_3(t_0)g_{12}(t_0)]$, so $\tilde{\Sigma}^{IC}$ is completely determined by the initial values of F_1 and g_{12}, $\tilde{\Sigma}_1^{IC}(t)g(t) \sim \int d3dqV^2(q)f_3(t_0)g(t_0)$.

Energy renormalization. Using these definitions, we rewrite the full equation of motion for the matrix elements of g_{12}, Eq. (7.6),

$$\left\{ i\hbar\frac{\partial}{\partial t} - (E_1^0 + E_2^0 - E_{1'}^0 - E_{2'}^0) \right\} g(12;1'2',t) =$$

[3]They do not have this property and, therefore, do not contribute to selfenergy diagrams, although they are of the same order in the interaction and the density

$$V(1-1')\left\{f_1'f_2'(1\pm f_1)(1\pm f_2)-f_1f_2(1\pm f_1')(1\pm f_2')\right\}-$$
$$\left\{\tilde{\Sigma}_1(t)+\tilde{\Sigma}_2(t)-\tilde{\Sigma}_1'^*(t)-\tilde{\Sigma}_2'^*(t)\right\}g(12;1'2',t). \tag{7.9}$$

Eq. (7.9) contains spin statistics effects (exchange and Pauli blocking) as derived in Ch. 3. To be fully consistent with that, spin statistics effects have to be incorporated into the selfenergy terms too. As was already shown in Sec. 3.4.1, the closure (7.1) must then be generalized according to

$$\boxed{V_{13}\,F_3\,g_{12}\longrightarrow(\hat{V}_{13}\,F_3\mp F_1F_3V_{13})\,g_{12}} \tag{7.10}$$

and we have further to take into account the exchange factor $1\pm P_{13}\pm P_{23}$ in the last term of the (anti-)symmetrized second hierarchy equation (3.22). The explicit result including Pauli blocking and exchange is then

$$\begin{aligned}\tilde{\Sigma}_1(t)g(12;1'2',t)=\frac{1}{i\hbar}\int_0^{t-t_0}d\tau\int\frac{d3}{(2\pi\hbar)^3}\frac{d\bar{1}}{(2\pi\hbar)^3}\frac{d\bar{3}}{(2\pi\hbar)^3}(2\pi\hbar)^3\delta(1+3-\bar{1}-\bar{3})\\ \times\Big(V(1-\bar{1})\pm V(1-\bar{3})\Big)e^{\frac{i}{\hbar}\int_{t_0}^{\tau}d\bar{\tau}\left(\bar{\epsilon}_1+\epsilon_2+\bar{\epsilon}_3-\epsilon^*_{1'}-\epsilon^*_{2'}-\epsilon^*_3\right)\big|_{\bar{\tau}}}g(12;1'2',t-\tau)\\ \times\Big(\bar{f}_3^<\bar{f}_1^<f_3^>\mp\bar{f}_3^>\bar{f}_1^>f_3^<\Big)\Big|_{t-\tau}\{i\hbar\delta(t-t_0-\tau)+V(1-\bar{1})\}\end{aligned} \tag{7.11}$$

where the initial correlation term is included under the time integral by means of the time delta function, and, in accordance with the definition of the operators $F^{\gtrless}$, Eq. (3.18), we defined $f^<=f$ and $f^>=1\pm f$.

Finally, Eq. (7.9) allows us to directly identify the renormalized one–particle and two–particle Hamiltonians, and also the so far unknown three-particle Hamiltonian, $\bar{H}_1$, $H_{12}^{0\text{eff}}$ and $H_{123}^{0\text{eff}}$:

$$\boxed{\begin{aligned}H_{12}^{0\text{eff}}&=\bar{H}_1+\bar{H}_2\\ H_{123}^{0\text{eff}}&=\bar{H}_1+\bar{H}_2+\bar{H}_3\\ \bar{H}_1\,g(12;1'2',t)&=\left\{H_1^0+\tilde{\Sigma}_1(t)\right\}g(12;1'2',t)\end{aligned}} \tag{7.12}$$

Eq. (7.12) shows that $\tilde{\Sigma}_1$ in fact renormalizes the bare one-particle Hamiltonian H_1^0, and thus is related to selfenergy effects. The new Hamiltonian $\bar{H}_1$ is, in general, complex, containing an energy shift Δ_1 with respect to the kinetic energy, which is related to $\text{Re}\tilde{\Sigma}_1$, and a broadening (damping) γ_1 which leads to a finite lifetime of the one-particle state reflecting the influence of the surrounding medium on the particle. This damping correction is given by $\text{Im}\tilde{\Sigma}_1$.

7.2 Renormalized binary correlation operator

Having determined the renormalized one–particle Hamiltonian and its eigenvalue ϵ_1 we now derive the correlation operator with renormalization effects. Instead of Eq. (6.5), we have to solve Eq. (7.9), with the initial condition $g_{12}(t_0) = g^0$,

$$i\hbar\frac{\partial}{\partial t}g_{12} - \left(\bar{H}_1 + \bar{H}_2\right) g_{12} + g_{12}\left(\bar{H}_1^\dagger + \bar{H}_2^\dagger\right) = \left\{\hat{V}_{12}F_1F_2 - F_1F_2\hat{V}_{12}^\dagger\right\}, \quad (7.13)$$

which differs from Eq. (6.5) only by the substitution $\bar{H}_1^0 \to \bar{H}_1$[4]. Thus, we can use the results of Ch. 6, where only the free propagators $U^{0\pm}$ are replaced by the renormalized propagators $U^\pm$ which follow from the equation

$$\left\{i\hbar\frac{\partial}{\partial t} - \bar{H}_1\right\} U_1^+(tt') = i\hbar\,\delta(t-t'), \quad (7.14)$$

and its adjoint, respectively, and $\bar{H}_1$ is given by Eq. (7.12) with $g_{12} \to U_1^+$. Inserting $U^\pm$ into Eq. (6.4), we obtain the renormalized two-particle correlation operator,

$$\begin{aligned} g_{12}(t) &= U_1^+(tt_0)\,U_2^+(tt_0)\;g^0\;U_1^-(t_0t)\,U_2^-(t_0t) \qquad (7.15)\\ &+ \frac{1}{i\hbar}\int_{t_0}^{\infty} d\bar{t}\,U_1^+(t\bar{t})\,U_2^+(t\bar{t})\left\{\hat{V}_{12}F_1F_2 - F_1F_2\hat{V}_{12}^\dagger\right\}|_{\bar{t}}\;U_1^-(\bar{t}t)\,U_2^-(\bar{t}t), \end{aligned}$$

where again g^0 denotes arbitrary initial binary correlations.

Local approximation for $U^\pm$. The physical properties of the renormalized propagators become particularly transparent in the local approximation, i.e. if we approximate $U^\pm(tt') \approx U^\pm(t-t')$. Then Eq. (7.14) has a simple solution (details of the derivation are given in Appendix D) which in momentum representation reads [here $U_1 = U(p_1)$]

$$U_1^\pm(\tau) = \Theta(\pm\tau)e^{-\frac{i}{\hbar}(E_1 \mp i\gamma_1)\tau}, \quad (7.16)$$

what generalizes the previous result (6.6). One clearly sees the difference to the free propagators $U^{0\pm}$: $U^\pm$ are *damped quasi-particle propagators* with a shifted one-particle energy $E_1 = E_1^0 + \Delta_1$ and a damping γ_1 that limits the life time of the one-particle states.[5] Shift and damping are consequences of the influence of the surrounding particles on particle "1". We have determined

[4]Since $\bar{H}_1$ is non-hermitean, it does not lead to a commutator on the l.h.s.
[5]See also Sec. 12.6, e.g. Fig. 12.12.

the energy shift and the damping selfconsistently by taking into account the relevant contributions from the third hierarchy equation. Using the local approximation, Eq. (7.16), we obtain for the matrix elements in the homogeneous case $(1+2=1'+2')$

$$\boxed{\begin{aligned} g(12;1'2',t) &= e^{\frac{i}{\hbar}(E_{12}-E'_{12})(t-t_0)}\, e^{-\frac{1}{\hbar}(\gamma_{12}+\gamma'_{12})(t-t_0)}\, g^0(12;1'2') \\ &+ \frac{1}{i\hbar}\int_{t_0}^{t} d\bar{t}\, e^{\frac{i}{\hbar}(E_{12}-E'_{12})(t-\bar{t})}\, e^{-\frac{1}{\hbar}(\gamma_{12}+\gamma'_{12})(t-\bar{t})} \\ &\times V(1-1')\left\{f_1^{<'}f_2^{<'}f_1^{>}f_2^{>} - f_1^{>'}f_2^{>'}f_1^{<}f_2^{<}\right\}|_{\bar{t}} \end{aligned}} \tag{7.17}$$

where $E_{12}=E_1+E_2$ and $\gamma_{12}=\gamma_1+\gamma_2$, and we used the definition (3.18).

With the renormalized propagators we have obtained the expected evolution behavior of the correlations: Due to the damping effects, the initial correlation term decays and vanishes for times larger than a characteristic time, the correlation time τ_{cor}. On the other hand, the scattering induced correlations (integral term) are zero close to the initial moment, and are being built up on the same scale τ_{cor}. Eq. (7.17) allows us to give a qualitative estimate for the *correlation time*[6]

$$\tau_{cor} \sim \frac{\hbar}{\gamma_{12}}, \tag{7.18}$$

as the lifetime of one-particle states (see also Fig. 7.3.).

This is a rather general result, which remains qualitatively the same if one goes beyond the local approximation. In general, the solution for the matrix elements of g_{12} can be written in the following form which replaces Eq. (7.17),

$$\begin{aligned} g(12;1'2',t) &= U_1^+(tt_0)U_2^+(tt_0)U_{1'}^-(t_0t)U_{2'}^-(t_0t)\; g^0(12;1'2') \\ &+ \frac{1}{i\hbar}\int_{t_0}^{\infty} d\bar{t}\, U_1^+(t\bar{t})U_2^+(t\bar{t})U_{1'}^-(\bar{t}t)U_{2'}^-(\bar{t}t) \\ &\times V(1-1')\left\{f_1^{<'}f_2^{<'}f_1^{>}f_2^{>} - f_1^{>'}f_2^{>'}f_1^{<}f_2^{<}\right\}|_{\bar{t}}. \end{aligned} \tag{7.19}$$

Moreover, this general behavior of the binary correlations remains valid also if one considers more complex hierarchy closures, as we will see in the next chapters. The important result is that we are now able to describe the dynamics of a many-particle systems on the whole time range, starting from arbitrary short times where initial correlations dominate to long times, where the system behavior is governed by scattering induced correlations.

[6]Strictly speaking, we have to write $\gamma_{12} \to \tilde{\gamma}_{12} = \gamma(\tilde{p}_1,\tilde{t}) + \gamma(\tilde{p}_2,\tilde{t})$, where $\tilde{p}_{1,2}$ are some average momentum values, but we will see below (Fig. 7.6.) that the momentum dependence is rather weak. Also, γ depends on time. But it turns out that for times larger than the correlation time, there is only small further change.

7.3 Non-Markovian quantum Landau equation with selfenergy

For the derivation of the renormalized non-Markovian Landau equation we again consider the homogeneous case and omit the Hartree-Fock contribution[7]. Proceeding exactly as in Ch. 6, we insert the solution $g_{12}(t)$ on the r.h.s. of the first hierarchy equation (6.1) and obtain

$$\frac{d}{dt} f(\mathbf{p}_1, t) = I(\mathbf{p}_1, t) + I^{IC}(\mathbf{p}_1, t), \tag{7.20}$$

with the renormalized collision integrals

$$\begin{aligned} I(\mathbf{p}_1, t)(t) = \frac{2}{\hbar^2} \int_0^{t-t_0} d\tau \int \frac{d\mathbf{p}_2}{(2\pi\hbar)^3} \int \frac{d\bar{\mathbf{p}}_1}{(2\pi\hbar)^3} \int \frac{d\bar{\mathbf{p}}_2}{(2\pi\hbar)^3} (2\pi\hbar)^3 \delta(\mathbf{p}_{12} - \bar{\mathbf{p}}_{12}) \\ \times V(\bar{\mathbf{p}}_1 - \mathbf{p}_1)\Big(V(\mathbf{p}_1 - \bar{\mathbf{p}}_1) \pm V(\mathbf{p}_1 - \bar{\mathbf{p}}_2)\Big) U_1^+(t\tau)\, U_2^+(t\tau)\, \bar{U}_1^-(\tau t)\, \bar{U}_2^-(\tau t) \\ \times \Big\{\bar{f}_1 \bar{f}_2 [1 \pm f_1][1 \pm f_2] \; - \; f_1 f_2 [1 \pm \bar{f}_1][1 \pm \bar{f}_2]\Big\}|_{t-\tau}, \end{aligned} \tag{7.21}$$

and, the contribution from the initial correlations,

$$\begin{aligned} I^{IC}(\mathbf{p}_1, t - t_0) \;&=\; -\frac{2}{\hbar} \int \frac{d\mathbf{p}_2}{(2\pi\hbar)^3} \int \frac{d\bar{\mathbf{p}}_1}{(2\pi\hbar)^3} \int \frac{d\bar{\mathbf{p}}_2}{(2\pi\hbar)^3} (2\pi\hbar)^3 \delta(\mathbf{p}_{12} - \bar{\mathbf{p}}_{12}) \\ &\times \;\; \Big(V(\mathbf{p}_1 - \bar{\mathbf{p}}_1) \pm V(\mathbf{p}_1 - \bar{\mathbf{p}}_2)\Big) \\ &\times \;\; \mathrm{Im}\Big\{U_1^+(tt_0)\, U_2^+(tt_0)\, \bar{U}_1^-(t_0 t)\, \bar{U}_2^-(t_0 t) g^0(\mathbf{p}_1, \mathbf{p}_2; \bar{\mathbf{p}}_1, \bar{\mathbf{p}}_2)\Big\}, \end{aligned} \tag{7.22}$$

where U^+ obeys equation (7.14).

In the *local approximation* (7.16), we obtain the simpler collision integrals

$$\begin{aligned} I(\mathbf{p}_1, t) = \frac{2}{\hbar^2} \int_0^{t-t_0} d\tau \int \frac{d\mathbf{p}_2}{(2\pi\hbar)^3} \int \frac{d\bar{\mathbf{p}}_1}{(2\pi\hbar)^3} \int \frac{d\bar{\mathbf{p}}_2}{(2\pi\hbar)^3} (2\pi\hbar)^3 \delta(\mathbf{p}_{12} - \bar{\mathbf{p}}_{12}) \\ \times V(\bar{\mathbf{p}}_1 - \mathbf{p}_1)\Big(V(\mathbf{p}_1 - \bar{\mathbf{p}}_1) \pm V(\mathbf{p}_1 - \bar{\mathbf{p}}_2)\Big) \cos\Big\{\frac{E_{12} - \bar{E}_{12}}{\hbar}\tau\Big\} e^{-\frac{\gamma_{12}+\bar{\gamma}_{12}}{\hbar}\tau} \\ \times \Big\{\bar{f}_1 \bar{f}_2 [1 \pm f_1][1 \pm f_2] \; - \; f_1 f_2 [1 \pm \bar{f}_1][1 \pm \bar{f}_2]\Big\}|_{t-\tau}, \end{aligned} \tag{7.23}$$

and, the contribution from the initial correlations,

$$I^{IC}(\mathbf{p}_1, t - t_0) \;=\; -\frac{2}{\hbar} \int \frac{d\mathbf{p}_2}{(2\pi\hbar)^3} \int \frac{d\bar{\mathbf{p}}_1}{(2\pi\hbar)^3} \int \frac{d\bar{\mathbf{p}}_2}{(2\pi\hbar)^3} (2\pi\hbar)^3 \delta(\mathbf{p}_{12} - \bar{\mathbf{p}}_{12})$$

[7] It is no problem to consider the inhomogeneous case. Then the Hartree-Fock term has to be included in the equation of motion for $U^{\pm}$, Eq. (7.14), and one has to use the coordinate or the Wigner representation, Sec. 2.3.

$$\times \quad \left(V(\mathbf{p}_1 - \bar{\mathbf{p}}_1) \pm V(\mathbf{p}_1 - \bar{\mathbf{p}}_2)\right) e^{-(\gamma_{12}+\bar{\gamma}_{12})(t-t_0)/\hbar}$$
$$\times \quad \mathrm{Im}\left\{e^{-\frac{i}{\hbar}(E_{12}-\bar{E}_{12})(t-t_0)} g^0(\mathbf{p}_1, \mathbf{p}_2; \bar{\mathbf{p}}_1, \bar{\mathbf{p}}_2)\right\}, \qquad (7.24)$$

which are the generalization of Eqs. (6.8), (6.9). Eq. (7.23) agrees with the result which was derived from the Kadanoff-Baym equations by applying the generalized Kadanoff-Baym ansatz (GKBA) [LvV86] in [Kuz91, HE92], cf. Sec. 12.6. Here, we obtained the result *without postulating the GKBA*, based on the BBGKY-hierarchy with the generalized closure relation (7.1).

Thus, the problem is solved. The generalized Landau equation is given by a closed system of equations: the kinetic equation (7.20) with the non-Markovian collision integral (7.21), the additional integral (7.22) and the expression for the selfenergy (7.11), which are coupled in a complicated way. The energy shift and the damping coefficient are momentum and time dependent and appear under the retardation integral in the collision terms and in $\tilde{\Sigma}$ itself and have to be calculated selfconsistently with the distribution function.

7.3.1 Properties of the Landau equation. Memory effects

Let us summarize the properties of the generalized Landau equation with the collision integrals (7.21,7.22) or (7.23,7.24), respectively.

i) The integrals conserve density and momentum.

ii) *Retardation (memory) effects*: The collision integrals (7.21) and (7.23) contain retardation in the distribution functions, i.e. $I(t)$ depends on the values of f at all previous times. The relative weight of the previous values decreases with growing retardation. The reason is the damping effect contained in the renormalized propagators which leads to a finite effective duration of the memory. This non-local (non-Markovian) time dependence is, however, only a formal consequence of the solution procedure: The original coupled equations for F_1 and g_{12} are local in time, only the formal solution for g_{12} introduces the non-locality. We will consider the properties of the coupled equations below in this section.

iii) *Energy broadening:* The collision integrals (7.21) and (7.23) contain collisional energy broadening of the form $\cos\{(E_{12} - \bar{E}_{12})t/\hbar\}$ instead of the energy delta function which appears in conventional Markovian kinetic equations. This is an effect which is related to the finite collision duration which also exists in classical systems where it is given by $\cos\{(kv - \bar{k}\bar{v})t/\}$ [Kli75]. This effect is important on short times, $t < \tau_{cor}$, when it allows scattering events which do not conserve the one-particle energies, for example scattering into high momentum states.

iv) *Time Reversibility:* (1) The non-Markovian Landau equation without selfenergy ($\Delta = \gamma = 0$) is time-reversible. (2) With energy renormalization effects included, Eq. (7.21) remains reversible as long as no approximations to the time structure have been made: In particular, if the propagators are exact solutions of Eq. (7.14). Any approximation to $\tilde{\Sigma}$ which is based on a retardation expansion or on a local approximation for the propagators immediately breaks the time symmetry and introduces irreversibility.

v) *Energy conservation:* (1) If selfenergy effects are neglected, the collision integrals (7.23), (7.24) conserve total energy. (2) The local approximation for the selfenergy is not conserving. (3) The integrals with renormalization effects included conserve energy only if the exact propagators are used (this follows from the general properties of the hierarchy closure, cf. Sec. 2.6.1). This question will be investigated in more detail in Sec. 7.3.2.

vi) *Slowing down of the relaxation:* The non-Markovian collision integrals (7.23) and (7.21) cause a slowing down of the relaxation compared to conventional Markovian equations, since for early moments the time integration interval is small, cf. Fig. 7.1. On the other hand, the collision terms which arise from initial correlations start with non-zero values (except for $g^0 = 0$) and compensate this effect.

vii) *Short-time behavior:* The behavior of the system on short times ($t < \tau_{cor}$) deviates strongly from the conventional kinetic relaxation. It corresponds to the coupled simultaneous evolution of one-particle and two-particle quantities. Thus, the equations of motion effectively contain second derivatives in time in contrast to Markovian equations which are of first order. One, therefore, can expect oscillatory effects in the time dependence, which, are indeed observed if selfenergy effects are neglected, see Ch. 6, Fig. 6.2. The proper inclusion of the latter, however, suppresses oscillatory behavior in carrier-carrier scattering, see below.

viii) *Long-time behavior:* For long times $t \gg \tau_{cor}$ memory effects are negligible. Nevertheless, the relaxation may strongly deviate from the conventional Markovian description. The reason is that correlations built up at early times remain in the system and, in the case of strong interaction between the particles, may cause significant changes in the macroscopic behavior, including the equilibrium distribution and correlation functions. This is most easily seen from a retardation expansion, see Sec. 7.3.3.

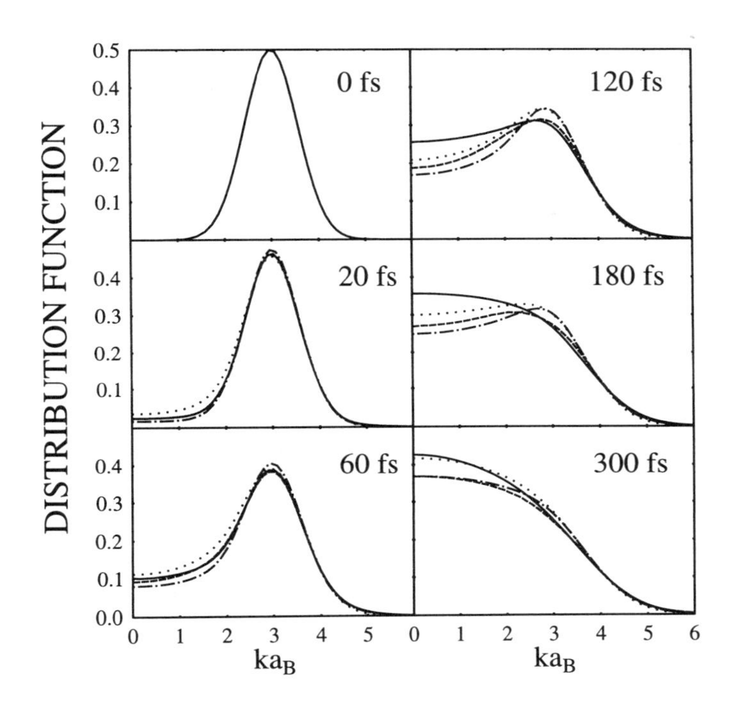

Fig. 7.1.: Relaxation of the Wigner distribution function for a bulk semiconductor with statically screened Coulomb interaction ($n = 3.64 \times 10^{17} cm^{-3}, \kappa = 1.16/a_B$) for different scattering models: Markovian Landau integral, cf. Eq. (7.48), (dotted line); zeroth order retardation approximation, cf. Eq. (7.43), (full line); non-Markovian Landau integral (7.21) with full retardation but no selfenergy (dashes), and Kadanoff-Baym equations, cf. Sec. 12.4, (dash-dotted line), [BKS+96].

Figs. 7.1 and 7.2. illustrate these properties comparing the relaxation calculated with different approximations for the collision integral in static Born approximation. The interaction potential is a Debye (Yukawa) potential $V(q) = 4\pi e^2/\epsilon_b(q^2 + \kappa^2)$. The differences are small if one takes for κ its equilibrium value. This of course overestimates screening neglecting the fact that screening is built up on the same time scale. These problems will be discussed in Chs. 9 and 12. Fig. 7.2. compares the relaxation of kinetic energy for the four models. While kinetic energy is conserved by the Markovian equation, it increases in the other cases. This reflects the buildup of correlations which leads to an increase of correlation energy in the system (which is negative due to the attractive interaction),[8] therefore, due to conservation of total energy, we observe an increase of kinetic energy. Interestingly, the zeroth order retardation approximation (b), where *memory and selfenergy are neglected*, but energy broadening effects are retained, cf. Eq. (7.43), is very close to the Kadanoff-Baym equations (which will be discussed in Ch. 12). Notice that for the full non-Markovian calculation without selfenergy (c) a continuous increase is observed. This is related to scattering of particles into high momentum states and a broadening of the distribution.

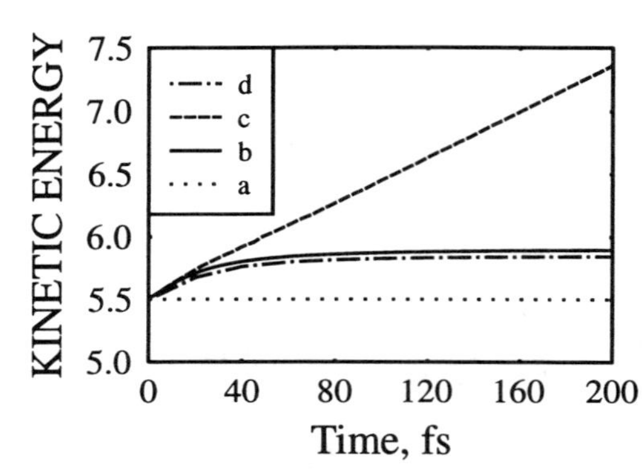

Fig. 7.2. Relaxation of kinetic energy density (in E_R/a_B^3) for the calculations of Fig. 7.1. a: Markov limit, b: Non–Markovian Landau equation with selfenergy (cf. Fig. 7.6), c: non–Markovian Landau equation without selfenergy, d: Kadanoff-Baym equations (see Ch. 12), [BKS+96].

In the following, we discuss the relaxation properties of the non-Markovian Landau equation more in detail. To simplify the notation, we introduce for the spectral kernel in the collision integral (7.23) the abbreviation

$$D(12;1'2',\omega,\Gamma,t) = \cos\{\omega t\}e^{-\Gamma t}, \tag{7.25}$$

where we denoted

$$\omega(12;1'2',t) = (E_{12} - E'_{12})/\hbar, \tag{7.26}$$

$$\Gamma(12;1'2',t) = (\gamma_{12} + \gamma'_{12})/\hbar. \tag{7.27}$$

Notice that ω and Γ are (weakly) time dependent via the energy renormalization. Further we write for the combination of the distributions in Eq. (7.23),

$$\Phi(12;1'2',t) = \left\{f'_1 f'_2[1 \pm f_1][1 \pm f_2] \;-\; f_1 f_2[1 \pm f'_1][1 \pm f'_2]\right\}\Big|_t. \tag{7.28}$$

Correlation Dynamics. We now can analyze the evolution of the correlations more in detail, completing the qualitative analysis of Ch. 5. For

[8] In a multi-component plasma, despite repulsive interaction of identical charges, the net interaction energy is negative, assuring stability. This applies also to a "one-component" plasma, where one has to account for the neutralizing background of oppositely charged carriers.

this, it is instructive to consider, instead of the formal solution for the complex correlation matrix element (7.17), closed equations for its imaginary and real part. Separating in Eq. (7.9) real and imaginary part and denoting the term with the distribution functions on the r.h.s. of Eq. (7.9) by $\hbar\Psi(12;1'2',t) = V(1-1')\Phi(12;1'2',t)$, we obtain for fixed momenta

$$\frac{d}{dt}\mathrm{Im}\, g = \omega\,\mathrm{Re}\, g - \Gamma\,\mathrm{Im}\, g - \Psi, \tag{7.29}$$

$$\frac{d}{dt}\mathrm{Re}\, g = -\Gamma\,\mathrm{Re}\, g - \omega\,\mathrm{Im}\, g. \tag{7.30}$$

Differentiating Eqs.(7.29) with respect to time, there follow closed equations of motion for the real and imaginary part, respectively:

$$\frac{d^2}{dt^2}\mathrm{Im}\, g + 2\Gamma\,\frac{d}{dt}\mathrm{Im}\, g + \left(\omega^2+\Gamma^2\right)\mathrm{Im}\, g = -\Gamma\Psi + \delta_I, \tag{7.31}$$

$$\frac{d^2}{dt^2}\mathrm{Re}\, g + 2\Gamma\,\mathrm{Re}\, g + \left(\omega^2+\Gamma^2\right)\mathrm{Re}\, g = \omega\Psi + \delta_R, \tag{7.32}$$

$$\text{with}\qquad \delta_I = -\frac{d}{dt}\Gamma\,\mathrm{Im}\, g + \frac{d}{dt}\omega\mathrm{Re}\, g - \frac{d}{dt}\Psi,$$

$$\delta_R = -\frac{d}{dt}\Gamma\,\mathrm{Re}\, g - \frac{d}{dt}\omega\mathrm{Im}\, g,$$

which allow for a clear interpretation of our theoretical model. These equations are fully equivalent to the non-Markovian Landau equation, but they are local in time, i.e. Markovian. If we neglect the, in most cases, weak time dependence of Γ, ω and Ψ, the corrections δ_I and δ_R vanish, and we have two equations of damped quasi-harmonic oscillations which are subject to a slowly varying external force defined by the distribution functions. The oscillation frequency is given by $\Omega = [(\omega^2+\Gamma^2)-\Gamma^2]^{1/2} = (E_{12}-E'_{12})/\hbar$, whereas the damping coefficient is just the sum of the one-particle damping coefficients Γ. Both are weakly time-dependent via their functional dependencies on the distribution functions. The solution of Eqs. (7.31,7.32) is

$$\begin{aligned}\mathrm{Re}\, g(t,\omega,\Gamma) &= \Psi(t)C_0(\omega,\Gamma,t) + \\ &\quad \left\{\mathrm{Re}\, g^0\,\cos[\omega(t-t_0)] - \mathrm{Im}\, g^0\,\sin[\omega(t-t_0)]\right\} e^{-\Gamma(t-t_0)},\end{aligned} \tag{7.33}$$

$$\begin{aligned}\mathrm{Im}\, g(t,\omega,\Gamma) &= -\Psi(t)D_0(\omega,\Gamma,t-t_0) + \\ &\quad \left\{\mathrm{Im}\, g^0\,\cos[\omega(t-t_0)] + \mathrm{Re}\, g^0\,\sin[\omega(t-t_0)]\right\} e^{-\Gamma(t-t_0)},\end{aligned} \tag{7.34}$$

where D_0 and C_0 are given by

$$D_0(\omega,\Gamma,t-t_0) = \frac{\Gamma}{\omega^2+\Gamma^2} + \frac{e^{-\Gamma t}}{\omega^2+\Gamma^2}\left(\omega\sin\omega t - \Gamma\cos\omega t\right), \tag{7.35}$$

$$C_0(\omega,\Gamma,t) \;=\; \frac{\omega}{\omega^2+\Gamma^2} - \frac{e^{-\Gamma t}}{\omega^2+\Gamma^2}\left(\Gamma\sin\omega t + \omega\cos\omega t\right), \qquad (7.36)$$

and correspond to the approximation where the distribution function does not change over an oscillation period. The next corrections to $\mathrm{Re}\, g$ and $\mathrm{Im}\, g$ follow from assuming $\Psi \sim t^1, t^2$ and so on, resulting in new functions $D_1, D_2, \ldots$ and $C_1, C_2 \ldots$, respectively. This expansion is identical to a retardation expansion of the non-Markovian solution for g_{12}, cf. Sec. 7.3.3.

The solution (7.33,7.34) allows for a transparent analysis of the dynamics of binary correlations: If selfenergy is neglected ($\gamma_{12} = \Delta_{12} = 0$), the oscillations would be undamped, with the frequency $\hbar^2\Omega_0^2 = E_1^0 + E_2^0 - E_{1'}^0 - E_{2'}^0$. This means, the oscillations are more rapid the more the scattering event violates kinetic energy conservation. The matrix element on the "energy shell" (where the one-particle energy is conserved) does not change in time at all. During the relaxation, the oscillations of the off-shell matrix elements become increasingly more rapid. Their contribution to the collision integral which contains a momentum average over $\mathrm{Im}\, g$ vanish due to destructive superposition, and only the on-shell terms remain. This means, the relaxation enters the "classical" kinetic stage (Markovian or Bogolyubov regime).

The account of one-particle damping gives rise mainly to damping of the oscillations of the off-shell correlation matrix elements with a characteristic decay time $1/\Gamma$. Furthermore, the functions D_o and C_o obtain a finite spectral width. For the on-shell components we find

$$\mathrm{Re}\, g(t,\omega=0,\Gamma) = \mathrm{Re}\, g^0(\omega=0,\Gamma)\, e^{-\Gamma(t-t_0)}, \qquad (7.37)$$

$$\mathrm{Im}\, g(t,\omega=0,\Gamma) = \mathrm{Im}\, g^0(\omega=0,\Gamma)\, e^{-\Gamma(t-t_0)} - \frac{\Psi(t-t_0)}{\Gamma}[1-e^{-\Gamma(t-t_0)}]. \qquad (7.38)$$

This clearly shows the damping of initial correlations (first term) and the correlation build-up (second term). With this "on–shell" approximation we exactly recovered the correlation time approximation [Bon96] which was introduced phenomenologically in Ch. 5.

7.3.2 Dynamics of physical observables. Energy conservation

The time evolution of physical quantities on arbitrary time scales is readily computed from the non-Markovian kinetic equation and the solution for $g_{12}(t)$.

Dynamics of one-particle quantities. For example, we have for the time derivative of the average of a one-particle quantity

$$\frac{d}{dt}\langle \mathcal{A}_1\rangle = 2\int \frac{d\mathbf{p}_1}{(2\pi\hbar)^3}\, \frac{\partial \mathcal{A}(\mathbf{p}_1,t)}{\partial t}\, f(\mathbf{p}_1,t) + 2\int \frac{d\mathbf{p}_1}{(2\pi\hbar)^3}\, \mathcal{A}(\mathbf{p}_1,t)\, \frac{d}{dt} f(\mathbf{p}_1,t),$$

where the first term vanishes, if $\mathcal{A}$ is not explicitly time dependent,[9] and in the second, one can either insert the r.h.s. of the kinetic equation (collision integrals) or its expression in terms of Img,

$$\begin{aligned}\frac{d}{dt}f(\mathbf{p}_1,t) &= I(\mathbf{p}_1,t)=\frac{2}{\hbar}\int\frac{d\mathbf{p}_2}{(2\pi\hbar)^3}\int\frac{d\mathbf{p}_1'}{(2\pi\hbar)^3}\int\frac{d\mathbf{p}_2'}{(2\pi\hbar)^3}\\ &\times\ \{V(\mathbf{p}_1-\mathbf{p}_1')\pm V(\mathbf{p}_1-\mathbf{p}_2')\}\\ &\times\ (2\pi\hbar)^3\delta(\mathbf{p}_1'+\mathbf{p}_2'-\mathbf{p}_1-\mathbf{p}_2)\,\mathrm{Im}\,g(\mathbf{p}_1',\mathbf{p}_2';\mathbf{p}_1,\mathbf{p}_2). \quad (7.39)\end{aligned}$$

In this form, the expression is valid for arbitrary approximations to the BBGKY-hierarchy. Also, memory effects, initial correlations and selfenergy are included via the formal solution for Im g.

Two-particle quantities. Correlation energy. Two-particle averages are given by Eq. (2.14) and contain a mean-field and a correlation contribution, $\langle\mathcal{A}_{12}\rangle=\frac{n^2}{2}\mathrm{Tr}_{12}\mathcal{A}_{12}F_1F_2\Lambda_{12}^{\pm}+\frac{n^2}{2}\mathrm{Tr}_{12}\mathcal{A}_{12}g_{12}$. The correlation term is readily obtained from the solution $g_{12}(t)$. For example, using $\mathcal{A}_{12}\rightarrow V_{12}$, yields the correlation energy density, which, is determined by the real part of the correlation matrix, and in Born approximation, with solution (7.19), is given by

$$\begin{aligned}\langle V\rangle(t) &= -\frac{N}{2\hbar}\int\frac{d\mathbf{p}_1}{(2\pi\hbar)^3}\int\frac{d\mathbf{p}_2}{(2\pi\hbar)^3}\int\frac{d\bar{\mathbf{p}}_1}{(2\pi\hbar)^3}\int\frac{d\bar{\mathbf{p}}_2}{(2\pi\hbar)^3}(2\pi\hbar)^3\delta(\mathbf{p}_{12}-\bar{\mathbf{p}}_{12})\\ &\times\Big(V(\bar{\mathbf{p}}_1-\mathbf{p}_1)\pm V(\bar{\mathbf{p}}_2-\mathbf{p}_1)\Big)\,\mathrm{Re}\left[\int_{t_0}^{\infty}d\tau U_1^+(t\tau)\,U_2^+(t\tau)\,\bar{U}_1^-(\tau t)\,\bar{U}_2^-(\tau t)\right.\\ &\quad\times\ V(\bar{\mathbf{p}}_1-\mathbf{p}_1)\left\{2\,\bar{f}_1\bar{f}_2[1\pm f_1][1\pm f_2]\right\}|_\tau\\ &\quad-\ \left.\hbar\,g^0(\bar{\mathbf{p}}_1,\bar{\mathbf{p}}_2;\mathbf{p}_1,\mathbf{p}_2)U_1^+(tt_0)\,U_2^+(tt_0)\,\bar{U}_1^-(t_0t)\,\bar{U}_2^-(t_0t)\right], \quad (7.40)\end{aligned}$$

where N is the particle number. In the special case of the local approximation for the propagators, the potential energy density simplifies to [BKS+96]

$$\begin{aligned}\langle V\rangle(t) &= -\frac{N}{2\hbar}\int\frac{d\mathbf{p}_1}{(2\pi\hbar)^3}\int\frac{d\mathbf{p}_2}{(2\pi\hbar)^3}\int\frac{d\bar{\mathbf{p}}_1}{(2\pi\hbar)^3}\int\frac{d\bar{\mathbf{p}}_2}{(2\pi\hbar)^3}(2\pi\hbar)^3\delta(\mathbf{p}_{12}-\bar{\mathbf{p}}_{12})\\ &\times\Big(V(\bar{\mathbf{p}}_1-\mathbf{p}_1)\pm V(\bar{\mathbf{p}}_2-\mathbf{p}_1)\Big)\left[\int_0^{t-t_0}d\tau e^{-(\gamma_{12}+\bar{\gamma}_{12})\tau/\hbar}\sin\Big(\frac{E_{12}-\bar{E}_{12}}{\hbar}\tau\Big)\right.\\ &\quad\times\ V(\bar{\mathbf{p}}_1-\mathbf{p}_1)\left\{2\,\bar{f}_1\bar{f}_2[1\pm f_1][1\pm f_2]\right\}|_{t-\tau}\\ &\quad-\ \left.\hbar\,e^{-(\gamma_{12}+\bar{\gamma}_{12})(t-t_0)/\hbar}\mathrm{Re}\Big\{g^0(\bar{\mathbf{p}}_1,\bar{\mathbf{p}}_2;\mathbf{p}_1,\mathbf{p}_2)e^{-\frac{i}{\hbar}(E_{12}-\bar{E}_{12})(t-t_0)}\Big\}\right]. \quad (7.41)\end{aligned}$$

[9] Otherwise, this term can be calculated using e.g. the numerically obtained solution $f(t)$.

This is the correlation energy density of a quantum system in second Born approximation, including exchange, selfenergy and initial correlations. This expression is valid for all times, including the short-time behavior and the correct asymptotic result and reduces, for classical systems without initial correlations and selfenergy, to the result of Ref. [Mor95].

Total energy conservation. Conservation of kinetic plus potential energy is readily verified explicitly for the non-Markovian Landau equation. To this end, one calculates the time derivative of kinetic energy, using the above results for one-particle averages with $\mathcal{A} \to T = p^2/2m$. We do not reproduce the lengthy calculations here[10] and mention that one recovers the properties listed in point v) of Sec. 7.3.1. In general, the explicit analysis becomes complicated by the time dependence of the selfenergy terms. While the result with the full propagators is conserving[11], additional approximations to the selfenergy terms, such as the local approximation, cf. Sec. 7.3.5, violate the exact conservation[12].

Nevertheless, it must be stressed that the (exact or approximate) recovery of total energy conservation is a tremendous progress in kinetic theory. Recall that conventional Markovian kinetic equations of the Boltzmann type conserve only kinetic energy which leads to wrong results for correlated many-particle systems. The idea that the conservation properties are related to non-Markovian behavior, was apparently first pointed out by Bärwinkel and Grossmann [BG67, Bär69] and studied in detail by Klimontovich and Ebeling [KE72], see also [Kli75]. Their main result was that total energy conservation is recovered already if the Boltzmann type collision integrals are supplemented by the first retardation correction. This leads us to the concept of *retardation expansion* which turns out to be a straightforward and intuitive concept for the derivation of approximations to non-Markovian equations and which will be studied in Sec. 7.3.3.

---

[10] Inserting into Eq. (7.39) Img in local approximation, Eq. (7.17), one can in $d\langle T\rangle/dt$ take the time derivative out of the whole expression. This has to be compensated by terms arising from the time dependence of the integrand. If selfenergy is neglected completely, all these additional terms can be shown to cancel, and the result for $\langle T\rangle$ is just minus $\langle V\rangle$ of Eq. (7.41) plus a constant, which is given by the initial value of total energy.

[11] This follows from the general conservation properties of the hierarchy studied in Secs. 2.2.2, 2.5.2: the corresponding hierarchy closure (7.1) involves only terms of commutator form.

[12] For example, a numerical treatment requires simplifications. The phenomenological choice of a constant damping γ leads to correction to total energy which is of the order $O(V^4)$, whereas the mean potential energy (7.41) is of the order $O(V^2)$. With selfconsistent approximations, one can achieve an improvement to $O(V^6)$, [BKS+96, BBK97]. On the other hand, one can use the requirement of total energy conservation directly to construct new approximations for the selfenergy.

Kinetic energy relaxation. Conservation of total energy has interesting consequences for the dynamics of physical observables, in particular for the kinetic energy $\langle T \rangle$. During the build-up of correlations also correlation energy increases (its absolute value). On the contrary, if the initial state is "overcorrelated" by some external mechanism which allows to "arrange" particles, correlation energy may decrease towards its stationary value. As a result of total energy conservation, the dynamics of the potential energy has to be compensated by the opposite trend of kinetic energy. This is, in fact, observed in the numerical analysis, as we have seen above in Fig. 7.2. On the other hand, in the long-time limit, where the system approaches the kinetic regime where conventional Markovian kinetic equations are applicable, kinetic energy is conserved. Therefore, when the kinetic energy change saturates, the relaxation switches from the initial to the kinetic stage, cf. Ch. 5. Thus, the saturation point of kinetic (or potential) energy yields a natural definition of the correlation time of nonequilibrium many-body systems [BK96], see Fig. 7.3.

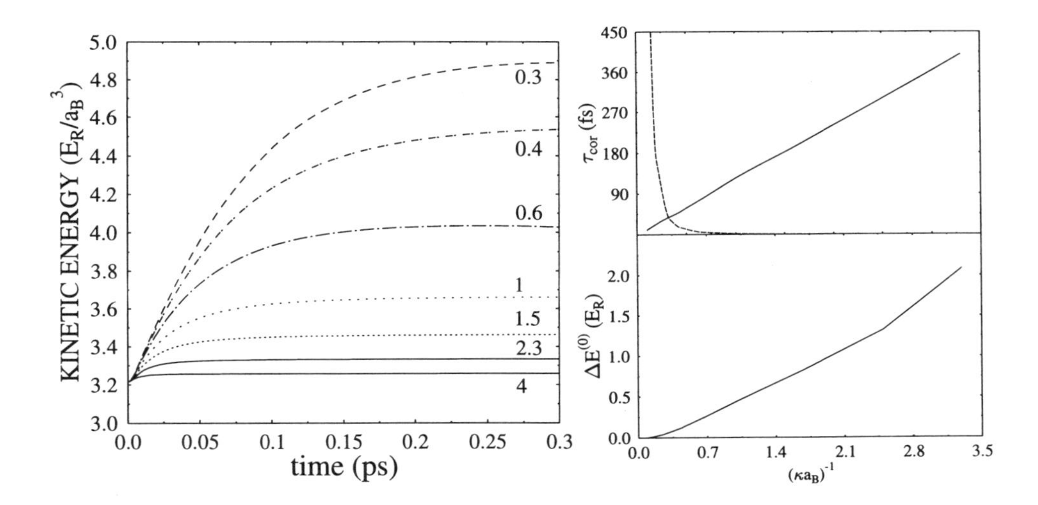

Fig. 7.3.: **(a)**: Kinetic energy density according to the zeroth retardation approximation Eq. (7.43) for different screening parameters, [BK96]. **(b)**: Nonequilibrium correlation time obtained graphically from (a) as the time of saturation of kinetic energy. The dashed line uses the alternative definition, Eq. (7.18), $\tau_{cor} = \hbar/4\gamma^{(3)}(p_0, t_0)$ with the Markovian selfenergy (see Sec. 7.3.5, Eq. (7.67)) taken at the peak momentum of the initial distribution [BK96]. **(c)**: Absolute amount of kinetic energy increase. Parameters for bulk GaAs, $n = 1.49 \times 10^{18} cm^{-3}$, initial distribution peaked around $p = 2\hbar/a_B$.

Influence of the initial correlations. In the presence of nonzero initial correlations, the relaxation behavior for $t < \tau_{cor}$ will be essentially influenced by the collision integral I^{IC}, cf. Eqs. (7.22) or (7.24). This is clearly seen on the behavior of macroscopic observables. As an example, we have shown in Fig. 7.4 the relaxation of the correlation energy for the case of zero initial correlations compared with the case of equilibrium correlations, $g^0 \to g_0^{EQ}$ given by Eq. (7.62). In the latter case, correlations are already in the system at the beginning. For more details, we refer to Ref. [KSB][13].

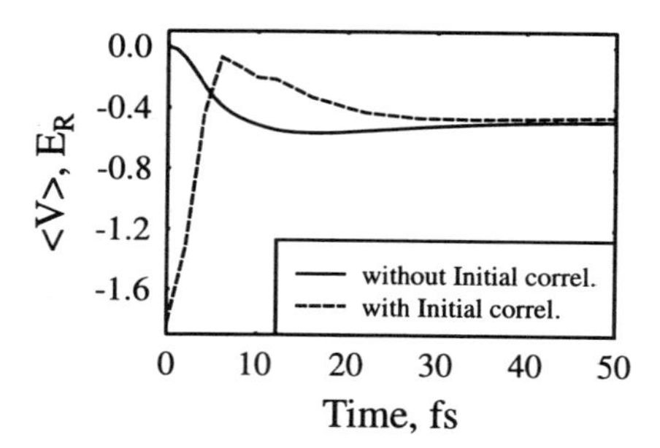

Fig. 7.4. Time evolution of correlation energy for an uncorrelated initial stated compared to the case of equilibrium initial correlations given by the first term of Eq. (7.62).

7.3.3 Markov limit and corrections. Retardation expansion

The idea of a retardation expansion is based on Bogolyubov's functional hypothesis which states that the one-particle quantities (distribution functions) change only weakly during the equilibration of the binary correlations, cf. Ch. 5. This means that the one-particle function $\Phi(t-\tau)$, Eq. (7.28), will change only little over the time interval $[t_0, \tau_{cor}]$, and under the $\tau-$integral, we may expand $\Phi(t-\tau)$ with respect to τ, around $\tau = 0$, according to

$$\Phi(t-\tau) = \Phi(t) - \tau\Phi'(t) + \ldots + \frac{(-1)^n}{n!}\tau^n\Phi^{(n)}(t) + \ldots. \tag{7.42}$$

This gives rise to a series of integrals, where in each the $\tau-$integration can be carried out easily. The full series is equivalent to the original non-Markovian integral, but, of course, the idea is to approximate it by a few terms only. The success of the retardation expansion is just based on the fact that this is indeed possible in many situations. In particular, we will see in a moment that (numbering individual terms by their respective power in τ)

a) the short-time behavior ($t \ll \tau_{cor}$) is dominated by the zeroth order term,

[13]We mention that the numerical calculation of the correlation energy in a quantum plasma is very complicated due to the large number of arguments of g_{12}. A much simpler method is given by formula (12.85) of the Green's functions approach, which is discussed in Ch. 12, and which was used to compute the data of Fig. 7.4.

b) the long-time limit of the zeroth order term yields the conventional Markovian collision integral I_0^M,

c) in most cases, the first order term already yields the dominating correlation correction,

d) in the long-time limit, $I_0^M + I_1^M$ are sufficient to guarantee total energy conservation and the correct thermodynamics of nonideal systems.

We briefly outline the derivation of the main results. Since all observables, including the collision integral are completely determined by g_{12}, for their retardation expansion it is sufficient to find the expansion of Reg and Img. In the local approximation (7.17), Img involves the retardation integral $\int_0^{t-t_0} d\tau \cos\omega\tau\, e^{-\Gamma\tau}\, \Phi(t-\tau)$, whereas in Re$g$ appears the same integral except for the substitution cos $\to$ sin. These integrals are denoted by B and A, respectively, the corresponding integrals for the case without selfenergy are denoted by B^0 and A^0. These four integrals are calculated in Appendix E, where also their limits for short and long times and the lowest order expansion terms are given. The initial correlation contribution to g_{12} remains unchanged and will not be considered here.

Retardation Expansion of Im g**.** We consider first the expansion of Im g, since it determines the non-Markovian collision integral, cf. Eq. (7.39). The imaginary part of the matrix elements of the correlation operator, Eq. (7.9), are related to the integral B of App. E by $\mathrm{Im}\, g(121'2',t) = \frac{V(1-1')}{\hbar} B(\omega,\Gamma,t)$, where $\omega(12;1'2')$ and $\Gamma(12;1'2')$ are given by Eqs. (7.26), (7.27), and we will also use the definition for Φ, Eq. (7.28). Further, we will drop the momentum arguments in ω, Γ, Φ because they are identical to those of g, and we will denote $V(1-1') \to V$.

Making use of the results for the integral B^0 of App. E, we write down the first two expansion terms for the case without selfenergy (labeled by the subscript "0" and "1"),

$$\mathrm{Im}\, g_0^0(t) = \frac{V}{\hbar}\frac{\sin\omega t}{\omega}\,\Phi(t), \tag{7.43}$$

$$\mathrm{Im}\, g_1^0(t) = \frac{V}{\hbar\omega^2}\Big[-1+\cos\omega t+\omega t\sin\omega t\Big]\frac{d}{dt}\Phi(t). \tag{7.44}$$

For *short times*, close to the initial moment t_0, the first order contribution vanishes more rapidly than the zeroth order term, and the asymptotic is given by

$$\lim_{t\to t_0} \mathrm{Im}\, g^0(t) = \lim_{t\to t_0} \mathrm{Im}\, g_0^0(t) = \frac{V}{\hbar}\, t\, \Phi(t_0) + O(t^3). \tag{7.45}$$

For larger times, but still $t < \tau_{cor}$, one has to use the full expressions (7.43), (7.44), but still the zeroth order term is expected to dominate. Eq. (7.43) is usually called *energy broadening approximation*, since it neglects retardation effects but retains the broadening of the energy delta function, or *completed collision approximation*. In particular, one can use this approximation to define the nonequilibrium correlation time [BK96], as shown in Fig. 7.3. There, the kinetic energy relaxation is shown for different values of the inverse screening length (range of the potential) κ. The saturation time of kinetic energy corresponds to the end of the initial stage and the crossover to the kinetic regime, see Ch. 5.

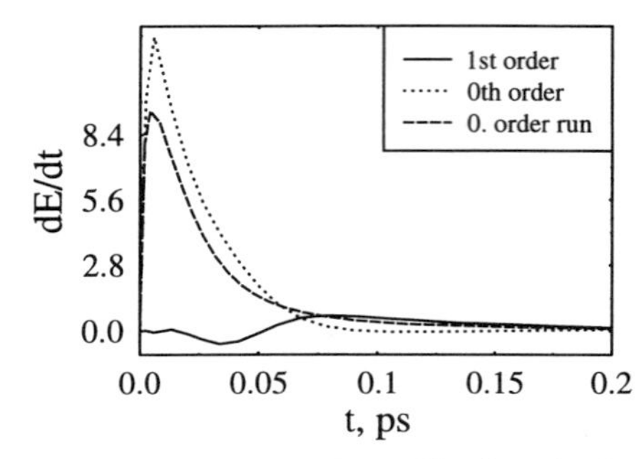

Fig. 7.5. Comparison of the zeroth and first order retardation contributions to kinetic energy density. Also, results from a pure 0th order calculation are shown. Parameters as in Fig. 7.1.

Consider now the **Markov limit** which is defined as

$$\boxed{\lim_{(t-t_0)\to-\infty} g(t) = g^M(t), \quad \text{and} \quad \lim_{(t-t_0)\to-\infty} g(t_0) = 0} \tag{7.46}$$

This means, that the initial moment is shifted back to minus infinity[14] and, at the same time, initial correlations are required to vanish, since correlations from the distant past should not influence the evolution[15]. This is also called *principle of weakening of initial correlations*[16]. However, shifting the initial time t_0 is only a formal mathematical procedure which is used to derive the conventional Markovian kinetic equations. In reality we expect, that the evolution of the system transforms from an initial stage into a (Markovian) later stage, exclusively due to internal interaction processes. Therefore, a kinetic

[14] Moving t_0 to $-\infty$, the problem changes qualitatively, it is transformed from an initial value to a boundary (on the time arrow) value problem.

[15] As mentioned in Ch. 5, exceptions are long-living correlations, such as bound states. In that case, g_{12} is split in a short-living and a long-living contribution, and only the first is required to decay, $\lim_{(t-t_0)\to-\infty} g(t_0) = g^{long}(t_0)$.

[16] It has been introduced into the density operator approach by Bogolyubov [Bog46] and has been generalized to real time Green's functions by Kremp et al. [KSB85]. While these and related works *postulated* this principle, in our approach it follows directly, since the dynamical evolution of the correlations (i.e. non–Markovian behavior) is included.

theory which is applicable to times $t < \tau_{cor}$, must yield this transition "automatically". Moreover, initial correlations have to decay during the evolution, and the second condition (7.46) transforms into

$$\boxed{\lim_{(t-t_0)\to-\infty} g^{IC}(t) = 0, \quad \text{where} \quad g^{IC}(t) = U^+_{12}(tt_0)g(t_0)U^-_{12}(t_0t)} \tag{7.47}$$

It is readily verified that our approach satisfies these requirements. First we notice from Eq. (7.46) that the limit $t_0 \to -\infty$ is equivalent to $t \to \infty$, and is thus realized by the evolution itself. In this limit, expressions (7.43) and (7.44) yield

$$\operatorname{Im} g_0^{0M}(t) = \frac{V}{\hbar}\,\delta(\omega)\,\Phi(t), \tag{7.48}$$

$$\operatorname{Re} g_1^{0M}(t) = \frac{V}{\hbar}\,\frac{d}{d\omega}\frac{\mathcal{P}}{\omega}\,\frac{d}{dt}\Phi(t), \tag{7.49}$$

where Eq. (7.48) is the well-known Markovian correlation function which gives rise[17] to the conventional Landau collision integral containing the familiar kinetic energy conserving delta function. Notice that g_0^{0M} still depends on time, but only functionally, via the distributions. This means, the relaxation described by g_0^{0M} corresponds already to the kinetic stage, cf. Ch. 5. When the distributions have reached their equilibrium too, g_0^{0M} goes over (in the absence of g_1^{0M}, g_2^{0M} etc.) to the equilibrium pair correlation function of an ideal quantum system. However, in a correlated system, the higher order expansion terms, in particular g_1^0 do not vanish. Even in the Markov limit, this term is nonzero, but approaches Eq. (7.49). As a result, the correlation function is given as the sum of the zeroth + first order terms[18] and may significantly deviate from that of an ideal system. For the same reason, the equilibrium one-particle distribution functions will, in this case, deviate from Fermi or Bose functions. Therefore, indeed, the higher order retardation terms yield nonideality corrections. Fig. 7.5. shows the effect of the first order retardation term on the kinetic energy relaxation. Typically, its effect at short times is weak.

However, there are also problems with the long-time limit of the expansion (7.43), (7.44), ..., and the underlying form of the correlation function. In particular, due to the lack of damping, the initial correlation term does not decay and condition (7.47) is violated. Therefore, we return to the more general

[17] if inserted into Eq. (7.39)

[18] plus possibly higher orders, but, in most cases, the series converges very fast.

case, which includes energy renormalization, and we have from Appendix E

$$\mathrm{Im}\, g_0(t) = \frac{V}{\hbar}\,\Phi(t)\left\{\frac{\Gamma}{\omega^2+\Gamma^2} + \frac{e^{-\Gamma t}}{\omega^2+\Gamma^2}\left(\omega\sin\omega t - \Gamma\cos\omega t\right)\right\}, \tag{7.50}$$

$$\begin{aligned}\mathrm{Im}\, g_1(t) &= \frac{V}{\hbar}\,\frac{d}{dt}\Phi(t)\\ &\times \left\{-\frac{\omega^2-\Gamma^2}{(\omega^2+\Gamma^2)^2} + \frac{e^{-\Gamma t}}{(\omega^2+\Gamma^2)^2}\Big[\left(2\Gamma\omega + (\omega^2+\Gamma^2)\,\omega t\right)\sin\omega t\right.\\ &+ \left.\left(\omega^2-\Gamma^2-(\omega^2+\Gamma^2)\Gamma t\right)\cos\omega t\Big]\right\}.\end{aligned} \tag{7.51}$$

Details and further expansion terms can be found in Appendix E, where also the short-time behavior is analyzed. Here, we provide only the result for the *Markov limit*,

$$\mathrm{Im}\, g_0^M(t) = \frac{V}{\hbar}\,\frac{\Gamma}{\omega^2+\Gamma^2}\,\Phi(t), \tag{7.52}$$

$$\mathrm{Im}\, g_1^M(t) = \frac{V}{\hbar}\,\frac{\omega^2-\Gamma^2}{(\omega^2+\Gamma^2)^2}\,\frac{d}{dt}\Phi(t). \tag{7.53}$$

Let us summarize the main consequences of these results:

1. *Short-time behavior:* For times close to the initial time t_0, the retardation is weak. Therefore, the zeroth order term (7.43) or (7.50), is the exact asymptotic of the full non-Markovian expression (this is correct, if no correlations exist in the system at $t = t_0$. Otherwise, at very short times, the initial correlation term in $g(t)$ may be dominant.)

2. *Damping at short times:* At early times, energy renormalization only begins to build up, therefore the result of Eq. (7.50) will be close to that of Eq. (7.43). Again, if there are non-zero initial correlations in the system, there may already exist a finite value of γ and Δ at $t = t_0$ (resulting from the initial value contribution in $\tilde{\Sigma}$). In this case, the two results (7.50,7.43) are different even at early times.

3. *Long-time behavior:* Expression (7.48) is just the familiar equilibrium result for the quantum binary correlation function in Born approximation. The thermodynamic equilibrium result follows from introducing Bose/Fermi functions in Φ. Its characteristic feature is the delta function, reflecting conservation of the one-particle energy in each scattering event.

4. *Correlation corrections at long times:* In addition to Eq. (7.48) other terms contribute to $\mathrm{Im}\, g$, most importantly, the first order term (7.49). This term also contributes to the macroscopic observables of the system and guarantees total energy conservation in the long-time limit.

5. *Selfenergy effects at long times:* With energy renormalization included, the Markov limit of the zeroth order retardation term differs from Eq. (7.48). The reason is that the energy shift and damping do not vanish at long times, (they are another consequence of correlations in the system), leading to Eq. (7.52) instead. This expression contains a Lorentzian spectral function with a finite width Γ. Though we have to expect modifications of the sharp spectral function (finite life time effects), this particular form is problematic, since it does not lead to the correct correlated equilibrium state. The reason is the local approximation for the propagators which lead to the exponential damping term. Thus, for long-time studies, this approximation has to be improved [HB96], see Sec. 12.6.

Retardation Expansion of $\mathrm{Re}\, g$**.** We complete the investigation of retardation expansion by providing the results for $\mathrm{Re}\, g$ which determines the correlation energy in the system. The real part of the correlation matrix elements, Eq.(7.9), is related to the integral A of App. E via $\mathrm{Re}\, g(121'2', t) = \frac{V(1-1')}{\hbar} A(\omega, \Gamma, t)$. With the results for the integral A^0, we write down the first two expansion terms for the case without selfenergy,

$$\mathrm{Re}\, g_0^0(t) = \frac{V}{\hbar}\, \Phi(t)\, \frac{1-\cos\omega t}{\omega}, \tag{7.54}$$

$$\mathrm{Re}\, g_1^0(t) = \frac{V}{\hbar\omega^2}\, \frac{d}{dt}\Phi(t) \Big[-\omega t \cos\omega t + \sin\omega t \Big], \tag{7.55}$$

and with energy renormalization included, we have from the integral A,

$$\mathrm{Re}\, g_0(t) = \frac{V}{\hbar}\, \Phi(t) \left\{ \frac{\omega}{\omega^2+\Gamma^2} - \frac{e^{-\Gamma t}}{\omega^2+\Gamma^2} \left(\Gamma \sin\omega t + \omega \cos\omega t\right) \right\}, \tag{7.56}$$

$$\begin{aligned} \mathrm{Re}\, g_1(t) &= \frac{V}{\hbar}\, \frac{d}{dt}\Phi(t) \left\{ \frac{\omega\Gamma}{(\omega^2+\Gamma^2)^2} + \frac{e^{-\Gamma t}}{(\omega^2+\Gamma^2)^2} \right. \\ &\times \Big[\Big(\omega^2 - \Gamma^2 - (\omega^2+\Gamma^2)\, \Gamma t\Big) \sin\omega t \\ &+ \left. \Big(2\omega\Gamma - (\omega^2+\Gamma^2)\, \omega t\Big) \cos\omega t \Big] \right\}. \end{aligned} \tag{7.57}$$

In the short-time limit, again the zeroth order terms (7.54,7.56) dominate the behavior. Interestingly, the real part of the correlations is being built up more slowly then the imaginary part ($\sim t^2$ versus t). In the long-time limit (Markov limit) we find for the case without selfenergy,

$$\mathrm{Re}\, g_0^{0M}(t) \;=\; \frac{V}{\hbar}\, \Phi(t)\, \mathcal{P}\, \frac{1}{\omega}, \tag{7.58}$$

and with renormalization taken into account

$$\mathrm{Re}\, g_0^{M}(t) \;=\; \frac{V}{\hbar}\, \Phi(t)\, \frac{\omega}{\omega^2+\Gamma^2}. \tag{7.59}$$

Higher order expansion terms can be found in Appendix E.

With these results one can readily perform a retardation expansion of the mean value of two-particle observables where we may directly apply the results of the expansion of $\mathrm{Re}\, g$. Of course, in the presence of initial correlations, we have to include the initial correlation contribution to $\mathrm{Re}\, g$ in all expressions.

7.3.4 Equilibrium correlations

Although the main focus of this book are short–time phenomena, to be consistent, any dynamical theory must contain the familiar equilibrium results as its long time limit. We briefly demonstrate this in this Section on the example of the equilibrium binary correlations[19].

The simplest case to consider is the zeroth order retardation term of the correlation function where, furthermore, the Markov limit has been taken, cf. Eqs. (7.48,7.58),

$$g_0^{0M}(12;1'2',t) = V(1-1')\,\Phi(12;1'2',t)\left[\mathcal{P}\frac{1}{\Delta E_{12}^0} + i\,\delta(\Delta E_{12}^0)\right], \tag{7.60}$$

where $1 \equiv \mathbf{p}_1$ and $\Delta E_{12}^0 \equiv E_1^0 + E_2^0 - E_{1'}^0 - E_{1'}^0$. The equilibrium limit follows, in lowest order in the interaction, if the distribution functions in Φ are replaced by Bose/Fermi functions, $f_1^{-1} = \exp[\beta(p_1^2/2m-\mu)] \mp 1$, where β and μ are the inverse temperature and chemical potential, respectively. For example, in the homogeneous case follows, with $1' \to p_1+q$ and $2' \to p_2-q$ and, with help of the identity,

$$\left[f_{1+q}^{-1} \pm 1\right]\left[f_{2-q}^{-1} \pm 1\right] = \left[f_1^{-1} \pm 1\right]\left[f_2^{-1} \pm 1\right]\, e^{-\beta\Delta E_{12}^0},$$

for the function Φ

$$\Phi_0^{\mathrm{EQ}}(12;1+q,2-q) = \left[1 - e^{-\beta\Delta E_{12}^0}\right] f_{1+q} f_{2-q}\,[1\pm f_1]\,[1\pm f_2]\,. \tag{7.61}$$

[19] These results have been obtained in collaboration with D. Semkat.

Expanding the exponential into a Taylor series and inserting the result (7.61) into Eq. (7.60), the equilibrium binary correlation function follows to[20]

$$g_0^{EQ}(12; 1+q, 2-q) = -\frac{V_q}{kT}\, f_{1+q} f_{2-q}\,[1 \pm f_1]\,[1 \pm f_2] \times \left\{ 1 - \sum_{l=2}^{\infty} \frac{(-1)^l}{l!} \left[\frac{\Delta E_{12}^0}{kT}\right]^{l-1} \right\}. \quad (7.62)$$

The first line is the well-known equilibrium correlation function in second Born approximation[21], see also the discussion of equilibrium correlations in Sec. 2.7, whereas the sum in the second line contains off-shell contributions due to collisional broadening. As mentioned above, Eq. (7.62) is only the lowest order result for g. Higher orders are straightforwardly obtained by including the first, second etc. retardation corrections to g and also the correlation corrections to the equilibium distributions (deviations from the Bose/Fermi function).

7.3.5 ∗Approximations for the selfenergy

We now consider the explicit result for the selfenergy contribution $\tilde{\Sigma}$ (7.11) and its properties more in detail. Also, to make it feasible for numerical analysis, we discuss simplifications. First, we mention that the exact result (7.11) leads to a time reversible dynamics[22]. Irreversibility will be introduced only if additional approximations to the time structure of Eq. (7.11) are applied (see below). Notice that Eq. (7.11) does not define $\tilde{\Sigma}$ explicitly, since it appears also under the time integral (recall that $\epsilon_1 = E_1^0 + \Delta_1 + i\gamma_1$, where $\tilde{\Sigma}_1(t)\, g(t) = [\Delta_1(t) + i\gamma_1(t)]\, g(t)$). A further complication is that the binary correlation matrix appears under the time integral too, and the internal time structure is complicated. For these reasons, additional simplifications are necessary, and we list in the following useful approximations[23].

---

[20]The imaginary part of the correlation function vanishes, because it contains a factor $x\delta(x)$ which is zero.

[21]This is the weak coupling limit of the binary collision (T-matrix) approximation, cf. Ch. 8, where $g_0^{EQ} \sim e^{-V/kT} - 1$.

[22]Recall that the result (7.11) was obtained from the second and third hierarchy equations with the closure (7.1) without any approximation with respect to the time dependencies. Based on the discussion in Sec. 2.2.1, we, therefore, conclude that the time-reversibility of the full hierarchy is retained.

[23]We briefly outline the main steps in the derivation of approximation (7.63): the selfenergy terms are small corrections to the free two-particle dynamics (in Born approximation, they are of the order V^2, constituting a small perturbation of the free energy). Therefore, one may apply perturbation theory to treat ϵ under the time integral. In particular, the renormalized energy may often be approximated by the zeroth order retardation term

i) A simpler, but still rather complicated approximation is given by

$$\tilde{\Sigma}_1^{(1)}(t) = \frac{1}{i\hbar}\int_{t_0}^{t} d\tau \int \frac{d3}{(2\pi\hbar)^3}\frac{d\bar{1}}{(2\pi\hbar)^3}\frac{d\bar{3}}{(2\pi\hbar)^3}(2\pi\hbar)^3\delta(1+3-\bar{1}-\bar{3})$$
$$\times V(1-\bar{1})\Big(V(1-\bar{1}) \pm V(1-\bar{3})\Big)e^{\frac{i}{\hbar}\left(\bar{E}_1^0+\bar{E}_3^0-E_1^0-E_3^0\right)(t-\tau)} \quad (7.63)$$
$$\times e^{\frac{i}{\hbar}\left(\bar{\Delta}_1+\bar{\Delta}_3-\Delta_1-\Delta_3\right)(t-t_0)}e^{-\frac{1}{\hbar}(\bar{\gamma}_1+\bar{\gamma}_3+\gamma_1+\gamma_3)(t-t_0)}\Big(\bar{f}_3^<\bar{f}_1^<f_3^> \mp \bar{f}_3^>\bar{f}_1^>f_3^<\Big)\Big|_\tau,$$

where Δ and γ are to be taken at time t. Together with the relation $\tilde{\Sigma}_1^{(1)}(t) = [\Delta_1(t) + i\gamma_1(t)]$, expression (7.63) selfconsistently defines the energy renormalization and the damping, including its momentum and time dependence. This approximation is well suited for numerical use as shown in Fig. 7.6., [KBKS97].[24]

ii) As we have seen in Sec. 7.3.3, a convenient way of generating approximate expressions for non-Markovian integrals is to perform a retardation expansion. We can do the same with expression (7.63) and obtain, in zeroth order, results similar to Img_0^0 and Reg_0^0, Eqs. (7.43) and (7.54), respectively. Under the momentum integrals, we have

$$\Delta_1^{(2)}(t) \sim \frac{1-\cos\{(\bar{E}_1^0+\bar{E}_3^0-E_1^0-E_3^0)(t-t_0)/\hbar\}}{\bar{E}_1^0+\bar{E}_3^0-E_1^0-E_3^0}, \quad (7.64)$$
$$\gamma_1^{(2)}(t) \sim \frac{\sin\{(\bar{E}_1^0+\bar{E}_3^0-E_1^0-E_3^0)(t-t_0)/\hbar\}}{\bar{E}_1^0+\bar{E}_3^0-E_1^0-E_3^0}, \quad (7.65)$$

which is readily generalized to higher orders in the retardation.

iii) A further simplification is to take the long–time limit (Markov limit) of Eqs. (7.64,7.65):

$$\Delta_1^{(3)}(t) \sim \frac{\mathcal{P}}{\bar{E}_1^0+\bar{E}_3^0-E_1^0-E_3^0}, \quad (7.66)$$
$$\gamma_1^{(3)}(t) \sim \delta(\bar{E}_1^0+\bar{E}_3^0-E_1^0-E_3^0), \quad (7.67)$$

$\epsilon(t-\tau) \approx \epsilon(t)$. Also, under the time integral, selfenergy corrections may often be neglected compared to the free energy E^0. Using these arguments, one can approximate under the time integral g by its Markovian limit g_0^M, Eq. (7.48). As a result, g can be taken out of the integral, turning $\tilde{\Sigma}$ into a function instead of an operator. Furthermore, the delta function in g^{M_0} leads to cancellation in the difference of six energies under the integral which becomes $\bar{E}_1^0+\bar{E}_3^0-E_1^0-E_3^0 \approx \bar{\epsilon}_1^0+\bar{\epsilon}_3^0-\epsilon_1^0-\epsilon_3^0$. Neglecting further the retardation in ϵ, we arrive at Eq. (7.63.)

[24] Although Eq. (7.63) defines $\Delta_1(t)$ and $\gamma_1(t)$ in a complicated way by an integral equation, this poses no problem for a numerical solution, and these integrals may be treated in the same way as the collision integrals of kinetic equations, cf. Appendix F.

where $\mathcal{P}$ denotes the principal value. Again, this result is readily generalized to higher order retardation terms (cf. Sec. 7.3.3).

The last approximation is qualitatively different from the preceding ones. With the *Markov approximation* used to eliminate the time integral in $\tilde{\Sigma}$, the energy renormalization is "switched on" instantaneously, whereas in the other approximations, Δ and γ are being built up steadily. Therefore, the Markov limit leads to an overestimate of damping effects on the early stage of relaxation. In Fig. 7.3.b, the dashed line shows the correlation time calculated from the imaginary part of the Markovian selfenergy. Notice the opposite dependence on the potential range $1/\kappa$ compared to τ_{cor} obtained from the kinetic energy relaxation [BK96, Köh96]. The reason is that with decreasing potential range (increasing screening and κ), energy saturates faster because particles cross the Debye sphere faster, after which they are correlated. On the other hand, with increasing potential strength (decreasing κ), Σ (or, equivalently, the scattering rates) grow rapidly. Interestingly, both curves cross around the equilibrium value for κ.[25]

What approximation to use is dictated by consistency requirements. The numerical analysis shows that the best results (e.g. in terms of energy conservation) are obtained if the collision integral and the selfenergy terms are treated on the same level of approximation. This applies also to the treatment of initial correlations.

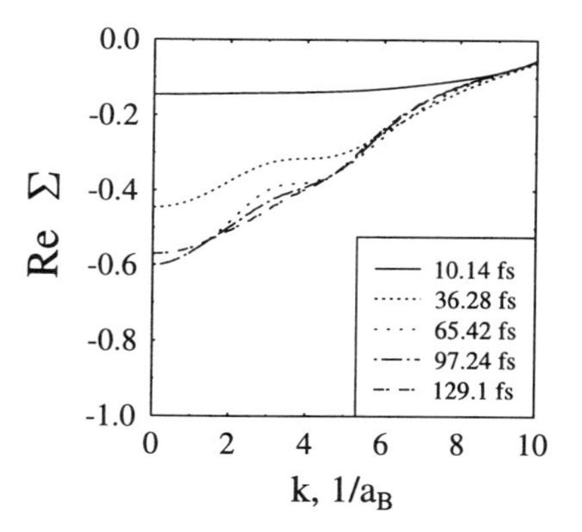

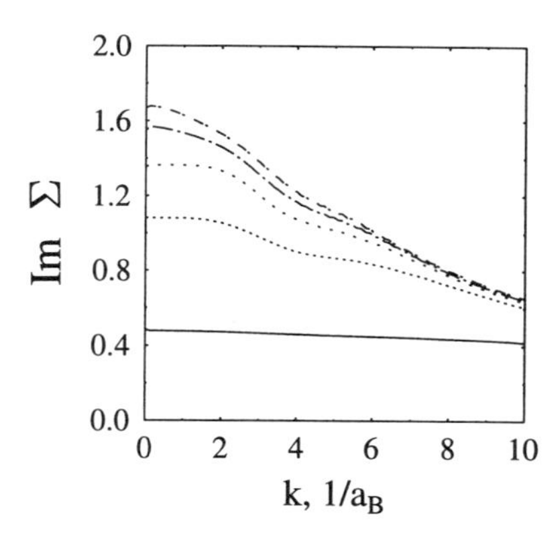

Fig. 7.6.: Real and imaginary part of the energy renormalization $\tilde{\Sigma}$ as a function of momentum and time according to the selfconsistent approximation (7.63). Same parameters as in Fig. 7.1., [BBK97].

[25]This comparison indicates that a better choice of the coefficient in Eq. (7.18) is 1/2 instead of 1/4 which was used in Fig. 7.3.b.

Initial correlations. Recall that the exact result for the selfenergy corrections (7.11) contains, in addition to the integral term, also a contribution from initial correlations. Obviously, if the kinetic equation contains an initial correlation collision integral (if $g(t_0 \neq 0)$, this term must be included also in the selfenergy. In that case, at the early stage of relaxation, the system already is correlated which must affect the selfenergy too, and the evolution has to start with nonzero γ and Δ, which are determined by $f(t_0)$ and $g(t_0)$.

Fig. 7.6. shows numerical results for the evolution of $\tilde{\Sigma}$ according to the selfconsistent approximation (i), i.e. the non-Markovian Landau equation was solved simultaneously with Eq. (7.63). The figure shows the damping and energy shift as a function of the wave number for different times during the initial stage, beginning from an uncorrelated state. The correlation time in this system is about 50 femtoseconds. After this time, correlations are basically formed, and, around this time, also the selfenergy is built up, further changes are only gradual (Bogolyubov regime).

We note again the problems related to the *local approximation* for the selfenergy correction $\tilde{\Sigma}$. In the long-time limit it leads to a broadened spectral kernel D which has a finite width Γ rather than to a sharp energy delta function. This does not yield the correct equilibrium distribution of correlated systems and also slows the relaxation down. Therefore, for selfconsistent calculations of the selfenergy which go beyond the correlation time, it is necessary to use improved expressions. One approach is to use phenomenological analytical expressions with non-Lorentzian tails [HB96]. An exact approach requires to solve the full Dyson equation for the two-time propagators (7.3) or, alternatively, the two–time Kadanoff-Baym equations [BKS+96]. These questions are discussed more in detail in Sec. 12.6.

7.4 *Discussion of the selfenergy concept. Relation to Green's functions results

In this Section we give a more general and strict discussion of the renormalization concept and identify the retarded and advanced selfenergies. Also, we show that these quantities are the same as are introduced in Green's functions theory. We will use a more compact operator notation and also retarded and advanced propagators $U^{\pm}$.

We reconsider the third hierarchy equation in its (anti-)symmetrized version (3.24) which was derived in Sec. 3.4. Using the decoupling approximation that lead us to the renormalization effects in Sec. 7.1, we can write

$$i\hbar\frac{\partial}{\partial t}g_{123} = H_{123}^{0\text{eff}}\,g_{123} + \left(S_{13}^{\gtrless} \mp S_{13}^{\lessgtr} + S_{23}^{\gtrless} \mp S_{23}^{\lessgtr} \mp G_{13} \mp G_{23}\right) g_{12} \; - \text{h.c.}, \quad (7.68)$$

where the hermitean conjugation refers to the whole r.h.s. and $g_{123}(t_0) = g_{123}^0$. Furthermore, $G_{ab} = ng_{ab}V_{ab}$, and we made use of the result $S_{ab}^{\gtrless} = F_a^{\gtrless} F_b^{\gtrless}\, V_{ab}\, F_b^{\lessgtr}$ of Sec. 3.4.1. We obtain the formal solution of Eq. (8.3) for g_{123},

$$\begin{aligned} g_{123}(t) &= \frac{1}{i\hbar}\int_{t_0}^{\infty} d\bar{t}\, U_{123}^{0+}(t\bar{t})\, K_{123}(\bar{t})\, U_{123}^{0-}(\bar{t}t), \\ \text{with } K_{123}(t) &= i\hbar\delta(t_0 - t)\, g_{123}^0 + \\ & \left\{\left(S_{13}^{\gtrless} \mp S_{13}^{\lessgtr} + S_{23}^{\gtrless} \mp S_{23}^{\lessgtr} \mp G_{13} \mp G_{23}\right) g_{12} \; - \text{h.c.}\right\}_t, \end{aligned} \quad (7.69)$$

where the first term on the r.h.s. arises from initial three-particle correlations. In consistency with the hierarchy closure (7.68), only initial correlations of the same form as the r.h.s. of Eq. (7.68) are permitted which are obtained from the second term in Eq. (7.69) by the substitution $V_{ab} \longrightarrow i\hbar\delta(t_0 - t)$, (as it was done in Eq. (7.11)). We, therefore, will remember, that each term $S^{\gtrless}$ and G contains a corresponding initial correlation contribution.

The retarded propagator U_{123}^{0+} is the solution of the equation

$$i\hbar\frac{\partial}{\partial t}U_{123}^{0+}(tt') - H_{123}^{0\text{eff}}\, U_{123}^{0+}(tt') = i\hbar\delta(t - t'), \quad (7.70)$$

and the advanced propagator U_{123}^{0-} obeys the adjoint equation and is related to the retarded propagator by $U_{123}^{0-}(t,t') = [U_{123}^{0+}(t',t)]^{\dagger}$. Furthermore, due to the neglect of the interaction potentials in the Hamiltonian (Born approximation), we expect $H_{123}^{0\text{eff}} = \bar{H}_1 + \bar{H}_2 + \bar{H}_3$ and $U_{abc}^{0\pm} = U_a^{\pm}U_b^{\pm}U_c^{\pm}$. Expression (7.69) for g_{123} is again inserted into the equation for g_{12}, Eq. (3.22)

$$\begin{aligned} & i\hbar\frac{\partial}{\partial t}g_{12} - \left[H_1 + H_1^{HF} + H_2 + H_2^{HF}, g_{12}\right] - (I_{12}^{>} - I_{12}^{<}) \\ & = \int_{t_0}^{\infty} d\bar{t}\,\frac{n}{i\hbar}\text{Tr}_3\left\{\left[V_{13} + V_{23}, U_{123}^{0+}(t\bar{t})K_{123}(\bar{t})U_{123}^{0-}(\bar{t}t)\right]\right\}\Lambda_{3,12}^{\pm}, \end{aligned} \quad (7.71)$$

where $I^{\gtrless}$ are defined by Eq. (3.18) and $\Lambda_{c,ab}^{\pm} = (1 \pm P_{ac} \pm P_{bc})$.

Retarded and advanced selfenergy. We now rewrite the integral term as an operator acting on g_{12}

$$\tilde{\Sigma}_{12}(t)\, g_{12}(t) = \int_{t_0}^{\infty} d\bar{t}\, \left\{\Sigma_{12}^{+}(t\bar{t})\, g_{12}(\bar{t})\, U_{12}^{-}(\bar{t}t) - \text{ h.c.}\right\}, \quad (7.72)$$

which, obviously, may be decomposed into one-particle and two-particle contributions,

$$\tilde{\Sigma}_{12} = \tilde{\Sigma}_1 + \tilde{\Sigma}_2 + \tilde{\Sigma}_{12}^{\rm cor} \quad \text{and} \quad \Sigma_{12}^{\pm} = \Sigma_1^{\pm} U_2^{\pm} + \Sigma_2^{\pm} U_1^{\pm} + \Sigma_{12}^{\rm cor\pm}, \tag{7.73}$$

which follow from comparison with Eq. (7.71),

$$\begin{aligned} \Sigma_1^{+}(tt') &= \frac{n}{i\hbar} \mathrm{Tr}_3 \left\{ V_{13} U_{13}^{0+}(t\bar{t}) \left(S_{13}^{>} \mp S_{13}^{<} \mp G_{13} \right)\Big|_{\bar{t}} U_3^{-}(\bar{t}t) \right\} \Lambda_{13}^{\pm}, && (7.74) \\ \Sigma_{12}^{cor\pm}(tt') &= \frac{n}{i\hbar} \mathrm{Tr}_3 \left\{ V_{23} U_{123}^{0+}(t\bar{t}) \left(S_{13}^{>} \mp S_{13}^{<} \mp G_{13} \right)\Big|_{\bar{t}} U_3^{-}(\bar{t}t) \right\} \\ &\times \; \Lambda_{3,12}^{\pm} + 1 \longleftrightarrow 2. && (7.75) \end{aligned}$$

We will see in a moment, that $\Sigma_{12}^{\pm}$ are in fact the retarded and advanced selfenergies familiar from Green's functions theory. But first, we rewrite the generalized second hierarchy equation (7.71) using the definition of $\Sigma_{12}^{\pm}$.

$$i\hbar \frac{\partial}{\partial t} g_{12} - \left\{ H_{12}^{0\rm eff}\, g_{12} - g_{12}\, H_{12}^{0\rm eff\dagger} \right\} = I_{12}^{>} - I_{12}^{<}, \tag{7.76}$$

where the effective two–particle Hamiltonian is given by

$$\begin{aligned} H_{12}^{0\rm eff}\, g_{12} &= \left(H_1 + H_1^{HF} + H_2 + H_2^{HF} \right) g_{12} + \int_{t_0}^{\infty} d\bar{t}\, \Sigma_{12}^{+}(t\bar{t})\, g_{12}(\bar{t})\, U_{12}^{0-}(\bar{t}t) \\ &= \left(\bar{H}_1 + \bar{H}_2 + H_{12}^{0\rm cor} \right) g_{12}, \end{aligned} \tag{7.77}$$

and in the last line we used the decomposition (7.73) and obtain

$$\begin{aligned} \bar{H}_1\, g_{12} &= H_1\, g_{12} + H_1^{HF}\, g_{12} + \int_{t_0}^{\infty} d\bar{t}\, \Sigma_1^{+}(t\bar{t})\, U_2^{+}(t\bar{t})\, g_{12}(\bar{t})\, U_{12}^{-}(\bar{t}t), && (7.78) \\ H_{12}^{0\rm cor} &= \int_{t_0}^{\infty} d\bar{t}\, \Sigma_{12}^{\rm cor+}(t\bar{t})\, g_{12}(\bar{t})\, U_{12}^{0-}(\bar{t}t). && (7.79) \end{aligned}$$

We see that $\Sigma_{12}^{\rm cor+}$ gives rise to an additional term in $H_{12}^{0\rm eff}$ that destroys the additivity. This is in contrast to our expectation, at least within the Born approximation. We, therefore, conclude that in this approximation, $\Sigma_{12}^{\rm cor+}$ has to be neglected[26]. Then, we indeed obtain additive results for the Hamiltonians,

$$\begin{aligned} H_{12}^{0\rm eff}\, g_{12} &= \left(\bar{H}_1 + \bar{H}_2 \right) g_{12}, && (7.80) \\ H_{123}^{0\rm eff}\, g_{12} &= \left(\bar{H}_1 + \bar{H}_2 + \bar{H}_3 \right) g_{12}. && (7.81) \end{aligned}$$

[26] In general, it has to be expected that Σ contains two-particle contributions. How they have to be introduced in the BBGKY-hierarchy is still an open problem.

Dyson equation. With this result, we now obtain the equation of motion for the renormalized one-particle propagator. Solving Eq. (7.70), using the factorization property of $U^{0\pm}_{123}$, we find the following equation for the quasi-particle propagators,

$$\left(i\hbar\frac{\partial}{\partial t} - H_1 - H_1^{HF}\right) U_1^+(tt') - \int_{t_0}^{\infty} d\bar{t}\, \Sigma_1^+(t\bar{t})\, U_1^+(\bar{t}t') = i\hbar\delta(t-t')\,, \quad (7.82)$$

which is nothing else than the *Dyson equation* for the retarded propagator (Green's function), for details cf. Appendix D. This again confirms that $\Sigma_1^{\pm}$ is indeed the retarded/advanced selfenergy. We can make this agreement more obvious by writing down the explicit result for the selfenergies in Born approximation using the definitions for $S^{\gtrless}$ and G,

$$\begin{aligned}\Sigma_1^+(tt') &= \frac{n}{i\hbar}\mathrm{Tr}_3\Big\{V_{13}^{\pm}U_1^+(tt')U_3^+(tt')\Big[F_1^> F_3^> V_{13}F_3^< \mp F_1^< F_3^< V_{13}F_3^> \mp n g_{13}V_{13} \\ &+ i\hbar\delta(t_0-t')\big(F_1^> F_3^> F_3^< \mp F_1^< F_3^< F_3^> \mp n g_{13}\big)\Big]\Big|_{t'} U_3^-(t't)\Big\}. \quad (7.83)\end{aligned}$$

The structure of this expression suggests to define

$$\begin{aligned} g_a^>(tt') &= U_a^+(tt')\,F_a^>(t') - F_a^>(t)\,U_a^-(tt'), \quad (7.84)\\ g_a^<(t't) &= U_a^+(tt')\,F_a^<(t') - F_a^<(t)U_a^-(tt'), \quad (7.85)\end{aligned}$$

which lets us rewrite Eq. (7.83) in a more compact form where we also absorb the initial correlations by defining $\tilde{V}_{ab}(t) = V_{ab} + i\hbar\delta(t-t_0)$,

$$\begin{aligned}\Sigma_1^+(tt') &= \frac{n}{i\hbar}\mathrm{Tr}_3\, V_{13}^{\pm}\Big\{g_1^>(tt')\,g_3^>(tt')\,\tilde{V}_{13}\,g_3^<(t't) \mp g_1^<(tt')\,g_3^<(tt')\,\tilde{V}_{13}\,g_3^>(t't) \\ &\mp U_1^+(tt')\,U_3^+(tt')\,n g_{13}(t')\,\tilde{V}_{13}(t')\,U_3^-(t't)\Big\}. \quad (7.86)\end{aligned}$$

This is the retarded selfenergy in second Born approximation, which is familiar from nonequilibrium Green's functions theory, e.g. [KB89], where $g^{\gtrless}$ are two-time one–particle correlation functions, see Ch. 12. Also, relations (7.84) are nothing but the generalized Kadanoff–Baym ansatz of Lipavský et al. [LvV86][27]. Interestingly, our approach includes arbitrary initial two-particle correlations in the definition of the selfenergy in a quite natural way.

[27] Strictly speaking, the last term in the selfenergy expressions, i.e. the term $\sim g_{13}$ is missing in the Green's functions results. In the BBGKY-hierarchy, it appears straightforwardly from the (anti-)symmetrization of the third hierarchy equation, cf. Sec. 3.4. A possible approach is to iteratively replace g_{13} by the solution of the second hierarchy equation. Most likely, the resulting terms can be "absorbed" by a generalization of the relations (7.84), which should agree with the additional terms (beyond the generalized Kadanoff-Baym ansatz) in the *exact* reconstruction formula of Ref. [LvV86], see Eq. (12.106).

We will see in the next Chapters, that the density operator approach to selfenergy is straightforwardly extended to more complex closure approximation to the hierarchy, including the ladder approximation, the RPA and the screened ladder approximation.

Chapter 8

Strong Coupling Effects. Ladder (T-Matrix) Approximation

We again consider a quantum many-body system, but this time with short-range binary interactions $V(r_i - r_j)$. We permit the interaction to be of arbitrary strength, so that the mean value of the potential energy $\langle V \rangle$ may be of the same order or even larger than that of the kinetic one $\langle T \rangle$, which is usually refered to as "strong coupling", cf. Ch. 1. Moreover, in case of an attractive potential, bound states are possible. To simplify the notation below, we consider a one-component system and the spatially homogeneous case only. Generalizations to several species and to inhomogeneous systems are straightforward.

Our goal in this Chapter is to derive the non-Markovian kinetic equation which - as in Ch. 7 - includes spin statistics effects, selfenergy and initial correlations, and which is valid on ultra-short and at long times as well. The difference here is the full inclusion of strong coupling effects[1] which will lead us to the non-Markovian Boltzmann equation [KBKS97], which essentially extends the validity of the kinetic equations. Therefore, our derivations will focus on the treatment of these phenomena, while for less specific questions, which are related to non-Markovian effects, we will reference the corresponding detailed results of Ch. 7. Furthermore, effects of the dynamics, in the case of Coulomb systems, are excluded here (see footnote 1 of Ch. 6). They will be considered in Chs. 9 and 10 for the case of weak and strong coupling, respectively.

[1] We mention interesting related work of Dufty and co-workers, see [BD81] and references therein.

8.1 Generalized binary collision approximation

The first step in deriving the non-Markovian Boltzmann equation is to obtain the appropriate closure to the hierarchy equations. As discussed in Sec. 2.6, we have to use the binary collision approximation (ladder or T-matrix approximation), which is obtained by neglecting the polarization terms in the second hierarchy equation, i.e. the terms $\Pi_{12}^{(1)}$ and $\Pi_{12}^{(2)}$ in Eq. (3.22), leading us to

$$i\hbar\frac{\partial}{\partial t}F_1 - [H_1 + H_1^{HF}, F_1] = n\mathrm{Tr}_2[V_{12}^{\pm}, g_{12}] \tag{8.1}$$

$$\begin{aligned} & i\hbar\frac{\partial}{\partial t}g_{12} - [H_1 + H_1^{HF} + H_2 + H_2^{HF}, g_{12}] - \Big(N_{12}V_{12}g_{12} - g_{12}V_{12}N_{12}\Big) \\ = & \Big(N_{12}V_{12}F_{12}^{<} - F_{12}^{<}V_{12}N_{12}\Big) + n\mathrm{Tr}_3[V_{13} + V_{23}, g_{123}](1 \pm P_{13} \pm P_{23}), \end{aligned} \tag{8.2}$$

where the Hartree-Fock potential H^{HF} and the potential with the exchange term included, $V^{\pm}$, were defined in Ch. 3. Here, we will write the Pauli blocking factor $N_{ab} = (1 \pm nF_a \pm nF_b)$ explicitly and also use the short notation $F_{ab}^{\gtrless} = F_a^{\gtrless} F_b^{\gtrless}$, cf. Eq. (3.18), and "$\pm$" refers to bosons/fermions. The main difference to the Born approximation so far is the appearance of the ladder terms (term with the potential) on the l.h.s. of Eq. (8.2). It is this term that is responsible for strong coupling effects, in analogy to the potential term in the Schrödinger equation for interacting particles. In the case of a pair "1-2" of oppositely charged particles, this potential is attractive and may allow for the formation of bound states. On the other hand, the immediate "technical" consequence of this term is that the solution for g_{12} will be more complicated than the one obtained in the case of the Born approximation in Ch. 7.

But before solving Eq. (8.2), we again include selfenergy effects, for the same reasons as were discussed in Ch. 7. (Without selfenergy, we would simply set $g_{123} = 0$). To be consistent with the ladder approximation for g_{12}, we will need to include the selfenergy in binary collision approximation too, which requires the equation of motion for g_{123} to be taken in the approximation (cf. Eq. (3.24):

$$\begin{aligned} & i\hbar\frac{\partial}{\partial t}g_{123} - \Big\{H_{123}^{\mathrm{eff}}\, g_{123} - g_{123}\, H_{123}^{\mathrm{eff}\dagger}\Big\} \\ = & \Big(S_{13}^{>} \mp S_{13}^{<} + S_{23}^{>} \mp S_{23}^{<} \mp G_{13} \mp G_{23}\Big)\, g_{12} \; - \mathrm{h.c.}, \end{aligned} \tag{8.3}$$

where the hermitean conjugation is applied to the whole r.h.s. Furthermore, $G_{ab} = ng_{ab}V_{ab}$, and we made use of the result $S_{ab}^{\gtrless} = F_{ab}^{\gtrless}\, V_{ab}\, F_b^{\lessgtr}$ of Sec. 3.4.1. Compared to the exact third hierarchy equation with spin statistics, Eq. (3.24), here we neglected 4–particle contributions related to higher orders in the density ($n\mathrm{Tr}_4$–terms) and accounted only for interactions between particle 1 with particle 3 and 2 with 3, respectively, retaining the full correlation g_{12} between particles 1 and 2. As in Ch. 7, we use an effective (renormalized) Hamiltonian,

$$H_{123}^{\mathrm{eff}} = \bar{H}_1 + \bar{H}_2 + \bar{H}_3 + N_{12}V_{12} + N_{13}V_{13} + N_{23}V_{23}, \tag{8.4}$$

which contains the renormalized (yet unknown) one-particle Hamiltonians. Consistently with the binary collision approximation, in addition, the three-particle ladder terms (potential terms in (8.4)) are included in H_{123}^{eff}. Thus, we obtain the same equation of motion for g_{123} as previously, cf. Eq. (7.68), except that we have to replace the Hamiltonian $H_{123}^{0\mathrm{eff}} \longrightarrow H_{123}^{\mathrm{eff}}$. We mention that with the decomposition (8.4), we assumed that energy renormalization enters only via the one-particle Hamiltonians, thus neglecting correlation contributions Σ_{12}^{cor}.

8.2 *Selfenergy in ladder (T-matrix) approximation

Proceeding further as in Sec. 7.4, we now solve Eq. (8.3) for g_{123}. One readily checks that the solution is the same as Eq. (7.69),

$$g_{123}(t) = \frac{1}{i\hbar}\int_{t_0}^{\infty} d\bar{t}\, U_{123}^{+}(t\bar{t})\, K_{123}(\bar{t})\, U_{123}^{-}(\bar{t}t), \tag{8.5}$$

where now, due to the ladder terms in the Hamiltonian, instead of $U_{123}^{0\pm}$ we have more general propagators which are the solution of the equation

$$i\hbar\frac{\partial}{\partial t}U_{123}^{+}(tt') - H_{123}^{\mathrm{eff}}U_{123}^{+}(tt') = i\hbar\delta(t-t'), \tag{8.6}$$

and its adjoint (U^-), respectively, where $U_{123}^{-}(t,t') = [U_{123}^{+}(t',t)]^{\dagger}$. The inhomogeneity K_{123} in Eq. (8.5) is again given by Eq. (7.70) and includes the initial correlation term. We mention, however, that there is one important difference compared to the Born approximation: Due to the interaction potentials in the Hamiltonian in Eq. (8.6), the propagators do not factorize into products of one-particle propagators.

The further derivation parallels that of Sec. 7.4 except for the more general propagators: After inserting $g_{123}(t)$ into the second hierarchy equation (8.2), the selfenergy contribution can be incorporated to yield a renormalized Hamiltonian H_{12}^{eff} which includes the retarded selfenergies Σ_1^+ and Σ_2^+,

$$i\hbar\frac{\partial}{\partial t}g_{12} - \left\{H_{12}^{\mathrm{eff}}\, g_{12} - g_{12}\, H_{12}^{\mathrm{eff}\dagger}\right\} = \left(N_{12}V_{12}F_{12}^{<} - F_{12}^{<}V_{12}N_{12}\right), \tag{8.7}$$

where the Hamiltonians are given by

$$H_{12}^{\text{eff}}\, g_{12} = \left(\bar{H}_1 + \bar{H}_2 + N_{12}V_{12}\right) g_{12}, \tag{8.8}$$

$$\bar{H}_1\, g_{12} = \left(H_1 + H_1^{HF}\right) g_{12} + \int_{t_0}^{\infty} d\bar{t}\, \Sigma_1^+(t\bar{t})\, U_2^+(t\bar{t})\, g_{12}(\bar{t})\, U_{12}^-(\bar{t}t). \tag{8.9}$$

Some care has to be taken for the evaluation of the selfenergies $\Sigma_1^{\pm}$, since the propagators do not factorize. We, therefore, start with the full renormalization integral, cf. Eq. (7.71), where there are four pairwise adjoint one-particle terms that contain squares of the potentials V_{13} and V_{23}, respectively. For example, for the term with V_{13}^2, we have

$$\tilde{\Sigma}_1^+(t) g_{12}(t) = \int_{t_0}^{\infty} d\bar{t} \frac{n}{i\hbar} \text{Tr}_3 \left\{ V_{13}^{\pm} U_{123}^+(t\bar{t}) \left(S_{13}^{>} \mp S_{13}^{<} \mp G_{13} \right)\Big|_{\bar{t}} U_{123}^-(\bar{t}t) \right\}. \tag{8.10}$$

An important conclusion is now that, within the binary collision approximation, a *partial factorization* of $U_{123}^{\pm}$ can be performed. Taking binary correlations into account but neglecting three-particle correlations, yields the following factorizations

$$V_{ac}\, U_{abc}^{\pm} = V_{ac}\, U_{ac}^{\pm}\, U_b^{\pm}, \tag{8.11}$$

$$g_{ab}\, U_{abc}^{\pm} = g_{ab}\, U_{ab}^{\pm}\, U_c^{\pm}, \tag{8.12}$$

what allows to introduce the retarded one-particle selfenergy in Eq. (8.10),

$$\tilde{\Sigma}_1^+(t) g_{12}(t) = \int_{t_0}^{\infty} d\bar{t}\, \Sigma_1^+(t\bar{t})\, U_2^+(t\bar{t})\, g_{12}(\bar{t})\, U_{12}^-(\bar{t}t), \tag{8.13}$$

where we identify

$$\Sigma_1^+(tt') = \frac{n}{i\hbar} \text{Tr}_3\, V_{13}^{\pm} U_{13}^+(tt') \left[F_{13}^{>} \tilde{V}_{13} F_3^{<} \mp F_{13}^{<} \tilde{V}_{13} F_3^{>} \mp n g_{13} \tilde{V}_{13} \right]\Big|_{t'} U_3^-(t't), \tag{8.14}$$

and, as before, used the potential that includes the initial correlations $\tilde{V}_{ab}(t) = V_{ab} + i\hbar\delta(t - t_0)$. In Sec. 8.3, this expression will be transformed to the familiar selfenergy in binary collision (ladder) approximation, introducing the concepts of scattering theory and the $T-$operator.

Thus, we have concluded the derivation of the generalized binary collision approximation. The problem is now well formulated, given by a closed system of equations for F_1, Eq. (8.1) and g_{12}, Eq. (8.7), together with the expressions for the renormalized Hamiltonians and the retarded selfenergy, Eqs. (8.8), (8.9), (8.14) and the equation for the propagators (8.6). This system is an important extension of the conventional binary collision scenario by the inclusion of many-particle effects, such as selfenergy and Pauli blocking. Moreover, no approximations with respect to the times have been made which would restrict the applicability of this system.

What is left now is to find equations of motion for the one-particle and two-particle propagators which enter the Hamiltonians and the selfenergy. Furthermore, it will be convenient to find a formal solution for the binary correlation operator which, inserted into the equation for F_1, yields a closed kinetic equation, the generalized Boltzmann equation.

8.3 Correlation operator in binary collision approximation

Formal solution for the binary correlation operator. We now solve the equation of motion for g_{12}, Eq. (8.7) together with the initial condition $g_{12}(t)|_{t=t_0} = g_{12}^0$, and obtain, in full analogy to the solution for g_{123}, Eq. (8.5),

$$\begin{aligned} g_{12}(t) &= U_{12}^+(tt_0)\, g_{12}^0\, U_{12}^-(t_0t) \qquad (8.15)\\ &+ \frac{1}{i\hbar}\int_{t_0}^{\infty} d\bar{t}\, U_{12}^+(t\bar{t})\, \left[N_{12}(\bar{t})V_{12}F_{12}^<(\bar{t}) - F_{12}^<(\bar{t})V_{12}N_{12}(\bar{t})\right]\, U_{12}^-(\bar{t}t). \end{aligned}$$

The first term on the r.h.s. follows from the homogeneous part of Eq. (8.7) and describes the dynamics of initial correlations, according to the effective two–particle Hamiltonian H_{12}^{eff}, whereas the second gives the contribution of two–particle correlations which are being built up after the initial time t_0, during the relaxation process.

8.3.1 Propagators and scattering quantities

The solution (8.15) has the same form as in the Born approximation, cf. Eq. (7.15), with the only difference, that the propagators are more complicated and are defined by the equation (U^- obeys the adjoint equation)

$$\begin{aligned} &\left(i\hbar\frac{\partial}{\partial t} - H_1 - H_1^{HF} - H_2 - H_2^{HF} - N_{12}(t)V_{12}\right) U_{12}^+(tt') \\ &- \int_{t_0}^{\infty} d\bar{t}\left[\Sigma_1^+(t\bar{t})U_2^+(t\bar{t}) + \Sigma_2^+(t\bar{t})U_1^+(t\bar{t})\right] U_{12}^+(\bar{t}t') = i\hbar\delta(t-t'), \quad (8.16) \end{aligned}$$

with the self energies Σ_1^+, Σ_2^+ given by Eq. (8.14). The propagators $U_{12}^+(tt')$ and $U_{12}^-(tt')$ are related by $\left[U_{12}^+(tt')\right]^\dagger = U_{12}^-(t't)$. Obviously, Eq. (8.16) represents a Schrödinger–like equation of motion for interacting quasiparticles with an effective two–particle Hamiltonian (8.8), that is modified by the one–particle selfenergy contributions Σ_1^+ and Σ_2^+, and contains the interaction potential with Pauli blocking corrections. In case of an attractive binary interaction, it includes the possibility of bound states which are modified by medium effects.

One-particle propagators and renormalized energy spectrum. Before continuing the analysis of Eq. (8.16), let us consider the free propagators $U_{12}^{0\pm}$ which obey Eq. (8.16) without the interaction potential,

$$\left(i\hbar\frac{\partial}{\partial t} - H_1 - H_1^{HF} - H_2 - H_2^{HF}\right) U_{12}^{0+}(tt') \\ - \int_{t_0}^{\infty} d\bar{t} \left[\Sigma_1^+(t\bar{t})\, U_2^+(t\bar{t}) + \Sigma_2^+(t\bar{t})\, U_1^+(t\bar{t})\right] U_{12}^{0+}(\bar{t}t') = i\hbar\delta(t-t'). \quad (8.17)$$

The structure of this equation suggests the ansatz $U_{12}^{0+}(tt') = U_1^+(tt')\, U_2^+(tt')$, which leads to the following equation of motion for the one-particle propagators,

$$\left(i\hbar\frac{\partial}{\partial t} - H_1 - H_1^{HF}\right) U_1^+(tt') - \int_{t_0}^{\infty} d\bar{t}\, \Sigma_1^+(t\bar{t})\, U_1^+(\bar{t}t') = i\hbar\delta(t-t'). \quad (8.18)$$

This equation corresponds to the well–known Dyson equation of Green's function theory. Formally, it is the same equation as was obtained in Sec. 7.4, cf. Eq. (7.82). However, here we have a much more general selfenergy function, given in generalized binary collision approximation, Eq. (8.14). Eq. (8.18) closes our system of equations which determine the binary collision approximation. Now, also the two-particle and three-particle propagators and the renormalized Hamiltonians are defined.

Example: propagators in local approximation. For practical reasons, in many cases it is desirable to have simplifying approximations which may be used as a first qualitative approach to the full problem. An important simplification is obtained if the propagators are taken in the local approximation, cf. Appendix D,

$$U_1^{\pm}(\tau) = \Theta(\pm\tau) e^{-\frac{i}{\hbar}(E_1 \mp i\gamma_1)\tau}. \quad (8.19)$$

This corresponds to quasi-particle propagators with an effective one–particle energy, approximately given by

$$E_1(pRT) = \frac{p^2}{2m} + \mathrm{Re}\Sigma_1^+(p\omega RT)\bigg|_{\omega=E_1(pRT)/\hbar}, \quad (8.20)$$

and a damping of the one-particle states defined by the imaginary part of the selfenergy, γ_1, here in binary collision approximation (8.10).

Integral equations for $U^{\pm}_{12}$. Let us now return to the equations of motion for the two-particle propagators, Eqs. (8.16), (8.17). There exists a well-developed theoretical formalism to describe binary collisions, which is given by quantum scattering theory, see for example the monographs [Tay72, New82]. Here, we have to generalize the conventional scattering theory to scattering in a medium, i.e. to include selfenergy and spin statistics effects. Therefore, we have to find a way to relate the propagators $U^{\pm}_{12}$ to the familiar quantities of scattering theory, such as the Møller operator and the scattering operator (T-operator), and to generalize these quantities by inclusion of the influence of the surrounding medium.

To do this, it is convenient to transform the differential equations for the two-particle propagators $U^{\pm}_{12}$ into integral equations. It is easy to proof that the effective two–particle propagator obeys the following integral equations[2]

$$\boxed{\begin{aligned} U^{+}_{12}(tt') &= U^{0+}_{12}(tt') - \frac{i}{\hbar}\int_{-\infty}^{\infty} d\bar{t}\, U^{0+}_{12}(t\bar{t})\, N_{12}(\bar{t})\, V_{12}\, U^{+}_{12}(\bar{t}t') \\ &= U^{0+}_{12}(tt') - \frac{i}{\hbar}\int_{-\infty}^{\infty} d\bar{t}\, U^{+}_{12}(t\bar{t})\, N_{12}(\bar{t})\, V_{12}\, U^{0+}_{12}(\bar{t}t') \end{aligned}} \tag{8.21}$$

Again, U^{-}_{12} satisfies the adjoint equations. Eq. (8.21) has a clear physical meaning. While the first term on the r.h.s. is related to "free" quasi-particles (which is exact in the case of weak coupling or Born approximation), the integral terms account for the coupling between the two particles via the binary interaction. Notice that the first term describes more than the propagation of a free particle pair, since the propagator equation (8.17) contains selfenergy effects on the T-matrix level. Due to this renormalization of the one-particle behavior, we will continue to use the notion "quasi-particle" propagators for $U^{0\pm}$. Eq. (8.21) is a many-particle generalization of the propagator equation of scattering theory [Tay72, New82]. In particular, it contains, in addition, the Pauli blocking factors $N_{12}(t)$.

Generalized propagators. To establish a closer relation to standard scattering theory, it is useful to consider, instead of $U^{\pm}_{12}$, new retarded and advanced propagators which are defined as

$$G^{\pm}_{12}(tt') = \pm\Theta\left[\pm(t-t')\right]\left\{\mathcal{G}^{>}_{12}(tt') - \mathcal{G}^{<}_{12}(tt')\right\}, \tag{8.22}$$

where we denoted

$$\mathcal{G}^{\gtrless}_{12}(tt') = U^{+}_{12}(tt')F^{\gtrless}_{12}(t') + F^{\gtrless}_{12}(t)U^{-}_{12}(tt'), \tag{8.23}$$

[2]To this end, one acts with the operator of the l.h.s. of Eq. (8.16) on the whole equation using also the equation of motion for $U^{0\pm}_{12}$, Eq. (8.17).

allowing us to rewrite Eq. (8.22) as

$$G^{\pm}_{12}(tt') = \pm\Theta[\pm(t-t')]\left\{U^{+}_{12}(tt')N_{12}(t') + N_{12}(t)U^{-}_{12}(tt')\right\}. \qquad (8.24)$$

We emphasize that the definition of the new quantities $\mathcal{G}^{\gtrless}_{12}$ does not contain any approximations. Thus the equations for $\mathcal{G}^{\gtrless}_{12}$ are fully equivalent to the equations for $U^{\pm}_{12}$. The same properties hold for the generalized quasiparticle propagators $G^{0\pm}_{12}$.[3] The propagators $G^{\pm}_{12}$ and $G^{0\pm}_{12}$ have the advantage to "absorb the spin statistics", contained in the Pauli blocking factors N_{12}, so that the structure of the resulting equations becomes simpler and similar to that of quantum scattering theory for spinless particles. However, in contrast to conventional scattering theory, $G^{\pm}_{12}$ describe *in–medium scattering*, accounting for Bose or Fermi statistics and selfenergy effects. Using Eq. (8.21) and the definition (8.24), we obtain an integral equation for G^{+}_{12}:

$$G^{+}_{12}(tt') = G^{0+}_{12}(tt') - \frac{i}{\hbar}\int_{-\infty}^{+\infty} d\bar{t}\,G^{0+}_{12}(t\bar{t})V_{12}G^{+}_{12}(\bar{t}t'). \qquad (8.28)$$

As before, the adjoint equation yields the integral equation for the advanced propagator G^{-}_{12}.

Møller operators. It is useful to rewrite (8.28) as

$$\begin{aligned} G^{+}_{12}(tt') &= \int_{-\infty}^{+\infty} d\bar{t}\,G^{0+}_{12}(t\bar{t})\left\{\delta(t'-\bar{t}) - \frac{i}{\hbar}V_{12}G^{+}_{12}(\bar{t}t')\right\} \\ &= \int_{-\infty}^{+\infty} d\bar{t}\,G^{+}_{12}(t\bar{t})\left\{\delta(t'-\bar{t}) - \frac{i}{\hbar}V_{12}G^{0+}_{12}(\bar{t}t')\right\}. \qquad (8.29) \end{aligned}$$

To make use of the methods of quantum scattering theory, we now introduce a generalized Møller operator Ω^{+}_{12} by[4]

$$\boxed{\Omega^{+}_{12}(tt') = \delta(t-t') - \frac{i}{\hbar}G^{+}_{12}(tt')V_{12} = \delta(t-t') - \frac{i}{\hbar}U^{+}_{12}(tt')N_{12}\,V_{12}} \qquad (8.30)$$

[3] For completeness, we give the relations for the quasiparticle propagators,

$$G^{0\pm}_{12}(tt') = \pm\Theta\left[\pm(t-t')\right]\left\{\mathcal{G}^{0>}_{12}(tt') - \mathcal{G}^{0<}_{12}(tt')\right\}, \qquad (8.25)$$

with $$\mathcal{G}^{0\gtrless}_{12}(tt') = U^{0+}_{12}(tt')F^{\gtrless}_{12}(t') + F^{\gtrless}_{12}(t)U^{0-}_{12}(tt'), \qquad (8.26)$$

and also $$G^{0\pm}_{12}(tt') = \pm\Theta[\pm(t-t')]\left\{U^{0+}_{12}(tt')N_{12}(t') + N_{12}(t)U^{0-}_{12}(tt')\right\}. \qquad (8.27)$$

[4] The adjoint operator follows from $\Omega^{-}_{12}(tt') = [\Omega^{+}_{12}(t't)]^{\dagger}$.

which allows us to write Eq. (8.29) and its adjoint in the compact form

$$G^+_{12}(tt') = \int\limits_{-\infty}^{+\infty} d\bar{t}\, \Omega^+_{12}(t\bar{t})\, G^{0+}_{12}(\bar{t}t'); \quad G^-_{12}(tt') = \int\limits_{-\infty}^{+\infty} d\bar{t}\, G^{0-}_{12}(t\bar{t})\, \Omega^-_{12}(\bar{t}t'), \quad (8.31)$$

and similarly, for the propagators $U^\pm_{12}$

$$\boxed{U^+_{12}(tt') = \int\limits_{-\infty}^{+\infty} d\bar{t}\, \Omega^+_{12}(t\bar{t})\, U^{0+}_{12}(\bar{t}t'); \quad U^-_{12}(tt') = \int\limits_{-\infty}^{+\infty} d\bar{t}\, U^{0-}_{12}(t\bar{t})\, \Omega^-_{12}(\bar{t}t')} \quad (8.32)$$

These equations clearly show the physical meaning of the generalized Møller operators: They transform the "free (quasi-particle) trajectory of the pair 1-2" into a fully correlated one, or, more precisely, the action of the Møller operator on the quantum state of a pair of quasi-particles generates the correlated pair state.

Scattering (T)-operators. Finally, let us introduce the central quantity of the binary collision approximation, the T–operator, by defining

$$\boxed{T^+_{12}(tt') = V_{12}\Omega^+_{12}(tt')\,; \qquad T^-_{12}(tt') = \Omega^-_{12}(tt')V_{12}} \quad (8.33)$$

With Eqs. (8.31) and (8.33), we can express the propagators $G^\pm$ by the T–operators

$$\boxed{V_{12}G^+_{12}(tt') = \int\limits_{-\infty}^{\infty} d\bar{t}\, T^+_{12}(t\bar{t})\, G^{0+}_{12}(\bar{t}t'); \quad G^-_{12}(tt')V_{12} = \int\limits_{-\infty}^{\infty} d\bar{t}\, G^{0-}_{12}(t\bar{t})\, T^-_{12}(\bar{t}t')} \quad (8.34)$$

and, with Eqs. (8.28) and (8.30), the Møller operator can be expressed in terms of the T–operator too

$$\Omega^+_{12}(tt') = \delta(t-t') - \frac{i}{\hbar}\int_{-\infty}^{\infty} d\bar{t}\, G^{0+}_{12}(t\bar{t})\, T^+_{12}(\bar{t}t'). \quad (8.35)$$

Eq. (8.34) allows for a clear understanding of the T-operator: Strong coupling effects which are contained in $G^\pm$ are taken over by $T^\pm$, leaving a propagator $G^{0\pm}$ free of interaction between the particles.

What is left now, is to derive the two fundamental equations for the generalized T–operator, the Lippmann–Schwinger equation and the optical theorem. First, combining Eqs. (8.31), (8.33) and (8.34), we obtain the well–known

Lippmann–Schwinger equation:[5]

$$\begin{aligned} T^+_{12}(tt') &= V_{12}\delta(t-t') - \frac{i}{\hbar} V_{12} G^+_{12}(tt') V_{12} \\ &= V_{12}\delta(t-t') - \frac{i}{\hbar}\int_{-\infty}^{+\infty} d\bar{t}\, V_{12}\, G^{0+}_{12}(t\bar{t})\, T^+_{12}(\bar{t}t') \end{aligned} \tag{8.37}$$

Secondly, we derive from Eqs. (8.37) and (8.36) the **optical theorem** in time representation

$$T^+_{12}(tt') - T^-_{12}(tt') = -\frac{i}{\hbar}\int d\bar{t} d\bar{\bar{t}}\; T^+_{12}(t\bar{t}) \left\{ G^{0+}_{12}(\bar{t}\,\bar{\bar{t}}) - G^{0-}_{12}(\bar{t}\,\bar{\bar{t}}) \right\} T^-_{12}(\bar{\bar{t}}\;t') \tag{8.38}$$

The difference of the quasiparticle propagators entering Eq. (8.38), can be rewritten in terms of the propagators $U^{0\pm}$, according to Eq. (8.27)

$$G^{0+}_{12}(tt') - G^{0-}_{12}(tt') = U^{0+}_{12}(tt') N_{12}(t') - N_{12}(t) U^{0-}_{12}(tt')\,.$$

With the above equations, we have obtained a closed system for the scattering quantities and quasiparticle propagators. It should be mentioned again that, in comparison to scattering theory for an isolated pair of particles, here, in–medium effects are incorporated, cf. Eqs.(8.37) and (8.38). With the generalized two–particle propagators G^+_{12} and G^{0+}_{12}, many–body effects such as selfenergy and degeneracy due to Bose or Fermi statistics are taken into account providing for an important extension of conventional scattering theory.

Correlation operator in binary collision approximation. Let us now return to the determination of the two–particle correlation operator $g_{12}(t)$ which is given by the formal solution (8.15). Using relations (8.32) and (8.30), it is easy to show that $g_{12}(t)$ can be expressed by the Møller operators,

$$\begin{aligned} g_{12}(t) + F^<_{12}(t) &= \int_{t_0}^{\infty}\int_{t_0}^{\infty} d\bar{t} d\bar{\bar{t}}\; \Omega^+_{12}(t\bar{t})\, U^{0+}_{12}(\bar{t}t_0)\, g_{12}(t_0)\, U^{0-}_{12}(t_0\,\bar{\bar{t}})\, \Omega^-_{12}(\bar{\bar{t}}\;t) \\ &\quad + \int_{t_0}^{\infty}\int_{t_0}^{\infty} d\bar{t} d\bar{\bar{t}}\; \Omega^+_{12}(t\bar{t})\, \mathcal{G}^{0<}_{12}(\bar{t}\,\bar{\bar{t}})\, \Omega^-_{12}(\bar{\bar{t}}\;t) \end{aligned} \tag{8.39}$$

[5] We will further need the corresponding equation for the advanced operator,

$$\begin{aligned} T^-_{12}(tt') &= V_{12}\delta(t-t') - \frac{i}{\hbar} V_{12} G^-_{12}(tt') V_{12} \\ &= V_{12}\delta(t-t') - \frac{i}{\hbar}\int_{-\infty}^{+\infty} d\bar{t}\, T^-_{12}(t\bar{t})\, G^{0-}_{12}(\bar{t}t)\, V_{12}\,. \end{aligned} \tag{8.36}$$

where the operator $\mathcal{G}_{12}^{0<}$ is given by Eq. (8.26). Notice that the sum on the l.h.s of Eq. (8.39) is just the reduced two-particle density operator F_{12}. With Eq. (8.39) we obtained an exact solution of the Bogolyubov hierarchy on the level of the binary collision approximation. In particular, there is no restriction with respect to the time. All nonequilibrium properties of the many–particle system can be derived from this expression in well–known manner, cf. Sec. 7.3.2. So, we can determine the collision integral in the equations of motion for the single–particle density operator, Eq. (8.1), which will include strong coupling effects, and which allows us to calculate the time evolution of the distribution function and of all one-particle observables. Furthermore, it is possible to evaluate the dynamics of all two-particle properties from g_{12}, e.g., the mean potential energy, Eq. (2.26).

Properties of the solution $g_{12}(t)$. At this point it is instructive to discuss some properties of the solution given by the expression (8.39).

1. Recall that $U_{12}^{0\pm}$ are two–particle propagators of free damped quasiparticles which, in general, are to be determined from Eq. (8.17). For illustration, it is again convenient to consider the local approximation, Eq. (8.19), which yields

$$U_{12}^{0\pm}(\tau) = \Theta(\pm\tau)e^{-\frac{i}{\hbar}[E_{12}\mp i(\gamma_{12})]\tau}, \tag{8.40}$$

   where we introduced the short notation $E_{12} = E_1 + E_2$ and $\gamma_{12} = \gamma_1 + \gamma_2$.

2. The binary correlation operator is influenced by its value at $t = t_0$, i.e. $g_{12}(t)$ depends on the dynamics of a correlated initial state. This contribution follows from the first term on the r.h.s. of Eq. (8.39). As in Ch. 7, we find that the effect of the initial correlations is weakened in time due to the damping of the quasiparticle propagators. This is most easily seen in the local approximation (8.40), where the initial correlation term reads

$$\int_{t_0}^{\infty}\int_{t_0}^{\infty} d\bar{t}d\,\bar{\bar{t}}\;\Omega_{12}^{+}(t\bar{t})e^{-\frac{i}{\hbar}[E_{12}-i\gamma_{12}](\bar{t}-t_0)}g_{12}(t_o)e^{-\frac{i}{\hbar}[E_{12}+i\gamma_{12}](t_0-\bar{\bar{t}})}\Omega_{12}^{-}(\bar{\bar{t}}\;t).$$

   This gives us an estimate for the time scale on which the initial correlations decay

$$\tau_{cor} \sim \frac{\hbar}{\gamma_{12}}. \tag{8.41}$$

   Therefore, for $t \gg \tau_{cor}$, the Bogolyubov assumption of weakening of initial correlations holds, and the nonequilibrium properties of the many–particle system can be described by simpler "conventional" Markovian equations, cf. Ch. 5. Our result is of interest for the understanding of the short-time behavior of the system. It shows that the Bogolyubov (kinetic) regime is established dynamically after relaxation of the correlations. In our approach, we do not need to postulate the Bogolyubov condition, it follows directly as a result of the dynamics of the system.[6]

3. Another property of the solution (8.39) is that the binary correlation operator is given by an expression which is nonlocal in time. At the actual time t, the operator is determined not only by its actual value, but also by its values in the past, that means, there are *memory effects* which can essentially influence the relaxation behavior of the system. This nonlocality can be seen from the second integral in Eq. (8.39) which describes the correlations built up from the initial time t_0 until the actual time t due to binary collisions. Interestingly, an analysis of the correlation function $\mathcal{G}_{12}^{0<}$, Eq. (8.26), shows, (e.g. with the local approximation (8.40)), that the memory has a "finite depth" which is again determined by the damping.

4. This non-Markovian behavior is completely analogous to our observations in Ch. 7, where a more detailed discussion can be found. However, it is important to verify, that these phenomena can also be seen for much more complicated approximations of many-particle theory, such as the ladder approximation. However, compared to the simpler case of the Born approximation, Ch. 7, there are also important differences: even though the correlation time is formally given by the same expression (8.41), here the damping is given by an essentially improved expression: it follows from the imaginary part of the retarded selfenergy on the full T-matrix level.

Retarded selfenergy in terms of the T-operator. Having introduced the scattering quantities, we can now express the retarded selfenergy in terms

[6]We want to mention, however, that the damping of one and two-particle states is, in general, very complex. Only in Born approximation it is reduced to one-particle damping. Otherwise, the propagators $U_{12}^{0\pm}$, as derived in our approach, yield only qualitatively correct results for the damping. Furthermore it is clear that various types of correlations have different decay times. In particular, bound state correlations or large scale fluctuations may have a rather long lifetime. A correct treatment of the latter type of correlations again requires the inclusion of damping effects resulting from two-particle dynamics. We will not discuss this problem here but mention the main results of such analysis [KK81, KKER86]: It turns out that bound states are affected by the surrounding medium much less than continuum (scattering) states, i.e. the former are damped less than the latter. The reason is a rather complex compensation mechanism between different many-particle (damping) effects for bound states. This compensation does not occur for continuum states.

of the T-operator, too. We start from expression Eq. (8.14) for Σ^+ in binary collision approximation, where we eliminate the two-particle propagator U^+_{13}, expressing it by the T-operator, Eq. (8.33), which leads to the result

$$\boxed{\begin{aligned}\Sigma_1^+(tt') &= \frac{n}{i\hbar}\mathrm{Tr}_3 \int_{-\infty}^{+\infty} d\bar{t}\, T^+_{13}(t\bar{t})\, U^{0+}_{13}(\bar{t}t') \\ &\quad \times \left\{F^>_{13}\,\tilde{V}_{13}\,F^<_3 \mp F^<_{13}\,\tilde{V}_{13}\,F^>_3 \mp n g_{13}\tilde{V}_{13}\right\}\Big|_{t'} U^-_3(t'\bar{t})\end{aligned}} \tag{8.42}$$

We mention that the original expression (8.14) contained exchange contributions in the potential $V^\pm_{13}$. Therefore, the T-operator has to be understood as to contain an exchange term too.

Eq. (8.42) is the ladder approximation to the selfenergy which includes all ladder type diagrams except the one with one rung only and which are "closed" by a single–particle propagator U^-. In terms of Green's functions, this selfenergy expression is discussed in [KB89, KKER86, Mah81]. However, our expression is more general. Eq. (8.42) again contains the generalized potentials $\tilde{V}_{ab}(t) = V_{ab}+i\hbar\delta(t-t_0)$ and thus includes initial correlations. This is important if the system is initially correlated ($g^0_{12} \neq 0$). In this case, the conventional selfenergy contribution to the effective Hamiltonians would initially be zero and only build up during the relaxation. This is clearly unphysical, since with correlations already being present, also energy renormalization and finite lifetime effects exist. Furthermore, Eq. (8.42) contains an additional term which involves the binary correlation operator g_{13}[7].

8.3.2 *Gradient expansion of g_{12} and physical observables

The physical consequences of the memory effects in g_{12} can be conveniently studied if $g_{12}(t)$ is expanded with respect to the retardation in time.[8] In particular, this allows us to evaluate the first corrections to the local (Markovian) behavior in explicit form. Let us discuss this expansion making the following two simplifying assumptions:

(i) We consider the special case of complete weakening of initial correlations,

$$\lim_{t_0 \to -\infty} g_{12}(t_0) = 0.$$

[7]This term is of higher order in the density and would be small at low densities. For a discussion, see Sec. 7.4.

[8]For a detailed discussion of retardation expansions, cf. Sec. 7.3.3.

(ii) The time dependence of the Møller operator and the related retarded and advanced quantities is approximated by

$$\Omega^{\pm}_{12}(tt') = \Omega^{\pm}_{12}(t-t')\,,$$

what means that in the scattering quantities, Pauli blocking is neglected.

In order to perform the retardation (gradient) expansion, we introduce "center of mass" and relative variables, t and τ, respectively:

$$F_{12}(t) = \int_{-\infty}^{+\infty}\int_{-\infty}^{+\infty} d\tau d\bar{\tau}\, \Omega^{+}_{12}(\tau)\mathcal{G}^{0<}_{12}\left(-(\tau+\bar{\tau}), t+\frac{\bar{\tau}-\tau}{2}\right)\Omega^{-}_{12}(\bar{\tau}), \qquad (8.43)$$

where $\tau = t - t'$ and $\bar{\tau} = \bar{\bar{t}} - t'$. Taylor expansion up to the first order in the relative times around t and Fourier transformation with respect to τ yields

$$F_{12}(t) = \int \frac{d\omega}{2\pi}\, \Omega^{+}_{12}(\omega)\mathcal{G}^{0<}_{12}(\omega,t)\Omega^{-}_{12}(\omega) + \qquad (8.44)$$
$$\frac{i}{2}\int\frac{d\omega}{2\pi}\left[\frac{d}{d\omega}\Omega^{+}_{12}(\omega)\frac{\partial}{\partial t}\mathcal{G}^{0<}_{12}(\omega,t)\Omega^{-}_{12}(\omega) - \Omega^{+}_{12}(\omega)\frac{\partial}{\partial t}\mathcal{G}^{0<}_{12}(\omega,t)\frac{d}{d\omega}\Omega^{-}_{12}(\omega)\right].$$

Gradient expansion of $\mathcal{G}^{0<}_{12}$. To evaluate expression (8.44), we need the expansion of $\mathcal{G}^{0<}_{12}$ which follows from Eq. (8.26), and we have, up to first order in τ,

$$\mathcal{G}^{0<}_{12}(\tau,t) = \left[U^{0+}_{12}(\tau) + U^{0-}_{12}(\tau)\right] F^{<}_{12}(t) - \frac{\tau}{2}\left[U^{0+}_{12}(\tau) - U^{0-}_{12}(\tau)\right]\frac{d}{dt}F^{<}_{12}(t). \quad (8.45)$$

If, furthermore, the damping in the propagators is being neglected, we obtain, after Fourier transformation and using Eqs. (D.22) and (D.23), of Appendix D, the result for the gradient expansion of $\mathcal{G}^{0<}_{12}$

$$\mathcal{G}^{0<}_{12}(\omega,t) = -2\pi\hbar i\,\delta\left(\hbar\omega - E_{12}\right)F^{<}_{12}(t) + \hbar\frac{d}{d\omega}\frac{\mathcal{P}}{\hbar\omega - E_{12}}\frac{d}{dt}F^{<}_{12}(t)\,, \qquad (8.46)$$

where we recall that $E_{12} = E_1 + E_2$ is the two–particle energy ($\mathcal{P}$ denotes the principal value. Finally, we write down the **gradient expansion of the binary density operator**, up to the first order in the retardation,

$$\begin{aligned} F_{12}(t) \;=\; & -i\,\Omega^{+}_{12}(E_{12})\,F^{<}_{12}(t)\,\Omega^{-}_{12}\,(E_{12}) \\ & +\hbar\int\frac{d\omega}{2\pi}\,\Omega^{+}_{12}(\omega)\,\frac{d}{d\omega}\frac{\mathcal{P}}{\hbar\omega - E_{12}}\,\frac{dF^{<}_{12}(t)}{dt}\,\Omega^{-}_{12}(\omega) \\ & +\frac{\hbar}{2}\left\{\frac{d\Omega^{+}_{12}}{dE_{12}}\,\frac{dF^{<}_{12}(t)}{dt}\,\Omega^{-}_{12} - \Omega^{+}_{12}\,\frac{dF^{<}_{12}(t)}{dt}\,\frac{d\Omega^{-}_{12}}{dE_{12}}\right\} \end{aligned} \qquad (8.47)$$

The first term on the r.h.s. of Eq. (8.47) represents the so–called local approximation (zeroth order in the retardation). This contribution leads to the usual quantum Boltzmann collision integral in the equation of the one–particle density operator and includes selfenergy corrections and degeneracy due to Bose or Fermi statistics. The further contributions are the first order gradient expansion terms which yield the corrections to the usual collision term which are important in nonideal systems, cf. Sec. 7.3.3.

Potential energy. Having determined the binary density operator, we can calculate all macroscopic observables with correlation corrections in binary collision approximation included. For example, the mean value of the potential energy is, according to Eq. (2.14), $\langle V\rangle = \frac{n^2}{2}\mathrm{Tr}_{12}V_{12}F_{12}$. With the local approximation for F_{12}, it follows

$$\langle V\rangle = \frac{n^2}{2}\mathrm{Tr}_{12}\left\{V_{12}\,\Omega^+_{12}F^<_{12}\Omega^-_{12}\right\}. \tag{8.48}$$

Using the relation between Ω^+_{12} and T^+_{12} according to Eq. (8.35) and the invariance of the trace, Eq. (8.48) may be transformed to

$$\langle V\rangle = -i\frac{n^2}{4}Tr_{12}\left\{T^+_{12}(E)F^<_{12}(t)\Omega^-_{12}(E) + \Omega^+_{12}F^<_{12}(t)T^-_{12}(E)\right\}. \tag{8.49}$$

Using again Eq. (8.35), we obtain

$$\langle V\rangle = i\frac{n^2}{2}Tr_{12}\left\{\mathrm{Re}T^+_{12}(E)F^<_{12}(t) + T^+_{12}(E)\frac{\mathcal{P}}{E-\bar{E}}N_{12}F^<_{12}(t)T^-_{12}(E)\right\}. \tag{8.50}$$

This expression explains once more the character of the approximations used in our theory. From relation (8.50) we are able to determine the mean value of the potential energy for an arbitrary nonequilibrium situation once the single-particle density operator $F_1(t)$ is known. Eq. (8.50) is a rather general result which extends our previous results obtained in Sec. 7.3.2 to the case of strong interaction.

Potential energy and pressure in thermodynamic equilibrium. Further simplifications are possible in limiting cases only. In particular, in *thermodynamic equilibrium*, the density operator is known explicitly, and we can replace

$$n^2\,F^<_{12} = n^2\,F_1\,F_2 = n_{12}(1 \pm f_1 \pm f_2), \tag{8.51}$$

where $f_{1,2}$ are Fermi functions, and n_{12} is the Bose function $n_{12}(\omega) = \{\exp\left[\beta(\hbar\omega - \mu_1 - \mu_2)\right] - 1\}^{-1}$. Inserting expression (8.51) into Eq.

(8.50), we obtain, after partial integration,

$$\langle V\rangle = -\frac{kT}{2}\int\limits_{-\infty}^{\infty}\frac{d\omega}{\pi}\ln\left|1 - z_1 z_2 e^{-\beta\omega}\right| \mathrm{Tr}_{12}\left\{\frac{\partial}{\partial\omega}\mathrm{Im}\left(G_0^+(\omega^+)T^+(\omega^+)\right)\right\}, \quad (8.52)$$

where $\omega^+ = \omega + i\epsilon$, and $z_{1,2} = \exp[\beta\mu_{1,2}]$ are the fugacities (activities).

The corresponding expression for the *equation of state* may be derived from Eq. (8.50) using a "charging" procedure[9]

$$\mathcal{V}\,(p - p_0) = -\frac{kT}{2}\sum_n \ln\left|1 - z_1 z_2 e^{-\beta E_n}\right| + \frac{kT}{2}\mathrm{Tr}_{12}\Big\{N_{12}\,n_{12}(E)\,\mathrm{Re}T^+$$
$$+ \ \ln\left|1 - z_1 z_2 e^{-\beta E}\right| \pi\delta(E - H)i\left(\frac{dT^+}{dE}N_{12}T^- - T^+ N_{12}\frac{dT^+}{dE}\right)\Big\}, \quad (8.55)$$

where $\mathcal{V}$ is the volume, and $T^\pm = T^\pm(E)$. The first term on the r.h.s. gives the contribution of the bound states. In the special case of a non-degenerate (classical) system, this is just the second cluster coefficient of the fugacity expansion of the pressure. Formulas of this type can be found e.g. in the text book of Landau and Lifshits [LL62]. Finally, by using a partial wave expansion of the T-matrix, it follows the Beth-Uhlenbeck representation [BU48], as it was shown in Refs. [KKK+84, KKE71].

Equilibrium pair distribution function. Inserting equilibrium distributions into the first term of Eq. (8.47), one can derive the equilibrium pair distribution function in binary collision approximation[10]. More easily, the classical equilibrium result is obtained from the equilibrium hierarchy (2.125): $f_{12}(|\mathbf{r}_1 - \mathbf{r}_2|) = e^{-V(|\mathbf{r}_1-\mathbf{r}_2|)/kT}$, see Sec. 2.7.

---

[9] This procedure consists in an integration over the coupling parameter λ from an noninteracting ($\lambda = 0$) to an interaction ($\lambda = 1$) system, see e.g. [KKER86],

$$\mathcal{V}\,(p - p_0) = -\int_0^1 \frac{d\lambda}{\lambda}\langle V\rangle_\lambda, \quad (8.53)$$

where the $\lambda-$integration can be carried out with the help of the identity $N_{12}\frac{\partial}{\partial\omega}(G_0^+T^+) = -\frac{\partial}{\partial\lambda}(G_0^+T^+G_0^+)$. The result is

$$\mathcal{V}\,(p - p_0) = \frac{kT}{2}\int\limits_{-\infty}^{\infty}\frac{d\omega}{\pi}\,ln\left|1 - z_1 z_2 e^{-\beta\omega}\right| \mathrm{Im}\mathrm{Tr}_{12}\left\{\frac{\partial G_0^+}{\partial\omega}T^+(\omega + i\epsilon)\right\}. \quad (8.54)$$

Following the calculations given in Ref. [KKK+84], we arrive at the well-known result (8.55) for the pressure.

[10] The procedure is the same as was used in Sec. 7.3.4.

8.3.3 *Recovery of the Lipavský ansatz

It is interesting to compare the obtained expression for the binary density operator in binary collision approximation, Eq. (8.39), with the corresponding result which follows from the theory of nonequilibrium Green's functions, [KSB86], see also Secs. 12.4 and 12.6. We give a brief summary of this result. Using the Green's functions formalism, one can derive an expression of the same form as Eq. (8.39). The main difference is that $\mathcal{G}_{12}^{0<}$ in Eq. (8.39), which is derived from Green's functions, is given by a product of two one–particle two–time correlation functions, i.e. (we suppress the momentum arguments)

$$\mathcal{G}_{12}^{0<}(tt') = i\, g_1^<(tt')\, g_2^<(tt'), \tag{8.56}$$

while, in the density operator approach, $\mathcal{G}_{12}^{0<}$ was given by Eq. (8.26). An agreement between our result, and Eq. (8.56), can be achieved only if the one–particle correlation functions in Eq. (8.56) are "reconstructed" from their value on the time diagonal, $\mp i\hbar g^<(tt) = F(t)$, according to the generalized Kadanoff-Baym ansatz proposed by Lipavský et al. [LvV86],

$$\mp i\hbar g^<(tt') = g^+(tt')\, F(t') - F(t)\, g^-(tt'), \tag{8.57}$$

where "-(+)" refers to fermions (bosons). Using this ansatz for $g_1^<$ and $g_2^<$ in Eq. (8.56), we obtain

$$\mathcal{G}_{12}^{0<}(tt') = \frac{-i}{\hbar^2}\left\{g_1^+(tt')\, g_2^+(tt')\, F_1(t')\, F_2(t') + F_1(t)\, F_2(t)\, g_1^-(tt')\, g_2^-(tt')\right\}, \tag{8.58}$$

where we used the fact that products of retarded and advanced functions of the same arguments vanish. If we now recall that $U^{0\pm}(t,t') = i\hbar g^\pm(t,t')$, Eq. (8.58) coincides with Eq. (8.26), which was an result of exact transformations within the density operator approach.

We thus showed that the Green's functions formalism agrees with the density operator result only if the reconstruction ansatz of Lipavský et al. is used. We want to underline, however, that with the density operator approach we are not require to postulate this ansatz. It is the structure of the binary correlation operator, here in binary collision approximation, Eqs. (8.39), (8.26), which is exactly of the form of the product of two generalized Kadanoff-Baym ansatzes. Notice also that this agreement does not depend on the particular choice of the free propagators.

8.4 Collision integral with memory effects

Let us now consider the derivation of the non-Markovian kinetic equation in binary collision approximation. For this, we start from the first equation of the hierarchy Eq. (8.1), in which we can now insert the solution for the binary correlation operator, Eq. (8.39),

$$\frac{d}{dt}F_1(t) = n\mathrm{Tr}_2\left[V_{12}, g_{12}\right] = I_1^{IC}(t) + I_1(t), \tag{8.59}$$

where we will not write the index "±" on V_{12}.[11] In Eq. (8.59), the second term is the conventional collision term and the first one is an additional contribution coming from the initial correlations (second and first integrals on the r.h.s. of Eq. (8.39), respectively). These collision integrals are given by

$$\begin{aligned} I_1(t) &= n\mathrm{Tr}_2\left[V_{12}, \int_{t_0}^{\infty}\int_{t_0}^{\infty} dt_1 dt_2 \Omega_{12}^{+}(tt_1)\, \mathcal{G}_{12}^{0<}(t_1t_2)\, \Omega_{12}^{-}(t_2t)\right], \quad (8.60)\\ I_1^{IC}(t) &= n\mathrm{Tr}_2\Big[V_{12}, \int_{t_0}^{\infty}\int_{t_0}^{\infty} dt_1 dt_2 \Omega_{12}^{+}(tt_1)\, U_{12}^{0+}(t_1t_0)\, F_{12}(t_0)\\ &\times \quad U_{12}^{0-}(t_0t_2)\, \Omega_{12}^{-}(t_2t) - F_{12}^{<}(t)\Big]. \quad (8.61)\end{aligned}$$

After rather lengthy calculations, one arrives at the result for I,

$$\boxed{\begin{aligned} &I(p_1,t) = \frac{i}{\hbar^2}\int dp_2\, d\bar{p}_1\, d\bar{p}_2 \int dt_1 dt_2 dt_3 \times\\ &\Big\{ \left\langle p_1p_2\left|T_{12}^{+}(tt_1)\right|\bar{p}_2\bar{p}_1\right\rangle \bar{U}_{12}^{0+}(t_1t_2)\left\langle \bar{p}_1\bar{p}_2\left|T_{12}^{-}(t_2t_3)\right|p_2p_1'\right\rangle U_{12}^{0-}(t_3t)\\ &\qquad\times\left[\bar{N}_{12}(t_2)\, f_{12}^{<}(t_3) - \bar{f}_{12}^{<}(t_2)N_{12}(t_3)\right]\\ &+\left\langle p_1p_2\left|T_{12}^{+}(tt_1)\right|\bar{p}_2\bar{p}_1\right\rangle \bar{U}_{12}^{0-}(t_1t_2)\left\langle \bar{p}_1\bar{p}_2\left|T_{12}^{-}(t_2t_3)\right|p_2p_1'\right\rangle U_{12}^{0-}(t_3t)\\ &\qquad\times\left[\bar{N}_{12}(t_1)f_{12}^{<}(t_3) - \bar{f}_{12}^{<}(t_1)N_{12}(t_3)\right]\\ &-U_{12}^{0+}(tt_3)\left\langle p_1p_2\left|T_{12}^{+}(t_3t_2)\right|\bar{p}_2\bar{p}_1\right\rangle \bar{U}_{12}^{0-}(t_2t_1)\left\langle \bar{p}_1\bar{p}_2\left|T_{12}^{-}(t_1t)\right|p_2p_1'\right\rangle\\ &\qquad\times\left[\bar{N}_{12}(t_2)f_{12}^{<}(t_3) - \bar{f}_{12}^{<}(t_2)N_{12}(t_3)\right]\\ &-U_{12}^{0+}(tt_3)\left\langle p_1p_2\left|T_{12}^{+}(t_3t_2)\right|\bar{p}_2\bar{p}_1\right\rangle \bar{U}_{12}^{0+}(t_2t_1)\left\langle \bar{p}_1\bar{p}_2\left|T_{12}^{-}(t_1t)\right|p_2p_1'\right\rangle\\ &\qquad\times\left[\bar{N}_{12}(t_1)f_{12}^{<}(t_3) - \bar{f}_{12}^{<}(t_1)N_{12}(t_3)\right]\Big\}\end{aligned}} \tag{8.62}$$

which was obtained in Ref. [KBKS97].[12] Here we used the following short notations for the momentum dependence of the different quantities: $N_{12} =$

[11] The exchange term yields an additive contribution to the T-operators and to the collision integral and is treated analogously.

[12] In the integral I_1, Eq. (8.60), one introduces the T–operator, using Eqs. (8.33), (8.35),

$$I_1(t) \quad = \quad n\mathrm{Tr}_2 \int_{t_0}^{t} dt_1 \Big\{ T_{12}^{+}(tt_1) - V_{12}\delta(t-t_1)\mathcal{G}_{12}^{0<}(t_1t) \tag{8.63}$$

$1 \pm f_1 \pm f_2$, $\bar{N}_{12} = 1 \pm \bar{f}_1 \pm \bar{f}_2$, $U_{12}^{0\pm} = U_1^{0\pm} U_2^{0\pm} = U^{0\pm}(p_1) U^{0\pm}(p_2)$, $\bar{U}_{12}^{0\pm} = \bar{U}_1^{0\pm} \bar{U}_2^{0\pm}$, $f_{12}^{\lessgtr} = f_1 f_2$, $\bar{f}_{12}^{\lessgtr} = \bar{f}_1 \bar{f}_2$, $f_1 = f(p_1)$ and $\bar{f}_1 = f(\bar{p}_1)$. Notice that the single-particle propagators $U^{0\pm}$ are to be determined selfconsistently from Eq. (8.17) and the adjoint equation, respectively.

Initial correlation integral. The collision integral arising from initial correlations is obtained in complete analogy. We apply Eqs.(8.33) and (8.35), and the final form is

$$
\begin{aligned}
I^{IC} &= n\mathrm{Tr}_2 \int dt' \left[T_{12}^{+}(tt')\mathcal{K}_{12}^{0}(t't) - \mathcal{K}_{12}^{0}(tt')\, T_{12}^{-}(t't)\right] - n\mathrm{Tr}_2 \left[V_{12}, F_{12}^{\lessgtr}(t)\right] \\
&- n\frac{i}{\hbar}\mathrm{Tr}_2 \int dt_1 dt_2 dt_3 \Big[G_{12}^{0+}(tt_3)\, T_{12}^{+}(t_3t_1)\mathcal{K}_{12}^{0}(t_1t_3) T_{12}^{-}(t_2t) \\
&\qquad + \quad T_{12}^{+}(tt_1)\mathcal{K}_{12}^{0}(t_1t_2) T_{12}^{-}(t_2t_3) G_{12}^{0-}(t_3t)\Big] , \qquad (8.65)
\end{aligned}
$$

where we denoted $\mathcal{K}_{12}^{0}(tt') = U_{12}^{0+}(tt_0)\, g_{12}(t_0)\, U_{12}^{0-}(t_0t')$. Further simplifications are only possible if $g_{12}(t_0)$ is given explicitly.

Eqs. (8.62) and (8.65) are the collision integral of the *generalized non-Markovian Boltzmann equation.* They are very general and go far beyond the usual Markovian Boltzmann equation. They include the full retardation (memory effects), as well as selfenergy and damping. These results are in full agreement with those obtained with Greens's function methods, see [KSB86, BKKS96].

Properties of the non-Markovian Boltzmann equation. We want to underline the fact that the kinetic equation with the collision integrals (8.62) and (8.65) follows from the solution of the initial value problem for g_{12}, Eq. (8.39), without any additional approximations. In particular, the time dependence is treated exactly. Therefore, this equation is valid without restrictions with respect to the time. This equation has the following remarkable properties:

$$
+ \quad \frac{i}{\hbar}\int_{t_0}^{t} dt_2 dt_3 T_{12}^{+}(tt_1)\mathcal{G}_{12}^{0<}(t_1t_2)\, T_{12}^{-}(t_2t_3)\, G_{12}^{0-}(t_3t)\Bigg\} \; - \text{h.c.}
$$

The first term of Eq. (8.63) is transformed with the help of the optical theorem (8.38) into

$$
\begin{aligned}
I_1(t) &= \frac{i}{\hbar} n\mathrm{Tr}_2 \int_{t_0}^{t} dt_1 dt_2 dt_3 \left\{T_{12}^{+}(tt_1)\left[G_{12}^{0+}(t_1t_2) - G_{12}^{0-}(t_1t_2)\right] T_{12}^{-}(t_2t_3)\mathcal{G}_{12}^{0<}(t_3t)\right. \\
&- \mathcal{G}_{12}^{0<}(tt_1) T_{12}^{+}(t_1t_3)\left[G_{12}^{0+}(t_2t_3) - G_{12}^{0-}(t_1t_2\right] T_{12}^{-}(t_3t) \qquad (8.64) \\
&+ \left. T_{12}^{+}(tt_1)\mathcal{G}_{12}^{0<}(t_1t_2) T_{12}^{-}(t_2t_3) G_{12}^{0-}(t_3t) - G_{12}^{0+}(tt_1) T_{12}^{+}(t_1t_2)\mathcal{G}_{12}^{0<}(t_2t_3) T_{12}^{-}(t_3t)\right\} .
\end{aligned}
$$

Here we used $G_{12}^{0+} - G_{12}^{0-} = \mathcal{G}_{12}^{0>} - \mathcal{G}_{12}^{0<}$. We finally transform the collision integral into the momentum representation, With (8.38) and using the T-matrices we obtain Eq. (8.62).

(i) The equation is nonlocal in time, i.e., the distribution function at time t is determined by its values for the preceding times too. We have a memory effect with the "memory depth" of the order of $\hbar/\gamma_{12}$.

(ii) For times $t < \hbar/\gamma_{12}$, initial correlations influence the behavior of the system significantly, while for $t \gg \hbar/\gamma_{12}$, they are being completely weakened.

(iii) The collision integral contains T-matrices which depend on two times. Translated into energy space, this means, there appear not only the conventional "on-shell" T-matrices (where the kinetic energy of the particle pair remains unchanged during the collision), but also "off-shell" T-matrices.

(iv) *Conservation of total energy* can be proved for the same three approximations as in case of the Born approximation, Ch. 7:

  (a) For the case with the exact expression (8.42) for the selfenergy (no approximations for the propagators), energy conservation can be proved for the case of distance dependent potentials.[13]

  (b) For the case without selfenergy, $\gamma_{12} \to 0$, we have a symmetric hierarchy closure, i.e. $P_{123}F_{123} = F_{123}$, which, according to Eq. (2.36) guarantees energy conservation (see also Sec. 2.5.2).

  (c) With selfenergy neglected and the Markov limit taken, energy conservation can be proved in first order retardation approximation, see Sec. 8.5.

(iv) Conventional Boltzmann–type two-particle scattering integrals, which conserve kinetic energy only, do not include bound states because, in that case, energy and momentum cannot be conserved simultaneously [McL89b, KK81]. In our case of the non-Markovian collision integral (8.62), this restriction is removed, because kinetic energy is not a conserved quantity ("off-shell" scattering). This means, the nonlocality in time allows for the existence of bound states in the framework of a two–particle collision approximation already. In particular, bound states may exist in the system already at the initial moment t_0, what can be accounted for by the choice of the initial correlations.

[13] This is based on the fact that the hierarchy closure (8.3) neglects only full commutators, cf. Sec. 2.2.2.

From the results of the binary collision approximation, we easily recover the Born approximation by taking the weak coupling limit. This is equivalent to using the first Born approximation for the T-matrix,

$$\left\langle p_1 p_2 \left| T_{12}^{\pm}(tt') \right| \bar{p}_2 \bar{p}_1 \right\rangle = \frac{V\left(p_1 - \bar{p}_1\right)}{(2\pi\hbar)^3} \delta\left(p_1 + p_2 - \bar{p}_1 - \bar{p}_2\right) \delta(t - t').$$

Inserted into the expression for the binary correlation operator or the collision integral, as well as into the selfenergy, two time integrations are trivially performed, and we recover the non-Markovian Landau equation of Ch. 7.

8.5 Kinetic equation in first order gradient expansion

Gradient expansion of the collision integral. In deriving Eq. (8.62) we have obtained a very general equation which is nonlocal in time (non–Markovian) and which is valid on arbitrary time scales. An essential question is now to investigate how the usual Boltzmann equation, which is local in time, is recovered from Eq. (8.62). As we have seen in Ch. 7, a very fruitful concept is to expand the correlation operator or the collision integral (8.62) with respect to the retardation. As a result, we will recover the Markovian Boltzmann equation, but in addition, we will obtain (Markovian too) correction terms which account for the effect of correlations.

Unfortunately, a retardation expansion of the non-Markovian Boltzmann collision integral is essentially more complicated than the expansion of the Landau integral, which is due to the complicated internal time structure of the T–operators and the three retardation (time) integrations in Eq. (8.62). For an expansion to exist, it is reasonable to assume that the scattering quantities T and Ω depend only on time differences, e.g., $T(t, t') = T(t - t')$. Furthermore, we perform the Markov limit (7.46) in the collision integrals. Here, we will not reproduce the lengthy calculations, but provide the main ideas and the result (more details can be found in [KBKS97]). The result up to first order in the retardation has the form

$$\frac{d}{dt} f(p_1) = I^0(p_1) + I^{(1)}(p_1) \,. \tag{8.66}$$

Markovian Boltzmann equation. From the zeroth order integral one recovers, after Fourier transform with respect to the retardation with application of the convolution theorem, the usual Boltzmann collision integral

$(E^+ = E + i\epsilon)$

$$\boxed{I^0 = \frac{2}{\hbar}\int \frac{dp_2 d\bar{p}_1 d\bar{p}_2}{(2\pi\hbar)^9} \left|\langle p_1 p_2 | T_{12}^+(E^+) | \bar{p}_2 \bar{p}_1\rangle\right|^2 \delta\left(E_{12} - \bar{E}_{12}\right) \left\{\bar{f}_{12}^> f_{12}^< - \bar{f}_{12}^< f_{12}^>\right\}} \tag{8.67}$$

which is local in time and, furthermore, contains only the T-matrices on the energy shell, i.e. for $E_{12} = \bar{E}_{12}$. The integral (8.67) describes the relaxation on the asymptotic stage, i.e., for times t sufficiently long after the initial moment, with $t \gg \tau_{\rm cor}$. As discussed in Ch. 7, this term yields the equilibrium state of an ideal (noninteracting) system. In contrast, the higher order retardation terms, most importantly, the first order term $I^{(1)}(p_1)$ in Eq. (8.66), are corrections accounting for correlation effects, which are of importance for higher orders of the density expansion of the collision integral. Furthermore, they guarantee the correct asymptotic value of the total energy and other thermodynamic and transport quantities.

First order retardation integral $I^{(1)}$. The first order terms follow from the retardation expansion of I, Eq. (8.62), by collecting all contributions linear in one of the retardation times. The final result is obtained after Fourier transformation (details can be found in [KBKS97])[14]

$$I^{(1)} = -\frac{1}{\hbar}\int \frac{dp_2 d\bar{p}_1 d\bar{p}_2}{(2\pi\hbar)^9} \left\{2\left|T_{12}^+(E)\right|^2 \frac{\mathcal{P}'}{\bar{E}-E}\left[\frac{df_{12}^<}{dt}\bar{f}_{12}^> - \bar{f}_{12}^<\frac{df_{12}^>}{dt}\right] + \right.$$
$$\left(\left|T_{12}^+(E)\right|^2 \frac{\mathcal{P}'}{\bar{E}-E} + \left|T_{12}^+(\bar{E})\right|^2 \frac{\mathcal{P}'}{\bar{E}-E} - \pi i\delta(E-\bar{E})\left[\frac{dT_{12}^+}{dE}T_{12}^- - T_{12}^+\frac{dT_{12}^-}{dE}\right]\right)$$
$$\left. \times \left[f_{12}^<\frac{d\bar{f}_{12}^>}{dt} - f_{12}^>\frac{d\bar{f}_{12}^<}{dt}\right]\right\}. \tag{8.70}$$

[14] There are two contributions to Eq. (8.70). The first follows from expanding the integral (8.62) with respect to $\tau_2 = t - t_3$ and yields, after Fourier transform

$$I_1^{(1)} = \frac{2}{\hbar}\mathrm{Tr}_2\left\{\left|T_{12}^+(\bar{E}+i\epsilon)\right|^2 \frac{d}{dE}\frac{\mathcal{P}}{E-\bar{E}}\left[\bar{F}_{12}^>\frac{dF_{12}^<}{dt} - \bar{F}_{12}^<\frac{d}{dt}F_{12}^>\right]\right\}. \tag{8.68}$$

The second contribution to $I^{(1)}$ arises from expanding $\bar{F}^>(t-\tau_1)$ and $\bar{F}^>(t-\tau)$, where $\tau = t - t_1, \tau_1 = t - t_2$ with respect to τ and τ_1, respectively. After some algebra, which involves the convolution theorem and relations (D.16), (8.39), one arrives at

$$\begin{aligned} I_2^{(1)} &= \frac{1}{\hbar}\int \frac{dp_2 d\bar{p}_1 d\bar{p}_2}{(2\pi\hbar)^9}\left\{-\left|T_{12}^+(\bar{E})\right|^2 \frac{\mathcal{P}'}{\bar{E}-E} - \left|T_{12}^+(E)\right|^2 \frac{\mathcal{P}'}{\bar{E}-E}\right. \\ &+ \left.\pi i\delta(E-\bar{E})\left[\frac{dT_{12}^+}{dE}T_{12}^- - T_{12}^+\frac{dT_{12}^-}{dE}\right]\right\}\left\{f_{12}^<\frac{d\bar{f}_{12}^>}{dt} - f_{12}^>\frac{d\bar{f}_{12}^<}{dt}\right\}. \end{aligned} \tag{8.69}$$

The sum of the two integrals (8.68) and (8.69) yields the result (8.70).

This result may be written in various ways. A particular helpful relation for the analysis of Eq. (8.70) may be derived from by differentiating the optical theorem (8.38) and using the dispersion relation for T_{12}^{+}:

$$T_{12}^{+}(\bar{E})\frac{\mathcal{P}' N_{12}}{\bar{E}-E}T_{12}^{-}(E) - \left|T_{12}^{+}(E)\right|^{2}\frac{\mathcal{P}' N_{12}}{\bar{E}-E}$$
$$= \pi i\delta(E-\bar{E})\left\{\frac{dT_{12}^{+}}{dE}N_{12}T_{12}^{-} - T_{12}^{+}N_{12}\frac{dT_{12}^{-}}{dE}\right\}. \qquad (8.71)$$

Eq. (8.71) determines the difference between the off–shell T–matrices for the energies $\bar{E}$ and E.

Classical limit. Drastic simplification of Eq. (8.70) is possible in the case of nondegenerate systems, i.e., if $\frac{d}{dt}(1-f_1-f_2) \approx 0$ and $f_{12}^{>} \approx 1$. In this case, we may write

$$I^{(1)}(p_1) = -\frac{1}{\hbar}\int\frac{dp_2 d\bar{p}_1 d\bar{p}_2}{(2\pi\hbar)^9}\left\{\left|T_{12}^{+}(E)\right|^{2}\frac{\mathcal{P}'}{\bar{E}-E} + \left|T_{12}^{+}(\bar{E})\right|^{2}\frac{\mathcal{P}'}{\bar{E}-E}\right.$$
$$\left. - i\pi\delta(E-\bar{E})\left[\frac{dT_{12}^{+}}{dE}T_{12}^{-} - T_{12}^{+}\frac{dT_{12}^{-}}{dE}\right]\right\}\frac{d}{dt}\left[\bar{f}_{12}^{>}(t) - f_{12}^{<}(t)\right]. \qquad (8.72)$$

With the help of Eq. (8.71), this expression may be reduced to the simpler form

$$I^{(1)}(p_1) = -\frac{2}{\hbar}\frac{d}{dt}\int\frac{dp_2 d\bar{p}_1 d\bar{p}_2}{(2\pi\hbar)^9}|T_{12}^{+}(E)|^2\frac{\mathcal{P}'}{E-\bar{E}}\left[\bar{f}_{12}^{<}(t) - f_{12}^{<}(t)\right]. \qquad (8.73)$$

Relation between zeroth and first order retardation terms. There exists a close connection of expressions $I^{(1)}$ and $I^{(0)}$, Eqs. (8.72) and (8.67). To derive it, let us define

$$I^{0}(\varepsilon) = -\frac{1}{\hbar}\int\frac{dp_2 d\bar{p}_1 d\bar{p}_2}{(2\pi\hbar)^9}\int d\omega\, 2\,\delta^{\varepsilon}(E-\omega)\, 2\,\delta^{\varepsilon}(\bar{E}-\omega)$$
$$\times \left|\langle p_1 p_2|\, T_{12}^{+}(E+i\epsilon)\,|\bar{p}_2\bar{p}_1\rangle\right|^2\left\{\bar{f}_{12}^{<}(t) - f_{12}^{<}(t)\right\}, \qquad (8.74)$$

where, $\delta^{\varepsilon}(x)$ is a broadened delta function, defined in App. A, and, therefore, $\lim_{\varepsilon\to 0} I^{0}(\varepsilon) = I^{0}$. Using this definition, one can derive an interesting relation between the first order retardation term $I^{(1)}(p_1)$ and the Boltzmann collision integral I^{0} (in agreement with Akhiezer and Peletminski [AP77]), which reads

$$I^{(1)}(p_1) = \frac{1}{2}\frac{d}{dt}\frac{d}{d\varepsilon}I^{0}(\varepsilon)\bigg|_{\varepsilon\to 0}. \qquad (8.75)$$

Consequently, for nondegenerate systems, the kinetic equation in first order gradient expansion may be written in the following compact form [KBBS96]

$$\boxed{\frac{d}{dt}f(p_1) = \left(1 + \frac{1}{2}\frac{d}{dt}\frac{d}{d\varepsilon}\right) I^0(\varepsilon)\Big|_{\varepsilon\to 0}} \tag{8.76}$$

As compared to Eq. (8.62), the result of Eq. (8.76) is less general, especially the non–Markovian character is lost, and no energy-broadening effects are retained. Nevertheless, retardation effects (and thereby correlation effects) are included up to the first order. Therefore, Eq. (8.76), goes far beyond the usual Boltzmann equation with the collision term (8.67). We will demonstrate this by considering the conservation laws.

Conservation of total energy. As already mentioned, the usual Boltzmann equation leads only to conservation laws for an ideal many–particle system. This follows from the delta–function $\delta(E_1 + E_2 - \bar{E}_1 - \bar{E}_2)$ in Eq. (8.67) which allows only scattering events that do not change the single–particle energy of the two particles. As a result, also the mean single-particle energy of the system is conserved. The first order gradient terms do not include such a delta function, and one can show that these gradient contributions lead to the correct conservation laws of correlated many–particle systems. We will demonstrate this for the conservation of (total) energy.

To this end, we multiply the kinetic equation (8.66), together with Eqs. (8.67), (8.72), by the kinetic energy $p_1^2/(2m_1)$ and calculate the trace with respect to the free index 1. Furthermore, we symmetrize the resulting expression with respect to the variables $1, 2$ and $\bar{1}, \bar{2}$, which yields the relation

$$\begin{aligned} \frac{\partial}{\partial t}\Big\{ \langle T\rangle + \frac{1}{2} \mathop{Tr}_{12} \Big[& T_{12}^{+}(E)\frac{\mathcal{P}}{E-\bar{E}}\bar{F}_{\bar{1}\bar{2}}^{<}(t)\, T_{\bar{1}\bar{2}}^{-}(\bar{E}) \\ & + T_{\bar{1}\bar{2}}^{+}(E)\frac{\mathcal{P}}{E-\bar{E}}\, T_{\bar{1}\bar{2}}^{-}(\bar{E})\, F_{\bar{1}\bar{2}}^{<}(t)\Big]\Big\} = 0. \end{aligned} \tag{8.77}$$

Using the classical limit of the potential energy in binary collision approximation (8.50),

$$\langle V\rangle = \frac{n^2}{2} \mathop{Tr}_{12} \left\{ \mathrm{Re}T_{12}^{+}(E)F_{\bar{1}\bar{2}}^{<}(t) - T_{12}^{+}(E)\frac{\mathcal{P}}{E-\bar{E}}\,\bar{F}_{\bar{1}\bar{2}}^{<}(t)\, T_{\bar{1}\bar{2}}^{-}(\bar{E})\right\}, \tag{8.78}$$

and the dispersion relation, $2\mathrm{Re}T_{\bar{1}\bar{2}}^{+}(E) = T_{12}^{+}(E)\frac{\mathcal{P}}{E-\bar{E}}T_{\bar{1}\bar{2}}^{-}(\bar{E})$, we find from Eq. (8.77) that the time derivative of total energy vanishes, i.e. $\langle T\rangle + \langle V\rangle = \text{const}$. This property of the first order retardation approximation is quite general and valid not only for the ladder approximation, cf. Chs. 6 and 9.

Density conservation. Bound states. We now consider consider the question of density conservation in the generalized Markovian Boltzmann equation which contains zeroth and first order retardation terms. Integrating this kinetic equation over p_1 it follows

$$\frac{dn}{dt} = \int \frac{dp_1}{(2\pi\hbar)^3} \left\{ I^0(p_1,t) + I^{(1)}(p_1,t) \right\} = \frac{dn^0}{dt} + \frac{dn^{(1)}}{dt}, \tag{8.79}$$

where n is the total density of the correlated (nonideal) system which consists of one-particle and correlation contributions. It is well known that the Markovian Boltzmann integral I^0 conserves density, i.e. $dn^0/dt = 0$. Similarly one readily proofs that $dn^{(1)}/dt = 0$, hence total density is conserved [Kli75].

It was mentioned above that the T-matrix approximation accounts for strong coupling effects, which, in the case of attractive potentials, includes the possibility of bound states. We briefly discuss how bound states are treated in our theory. In general, the binary correlation operator contains scattering and bound state contributions [KKER86], $g_{12} = g_{12}^{sc} + g_{12}^{b}$. The coupled equations for F_1 and g_{12}, Eq. (8.1) and (8.2) describe the dynamics of *all* correlations, starting from an initial value g_{12}^0. So, if at the initial moment, the system already contains bound states, g_{12}^0 will contain a contribution g_{12}^{0b}. Since the full non-Markovian Boltzmann equation (8.62) is equivalent to the coupled equations for F_1 and g_{12}, it fully contains scattering and bound state correlations too. In the general nonequilibrium situation, a clear separation of both contributions is not possible.

The situation is different if the Markov limit is taken. This limit includes weakening of initial correlations and is sensitive to the existence of bound state correlations which usually do not decay during the relaxation of F_1. Then we have to expect that the evolution of F_1 will be influenced by bound states and their dynamics,[15] which requires to selfconsistently solve a kinetic equation for

[15]It can be shown that there appears an additional collision integral in the kinetic equation for F_1, beyond the integrals I^0 and $I^{(1)}$, which contains the bound state distributions N_j [Dan90, BKKS96]. It is of the form

$$I^{(1b)}(p_1,T) = \frac{\partial}{\partial T} \int \frac{dp_2}{(2\pi\hbar)^3} \sum_{jP} \langle p_1 p_2 | \Psi^{jP} \rangle \, \langle \tilde{\Psi}^{jP} | p_2 p_1 \rangle \, N_j(P), \tag{8.80}$$

where the wave function Ψ^K obeys the eigenvalue equation (with the eigenvalues E_K)

$$\{E_K - \epsilon(p_1) - \epsilon(p_2)\} \langle p_1 p_2 | \Psi^K \rangle - [1 - f(p_1) - f(p_2)] \int \frac{d\bar{p}_1 d\bar{p}_2}{(2\pi\hbar)^6} \langle p_1 p_2 | V | \bar{p}_2 \bar{p}_1 \rangle \, \langle \bar{p}_1 \bar{p}_2 | \Psi^K \rangle ,$$

and $\tilde{\Psi}^K$ obeys the hermitean adjoint equation (due to the time dependence of the Pauli blocking factor these equations are not hermitean and one has to construct a bi-orthonormal basis with Ψ and $\tilde{\Psi}$).

the bound state distribution N_j (j labels all quantum numbers), which is of the form [BKKS96]

$$\frac{d}{dT}N_j(P,T) = I_j^{scatt} + I_j^{rearr} + I_j^{react}. \tag{8.81}$$

This equation accounts for the change of the bound state population due to elastic scattering, rearrangement processes and inelastic processes (e.g. excitation of bound states or their ionization/recombination etc.).

8.6 Numerical results and discussion

Importance of T-matrix effects. The most interesting questions for numerical analysis are, (1) how strong are T-matrix effects, compared to the simpler Born approximation and, (2), in which situations are these effects relevant. In the Introduction, it was shown, that weak coupling is observed in the limit of very high or low densities, $n > a_B^{-3}$ or $n < (e^2/\epsilon kT)^{-3}$, where a_B denotes the Bohr radius of the dominating bound state, and ϵ the background dielectric constant. In between, and especially at low temperatures, T-matrix effects are to be taken into account. In this Section, we present recent results which quantitatively compare T-matrix scattering rates to the corresponding weak coupling limit (i.e. the Born approximation) [GKSB98, GKSB] on the example of bulk semiconductors. We will be interested only in the simplest scattering mechanism - carrier-carrier scattering in electron-hole plasmas.[16]

Markovian T-matrix scattering rates. We consider the Markovian Boltzmann collision integral (8.67). It involves only "on-shell" T-matrices, which can be expressed by the differential scattering cross section according to [LL62, Tay72]

$$\frac{d\sigma_{ab}(p,\Omega)}{d\Omega} = (2\pi)^2\hbar^2 m_{ab}^4 \left|\langle \boldsymbol{p}\,|\mathbf{T}_{ab}|\,\overline{\boldsymbol{p}}\,\rangle^{\pm}\right|^2_{|\boldsymbol{p}|=|\overline{\boldsymbol{p}}|}. \tag{8.82}$$

Here, $\boldsymbol{p}$ is the momentum of relative motion, Ω scattering angle and $m_{ab} = m_a m_b/(m_a + m_b)$ denotes the reduced mass.[17] In the case of non-degenerate carriers, the cross sections can be evaluated efficiently using a phase shift approach to solve the radial Schrödinger equation[18], for details see Appendix F.

[16] For a T-matrix treatment of other processes such as impurity scattering in solids, see [Mah81] and references therein.

[17] The index $\pm$ indicates that, for $a = b$, exchange is included in the T-matrix.

[18] In the calculations presented, the radial Schrödinger equation was solved for a statically screened Coulomb potential and self energy shifts in Debye approximation[Zim87].

The magnitude of the cross section depends essentially on the scattering partners. In the case of electron-hole scattering, the interaction is attractive, and close collisions with large angle scattering are important. On the other hand, for identical carriers, exchange effects effectively lead to "back scattering". In Fig. 8.1. we show numerical comparisons of T-matrix total cross sections (i.e. the angle integrated differential cross section) and the corresponding Born approximation result, both calculated with a statically screened Debye potential with the inverse screening length $\kappa = 1/a_B$. One clearly sees that both approximations merge at high momenta, because there the kinetic energy exceeds the correlation energy. Deviations occur at small momenta. Interestingly, the $e-e$ cross section in T-matrix approximation is smaller than the Born approximation result, and the exchange leads to further reduction. On the other hand, as shown in Fig. 8.1, there are situations where the $e-h$ T-matrix cross section is larger than the Born approximation result, what is due to resonances[19] [GKSB98]. For smaller values of κ (lower densities), the resonance vanishes, and the cross section decreases below the Born approximation.

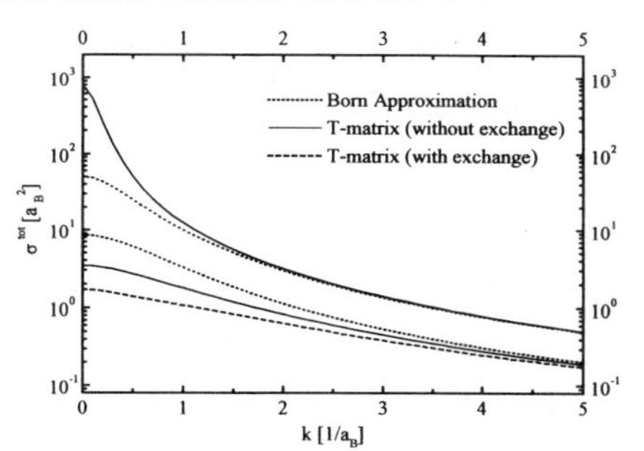

Fig. 8.1. Total cross section for electron-hole (upper two lines) and electron-electron scattering (lower three lines) in T-matrix approximation compared to the Born approximation. Bulk semiconductor with $\kappa = 1.0/a_B$.

Using the results for the cross sections, one can calculate the scattering rates $\Sigma^<$ and $\Sigma^>$[20] in the Markovian Boltzmann collision integral, i.e. $I^0(p,t) = \Sigma^<(p,t)f^>(p,t) - \Sigma^>(p,t)f^<(p,t)$. In the scattering rates, the cross section is averaged by the carrier distributions (explicit results for the T-matrix scattering rates are derived in Appendix F), and it is not clear if differences of the T-matrix cross section vs. the Born approximation will survive in the scattering rates. Therefore, we compare the sum of the scattering rates (the dephasing rate Γ) for the two cases in Fig. 8.2. We see that indeed T-matrix effects show up in the scattering rates at low momentum values. The effect is particularly strong at low temperatures and moderate densities - i.e. inside the *corner of correlations*, confirming our general discussion on many-body effects in Ch. 1.

---

[19] I.e. bound states merged into the continuum due to the Mott effect. Here, the effect comes from the vanishing of the 1s bound state.

[20] I.e. the probabilities of scattering events "into" and "out of" state p, respectively.

We mention that another quite efficient method in equilibrium is to solve the Lippmann-Schwinger equation by means of matrix inversion, e.g. [SSTH98].

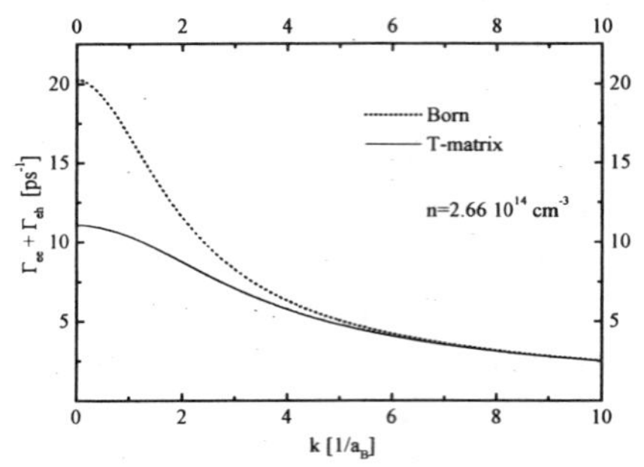

Fig. 8.2. Total electron dephasing rate $\Gamma_{ee} + \Gamma_{eh}$ ($\Gamma_{ab} = \Sigma^>_{ab} + \Sigma^<_{ab}$) for $n = 2.66 \times 10^{14} cm^{-3}$ and $T = 150K$.

It is possible to extend this analysis to nonequilibrium systems, for which explicit results fort the scattering rates can be given too (see Appendix F). Recent investigations have shown that the qualitative behavior remains the same as in equilibrium, but the magnitude of the strong coupling effects depends on the explicit form of the distribution function. For a detailed analysis and a discussion of dynamical interaction effects, see Ref. [GKSB]. Furthermore, the inclusion of bound states requires to consider also off-shell T-matrices (see above)[21].

Summary and comments. In this Chapter we presented a derivation of a quantum kinetic equation in binary collision approximation which generalizes the conventional quantum Boltzmann equation for the Wigner distribution function in several directions. 1., we extended the kinetic equation to the regime of ultrashort times by including the full dynamics of the binary correlations. 2., we included many-body (medium) effects on the two-particle interaction, by taking into account relevant contributions from three-particle correlations. 3., by avoiding any assumption about weakness of the interaction, our results are valid for systems with strong coupling, applying to strongly correlated systems, in particular to systems with bound states. Solving the BBGKY-hierarchy with the closure (8.3) allowed us to incorporate selfenergy effects on the level of the full ladder approximation.

Of course, there remain numerous open questions. It remains a challenging yet unfeasible task to solve the kinetic equation (8.59) with the integrals (8.62), (8.65), numerically. Further approximations for the T-matrices are necessary. Then it will eventually be possible to investigate the ultra-short-time dynamics of strongly correlated systems including the dynamics of bound state correlations.

---

[21]We mention that there have been developed efficient approximation techniques, such as expansion of the interaction in terms of separable potentials, for a recent study see [KK95].

Chapter 9

*Random Phase Approximation

In this Chapter, we continue the analysis of correlations and their buildup. While in the previous Chapters, we considered many-body systems with static interaction, in this Chapter, we consider the dynamics of charged particles and the build up of dynamical screening of the Coulomb interaction[1]. We will assume that the interaction is weak, i.e. that the ratio of potential to kinetic energy (coupling parameter Γ) is less than 1. This leads to the polarization approximation of the BBGKY–hierarchy which corresponds to the random phase approximation (RPA) of Green's functions theory. The resulting kinetic equation is the Balescu-Lenard equation [Bal60, Len60] which has been re-derived in various ways and discussed in great detail by many authors, e.g. [Kli75, Sil67, Ich73]. We will derive the non-Markovian generalization of this equation.[2] The extension to strong dynamic interaction is discussed in Ch. 10.

9.1 Generalized polarization approximation. Selfenergy

The polarization approximation to the hierarchy has been introduced in Sec. 2.6. With the modifications from spin statistics effects included, Sec. 3.3, this approximation is given by

$$i\hbar\frac{\partial}{\partial t}F_1 - [\bar{H}_1, F_1] \quad = \quad n\mathrm{Tr}_2[V_{12}^{\pm}, g_{12}], \tag{9.1}$$

[1]The results of this Chapter have been obtained together with D. Kremp and, in part, J.W. Dufty, see Refs. [BDK98, BKDK].

[2]Further references will be given below. Also, the Green's functions approach to screening is discussed in Ch. 12.

$$\begin{aligned}
i\hbar\frac{\partial}{\partial t}g_{12}[\bar{H}^0_{12}\,g_{12}] &= Q_{12} \\
+\Pi^{(1)}_{12}+\Pi^{(2)}_{12} &+ n\mathrm{Tr}_3[V_{13}+V_{23},g_{123}](1\pm P_{13}\pm P_{23}),
\end{aligned} \tag{9.2}$$

where $\bar{H}^0_{12}=\bar{H}_1+\bar{H}_2$, and the effective single–particle Hamiltonian is $\bar{H}_1 = H_1+H^{HF}_1 = H_1+n\mathrm{Tr}_2V_{12}F_2\Lambda^{\pm}_{12}$. The inhomogeneity Q_{12} and the polarization contributions are

$$Q_{12} = \hat{V}_{12}F_1F_2 - F_1F_2\hat{V}^{\dagger}_{12}, \tag{9.3}$$

$$\Pi^{(1)}_{12} = n\mathrm{Tr}_3[V^{\pm}_{13},F_1]g_{23}\Lambda^{\pm}_{23}, \tag{9.4}$$

and $\Pi^{(2)}_{12}$ follows from the substitution $1\longleftrightarrow 2$. The shielded potential $\hat{V}_{12}$, the (anti-)symmetrization factor $\Lambda^{\pm}_{12}$ and the permutation operator P_{12} account for Pauli blocking and exchange, respectively, and have been defined in Ch. 3.

In Eq. (9.2) we still included three–particle correlations what allows for some additional freedom in the approximation. As in Chs. 7 and 8, we will consider two cases:

I. $g_{123}=0$, this leads to the "usual" polarization approximation, and

II. $g_{123}\neq 0$, but it is defined by Eq. (9.5) or (9.6). This leads to a generalization of the polarization approximation by inclusion of selfenergy corrections. As a result, the single–particle Hamiltonian $\bar{H}_1$ will contain an additional selfenergy contribution. This approximation will be discussed below in this Section.

RPA–Selfenergy terms in the third hierarchy equation. Following our concept to introduce selfenergy in the BBGKY–hierarchy which was developed in Ch. 7, we obtain now the corresponding expression for the polarization approximation. We again consider the third hierarchy equation for the ternary correlations (2.99) where we keep the relevant part of the inhomogeneity (as in the Chapters before) and also the three–particle polarization terms [BDK98],

$$\begin{aligned}
i\hbar\frac{\partial}{\partial t}g_{123} - \left\{H^{\mathrm{eff}}_{123}\,g_{123} - g_{123}\,H^{\mathrm{eff}\dagger}_{123}\right\} = [V_{13}+V_{23},F_3g_{12}] \\
+n\mathrm{Tr}_4[V_{14},F_1\,g_{234}] + n\mathrm{Tr}_4[V_{24},F_2\,g_{134}] + n\mathrm{Tr}_4[V_{34},F_3\,g_{124}],
\end{aligned} \tag{9.5}$$

where $H^{\mathrm{eff}}_{123}=\bar{H}_1+\bar{H}_2+\bar{H}_3$. This equation does not contain spin statistics effects yet (the complete equation will be given below, Eq. (9.6)), but it allows for a better understanding of this approximation. The inhomogeneity, i.e. the first term on the r.h.s. of Eq. (9.5), is the familiar basic term which gives rise to selfenergy–type diagrams in the equation for g_{12}. The idea is again to solve Eq. (9.5) for g_{123} as a functional of g_{12} and to include the resulting term (which appears on the r.h.s. of Eq. (9.2)) into the Hamilton operator $\bar{H}^0_{12}$. The result will be a renormalized Hamiltonian $\bar{H}^0_{12} \to H^{\rm eff}_{12}$ which contains finite lifetime (damping) effects that are essential for the correct description of the short–time behavior (for details, see Ch. 7).

The specifics here is the inclusion of the three–particle polarization terms. This provides the necessary consistency with the polarization approximation for g_{12}. The explicit expressions obtained for the RPA–selfenergy below will justify this choice. Notice also, that the three-particle Hamilton operator is already renormalized, i.e. it contains the renormalized single–particle Hamiltonians, which are given below by Eq. (9.8). This selfconsistent scheme is equivalent to the summation of an infinite series of diagrams, which just yields the selfenergy expressions familiar from Green's functions theory [KB89]. We now complete Eq. (9.5) by including the additional terms which are due to the spin statistics, cf. Eq. (3.24)

$$i\hbar\frac{\partial}{\partial t}g_{123} - \left\{H^{\rm eff}_{123}\, g_{123} - g_{123}\, H^{\rm eff\dagger}_{123}\right\} =$$
$$\left\{(\hat{V}_{13} + \hat{V}_{23})F_3 g_{12} \mp nF_3(F_1 V_{13} + F_2 V_{23})g_{12} \mp n(g_{13}V_{13} + g_{23}V_{23})g_{12} - \text{h.c.}\right\}$$
$$+\Pi^{(1)}_{123} + \Pi^{(2)}_{123} + \Pi^{(3)}_{123}, \qquad (9.6)$$

where the three–particle polarization terms are defined as $\Pi^{(1)}_{123} = n\text{Tr}_4[V^{\pm}_{14}, F_1 g_{234}](1 \pm P_{24} \pm P_{34})$ and permutations (cf. Eq. (9.5)).

Eq. (9.6) closes the system (9.1), (9.2) and defines our generalized polarization approximation.[3] To derive explicit results for the selfenergy Σ^+, we will solve Eq. (9.6) and then transform the solution in analogy to the case of the ladder approximation, Sec. 8.2. The solution g_{123} will be expressed in terms of single–particle propagators which will turn out to be the same as in the solution of the second hierarchy equation. We, therefore, consider first the solution for g_{12}, since it is simpler, and return to the selfenergy problem below in this Section.

Binary correlation operator in RPA. Using Eqs. (9.2) and (9.6), we can rewrite the second hierarchy equation as

$$i\hbar\frac{\partial}{\partial t}g_{12} - \left\{H^{\rm eff}_{12}\, g_{12} - g_{12}\, H^{\rm eff\dagger}_{12}\right\} \;=\; Q_{12} + \Pi^{(1)}_{12} + \Pi^{(2)}_{12}, \qquad (9.7)$$

where the three–particle correlations are included into $H^{\rm eff}_{12} = \bar{H}_1 + \bar{H}_2$ via the

[3] As we will see below, this approximation is equivalent to the selfconsistent random phase approximation of Green's functions theory, where all propagators are renormalized by selfenergies in RPA.

renormalized single–particle Hamiltonians $\bar{H}_1$,

$$\bar{H}_1\, g_{12}(t) = \left(H_1 + H_1^{HF}\right)\, g_{12}(t) + \int_{t_0}^{t} d\bar{t}\, \Sigma_1^{+}(t\bar{t})\, U_2(t\bar{t})\, g_{12}(\bar{t}), \tag{9.8}$$

where Σ_1^+ and U_2 denote the retarded RPA selfenergy and the quasi–particle propagator which will be determined below. To simplify the solution, we will neglect exchange polarization effects, replacing $V_{13}^{\pm} \to V_{13}$ and $\Lambda_{23}^{\pm} \to 1$ in $\Pi_{12}^{(1)}$ and so on. We solve Eq. (9.7) with the initial condition $g_{12}(t_0) = g_{12}^0$ by making the ansatz

$$\boxed{g_{12}(t) = \mathrm{Tr}_{34} U_{13}(tt_0)\, U_{24}(tt_0)\, g_{34}^0 + \frac{1}{i\hbar}\mathrm{Tr}_{34} \int_{t_0}^{t} d\bar{t}\, U_{13}(t\bar{t})\, U_{24}(t\bar{t})\, Q_{34}(\bar{t})} \tag{9.9}$$

By construction, the solution contains retardation effects in time but also in the particle index which will allow to account for polarization effects. U_{13} is the generalization of the dielectric propagator (see e.g. [Ich73]), to quantum systems with selfenergy.[4] It obeys the linearized quantum Vlasov (Hartree) equation, cf. Ch. 4,

$$\boxed{\begin{aligned} i\hbar\frac{\partial}{\partial t} U_{13}(tt') - \left\{\bar{H}_1\, U_{13}(tt') - U_{13}(tt')\, \bar{H}_1^{\dagger}\right\} &= n\mathrm{Tr}_5\left[V_{15}, F_1\right] U_{53}(tt') \\ U_{13}(tt) &= \delta_{13} \end{aligned}} \tag{9.10}$$

where the second line defines the initial condition. One readily checks that the propagator defined by Eq. (9.10) indeed generates the correct solution $g_{12}(t)$.

Collision integral in generalized polarization approximation. Inserting the formal solution (9.9) into the first hierarchy equation (9.1), we obtain the collision integral of the kinetic equation in generalized polarization approximation (RPA),

$$\begin{aligned} I_1(t) &= n\mathrm{Tr}_{234} V_{12}^{\pm}\, U_{13}(tt_0)\, U_{24}(tt_0)\, g_{34}^0 + \frac{n}{i\hbar}\mathrm{Tr}_{234} \int_{t_0}^{t} d\bar{t} \\ &\times\ V_{\bar{1}\bar{2}}^{\pm}\, U_{13}(t\bar{t})\, U_{24}(t\bar{t})\, \{F_{34}^{>}\, V_{34}\, F_{34}^{<} - F_{34}^{<}\, V_{34}\, F_{34}^{>}\}\,|_{\bar{t}} - \mathrm{h.c.}, \end{aligned} \tag{9.11}$$

where we denoted $F_{ab}^{\gtrless} = F_a^{\gtrless} F_b^{\gtrless}$, $F_a^{>} = 1 \pm nF_a^{<}$ and $F_a^{<} = F_a$. The first integral is due to the initial correlations g^0, while the second one describes the correlation buildup including exchange terms.

[4] Alternatively, one can use retarded and advanced propagators, which are defined as $U^{\pm}(tt') = \Theta[\pm(t-t')]U(tt')$. Having this in mind, we will be able to go over to $U^{\pm}$ at any time later. Notice also, that U_{13} is different from the two-particle propagator in Ch. 8.

Result for the RPA–selfenergy. We now return to the third hierarchy equation (9.6). Since it has the same structure as the equation for g_{12}, accept for the additional particle index, we may use the same type of ansatz for $g_{123}(t)$, as for g_{12}, Eq. (9.9):

$$\begin{aligned} g_{123}(t) &= \mathrm{Tr}_{456}\, U_{14}(tt_0)\, U_{25}(tt_0)\, U_{36}(tt_0)\, g^0_{456} \\ &+ \frac{1}{i\hbar}\mathrm{Tr}_{456}\int_{t_0}^t d\bar{t}\, U_{14}(t\bar{t})\, U_{25}(t\bar{t})\, U_{36}(t\bar{t})\, R_{456}(\bar{t}), \end{aligned} \tag{9.12}$$

where g^0_{123} denotes initial three–particle correlations and R denotes the inhomogeneity in the third hierarchy equation, i.e. the whole term in brackets on the r.h.s. of Eq. (9.6). We begin the analysis by dropping the initial correlation term and taking R for the spinless case, $R_{123} = (V_{13} + V_{23})\, F_3\, g_{12} - \text{h.c.}$ and restore the full expression in the final result (9.17).

From this point, the derivation follows the same lines as in Sec. 7.4, and we may drop all the details. Inserting the solution (9.12) into Eq. (9.2), we obtain two terms of the form

$$S_1(t) = \frac{n}{i\hbar}\mathrm{Tr}_{3456}\int_{t_0}^t d\bar{t}\, V_{13}\, U_{14}(t\bar{t})\, U_{25}(t\bar{t})\, U_{36}(t\bar{t})\, V_{46}\, F_6(\bar{t})\, g_{45}(\bar{t}), \tag{9.13}$$

where S_2 follows from S_1 by replacing $V_{13} \to V_{23}$ and $V_{46} \to V_{56}$. There are two further terms, which are the hermitean conjugates of S_1 and S_2, respectively. Other combinations of potentials do not contribute to selfenergy. Since with the approximation (9.6) the dynamics of g_{123} is determined by correlations of the form $g_{12}F_3$, cf. the inhomogeneity R, the propagators, in combination with two potentials, must have a similar product structure,

$$V_{13}\, U_{14}\, U_{25}\, U_{36}\, V_{46} = V_{13}\, U_{14}\, U_{36}\, V_{46}\ \cdot U_2\, \delta_{2,5}. \tag{9.14}$$

This allows us to perform in Eq. (9.13) the trace over "5" and to rewrite this expression as

$$\begin{aligned} S_1(t) &= \frac{n}{i\hbar}\mathrm{Tr}_{346}\int_{t_0}^t d\bar{t}\, V_{13}\, U_{14}(t\bar{t})\, U_{36}(t\bar{t})\, V_{46}\, F_6(\bar{t})\, U_2(t\bar{t})\, g_{42}(\bar{t}) \\ &= \int_{t_0}^t d\bar{t}\, \Sigma_1^+(t\bar{t})\, U_2(t\bar{t})\, g_{12}(\bar{t}), \end{aligned} \tag{9.15}$$

where we introduced the retarded selfenergy

$$\Sigma_1^+(t\bar{t}) = \frac{n}{i\hbar}\mathrm{Tr}_{346}\, V_{13}\, U_{14}(t\bar{t})\, U_{36}(t\bar{t})\, V_{46}\, F_6(\bar{t}). \tag{9.16}$$

This is the retarded selfenergy in RPA, expressed in terms of generalized dielectric propagators. What remains now is to restore the tree–particle initial correlations and the spin statistics corrections in expression (9.16). The derivation is the same as discussed in the ladder approximation, Sec. 8.2, cf. Eq. (8.14),

$$\boxed{\Sigma_1^+(t\bar{t}) = \frac{n}{i\hbar}\mathrm{Tr}_{346}\, V_{13}^{\pm}\, U_{14}(t\bar{t})\, U_{36}(t\bar{t}) \left[F_{46}^{>}\tilde{V}_{46}F_6^{<} \mp F_{46}^{<}\tilde{V}_{46}F_6^{>} \mp n g_{46}\tilde{V}_{46}\right]\Big|_{\bar{t}}} \tag{9.17}$$

and, as before, we introduced an effective potential that includes the initial correlations $\tilde{V}_{ab}(t) = V_{ab} + i\hbar\delta(t - t_0)$.[5]

With these results, the kinetic equation in generalized polarization approximation, including selfenergy in RPA, has been found. We were able to reduce the dynamics of the binary correlations to that of simpler single–particle propagators, what is a result of the weak–coupling limit (neglect of ladder terms). In fact, the propagators U_{13} are effectively two–particle propagators too, which describe the transfer of correlations by long–range polarization effects. As is well known, these effects lead to screening of the long–range Coulomb interaction. In the following, we show that screening and screening buildup are indeed contained in the equations derived above. To this end, it is necessary to go over from a description in terms of the dielectric propagators to a more familiar one which involves typical plasma quantities, such as the screened potential V_s or the inverse dielectric function ϵ^{-1}.

9.2 Dynamical screening in nonequilibrium

We now return to the equation of motion of the dielectric propagator (9.10). Similar as for the ladder approximation, Ch. 8, it is advantageous to rewrite the differential equation for the propagators (9.10) as an integral equation,

$$\boxed{\begin{aligned} U_{13}(tt') &= U_{13}^0(tt') + \frac{1}{i\hbar}\mathrm{Tr}_7 \int_{t_0}^t d\bar{t}\, U_{17}^0(t\bar{t})\, n\mathrm{Tr}_5\left[V_{75}, F_7(\bar{t})\right] U_{53}(\bar{t}t') \\ &= U_{13}^0(tt') + \frac{1}{i\hbar}\mathrm{Tr}_7 \int_{t_0}^t d\bar{t}\, U_{17}(t\bar{t})\, n\mathrm{Tr}_5\left[V_{75}, F_7(\bar{t})\right] U_{53}^0(\bar{t}t') \end{aligned}} \tag{9.18}$$

which can be verified by direct differentiation. U^0 is the free propagator which obeys Eq. (9.10) without the polarization term.

Momentum representation. The relation to the screening properties becomes particularly transparent in the homogeneous case. We, therefore, transform the propagators and their equations of motion into the momentum representation. We first introduce the short notation[6] $1 = p_1$, $\sum_1 = 2\int dp_1/(2\pi\hbar)^3$,

[5] The last term in brackets is a higher order correction, for its discussion, see Sec. 7.4.

[6] All momenta are understood as vectors.

the momentum transfer $q = 1' - 1$ and the matrix elements of the density operator, the Coulomb potential and the renormalized Hamiltonian by $\langle 1|nF_1|1'\rangle = f_1\delta_{1,1'}$, $\langle 12|V_{12}|2'1'\rangle = V_q\delta_{1+2,1'+2'}/(2\pi\hbar)^3$, $\langle 1|\bar{H}_1|1'\rangle = \epsilon_1\delta_{1,1'}$, (for a basic discussion of the momentum representation of the BBGKY–hierarchy, see Sec. 2.3.3). The matrix elements of the dielectric propagator are denoted by

$$\begin{aligned} \langle 12|U_{12}(tt')|2'1'\rangle &= U(1,2;1',2',tt')\,\delta_{1+2,1'+2'}, && (9.19)\\ \text{with}\quad \langle 12|U_{12}(tt)|2'1'\rangle &= \delta_{1,2'}\,\delta_{2,1'}, \quad \text{or, equivalently,} \\ U(1,2;1+q,2-q,tt) &= \delta_{1,2-q}, && (9.20) \end{aligned}$$

where the last two lines represent the initial condition. In similar way, the matrix elements of the free propagators are defined which, moreover, have only two nontrivial arguments, $U^0(13;1'3'tt')\,\delta_{1+3,1'+3'} = U^0(11',tt')\,\delta_{1,3'}\delta_{1',3}$.[7] For illustration, we consider the *local approximation* for the propagators, cf. App. D. Then, we obtain from Eq. (9.22) with $\tau = t - t'$ and $\epsilon = E - i\gamma$,

$$U^0(1,1+q,\tau) = e^{-\frac{\gamma_1+\gamma_{1+q}}{\hbar}\tau}e^{-i\frac{E_1-E_{1+q}}{\hbar}\tau}. \tag{9.23}$$

The momentum representation of the integral equation (9.18) is readily obtained,

$$\begin{aligned} U(p_1,p_3;p_1+q,p_3-q,tt') &= U^0(p_1,p_1+q,tt')\,\delta(p_1-p_3+q) \\ &+\frac{V_q}{i\hbar}\int_{t'}^{t} d\bar{t}U^0(p_1,p_1+q,t\bar{t})\,\{f_{p_1+q}-f_{p_1}\}\,|_{\bar{t}}\,I(p_3,q,\bar{t}t'), \end{aligned} \tag{9.24}$$

where we introduced the abbreviation for the polarization integral

$$I(p_3,q,tt') = \sum_5 U(p_5-q,p_3;p_5,p_3-q,tt'). \tag{9.25}$$

With the integral equation (9.24) we have expressed the dielectric propagator in terms of the free propagator U^0. The first term corresponds to the free (without

[7]For completeness, we give the equation of the matrix elements of the free propagator

$$\left\{i\hbar\frac{\partial}{\partial t} - \left(\epsilon_{p_1} - \epsilon^*_{p_1+q}\right)\right\}U^0(p_1,p_1+q,tt') = 0; \quad U^0(p_1,p_1+q,tt) = 1. \tag{9.21}$$

The solution of this commutator equation is

$$U^0(1,1+q,tt') = \exp\left\{-\frac{i}{\hbar}\int_{t'}^{t} d\bar{t}\left[\epsilon_1(\bar{t}) - \epsilon^*_{1+q}(\bar{t})\right]\right\}, \tag{9.22}$$

which, obviously, is related to the quasiparticle propagators (cf. Appendix D) simply by $U^0(1,1+q,tt') = U_1(tt')\,U^*_{1+q}(tt')$.

polarization effects) propagation of a particle, while the integral describes the interaction with the medium. However, this is not an explicit equation for U, since it contains also the integral over U. We therefore, first derive an equation for the integral I by integrating Eq. (9.24) over p_1,

$$I(p_3, q, tt') = U^0(p_3 - q, p_3, tt') + V_q \int_{t'}^{t} d\bar{t}\, \Pi(q, t\bar{t})\, I(p_3, q, \bar{t}t'), \tag{9.26}$$

where we defined the **RPA polarization function**

$$\boxed{\Pi(q, tt') = \frac{1}{i\hbar} \sum_{p_1} U^0(p_1, p_1 + q, tt')\, \{f_{p_1+q} - f_{p_1}\}\,|_{t'}} \tag{9.27}$$

Equation (9.26) can be solved with the ansatz

$$I(p_3, q, tt') = \int_{t'}^{t} d\bar{t}\, \epsilon^{-1}(q, t\bar{t})\, U^0(p_3 - q, p_3, \bar{t}t'), \tag{9.28}$$

which, inserted into Eq. (9.26), yields, after straightforward calculations,[8]

$$\boxed{\epsilon^{-1}(q, tt') = \delta(t - t') + V_q \int_{t'}^{t} d\bar{t}\, \Pi(q, t\bar{t})\, \epsilon^{-1}(q, \bar{t}t')} \tag{9.31}$$

Equation (9.31) has the well–known form of a Dyson equation, what suggests to identify ϵ^{-1} with **the inverse dielectric function**.

[8]We outline the main steps. Inserting the ansatz (9.28) into the equation of motion for I, Eq. (9.26), we obtain

$$\int_{t'}^{t} dt_1\, \epsilon^{-1}(q, tt_1)\, U^0(3 - q, 3, t_1t') = U^0(3 - q, 3, tt')$$
$$+ V_q \int_{t'}^{t} dt_1\, \Pi(q, tt_1) \int_{t'}^{t_1} dt_2\, \epsilon^{-1}(q, t_1t_2)\, U^0(3 - q, 3, t_2t'). \tag{9.29}$$

Changing in the integral term the order of integrations $\int_{t'}^{t} dt_1 \int_{t'}^{t_1} dt_2 \rightarrow \int_{t'}^{t} dt_2 \int_{t_2}^{t} dt_1$ and then $t_1 \longleftrightarrow t_2$, we obtain

$$\int_{t'}^{t} dt_1\, \epsilon^{-1}(q, tt_1)\, U^0(3 - q, 3, t_1t') =$$
$$\int_{t'}^{t} dt_1 \left\{ \delta(t - t_1) + V_q \int_{t_1}^{t} dt_2\, \Pi(q, tt_2)\, \epsilon^{-1}(q, t_2t_1) \right\} U^0(3 - q, 3, t_1t'). \tag{9.30}$$

This equation has to be fulfilled for arbitrary functions U^0, and we conclude that the integrands are equal. Therefore, the expression in parentheses must vanish, what yields the Dyson equation (9.31).

Screening buildup. We now introduce the **dynamically screened potential**,

$$\boxed{V_s(q,tt') = V_q\,\epsilon^{-1}(q,tt')} \tag{9.32}$$

which, obviously, obeys the same equation as ϵ^{-1},

$$V_s(q,tt') = V_q\,\delta(t-t') + V_q \int_{t'}^{t} d\bar{t}\,\Pi(q,t\bar{t})\,V_s(q,\bar{t}t'). \tag{9.33}$$

Thus we have obtained important results, which are the basis for a description of the plasmon kinetics and screening buildup at short times. We briefly discuss these relations:

I. The system (9.31), (9.32) and (9.27) is the basis to describe the *dynamics of the screened Coulomb potential*, buildup of screening etc.[9] in the weak coupling limit, see also Sec. 9.3.3. This system has no restrictions with respect to the times and is, in particular, applicable to the initial stage of relaxation. These expressions are valid under arbitrary nonequilibrium conditions, what finds its expression in the dependence of V_s and ϵ^{-1} on two times, which is an immediate result of our calculations.

II. The dynamics of the screened potential is coupled to the dynamics of the particles via the distribution functions in Eq. (9.27), where the particles evolve according to the kinetic equation with the collision integral (9.11). This collision integral contains initial correlations, as well as screening effects via the dielectric propagators, which evolve according to Eq. (9.35). This system is closed by the dynamics of the free propagators U^0, Eq. (9.22), which fully selfconsistently includes selfenergy effects in RPA via Eq. (9.17.)

III. This system of equations followed from the BBGKY-hierarchy with the closure (9.6) without any further approximation.[10] In particular, the Dyson equation for V_s followed naturally from the equation of motion for the dielectric propagator. Furthermore, we have obtained the nonequilibrium polarization function Π which is defined by Eq. (9.27). As the Dyson equation, this expression is well known from Green's functions theory.[11]

[9] or, equivalently the longitudinal component of the electromagnetic field, see also Ch. 12

[10] The ansatz (9.26) was only used for identical transformations.

[11] There it is obtained using the selfenergy in RPA and by applying the generalized Kadanoff–Baym ansatz (GKBA), cf. Ch. 12. With Eq. (9.27) we have, in fact, derived the GKBA in RPA from the hierarchy.

IV. We have seen, that screening properties can be described by two alternative means - in terms of the dielectric propagator U or the screened potential V_s, respectively. Obviously, these quantities are closely related. Indeed, rewriting Eq. (9.28) in terms of U and V_s, we obtain the relation[12]

$$\boxed{V_q \sum_{p_5} U(p_5 - q, p_3; p_5, p_3 - q, tt') = \int_{t'}^{t} d\bar{t}\, V_s(q, t\bar{t})\, U^0(p_3 - q, p_3, \bar{t}t')} \tag{9.34}$$

This relation allows to transfer the correlation (polarization) effects from the propagators onto an effective interaction potential.

Using the result (9.28) for the integral over the propagator, I, we now obtain the solution for the propagator itself. Inserting expression (9.28) into Eq. (9.24), we obtain the solution

$$\boxed{\begin{aligned} U(p_1, p_3; p_1 + q, p_3 - q, tt') = U^0(p_1, p_1 + q, tt')\, \delta(p_1 - p_3 + q) + \frac{V_q}{i\hbar} \times \\ \int_{t'}^{t} dt_1 \int_{t'}^{t_1} dt_2 U^0(p_1, p_1 + q, tt_1) \left\{f_{p_1+q} - f_{p_1}\right\}\big|_{t_1} \epsilon^{-1}(q, t_1 t_2) U^0(p_3 - q, p_3, t_2 t') \end{aligned}} \tag{9.35}$$

which expresses the dielectric propagator in terms of the free propagator and the inverse dielectric function and thus, closes the system of equations. This equation generalizes the corresponding Markovian result obtained by various authors, e.g. [Ich73].

Density fluctuation function. For completeness, we mention that the dielectric propagator and the free propagator are directly related to the density fluctuation functions L and L^0, respectively:

$$L(1, 3; 1 + q, 3 - q, tt') = \frac{1}{i\hbar} U(1, 3; 1 + q, 3 - q, tt')\, \{f_{1+q} - f_1\}\big|_{t'}, \tag{9.36}$$

$$L^0(1, 1 + q, tt') = \frac{1}{i\hbar} U^0(1, 1 + q, tt')\, \{f_{1+q} - f_1\}\big|_{t'}, \tag{9.37}$$

and L^0 is closely connected with the RPA polarization function Π, Eq. (9.27), by $\Pi(q, tt') = \sum_1 L^0(1, 1 + q, tt')$. Multiplying Eq. (9.35) by $f_{p_3} - f_{p_3 - q}$, we can derive the corresponding representation of L in terms of L^0,

$$\begin{aligned} L(p_1, p_3; p_1 + q, p_3 - q, tt') = L^0(p_1, p_1 + q, tt')\, \delta(p_1 - p_3 + q) + \\ \frac{V_q}{i\hbar} \int_{t'}^{t} dt_1 \int_{t'}^{t_1} dt_2 L^0(p_1, p_1 + q, tt_1) \epsilon^{-1}(q, t_1 t_2) L^0(p_3 - q, p_3, t_2 t'), \end{aligned} \tag{9.38}$$

[12] Equivalently, Eq. (9.34) can be used to define V_s, replacing the ansatz (9.26). This relation is analogous to the relation (8.34) between propagators and the T-operators in case of the binary collision approximation, cf. Ch. 8.

which completely eliminates the distribution functions. These relations can also be derived on the abstract operator level, using the integral equations (9.18), and also for more general (not simply distance-dependent) interactions V_{12}. Due to the complex character of the function L, this is a very general relation, from which various other equations for screening quantities can be derived, see e.g. [SB87]. Again, we underline, that this result is obtained straightforwardly from the hierarchy in generalized polarization approximation without any additional assumptions.

Retarded and advanced quantities. The above equations which involve two-time quantities U, U^0, ϵ^{-1}, Π, V_s, L and L^0 are formally applicable to arbitrary times. With the initial values of the density operators F_1 and g_{12} given, we have obtained also the initial values for the dielectric propagator $U(tt)$, Eq. (9.10). Starting with these values, the evolution of the density operators can proceed either forward ($t > t' \geq t_0$) or backward ($t < t' \leq t_0$). All derived above relations are applicable to both cases. It is sometimes advantageous to formally distinguish the propagators for both cases by rewriting the dielectric propagator in terms of advanced and retarded propagators $U^{\pm}(tt') = \Theta[\pm(t-t')]U(tt')$, as we did in Ch. 8, see also App. D. As a result, all equations for the "ordinary" quantities split into a pair of equations for retarded and advanced quantities.[13] In many cases, this allows, in the end, to simplify complex expressions, such as the collision integral.

9.3 Non-Markovian Balescu-Lenard equation

The kinetic equation, i.e. the closed equation for the one-particle density operator, has already been derived above. It is given by Eq. (9.1) with the collision integral (9.11)

$$i\hbar\frac{\partial}{\partial t}F_1(t) - [\bar{H}_1, F_1(t)] \quad = \quad I_1(t). \tag{9.39}$$

As we have seen above, this equation is closed, if it is supplemented with the equations determining the inverse dielectric function and the dielectric

[13]Modifications appear only in case of multiple time integrations. For example, the integral $A(t_1t_2)$ splits in a sum of two according to

$$A(t_1t_2) = \int_{t_0}^{t_1} dt_3 \int_{t_0}^{t_2} dt_4 B(t_1t_3)C(t_3t_4)D(t_4t_2) = \int_{t_0}^{t_1} dt_3 \int_{t_0}^{t_2} dt_4 B^+(t_1t_3)C(t_3t_4)D^-(t_4t_2)$$
$$= \int_{t_0}^{t_1} dt_3 \int_{t_0}^{t_3} dt_4 B^+(t_1t_3)C^+(t_3t_4)D^-(t_4t_2) + \int_{t_0}^{t_1} dt_3 \int_{t_3}^{t_2} dt_4 B^+(t_1t_3)C^-(t_3t_4)D^-(t_4t_2),$$

where the resulting two integrals have a clear time ordering in all quantities.

propagator (9.31) and (9.35). Thus, in principle, the problem is solved.

On the other hand, we have seen that screening properties can be described in a more familiar way using, instead of the dielectric propagator, the screened potential. It is, therefore, often useful to eliminate the dielectric propagator from the the collision integrals expressing them in terms of ϵ^{-1} and U^0. For this, we have derived various relations, including Eqs. (9.26), (9.34) and (9.35). Further simplifications are possible using Eqs. (9.31), (9.32) and (9.33). However, a comparably simple result follows only in terms of retarded and advanced quantities $U^{0\pm}$ and $V_s^{\pm}$, see above. The main idea is that, according to Eq. (9.34), elimination of each dielectric propagator in Eq. (9.11) leads to elimination of one trace and addition of one time integration. We will not reproduce the rather lengthy transformations here and give the final result without the initial correlation term and without exchange[14]

$$
\boxed{
\begin{aligned}
I_k(t) \;=\;& \frac{1}{\hbar^4}\sum_{qk'}\int_{t_0}^{t} dt_2 \int_{t_0}^{t_2} dt_4 U^{0+}(k-q,k,tt_2)\Phi^{<}(k-q,k,t_2)V_s^{-}(qt_4t_2) \\
\times\;& \left[\int_{t_0}^{t_4} dt_3 V_s^{+}(qtt_3)U^{0-}(k'+q,k',t_3t_4)\Phi^{<}(k'+q,k',t_3)\right. \\
&\left. +\int_{t_4}^{t} dt_3 V_s^{+}(qtt_3)U^{0+}(k'+q,k',t_3t_4)\Phi^{<}(k'+q,k',t_4)\right] \\
+\;& \text{c.c.} - (> \longleftrightarrow <)
\end{aligned}
}
\tag{9.40}
$$

where the last line denotes complex conjugation of the whole expression and subtraction of an analogous term where $\Phi^{>}$ and $\Phi^{<}$ are interchanged. In Eq. (9.40) we denoted $\Phi^{\stackrel{>}{<}}(k_1k_2,t) = f^{\stackrel{<}{>}}_{k_1}(t)f^{\stackrel{>}{<}}_{k_2}(t)$ where $f^{<} = f$ and $f^{>} = 1 \pm f$, and the propagators are products of quasiparticle propagators $U^{\pm}$ (cf. App. D), i.e. $U^{0\pm}(k,k',tt') = U^{\pm}(k,tt')U^{\mp}(k',t't)$.[15]

The collision integral (9.40) is the non-Markovian generalization of the quantum Balescu-Lenard integral. It was first obtained by Kuznetsov [Kuz91] and Haug/Ell [HE92] from Green's functions theory. A similar integral without selfenergy effects was recently derived by Hohenester and Pötz [HP97], but its relation to the result (9.40) is still unclear.

[14]The transformations are analogous to the ones used to derive the non-Markovian Boltzmann collision integral, Ch. 8. The decomposition in the two integral terms in brackets arises from different time orderings in the t_3 and t_4 integrations as explained in footnote 13, see also Ref. [HE92]. The initial correlation term in Eq. (9.11) is treated analogously, for details, see Ref. [BKDK].

[15]The introduction of the retarded and advanced quantities was discussed at the end of Sec. 9.2. In particular, the explicit form of the propagators $U^{0\pm}(tt')$ follows immediately from Eq. (9.22) by multiplication with $\Theta[\pm(t-t')]$.

9.3.1 Properties of the non-Markovian Balescu-Lenard equation. Markov limit

(i) The collision integral (9.40) has a structure very similar to the non-Markovian Boltzmann integral (8.62), where, basically, the T-matrices $T^{\pm}$ are replaced by the screened potentials $V_s^{\pm}$. It contains complex retardation effects: first, the distribution functions enter at previous times and, second, the dynamical character (dependence on two times) of the screened potentials introduces additional "memory" effects.

(ii) The integral contains selfenergy effects, all propagators are renormalized by selfenergy in RPA, cf. Eq. (9.21) and contain damping effects. This becomes obvious in the local approximation (9.23) for the propagators. Thus, initial correlations in the system are being weakened during the evolution, and the memory in the integral (9.40) has a finite duration $\tau_{cor} \sim \hbar/2\gamma$, with $\gamma = -\text{Im}\epsilon$, and ϵ being the matrix element of the renormalized Hamiltonian $\bar{H}$. Formally the damping (or correlation) time is identical to the results of the Born approximation or the T-matrix approximation, but here γ is to be computed from the RPA selfenergy.

(iii) *Conservation of total energy* can be proved for the same three approximations as in case of the Born approximation, Ch. 7 and the ladder approximation Ch. 8:

1. For the case with the exact expression (9.17) for the selfenergy (no approximations for the propagators), energy conservation can be proved for the case of distance dependent potentials.
2. For the case without selfenergy, $\gamma \to 0$, we have a symmetric hierarchy closure, i.e. $P_{123}F_{123} = F_{123}$, which, according to Eq. (2.36) guarantees energy conservation (see also Sec. 2.5.2).
3. With selfenergy neglected and the Markov limit taken, energy conservation can be proved in first order retardation approximation, see Sec. 8.5.

(iv) In the limit of *static interaction*, $V_s^{\pm}(qtt') \to V_s(qt)\delta(t-t')$, the t_3 and t_4 integrations in Eq. (9.40) can be performed, and we immediately recover the non-Markovian Landau equation (7.21,7.22), with a statically screened potential, selfenergy and initial correlations.

(v) In the *Markov limit* (7.46), we straightforwardly recover the well-known results for the dielectric and screening properties. To this end, all dynamic quantities are taken in the local approximation, e.g. $\epsilon^{+-1}(tt') \to$

$\epsilon^{+-1}(T,\tau)$, where $T = (t+t')/2)$ and $\tau = t - t'$, and T is assumed to be constant on times $t \leq \tau_{cor}$. After Fourier transform with respect to τ and application of the convolution theorem, Eq. (9.31) yields the familiar inverse retarded RPA-dielectric function ($\delta \to +0$)

$$\boxed{\epsilon^{-1}(q,\omega+i\delta,T) = \left\{1 - V(q)\Pi(q,\omega+i\delta,T)\right\}^{-1}} \tag{9.41}$$

the inverse of which (i.e. the dielectric function) has already been obtained in Ch. 4, cf. Eq. (4.21).

$\Pi(q,\omega+i\delta)$ is the familiar retarded RPA polarization. If selfenergy is neglected, the Fourier transform of Eq. (9.27) together with Eq. (D.21) of App. D yields the Lindhard formula[16],

$$\boxed{\Pi(q,\omega+i\delta,T) = \sum_p \frac{f_p(T) - f_{p+q}(T)}{E_p^0 - E_{p+q}^0 + \hbar\omega + i\delta}} \tag{9.42}$$

which was derived in Ch. 4, cf. Eq. (4.14).[17]

(vi) Moreover, taking the *Markov limit* (7.46), of Eq. (9.40) we directly recover the conventional Balescu-Lenard collision integral. To see this, we first neglect initial correlations and neglect selfenergy in the propagators, i.e. $\epsilon \to E^0$. Introducing the Fourier transforms of $V_s^{\pm}$ and $U^{0\pm}$ in the collision integral and using the convolution theorem and the properties (D.21) and (D.22) of the propagators (cf. App. D), we obtain

$$\boxed{\begin{aligned} I_{0k}^M(t) = \frac{2}{\hbar}\sum_{qk'} \left|\frac{V_q}{\epsilon(q, E_k^0 - E_{k-q}^0 + i\delta, t)}\right|^2 \delta(E_k^0 + E_{k'}^0 - E_{k-q}^0 - E_{k'+q}^0)\times \\ \left\{f_{k-q}f_{k'+q}(1\pm f_k)(1\pm f_{k'}) - f_k f_{k'}(1\pm f_{k-q})(1\pm f_{k'+q})\right\}\Big|_t \end{aligned}} \tag{9.43}$$

This is the quantum version of the Markovian Balescu-Lenard equation [Bal60, Len60, Bal63], describing two-particle scattering on a dynamically screened potential. This equation is valid only on the kinetic stage, $t \gg \tau_{cor}$, and conserves only the single-particle energy. Consequently, it yields the thermodynamics of a dynamically screened, but *ideal* plasma.

[16] In principal, the derivation yields the renormalized energies ϵ in the denominator. However, this leads to inconsistencies, and the proper inclusion of renormalization effects is still being debated.

[17] Recall that all momenta have to be understood as vectors.

In Figs. 9.1. and 9.2. we show numerical solutions of the Markovian Balescu-Lenard equation for an electron beam penetrating a dense hydrogen plasma where quantum effects are important. Therefore, we used th quantum collision integral (9.43) and selfconsistently calculated the nonequilibrium dielectric function from Eqs. (9.41), (9.42).

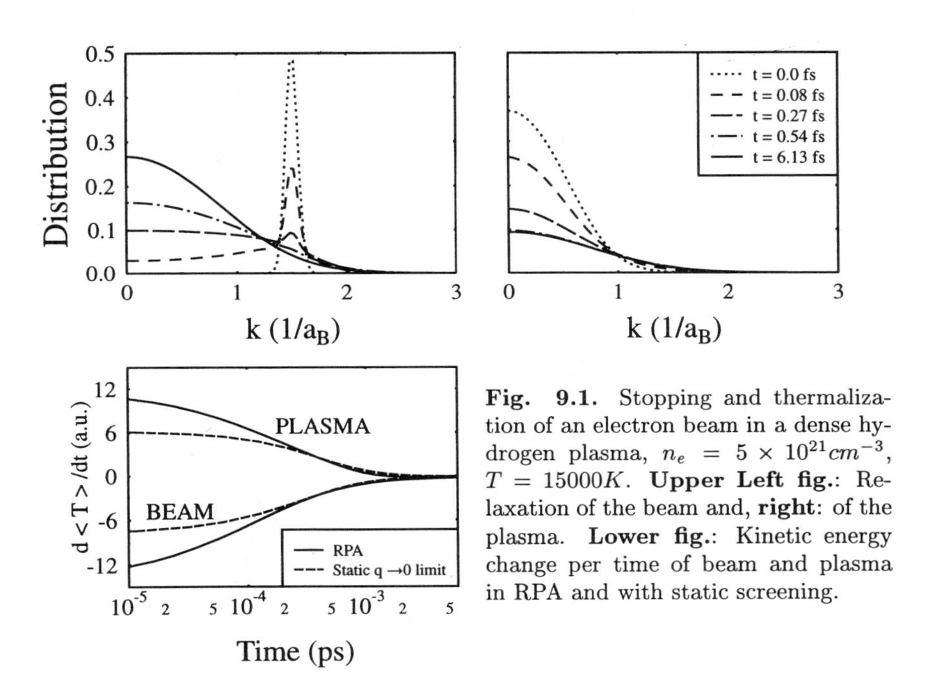

Fig. 9.1. Stopping and thermalization of an electron beam in a dense hydrogen plasma, $n_e = 5 \times 10^{21} cm^{-3}$, $T = 15000K$. **Upper Left fig.**: Relaxation of the beam and, **right**: of the plasma. **Lower fig.**: Kinetic energy change per time of beam and plasma in RPA and with static screening.

Clearly the RPA leads to a faster energy exchange between beam and plasma than the static screening approximation (static limit of the RPA). This is due to the account of collective plasma excitations bye the RPA. The same tendency is observed from a comparison of the total scattering rates $\Sigma^> + \Sigma^<$, which are related to the collision integral by $I = \Sigma^<(1 - f) - \Sigma^> f$, see Fig. 9.2. We mention that similar calculations including the investigation of the related plasmon un-damping have been performed in semiconductor theory, e.g. [SBK92, BSP+92, Col93].

(vii) A convenient way to extend the result (9.43) to nonideal plasmas is to perform a retardation (gradient) expansion of the full non-Markovian integral (9.40) up to first order [KE72, Kli75]. This was discussed in detail in Ch. 7 and can be performed in full analogy to the expansion of the non-Markovian Boltzmann equation in Ch. 8. Here, we only mention that for a nondegenerate plasma, one again can derive a simple relation between the zeroth order and first order gradient terms,

$$\boxed{I_1^M(k) = \frac{1}{2}\frac{d}{dt}\frac{d}{d\varepsilon}I_0^M(\varepsilon,k)\bigg|_{\varepsilon\to 0}} \tag{9.44}$$

where $I_0^M(\varepsilon,k)$ is the Markovian Balescu integral (9.43) where the delta function is replaced by a broadened delta function (cf. App. A) [AP77, KBBS96]. A Markovian kinetic equation with both integrals (9.43) and (9.44) included again satisfies total energy conservation and yields the correct thermodynamic limit of a correlated plasma [KE72, KBBS96]. We mention that the first order retardation term has recently been derived explicitly by Belyi et al. [BKW96].

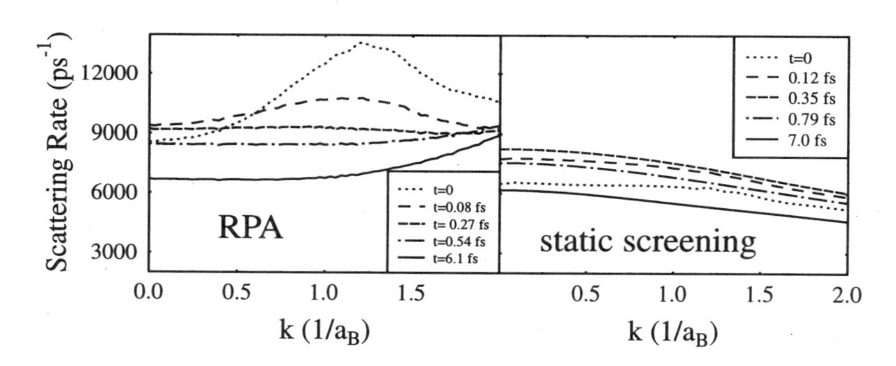

Fig. 9.2. Time evolution of the total scattering rates of the beam electrons, calculated in RPA (left) and with static screening (right fig.). Parameters same as in Fig. 9.1.

Scattering on collective excitations. Here we add a remark which is analogous to that at the end of Sec. 8.3.2. Correlations in plasmas give rise to short-range and long-range phenomena, where the latter refer to collective excitations (see Ch. 4). Both parts are fully included in the non-Markovian correlation operator (9.9) and Balescu-Lenard collision integral (9.40). However, with the Markov limit, where all initial correlations are weakened, the collective part is neglected (collective excitations correspond to a long living/long-range part of g_{12}. It these effects are important, this part should not be weakened. (Of course, there is not always a sharp separation possible). Plasmon effects give rise to an additional collision integral [Dan90]

$$I^{coll}(p) \sim \sum_K |V(q)\phi^{(K)}(q)|^2 \left\{N_K^> f_{p+q}^> f_p^< - N_K^< f_{p+q}^< f_p^>\right\} 2\pi\delta(E_p - E_{p+q} - \mathcal{E}_K),$$

where the N's are related to the number of plasmons with energy $\mathcal{E}_K$. The wave functions $\phi^{(K)}$ and the energies $\mathcal{E}_K$ satisfy the eigenvalue problem

$$\{\mathcal{E}_K + E_{p+q} - E_p\}\,\phi^{(K)}(q) - [f_{p+q} - f_p]\,V(q)\phi^{(K)}(q) = 0, \tag{9.45}$$

which is, basically, the homogeneous equation associated with the Fourier transformed Dyson equation $V_s^+(q,\omega) = V_q + V_q\Pi^+(q,\omega)V_s^+(q,\omega)$. Notice that, due to the time dependence of f, Eq. (9.45) is not hermitean, so, as in Sec. 8.3.2, a bi-orthonormal system has to be constructed. If the plasmons are in equilibrium, the N's are related to Bose distributions, otherwise, separate equations for the N's have to be solved simultaneously. This brings us back to the coupled carrier-plasmon kinetic equations of Pines/Schrieffer and Klimontovich which we discussed in Sec. 4.7. We mention that a much more general approach to this problem is given by plasma quantum electrodynamics which will be considered in Ch. 12.

9.3.2 Correlation energy in RPA

The correlation energy for the general case of full retardation is readily calculated from the formula $\langle V\rangle = \frac{n^2}{2}\mathrm{Tr}_{12}V_{12}g_{12}$, where for g_{12} the solution (9.9) is used. The trace gives rise to integrations over q, p_1 and p_2, which allows to apply Eq. (9.34) to both dielectric propagators. As a result, we obtain

$$\begin{aligned}\langle V\rangle(t) = -\frac{N}{\hbar}\int_{t_0}^{t} d\bar{t}\int_{\bar{t}}^{t} dt_1\int_{\bar{t}}^{t} dt_2 \int\frac{d\mathbf{q}}{(2\pi\hbar)^3}\int\frac{d\mathbf{p}_1}{(2\pi\hbar)^3}\int\frac{d\mathbf{p}_2}{(2\pi\hbar)^3}\,V(q)\\ \times\,\mathrm{Re}\Big\{\epsilon^{-1}(\mathbf{q},tt_1)\,\epsilon^{-1}(-\mathbf{q},tt_2)U^0(\mathbf{p}_1,\mathbf{p}_1+\mathbf{q},t_1\bar{t})\,U^0(\mathbf{p}_2,\mathbf{p}_2-\mathbf{q},t_2\bar{t})\\ \times\,\Big[V(q)\,2\,f_{\mathbf{p}_1+\mathbf{q}}(t_1)[1\pm f_{\mathbf{p}_1}(t_1)]f_{\mathbf{p}_2-\mathbf{q}}(t_2)[1\pm f_{\mathbf{p}_2}(t_2)]\\ +\,i\hbar\,\delta(\bar{t}-t_0)\,g^0(\mathbf{p}_1+\mathbf{q},\mathbf{p}_2-\mathbf{q};\mathbf{p}_1,\mathbf{p}_2,\bar{t})\Big]\Big\},\end{aligned} \tag{9.46}$$

where the second term in square brackets is due to initial correlations. This is a very general result for the correlation energy of a nonideal charged particle system. It includes retardation and energy broadening as well as selfenergy effects and is valid on all time scales. In particular, it applies to arbitrary short times since it fully contains the buildup of dynamical screening and the influence of initial correlations. If the effects of the dynamics are neglected, we recover from Eq. (9.46) the correlation energy in static Born approximation, cf. Eq. (7.40).[18]

Equilibrium correlation energy. In the limit of long times, Eq. (9.46) yields the equilibrium correlation energy of a weakly coupled plasma. Instead

[18] To this end, we again have to use the limiting result (9.41) for the dielectric function, where the static long wavelength limit ($\omega \to 0$ and, subsequently, $q \to 0$) has to be taken. As a result, in Eq. (9.46), the integrations over t_1 and t_2 are removed and the square of the statically screened Coulomb potential (Debye potential) appears.

of explicitly performing the Markov limit in Eq. (9.46), as it was demonstrated in Sec. 7.3.4, one can also use the equilibrium correlation function derived from the equilibrium hierarchy, cf. Sec. 2.7. Using there the polarization approximation for g_{12}, we obtain the solution for the classical pair correlation function in configuration space[19]

$$g_{12}^{eq}(r) = -\frac{e^2}{\epsilon_b kT} \frac{e^{-r/r_D}}{r} = -\frac{V_D(r)}{kT}, \tag{9.47}$$

where $r \equiv |\mathbf{r}_1 - \mathbf{r}_2|$, ϵ_b is the background dielectric constant, $r_D = \frac{1}{\kappa} = \sqrt{\frac{kT\epsilon_b}{4\pi n e^2}}$ is the Debye screening radius and V_D the Debye potential. The full correlation function (including the momentum dependence) of a classical homogeneous plasma follows by multiplying Eq. (9.47) with $f_{p_1}^{eq} f_{p_2}^{eq}$.

Using the result (9.47) it is now easy to compute the equilibrium correlation energy,

$$\langle V \rangle^{eq} = \frac{nN}{2} \int d\mathbf{r}\, \frac{e^2}{\epsilon_b r}\, g_{12}^{eq}(r) = -\frac{N}{2} \frac{\kappa e^2}{\epsilon_b}. \tag{9.48}$$

This is the well–known result for the correlation (or internal) energy of a one-component plasma (Debye–Hückel result).

9.3.3 Short-time behavior. Screening buildup

We now consider the early stage of relaxation, $t_0 \le t < \tau_{cor}$. Suppose, at the initial moment t_0, there are only very few free carriers in the system. At this time, some short excitation mechanism is turned on (a laser pulse, an external electric field etc.) which begins to generate carriers: depending on the system, this can be an ionization process (inside a solid or on its surface) or excitation of electrons from low lying energy bands into the conduction band, e.g. in a semiconductor. If these carriers are generated independently, there are no correlations among them initially.[20] Correlations between the particles only start to build up during the excitation process. This buildup of correlations is (in our case of weakly coupled plasmas) nothing but rearrangement of the carriers in the long-range Coulomb field of the others, i.e. the buildup of the screening cloud.

---

[19]This assumes that the one–particle distributions are equilibrium (Fermi/Bose) distributions.

[20]More generally, one may whish to be able to consider also situations where already carriers, and thus, also correlations exist in the system prior to the onset of the excitation.

These questions have been analyzed already long ago, especially for ionic fluids (electrolytes) and weakly ionized plasmas [DF28, KE62].[21] The well-known result is that the typical time scale for correlation buildup is just the period of the plasma oscillations, $\tau_{cor} \sim 2\pi/\omega_{pl}$. This is the same result as in equilibrium and has been discussed in Sec. 5.2, but here, ω_{pl} is a nonequilibrium quantity which may change during the relaxation as a result of the evolution of the plasma density, composition, degree of ionization and so on. These questions have also been studied for dense classical plasmas by means of Molecular dynamics simulations, e.g. [ZTR][22] which confirmed this time scale.

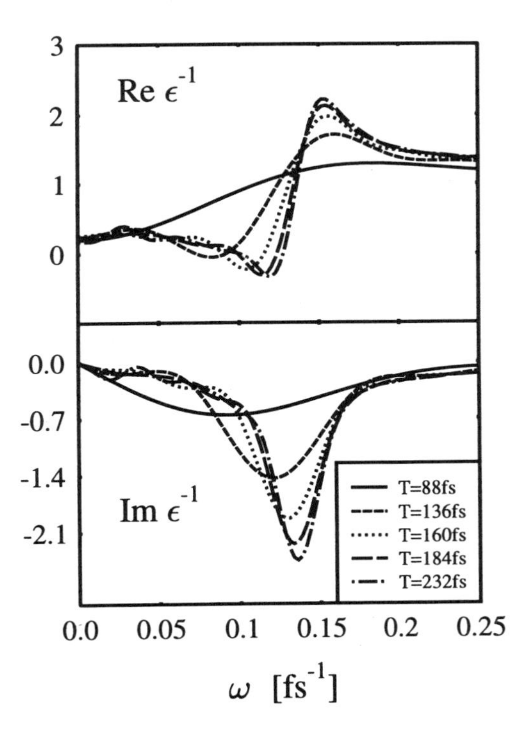

Fig. 9.3. Screening buildup in an optically excited bulk semiconductor. One clearly sees the formation of the optical plasmon peak in the spectral function. Calculated by solving Eq. (9.31) using Π from solutions of the Kadanoff-Baym equations in static Born approximation as an input [BKK$^+$] For more details, see Sec. 12.7.

Recently, the buildup of screening has attracted strong new interest in the context of semiconductors excited by femtosecond lasers [SBH94, SSH$^+$94, MHH$^+$95, BKK$^+$], where, using pulse durations shorter than the plasma period (see Ch. 1), one hopes to study these phenomena experimentally. Since Eq.

[21] Already in 1928 Debye and Falkenhagen calculated the relaxation time of the screening cloud around an ion if the ion is perturbed (removed), see also [Fal71].

[22] Starting from an uncorrelated (unscreened) initial state, the authors demonstrated the buildup of the correlation energy after about $t \sim 1/\omega_{pl}$ and also subsequent oscillations of the energy with this period, see also Ch. 13.

(9.40) is yet too complicated to solve, various models have been proposed. ElSayed et al. have proposed to use for the very first stage kinetic equations with an unscreened Coulomb potential which does not lead to divergencies, due to collisional energy broadening [SBH94]. Furthermore, a time-dependent plasmon pole approximation (cf. Sec. 4.4) was proposed to solve Eq. (9.31) [SSH$^+$94], for a textbook discussion, see [HJ96]. Finally, we mention first direct solutions of the Dyson equation for the screened potential [MHH$^+$95, BKK$^+$] which will be discussed further in Ch. 12. For numerical details, see App. F.

Chapter 10

*Dynamically Screened Ladder Approximation

In this Chapter we continue the discussion of dynamical screening. Here, we go beyond the polarization approximation of Ch. 9, extending the discussion to screening in the presence of strong correlations. As was discussed in Sec. 2.6, the dynamically screened ladder approximation (DSLA) includes both polarization and ladder terms and is given by (cf. Eq. (3.22),

$$i\hbar\frac{\partial}{\partial t}F_1 - [\bar{H}_1, F_1] = n\mathrm{Tr}_2[V_{12}^{\pm}, g_{12}], \tag{10.1}$$

$$i\hbar\frac{\partial}{\partial t}g_{12} - [\bar{H}_{12}^0, g_{12}] - (\hat{V}_{12}g_{12} - g_{12}\hat{V}_{12}^{\dagger}) = Q_{12}$$
$$+\Pi_{12}^{(1)} + \Pi_{12}^{(2)} + n\mathrm{Tr}_3[V_{13} + V_{23}, g_{123}](1 \pm P_{13} \pm P_{23}), \tag{10.2}$$

where $\bar{H}_{12}^0 = H_1 + H_2 + H_1^{HF} + H_2^{HF}$, and the effective single–particle Hamiltonian is

$$\bar{H}_1 = H_1 + H_1^{HF} = H_1 + n\mathrm{Tr}_2 V_{12} F_2 \Lambda_{12}^{\pm}. \tag{10.3}$$

The inhomogeneity Q_{12} and the polarization contributions are

$$Q_{12} = \hat{V}_{12}F_1F_2 - F_1F_2\hat{V}_{12}^{\dagger}, \tag{10.4}$$
$$\Pi_{12}^{(1)} = n\mathrm{Tr}_3[V_{13}^{\pm}, F_1]g_{23}\Lambda_{23}^{\pm}, \tag{10.5}$$

and $\Pi_{12}^{(2)}$ follows from the substitution $1 \longleftrightarrow 2$. The shielded potential $\hat{V}_{12}$ and the permutation operator P_{12} account for Pauli blocking and exchange, respectively, and have been defined in Ch. 3. The difference to the polarization approximation of Ch. 9, is the appearance of the shielded potential[1], (i.e. of the ladder terms) on the l.h.s. of Eq. (10.2).

[1]Recall that the shielded potential contains a Pauli blocking factor, $\hat{V}_{ab} = (1 \pm nF_a \pm nF_b)V_{ab}$, see Ch. 3.

10.1 Generalized screened ladder approximation. Selfenergy

Proceeding as in Ch. 9, we introduce selfenergy effects based on a consistent approximation for the ternary correlation operator. The hierarchy closure that corresponds to the generalized screened ladder approximation is (we immediately give the result with spin statistics fully included, cf. Sec. 9.1),

$$i\hbar\frac{\partial}{\partial t}g_{123} - \left\{H^{\text{eff}}_{123}\, g_{123} - g_{123}\, H^{\text{eff}\dagger}_{123}\right\} =$$
$$\left\{(\hat V_{13}+\hat V_{23})F_3 g_{12} \mp nF_3(F_1V_{13}+F_2V_{23})g_{12} \mp n(g_{13}V_{13}+g_{23}V_{23})g_{12} - \text{h.c.}\right\}$$
$$+\Pi^{(1)}_{123}+\Pi^{(2)}_{123}+\Pi^{(3)}_{123}, \qquad (10.6)$$

where the three–particle polarization terms are defined as $\Pi^{(1)}_{123} = n\text{Tr}_4[V^{\pm}_{14}, F_1 g_{234}](1 \pm P_{24} \pm P_{34})$. Again H^{eff}_{123} contains the still unknown renormalized single–particle Hamiltonians,

$$H^{\text{eff}}_{123}\, g_{123} = \left\{\bar H_1 + \bar H_2 + \bar H_3 + \hat V_{12} + \hat V_{13} + \hat V_{23}\right\} g_{123}, \qquad (10.7)$$

which are defined below in Eq. (10.10), but, in addition, the three-particle ladder terms. Eq. (10.6) closes the system (10.1), (10.2) and defines our generalized dynamically screened ladder approximation.[2]

We next derive the solution for the binary correlation operator in terms of generalized propagators. With these propagators, we then obtain the collision integral of the kinetic equation and also explicit results for the selfenergy Σ^+.

Binary correlation operator in DSLA. Using Eqs. (10.2) and (10.6), we can rewrite the second hierarchy equation as

$$i\hbar\frac{\partial}{\partial t}g_{12} - \left\{H^{\text{eff}}_{12}\, g_{12} - g_{12}\, H^{\text{eff}\dagger}_{12}\right\} \;=\; Q_{12} + \Pi^{(1)}_{12} + \Pi^{(2)}_{12}, \qquad (10.8)$$

where the three–particle correlations are included into H^{eff}_{12} via the renormalized single–particle Hamiltonians $\bar H_1$,

$$H^{\text{eff}}_{12}\, g_{12} \;=\; \left\{\bar H_1 + \bar H_2 + \hat V_{12}\right\} g_{12}, \qquad (10.9)$$
$$\bar H_1\, g_{12}(t) \;=\; \left(H_1 + H^{HF}_1\right) g_{12}(t) + \int_{t_0}^{t} d\bar t\, \Sigma^+_1(t\bar t)\, U_2(t\bar t)\, g_{12}(\bar t), \quad (10.10)$$

---

[2] This approximation is equivalent to the selfconsistent DSLA of Green's functions theory where all propagators are renormalized by selfenergies in DSLA, e.g. [KKER86].

and Σ^+ and U denote, respectively, the retarded DSLA selfenergy and the quasi–particle propagator which will be determined below. We solve Eq. (10.8) with the initial condition $g_{12}(t_0) = g^0_{12}$ by making the ansatz [BDK98]

$$\boxed{g_{12}(t) = \mathrm{Tr}_{34} U_{13,24}(tt_0)\, g^0_{34} + \frac{1}{i\hbar}\mathrm{Tr}_{34} \int\limits_{t_0}^{t} d\bar{t}\, U_{13,24}(t\bar{t})\, Q_{34}(\bar{t})} \tag{10.11}$$

This solution is similar to the one in RPA, cf. Ch. 9. However, here no factorization of the propagator into a product $U_{13}U_{24}$ is possible, what, of course, is due to the strong coupling effects (formally described by the interaction potential in the Hamiltonian). Correspondingly, the full propagator obeys an essentially more complicated equation,

$$\begin{aligned} i\hbar\frac{\partial}{\partial t} U_{13,24}(tt') - \left\{H^{\mathrm{eff}}_{12}\, U_{13,24}(tt') - U_{13,24}(tt')\, H^{\mathrm{eff}\dagger}_{12}\right\} &= \\ n\mathrm{Tr}_5\, [V^{\pm}_{15}, F_1]\, U_{53,24}(tt') + n\mathrm{Tr}_5\, [V^{\pm}_{25}, F_2]\, U_{13,54}(tt'), & \\ U_{13,24}(tt) &= \delta_{13}\, \delta_{24}. \end{aligned} \tag{10.12}$$

The third line defines the initial condition. One readily checks that the propagator defined by Eq. (10.12) indeed generates the correct solution $g_{12}(t)$ together with the initial condition.

Collision integral in DSLA. Inserting the formal solution (10.11) into the first hierarchy equation (10.1), we obtain the collision integral of the kinetic equation in screened ladder approximation,

$$I_1(t) = n\mathrm{Tr}_{234} V^{\pm}_{12} \left\{ U_{13,24}(tt_0)\, g^0_{34} + \frac{1}{i\hbar}\int_{t_0}^{t} d\bar{t}\; U_{13,24}(t\bar{t})\, Q_{34}(\bar{t}) - \mathrm{h.c.}\right\}. \tag{10.13}$$

The first integral is due to the initial correlations g^0, while the second one describes the correlation buildup including exchange terms.

Selfenergy in DSLA. We now return to the third hierarchy equation (10.6). It has a similar structure as the second hierarchy equation (10.2), and therefore, we use the same type of ansatz for $g_{123}(t)$ as Eq. (10.11),

$$g_{123}(t) = \mathrm{Tr}_{456}\, U_{123,456}(tt_0)\, g^0_{456} + \frac{1}{i\hbar}\mathrm{Tr}_{456} \int_{t_0}^{t} d\bar{t}\, U_{123,456}(t\bar{t})\, R_{456}(\bar{t}), \tag{10.14}$$

where g^0_{123} denotes initial three–particle correlations. Also, R denotes the inhomogeneity in the third hierarchy equation, i.e. the whole term in brackets on the r.h.s. of Eq. (10.6). We begin the anaysis by dropping the initial correlation term and taking R for the spinless case, $R_{123} = (V_{13} + V_{23})\, F_3\, g_{12} -$ h.c.. We will restore the full expression in the final result (10.19).

Now our derivation follows the same lines as in Chs. 7 and 9, and we only briefly list the necessary steps referring to these Chapters for details. Inserting the solution (10.14) into Eq. (10.2) we obtain one term of the form

$$S_1(t) = \frac{n}{i\hbar}\mathrm{Tr}_{3456}\int_{t_0}^{t} d\bar{t}\, V_{13}\, U_{123,456}(t\bar{t})\, V_{46}\, F_6(\bar{t})\, g_{45}(\bar{t}), \tag{10.15}$$

and a second analogous term S_2 where $V_{13} \to V_{23}$ and $V_{46} \to V_{56}$. Furthermore, there appear the hermitean conjugates of both. Other combinations of potentials do not contribute to selfenergy. The crucial point is now that in Eq. (10.15) the propagator may be simplified. Indeed, with the approximation (10.6) the dynamics of g_{123} is not determined by true three-particle correlations, but rather by lower order correlations of the form $g_{12}F_3$, cf. the inhomogeneity R. Therefore, the propagators, in combination with two potentials, must have a similar product structure,

$$V_{13}\, U_{123,456}\, V_{46} = V_{13}\, U_{13,46}\, V_{46} \;\cdot U_2\, \delta_{2,5}, \tag{10.16}$$

where, obviously, the indices "2, 5" are not affected by the potentials. This allows us to perform one trace in Eq. (10.15) and to rewrite this equation as

$$\begin{aligned} S_1(t) &= \frac{n}{i\hbar}\mathrm{Tr}_{346}\int_{t_0}^{t} d\bar{t}\, V_{13}\, U_{13,46}(t\bar{t})\, V_{46}\, F_6(\bar{t})\, U_2(t\bar{t})\, g_{42}(\bar{t}) \\ &= \int_{t_0}^{t} d\bar{t}\, \Sigma_1^{+}(t\bar{t})\, U_2(t\bar{t})\, g_{12}(\bar{t}), \end{aligned} \tag{10.17}$$

where we introduced the retarded selfenergy

$$\Sigma_1^{+}(t\bar{t}) = \frac{n}{i\hbar}\mathrm{Tr}_{346}\, V_{13}\, U_{13,46}(t\bar{t})\, V_{46}\, F_6(\bar{t}). \tag{10.18}$$

This is the retarded selfenergy in screened ladder approximation, expressed in terms of the DSLA propagator. What remains now is to restore the three–particle initial correlations and the spin statistics corrections in expression (10.18). The derivation is the same as discussed in the ladder approximation, Sec. 8.2, cf. Eq. (8.14), [3]

$$\Sigma_1^{+}(t\bar{t}) = \frac{n}{i\hbar}\mathrm{Tr}_{346}\, V_{13}^{\pm}\, U_{13,46}(t\bar{t}) \left[F_{46}^{>}\tilde{V}_{46}F_6^{<} \mp F_{46}^{<}\tilde{V}_{46}F_6^{>} \mp n g_{46}\tilde{V}_{46}\right]\Big|_{\bar{t}}, \tag{10.19}$$

and, as before, we introduced an effective potential that includes the initial correlations $\tilde{V}_{ab}(t) = V_{ab} + i\hbar\delta(t-t_0)$, and also $F_{ab}^{>} = F_a^{>}F_b^{>}$, $F_a^{>} = 1 \pm nF_a^{<}$ and $F_a^{<} = F_a$.

[3] Since the inhomogeneity in the third hierarchy equation is the same in all approximations for Σ, the approximations differ, basically, only in the actual propagators.

Limiting cases of the screened ladder approximation

The generalized screened ladder approximation is the most complex approximation which we considered. It includes strong coupling and polarization diagrams as well, and thus contains the simpler cases of the ladder approximation (Ch. 8), the RPA (Ch. 9) or the second Born approximation (Ch. 6) which are easily recovered as limiting cases of the DSLA–propagators:

i) The polarization approximation is recovered by neglecting the ladder term in the second hierarchy equation. This allows to factorize the DSLA–propagator: $U_{13,24}(tt') = U_{13}(tt')\, U_{24}(tt')$,

ii) The ladder approximation follows from neglecting the polarization terms in the second hierarchy equation. This leads to simpler propagators, $U_{13,24}(tt') = U_{12}(tt')\, \delta_{1,3}\, \delta_{2,4}$. [4]

iii) The Born approximation neglects both ladder and polarization terms and is thus obtained from $U_{13,24}(tt') = U_1(tt')\, U_2(tt')\, \delta_{1,3}\, \delta_{2,4}$, where $U_1(tt')$ is the simple single-particle propagator.

With these relations between the propagators, all results for the different approximations, including the binary correlation operator, collision integral and selfenergy, can be derived from the DSLA.

In principal, the coupled first and second hierarchy equations are well suitable for numerical solution, even on the level of the screened ladder approximation. Due to large requirements in computer memory, best candidates for a solution would initially be lower dimensional (2D or 1D) systems.

Unfortunately, the equation of motion for the DSLA–propagator is very complicated, and there is no direct relation of the DSLA–propagator to the physically important quantities, such as T-matrix or screened potential known yet.[5] Therefore, for practical purposes, simplifying approximations are of high interest, one of which we consider below.

10.2 Gould–DeWitt approximation

An interesting approximate treatment of the screened ladder approximation has been proposed by Gould, Williams and DeWitt [GD67, WD69]. It is based on the fact that the ladder and polarization terms (first and third terms on

---

[4] Notice that this propagator U_{12} differs from the one used in Ch. 8 by the fact that it obeys a commutator equation. But this equation can be easily transformed into the one used in Ch. 8. The same applies to the Born approximation.

[5] A detailed Green's functions discussion can be found in Ref. [KKER86].

the r.h.s. of Eq. (10.2), respectively) are contained either in the ladder or polarization approximation, cf. Chs. 8 and 9. On the other hand, the second term on the r.h.s. is included in both approximations. Thus, one may attempt to solve Eq. (10.2) using the ansatz

$$\boxed{g_{12}(t) \approx g_{12}^B(t) - g_{12}^L(t) + g_{12}^{LB}(t)} \tag{10.20}$$

where g_{12}^B, g_{12}^{LB} and g_{12}^L are the binary correlation operator in binary collision approximation, polarization approximation (RPA) and Born approximation[6], leading respectively, to the Boltzmann, Lenard-Balescu and Landau collision integral, cf. Chs. 6, 8, 9, i.e. they are defined by the equations

$$\begin{aligned}
i\hbar\frac{\partial}{\partial t}g_{12}^B - [\bar{H}_{12}^0, g_{12}^B] &= (\hat{V}_{12}g_{12}^B - g_{12}^B\hat{V}_{12}^\dagger) + (\hat{V}_{12}F_1F_2 - F_1F_2\hat{V}_{12}^\dagger), \\
i\hbar\frac{\partial}{\partial t}g_{12}^{LB} - [\bar{H}_{12}^0, g_{12}^{LB}] &= (\hat{V}_{12}F_1F_2 - F_1F_2\hat{V}_{12}^\dagger) \\
&+ n\mathrm{Tr}_3\left\{[V_{13}^\pm, F_1]\, g_{23}^{LB}\, \Lambda_{23}^\pm + 1 \leftrightarrow 2\right\}, \\
i\hbar\frac{\partial}{\partial t}g_{12}^L - [\bar{H}_{12}^0, g_{12}^L] &= (\hat{V}_{12}F_1F_2 - F_1F_2\hat{V}_{12}^\dagger).
\end{aligned} \tag{10.22}$$

Inserting the ansatz (10.20) into Eq. (10.2) and using the definitions (10.22), we find that all terms cancel except the following correction

$$\begin{aligned}
\Delta &= \hat{V}_{12}(g_{12}^{LB} - g_{12}^L) - (g_{12}^{LB} - g_{12}^L)\hat{V}_{12}^\dagger \\
&+ n\mathrm{Tr}_3\left\{[V_{13}^\pm, F_1]\,(g_{23}^B - g^L)\,\Lambda_{23}^\pm + 1 \leftrightarrow 2\right\}.
\end{aligned} \tag{10.23}$$

This scheme is very promising and has been used by various authors [Kli75, MKR+89] to compute transport properties of dense plasmas near equilibrium. Recently this scheme has been extended to nonequilibrium situations in plasmas and optically excited semiconductors for the calculation of the stopping power and of scattering rates, respectively [GSK96, GKSB]. Further extensions to the initial stage of relaxation in strongly coupled plasmas, including the buildup of correlations, seem possible.

---

[6]It is instructive to consider as an example the equilibrium limit of Eq. (10.20) for a non–degenerate plasma, cf. Sec. 56 of Ref. [Kli75]. For the distance dependent part, it follows (cf. Secs. 7.3.4, 8.3.2 and 9.3.2)

$$g_{ab}^{EQ}(r) = e^{-V_{ab}^C(r)/kT} - 1 + \frac{V_{ab}^C(r) - V_{ab}^D(r)}{kT}, \tag{10.21}$$

where V^C and V^D denote the Coulomb and Debye potential, respectively.

Chapter 11

Charged Many–Particle Systems in Electromagnetic Fields. Generalized Bloch Equations

In this Chapter, we restore the external field in the Hamilton operators H_1, H_{12} etc. Further, we will specify it to be an electromagnetic field since this is the most important case in the context of ultrafast relaxation. Our first encounter with an external field was in Ch. 4, where we focused mainly on longitudinal fields and the collective plasma response (plasmons) to them, thereby neglecting scattering. Here, our aim will be to derive quantum kinetic equations which include both, the effect of fields (collective effects) and of correlations. Correlations and their dynamics have been discussed in detail in the previous Chapters. We now will generalize these results by including the field–matter interaction. In particular, we will derive the generalized Bloch equations which include the electromagnetic field, spin statistics effects, i.e. exchange and Pauli blocking, as well as correlations (scattering contributions).[1]

As was discussed in Sec. 2.4.1 where, among other representations of the BBGKY–hierarchy, we considered also the Bloch representation, the Bloch equations are the quantum kinetic equations for a multiband system. In this case, the one-particle density operator is replaced by a matrix (Bloch matrix) that includes diagonal and off-diagonal elements (level populations and transition probabilities, respectively). So, our problem now is to derive the equations of motion for the populations of the bands f^{λ} and the *interband polarizations* $P^{\lambda\lambda'}, (\lambda \neq \lambda')$. The BBGKY-approach is very well suited for this task. Its

[1] Pioneering work in extending the atomic Bloch equations to semiconductors is due to Schäfer and Treusch [ST86] and Lindberg and Koch [LK88], who used, respectively, Green's functions and field operator (creation and annihiliation operators) techniques.

advantage is that all transformations and approximations can be performed at the compact operator level, and only at the end, one can transform the operator equation to the Bloch representation [2]. The procedure consists of the following steps [BDK98]:

1. Derivation of the first and second equation of the BBGKY-hierarchy for the operators F_1 and g_{12}, including external fields, (in some cases, also the third equation is needed), this was done in Sec. 2.5.1,

2. (anti-)symmetrization of these equations in the case of fermions/bosons, cf. Ch. 3,

3. decoupling of the BBGKY-hierarchy using an appropriate approximation for the correlation operators, cf. Sec. 2.6,

4. formal solution of the second hierarchy equation for $g_{12}(\{F_1\}, t)$, for the approximation of interest, cf. Chs. 6–10, and substitution of this solution into the collision term (r.h.s.) of the equation for F_1,

5. expansion of the resulting operator equation for F_1 into a basis of Bloch states. In particular, the collision term requires

6. expansion of the solution $g_{12}(t)$ in terms of Bloch states.

Points 1.-4. have been discussed in detail in the indicated Sections. What remains to be solved are points 5 and 6. We will consider these questions in detail in Secs. 11.4-11.7. But, before doing this, it is helpful to have a more general look on the effect of electro-magnetic fields on many-particle systems.

11.1 Field–matter interaction

The most general description of a charged particle system in an electromagnetic field is given by relativistic quantum electrodynamics, what will be discussed in Ch. 12. Here we consider only nonrelativistic systems, which sets an upper limit to the field intensities[3], and we will use a quasiclassical approach, where the field is treated classically.

[2]We mention that this is not possible for arbitrary types of hierarchy closures. If the truncation is different for different (Bloch)matrix elements, one first has to transform the hierarchy into the corresponding matrix representation. An example for such a decoupling is the χ^n–approach of Axt and Stahl. Here the decoupling is based on perturbation theory in powers of the field which, in general, includes different types of terms for the diagonal and off–diagonal matrix elements, [AS94a, AS94b], see also [LHBK94, BK95], cf. Sec.2.6.

[3]For the exciting but quite peculiar relativistic phenomena generated by ultra-intense lasers, the reader is referred to [PM94, Tea94, PtV96, Bea97] and references therein.

Maxwell's equations. The classical description of an electromagnetic field is based on a vector potential $\mathbf{A}$ and a scalar potential ϕ which are related to the electric field strength and the magnetic induction vectors by[4]

$$\boxed{\mathcal{E} = -\nabla\phi - \frac{1}{c}\frac{\partial \mathbf{A}}{\partial t}; \qquad \mathbf{B} = \nabla \times \mathbf{A}} \tag{11.1}$$

where $\mathcal{E}$ and B satisfy Maxwell's equations,

$$\begin{array}{|lcll|}\hline \nabla \times \mathbf{B} & = & \frac{1}{c}\frac{\partial \mathcal{E}}{\partial t} + \frac{4\pi}{c}\left(\mathbf{j} + \mathbf{j}^{ext}\right); & \nabla\,\mathbf{B} = 0 \\ \nabla \times \mathcal{E} & = & -\frac{1}{c}\frac{\partial \mathbf{B}}{\partial t}; & \nabla\,\mathcal{E} = 4\pi\left(\rho + \rho^{ext}\right) \\ \hline \end{array} \tag{11.2}$$

where $\rho, \mathbf{j}$ and $\rho^{ext}, \mathbf{j}^{ext}$ are charge density and charge current induced by the particles in the system[5] and by external sources, respectively. In some cases, the external source may be specified by an external field or external vector and scalar potentials $\mathbf{A}^{ext}, \phi^{ext}$. Examples are an externally controlled electric field (e.g. of a trap or an accelerator) or the electromagnetic field of a laser. Then, the total field acting on the particles is given by the sums $\mathbf{A}_{tot} = \mathbf{A} + \mathbf{A}^{ext}$ and $\phi_{tot} = \phi + \phi^{ext}$.

Charge and field energy conservation. From Eqs. (11.2), we immediately obtain the charge density balance and the field energy. Operating with ∇ on the first equation and eliminating $\nabla\mathcal{E}$ using the fourth equation, one derives

$$\frac{\partial}{\partial t}\left(\rho + \rho^{ext}\right) + \nabla\left(\mathbf{j} + \mathbf{j}^{ext}\right) = 0. \tag{11.4}$$

For the derivation of the field energy balance, we multiply the first Eq. (11.2) by $\mathcal{E}$ and the third by $\mathbf{B}$ and take the difference, with the result

$$\frac{\partial u}{\partial t} + \nabla\mathbf{S} = -\left(\mathcal{A} + \mathcal{A}^{ext}\right), \tag{11.5}$$

$$u = \frac{1}{8\pi}\left(\mathbf{B}^2 + \mathcal{E}^2\right); \quad \mathbf{S} = \frac{c}{4\pi}\,\mathcal{E} \times \mathbf{B}; \quad \mathcal{A} = \mathbf{j}\mathcal{E}; \quad \mathcal{A}^{ext} = \mathbf{j}^{ext}\mathcal{E}, \tag{11.6}$$

[4]Notice that in Eqs. (11.1) and (11.2), all symbols stand for operators which correspond to ensemble averages of the microscopic (fluctuating) observables. The microscopic quantities will be considered in Ch. 12.

[5]Alternatively, one can eliminate the induced current and charge density $\mathbf{j}$ and ρ from Eqs. (11.2) by introducing the electric induction $\mathbf{D}(t) = \mathcal{E}(t) + 4\pi\int_{-\infty}^{t} d\tau \mathbf{j}(\tau)$. Then, the first and fourth of Maxwell's equations become

$$\nabla \times \mathbf{B} = \frac{1}{c}\frac{\partial \mathbf{D}}{\partial t} + \frac{4\pi}{c}\mathbf{j}^{ext}; \qquad \nabla\,\mathbf{D} = 4\pi\rho^{ext}. \tag{11.3}$$

where u and $\mathbf{S}$ denote the energy density and the energy flux (Poynting vector), while $\mathcal{A}$ and $\mathcal{A}^{ext}$ are the work performed by the field on charged particles inside and outside the system.

Hamiltonian of the field–matter complex. The dynamics of the averaged fields is completely defined by Maxwell's equations together with the given $\mathbf{A}^{ext}(t)$ and $\phi^{ext}(t)$. For the statistical description of the particles in the framework of the BBGKY-hierarchy, we now need the Hamiltonian of the particles. We start with the Hamilton operator of the entire N-particle-field system which is given by

$$\boxed{\begin{aligned} H_{1\ldots N;f} &= \frac{1}{8\pi}\int_{\mathcal{V}} d^3r \left(B^2 + \mathcal{E}_t^2\right) + \sum_{s=1}^{N} H_s + \frac{1}{2}\sum_{i\neq j}^{N} V_{ij} \\ H_s &= H_s^0\left[\mathbf{p} - \frac{e_s}{c}\left(\mathbf{A} + \mathbf{A}^{ext}\right)\right] + e_s\left(\phi + \phi^{ext}\right) \end{aligned}} \tag{11.7}$$

where H_s^0 is the *field-free* Hamilton operator which depends e.g. on the band structure of a solid [6] and reduces, in the case of parabolic energy dispersion, to $H_s^0(\mathbf{p}) = p_s^2/2m_s$ with $\mathbf{p}_s = -i\hbar\nabla_s$. The first contribution in Eq. (11.7) corresponds to the energy of the transverse part of the field ($\mathcal{E}^2 = \mathcal{E}_l^2 + \mathcal{E}_t^2$), while the longitudinal part is due to the Coulomb interaction between the particles, which is contained in the last term of Eq. (11.7).[7] Compared to the conventional single-particle Hamiltonian, here the free momentum is replaced by the total (kinematic) momentum.[8] Notice that H_i contains the total field, i.e. internal + external contributions.

The coupling between particles and field is two–fold: First, the field affects the charged carriers via the single-particle Hamiltonian (11.7) of the particles. On the other hand, the field is modified by the carriers via charges or currents (polarization) acting as sources in Maxwell's equations. Here we focus on the first part.[9] Furthermore, we assume that the *field does not affect the elementary Coulomb interaction* between the charges.[10] Of course, this does not exclude field effects on the screening properties, i.e. on the rearrangement

[6] In that case, it includes the lattice periodic potential and, in low-dimensional semiconductor systems, also confinement potentials, e.g. [BK95, Jah96].

[7] This will become clear in the quantum electrodynamics approach in Ch. 12.

[8] This is a consequence of the gauge invariance requirement for the Dirac equation. The Hamiltonian (11.7) follows from it in the limit $v/c \ll 1$ and corresponds to the Pauli Hamiltonian, cf. Ch. 12.

[9] We thus imply, throughout this Chapter, that the fields and scalar and vector potential are obtained by solving Eqs. (11.1), (11.2) using techniques familiar from electrodynamics, e.g. [Jac75].

[10] This assumption breaks down only at ultra-high field intensities where the internal structure of the charges itself may become affected by the field.

of the carriers (polarization effects), cf. Ch. 9. Thus, the electromagnetic field contributes only to the single-particle Hamiltonian. Still, the field will appear in each equation of the BBGKY–hierarchy (2.16), in the commutator $[H_{1...s}, F_{1...s}]$ (drift term), what allows to account for a large variety of physical effects. In particular, with the field terms included into the first three hierarchy equations, one is able to describe transport in fields, field effects in the collision integrals and the field–induced modification of the selfenergy.

Examples of electromagnetic fields. The vector and scalar potential which are related to the electric field strength $\mathcal{E}$ and magnetic induction $\mathbf{B}$ according to Eq. (11.1), are defined only up to an additional gauge condition (the gauge problem and gauge invariance will be discussed in Sec. 11.2). This fact may be used to simplify Maxwell's equations for various important special cases, which include

i) For purely *longitudinal fields*, it follows $\mathbf{A} = 0$. Then, among the main effects are collective plasma excitations (cf. Ch. 4) or transport and charge acceleration in a static electric field. This case is relatively easy to treat. For the example of a weakly time and space dependent external electrical field $\mathcal{E}^{ext}$, one can use the scalar potential gauge, $\phi = -\mathbf{r}\,\mathcal{E}$, Eq. (11.18) which yields the following Hamilton operator, in coordinate and momentum representation, respectively, $(\mathcal{E}_{tot} = \mathcal{E} + \mathcal{E}^{ext})$

$$H_s \to H_s^0(-i\hbar\nabla) - e_s\mathbf{r}_s\mathcal{E}_{tot} = H_s^0(\mathbf{p}) - i\hbar e_s\mathcal{E}_{tot}\frac{\partial}{\partial \mathbf{p}_s}, \tag{11.8}$$

where the derivative is to be taken along the field direction.

ii) If there are no external charges, $\phi^{ext} = 0$, and the system is homogeneous, we have the case of a *purely transverse field*. Then, it is convenient to use the Coulomb gauge $\nabla\mathbf{A} = 0$, Eq. (11.19). As a result, the action of $\mathbf{p}$ on $\mathbf{A}$ gives zero, therefore, $\mathbf{p}\cdot\mathbf{A} = \mathbf{A}\cdot\mathbf{p}$,[11]. Thus, for example for parabolic energy dispersion, we have

$$H_s \longrightarrow \frac{\mathbf{p}_s^2}{2m_s} - \frac{e_s}{m_s c}\left(\mathbf{A} + \mathbf{A}^{ext}\right)\cdot\mathbf{p}_s + \frac{e_s^2}{m_s c^2}\left(\mathbf{A} + \mathbf{A}^{ext}\right)^2. \tag{11.9}$$

In many cases the last term is small, e.g. for moderate field intensities. Also, for periodic fields $\mathbf{A}(t) \sim \cos\omega_0 t$, the contribution of A^2 will oscillate with the frequency $2\omega_0$ which, under situations close to the resonance ω_0, will be far detuned. On the other hand, this term is important in experiments with high power lasers, e.g. [PM94, Tea94].

[11] In fact, this relation holds independently on the gauge if A varies sufficiently slowly in space, see e.g. [BK95].

iii) In the case of a transverse field, the field–matter interaction can often be treated within the *dipole approximation.* This correspond to situations where the particles responding to the electromagnetic wave are effectively point charges. The criterion for this to be satisfied is that the field does not vary noticeably over the spatial "extension" d_n of the particles,

$$\lambda_l \gg \bar{d}_n. \tag{11.10}$$

A measure for the average value $\bar{d}_n$ is the typical diameter of the electron orbit or of the wave function of the quantum state "n". For atomic systems, $\bar{d}_n$ is of the order of 1 Å, for semiconductors, of the order of the exciton Bohr radius (around 100 Å). For a harmonic field, λ_l is its wavelength. If higher harmonics are excited, we may use a Fourier expansion in terms of harmonics of some base oscillation with frequency ω_0 and wave vector $\mathbf{k}_0$,

$$\mathbf{A}(\mathbf{r},t) = \sum_{l=1}^{\infty} A_l(\mathbf{r},t) e^{-i(l\omega_0 t - \mathbf{k}_l \mathbf{r})} + c.c., \tag{11.11}$$

where A_l are slowly varying amplitudes, and $k_l = 2\pi/\lambda_l$. Then, the dipole approximation is applicable, if the criterion (11.10) is fulfilled for all nonzero components. This is usually the case, for example, for lasers in the visible regime, where $\lambda_1 > 3000$ Å. Only for high power lasers, there may be harmonics of very high order excited in the system, for which the criterion (11.10) may be violated. This condition may also break down for high energy photons above the visible spectral region (e.g. x-ray lasers).

Relation to the general results of Ch. 2. To make use of the general results regarding the BBGKY–hierarchy which were obtained in Ch. 2, we now have to establish a relation between the single-particle Hamiltonian (11.7) and Eq. (2.4). Obviously, we simply have to identify all terms in Eq. (11.7) which contain ϕ or A with the general potential $\mathcal{U}$. For example, in Eqs. (11.8) and (11.9), this corresponds to all terms except the kinetic energy part.

11.2 Field effects on the distribution and the propagators

Before deriving generalized kinetic equations for multiband systems, we summarize the main field effects in the case of single band (one–component) systems. As we have seen in the Chapters before, for the derivation of quantum kinetic equations, we need (I) the first hierarchy equation in the appropriate form and (II) explicit expressions for the collision integrals. In the case of external fields, we expect field-induced modifications in both parts. We, therefore, discuss next the changes in the quantum kinetic equations and in the quasiparticle propagators which determine the collision integral.

Kinetic equations with EM fields. We start from the first hierarchy equation in coordinate representation, Eq. (2.45)

$$\left\{ i\hbar\frac{\partial}{\partial t} + \frac{\hbar^2}{m}\nabla_R\nabla_r - \mathcal{U}(R+\frac{r}{2},t) + \mathcal{U}(R-\frac{r}{2},t)\right\} f(R,r,t) = I(R,r,t), \quad (11.12)$$

where we introduced the center of mass and difference coordinates R and r, and I is the collision integral.[12] If the system is only weakly inhomogeneous, we use the Wigner representation, Eq. (2.62), where in the difference of the potentials we keep only terms linear in r and use $r = i\hbar\nabla_{\mathbf{p}}$,

$$\left\{ \frac{\partial}{\partial t} + \frac{\mathbf{p}}{m}\nabla_{\mathbf{R}} - \nabla_{\mathbf{R}}\mathcal{U}(R,t)\cdot\nabla_{\mathbf{p}}\right\} f(R,p,t) = I(R,p,t). \quad (11.13)$$

This equation is the starting point for the investigation of field effects in kinetic equations for weakly inhomogeneous systems. Substituting for $\mathcal{U}$ the field in the respective approximation and gauge, this equation is applicable to all the situations discussed in Sec. 11.1 . For example, for a longitudinal electrical field, $\mathcal{U} \to e\phi^{ext}$, and $-\nabla_{\mathbf{R}}\mathcal{U} = e\mathcal{E}^{ext}$. Equations of this type have been investigated in great detail in almost any field of statistical physics. They allow to describe a large variety of field-matter interaction problems, including transport (electrical conductivity) or response to time-dependent fields. In the spatially homogeneous case, an elegant approach to Eq. (11.13) is to introduce a generalized momentum $\tilde{\mathbf{p}}$ by

$$\mathbf{p} = \tilde{\mathbf{p}} + \mathbf{p}^D = \tilde{\mathbf{p}} - \int_{-\infty}^{t} d\bar{t}\, \nabla_{\mathbf{R}}\mathcal{U}(\mathbf{R}\bar{t}), \quad (11.14)$$

which contains a field induced drift p^D, and the new distribution $F(\tilde{\mathbf{p}},t) = f(\mathbf{p},t) = f(\tilde{\mathbf{p}}+\mathbf{p}^D,t)$. Replacing in Eq. (11.13) $f(\mathbf{p})$ by $F(\tilde{\mathbf{p}})$, the field on the l.h.s. can be eliminated, and the equation can be treated as in the field-free case (in a drifting coordinate system). Only at the end, the field is restored according to Eq. (11.14). It then remains to express the collision integral in terms of $F(\tilde{\mathbf{p}})$ too.

[12]Notice that $I = I[F_{12}]$, i.e. it contains the mean-field term. Therefore, the potential $\mathcal{U}$ should not contain the longitudinal field $\mathcal{E}_l$.

An extensive analysis of this approach can be found in papers and monographs of Klimontovich and co-workers, [KP74, Kli75, Puc75], who derived a variety of generalized collision integrals which include constant and time-dependent electric fields, magnetic fields, field-dependent screening etc. Similar concepts have been developed in the field of quantum transport in solids, based on density matrix methods [KI86, KI87, IK88][13] or nonequilibrium Green functions [MH83, HM83a, HM83b, Jau83, JW84, HJ96], see below.

Longitudinal electric fields. In the following, we discuss the effect of longitudinal electric fields more in detail. Among them are

1. acceleration of carriers and correspondingly deformation of the distribution function, electron "run away" or "hot electrons";
2. modification of the collision integral by explicit appearance of the field in it, retardation effects [KL57, Kli75] and intra-collisional field effect, e.g. [Lev69, Bar73];
3. modification of the collective response of the plasma, modified plasmon spectra, instabilities etc. (see Ch. 4), for an overview, see also [ABR84, Sil73];
4. change of absorption/emission properties, e.g. of optical absorption in the presence of a longitudinal field (Franz-Keldysh effect [Kel57, Fra57]);
5. change of reaction rates by the enhancement of excitation or ionization processes (e.g. field-induced lowering of the ionization or dissociation energy) [KMSR93] or new ionization processes (field ionization etc.);
6. as a result, change of the chemical composition, population of atomic levels, energy bands in solids etc. [MSK93, SBRK96];
7. Modification of the energy level structure caused by the field, field-induced level splitting (Stark effect), modified tunnel probabilities etc.

Stationary carrier distribution in an electric field. We briefly comment on some important points regarding the carrier distribution function in the presence of an external field. If correlations are weak ("collisionless" plasma), the r.h.s. of Eq. (11.13) is small, and its details are not important. The plasma will relax towards a local equilibrium distribution which is shifted by the field induced drift momentum (11.14) $f_{\mathbf{p}}(\mathcal{E}, t) \longrightarrow f^{EQ}(\mathbf{p} - \mathbf{p}^D(t))$. This applies to classical and quantum systems as well, where f^{EQ} is a Maxwell

[13]This approach has been extended to spatially inhomogeneous fields in [Iaf].

or Fermi distribution respectively, to time-dependent fields and even to relativistic plasmas, see e.g. [ABR84]. In case of a *time-independent field* (slowly varying field or pulse of long duration), the drift momentum (11.14) grows continuously, and at some point the collisionless approximation will no longer be valid. The accelerated carriers (in particular electrons) encounter a variety of growing energy loss mechanisms, such as ionization of bound states, excitation of plasmons, generation of radiation or, in solids additionally, ionization of impurities, inter-valley scattering or excitation of phonons. This will eventually lead to a stationary but nonequilibrium distribution, the shape of which is determined by a balance of the field and scattering effects. Such stationary electron distributions in an external electric field ($d/dt = 0$ in Eq. (11.13)[14]) have been studied for many years in plasma physics, transport in solids and other fields. The standard approach is to expand the distribution into a series of polynomials, such as Sonine [ABR84] or Legendre polynomials, $f(\mathbf{p}) = f_0(\mathbf{p}) + A_1 f_1(\mathbf{p}) + A_2 f_2(\mathbf{p}) + \ldots$, and to transform Eq. (11.13) into a system of equations for the coefficients.

The most important example is a Legendre expansion, where in many cases, only the first two terms are used, i.e. $f(\mathbf{p}, \mathcal{E}) \approx f_0(p, \mathcal{E}) + \cos\Theta f_1(p, \mathcal{E})$, where Θ is the angle between $\mathbf{p}$ and $\mathcal{E}$. The dominating effect for weak fields is the generation of the non-isotropic component $\cos\Theta f_1(p, \mathcal{E})$ while f_0 remains close to an equilibrium distribution. At higher intensities, f_0 is modified too. The most significant deviation from f^{EQ} is a shift towards higher momenta. It may also occur that a fraction of highly energetic carriers is not slowed down efficiently giving rise to a "run away" effect which has been investigated in plasmas already a long time ago [Dre59, Dre60a, Dre60b, Gur60, KB60]. The precise form of the stationary distribution depends on the dominating energy loss mechanisms of the carriers. Analytical results have already been obtained in the classical works of Dawydov [Daw36] and Druyvesteyn, e.g. [Dru30, LL80a] and have been extended to a variety of fields. Corresponding results for semiconductors can be found e.g. in [Poz81].

Recent investigations focus on many-particle effects in dense plasmas and, in particular take into account impact ionization in partially ionized plasmas [KMSR93, Rie97]. Following Rietz [Rie97], we give a brief summary of some known analytical results in table 11.1. For very strong fields, the energy loss mechanisms may become insufficient to slow down the fast carriers, which leads to electrical break down. The formal consequence is that polynomial expansions are typically poorly converging and the distribution functions are no longer normalizable. Then, a numerical solution of the full anisotropic equation (11.13) is necessary.

---

[14] The stationary case is studied more easily using the original function $f(p, t)$.

Electron distributions in an electric field

$f(\mathbf{p}) = f^{EQ}(\mathbf{p} - \mathbf{p}^D(t))$	Eq. (11.14), $\mathcal{E}(t)$ arbitrary† "collisionless" plasma,
$f_0(p) \sim \exp\left(-\int_0^p \frac{dp'p'}{\frac{e^2\mathcal{E}^2 M}{3m_e\tau_E\tau_p} + m_e k_B T}\right)$	Dawydov [Daw36][1]
$f_0(E) \sim \exp\left(-\frac{E}{k_B\tilde{T}}\right),\ \tilde{T} = T_i + \frac{e^2\mathcal{E}^2 M}{3n_A^2 B^2 k_B}$	Field-dependent $\tilde{T}$ if τ_E, τ_p constant [2]
$f_0(E) \sim \exp\left(-\frac{3E^2 n^2 Q_{eA}^{T\,2} m_e}{e^2\mathcal{E}^2 M}\right),$	Druyvesteyn [Dru30][3] high field limit
$f_0(E) \sim \left(1 + \frac{CE}{k_B T}\right)^{-\frac{1}{C}},\ C = \frac{2e^2\mathcal{E}^2 m_e M}{3B^2 n^2}$	for $Q_{eA}^T = B/p^2$
$f(\mathbf{p}) = A(E)\,\delta(1 - \cos\Theta)$	limit of strong inelastic scattering [Poz81]

Table 11.1: Analytical results for the stationary electron (momentum (p) or energy (E)) distribution in a homogeneous electric field $\mathcal{E}$.
[1] τ_E, τ_p - energy and momentum relaxation times,
[2] $\tilde{T}, T_i$ - effective electron and ion temperature, respectively, B is a constant (Fokker-Planck coefficient, see [LL80a]);
[3] Q_{eA}^T - electron-atom transport cross section in a partially ionized plasma ($\tau_{E,p}^{-1} \sim p\,Q_{eA}^T$);
† drifting (time-dependent) local equilibrium distribution, see text.

The interaction of plasmas with time-dependent field is currently very actively studied. The effect of low-intensity laser fields in semiconductors will be considered below in Sec. 11.3. The situation is different if high intensity laser pulses are applied to the surface of the solid. For a discussion of the plasma creation, hot electrons, stimulated x-ray emission and related problems, see e.g. [CSU92, CKRU94] and references therein. For fields varying periodically in time sufficiently fast, obviously, the modification of the carrier distribution is weaker due to the variation of the direction of the acceleration. Here, the main effect of the field is carrier heating, excitation of plasma waves or instabilities and excitation or ionization of bound states. Very interesting phenomena occur in case of superposition of static and periodic fields, leading to complex nonlinear effects, e.g. [Cha90], Ch. 10.

Field effects on the quasiparticle propagators. Until now we have discussed the *direct* effect of the field (via the l.h.s. of Eqs. (11.12), (11.13)) on the carrier distributions. We now consider another, *indirect*, field effect, which arises from the modification of the r.h.s. of the kinetic equations, i.e. of the collision integrals. These scattering terms are determined by the binary correlation operator $I_1 = n\mathrm{Tr}_2[V_{12}^{\pm}, g_{12}]$, cf. Eq. (3.19). Our approach to the collision integrals was to explicitly solve the second hierarchy equation for g_{12}. Furthermore, the solutions $g_{12}(t)$, in various approximations (Chs. 6–10), were expressed in terms of quasiparticle propagators. Now, an important observation can be made: this procedure is directly applicable to the case of external fields also. The only change is that the propagator $U_1(tt')$ is to be determined from generalized equations of motion which contain the field [15], i.e.

$$\left\{ i\hbar\frac{\partial}{\partial t} - \bar{H}_1(t) \right\} U_1(tt') = 0, \qquad U_1(tt) = 1, \tag{11.15}$$

which has the solution $U_1(tt') = \mathrm{T}\, e^{-\frac{i}{\hbar}\int_{t'}^{t} d\bar{t}\, \bar{H}_1(\bar{t})}$. The effective single–particle Hamiltonian contains, beyond H_1 of Eq. (11.7), the Hartree-Fock term[16] and the retarded correlation selfenergy Σ_1^+, cf. Eq. (7.82),

$$\begin{aligned} \bar{H}_1\, U_1(tt') &= \left\{ H_1^0 \left[\mathbf{p} - \frac{e_1}{c}\left(\mathbf{A}(t) + \mathbf{A}^{ext}(t) \right) \right] + e_1\phi^{ext}(t) + H_1^{HF} \right\} U_1(tt') \\ &+ \int_{t_0}^{t} d\bar{t}\, \Sigma_1^+(t\bar{t})\, U_1(\bar{t}t'). \end{aligned} \tag{11.16}$$

[15] We could also use the retarded and advanced propagators $U^{\pm}$ instead of U, as it was done in Ch. 8. Then, the r.h.s. of Eq. (11.15) would be $i\hbar\delta(t-t')$, replacing the initial condition, see also Appendix D.

[16] The Hartree-Fock term accounts for the effect of the mean Coulomb potential, thus $\mathcal{E}_l$ should not appear in the field term.

The operator equation (11.15) together with Eq. (11.16) is completely general. No approximations have been made yet. It applies to inhomogeneous systems and fields $\mathbf{A}(\mathbf{r},t)$ and $\phi(\mathbf{r},t)$ of arbitrary strength and space and time dependence.[17] Eq. (11.15) is directly related to various fundamental equations:

I. If selfenergy terms and the Maxwell field are neglected, it is nothing but the time-dependent Schrödinger equation of a free particle in an external field $\mathbf{A}^{ext}, \phi^{ext}$, and the propagator is essentially the wave function, $U_1(tt') = \psi(t)\,\delta(t-t')$.

II. If the selfenergy is treated fully selfconsistently (i.e. it contains the same propagators also), Eq. (11.15) is directly related to the Dyson equation for the retarded Green's function, with $i\hbar\Theta(t-t')U(tt') = i\hbar g^{+}(tt')$, cf. Sec. 8.3.3.

This correspondence is extremely useful, as it allows one to use the abundant results from quantum theory of charged particles in electromagnetic fields on one hand[18], and, from nonequilibrium Green's functions, on the other. We list some important results in table 11.2 below.

U differs from the propagator of a free particle $U_p^0(tt') = \exp[-iE_p(t-t')/\hbar]$ in two ways: by the action of the field and by the influence of the surrounding

[17] For completeness, we mention that the scalar and vector potential are given only up to derivatives of an arbitrary real function $f(\mathbf{r},t)$, e.g. [LL62]. Obviously, this should not have an impact on measurable quantities, such as the electric and magnetic fields, averages (computed from the wave function of the system) etc, which should be *gauge invariant* (for a more fundamental discussion, see Ch. 12). This means, any transformation between two alternative definitions of $\mathbf{A}$ and ϕ (gauge transformations), obeys

$$\begin{aligned}
\mathbf{A}(\mathbf{r},t) &\longrightarrow \mathbf{A}(\mathbf{r},t) + \nabla f(\mathbf{r},t) \\
\phi(\mathbf{r},t) &\longrightarrow \phi(\mathbf{r},t) - \frac{\partial}{\partial t} f(\mathbf{r},t) \\
(\mathbf{E}(\mathbf{r},t), \mathbf{B}(\mathbf{r},t)) &\longrightarrow (\mathbf{E}(\mathbf{r},t), \mathbf{B}(\mathbf{r},t)) \\
\psi(\mathbf{r},t) &\longrightarrow e^{\frac{ie}{\hbar c} f(\mathbf{r},t)}\,\psi(\mathbf{r},t).
\end{aligned}$$

This gives some freedom to choose a specific function f to simplify the problem under consideration. Among these choices are

$$\mathbf{A}(\mathbf{r},t) = -t\,\mathbf{E}(\mathbf{r},t), \quad \text{vector potential gauge} \tag{11.17}$$

$$\phi(\mathbf{r},t) = -\mathbf{r}\,\mathbf{E}(\mathbf{r},t), \quad \text{scalar potential gauge} \tag{11.18}$$

$$\nabla\mathbf{A}(\mathbf{r},t) = 0, \quad \text{Coulomb gauge.} \tag{11.19}$$

On the other hand, it is often convenient to define the propagator U in a gauge invariant form, which will be given below.

[18] For nonrelativistic and relativistic quantum-mechanical results, see e.g. [LL62] and [LL80b], respectively.

particles (which is contained in Σ^{HF} and Σ^{+}). Despite its, at first sight, simple form, the solution of Eq. (11.15) may be extremely complicated and not even constitute an explicit solution at all. First, due to the selfenergy terms in $\bar{H}$, the propagator appears in the exponent also. Furthermore, $\mathbf{A}$ is not externally given, but has to be determined selfconsistently from Maxwell's equations (11.2). Therefore, the full problem can be solved only numerically.

Approximations for the propagators. There are important limiting cases, where analytical results for the the propagators can be found. If mean field and correlation effects (i.e. the selfenergy terms in the effective Hamiltonian $\bar{H}$) are small, one can apply perturbation theory, using e.g. the local approximation for Σ^{+} which leads to the free-particle propagator with an effective energy $E_p \to \epsilon_p = E_p + \Delta_p - i\gamma_p$, cf. Appendix D. If, on the other hand, the field is weak, one can use the propagator of the field-free system as the starting point of a perturbative approach.

Table 11.2 lists several important results for the propagators, both without and with field. (A) and (B) summarize the result for a free particle which can be generalized to more complicated energy dispersions (C), as is the case in solids. (D) and (E) indicate the effect of the surrounding particles (selfenergy).[19] In the following lines, the influence of homogeneous fields is considered (Extensions to non-uniform fields are discussed in [Iaf, HJ96]. (F) shows that all previous results remain valid if the momentum is replaced by the kinematic momentum $\tilde{p}$ as indicated. (This is consistent with our discussion of field effects in the distribution function above). Finally, lines (G)–(I) contain explicit results for free particles in a constant uniform and a harmonic field, respectively. Case (G) is familiar from quantum mechanics, the corresponding result for the electron wave function can be found e.g. in [LL62]. The quantum mechanical result for an electron in a periodic field was derived by Keldysh [Kel64b]. The generalization to the relativistic case is known too since the work of Volkov [Vol35]. Furthermore, the case of a uniform magnetic field can be treated in analogy to Landau's solution of the Schrödinger equation [LL62]. The results (G)-(I) can also be generalized to include the dependence on the full momentum vector, for case (G) see [HJ96], case (I) follows from straightforward computation of the time integral in (F). We mention that the explicit expression (I) in terms of Bessel functions is based on relation (A.4) of App. A. Notice that for an efficient computation of the propagators it is important to choose the appropriate representation, which very much depends on the system under consideration. Lines (A)–(E) of Table 11.2 are written

[19] However, this simple quasiparticle approximation (D) leads to a Lorentzian spectrum (E), which does not describe the long-time behavior correctly [BKS+96, HB96]. We will consider improvements in Ch. 12.

Free particle $\mathbf{A} = 0,\ \Sigma = 0$	$U^0_{\mathbf{p}}(\mathbf{R}, tt') = e^{-\frac{i}{\hbar}\left[\frac{p^2}{2m}\tau - \mathbf{pR}\right]}$ $U^0_{\mathbf{p}}(\mathbf{R} = 0, \omega, T) = \delta(\omega - \frac{p^2}{2m\hbar})$	(A) (B)
General dispersion[1]	$p^2/2m \to E^0_{\mathbf{p}}, \quad e^{\frac{i}{\hbar}\mathbf{pR}} \to e^{\frac{i}{\hbar}\mathbf{pR}}\mathcal{U}_{\mathbf{p}}(\mathbf{R})$	(C)
$\mathbf{A} = 0$, Quasiparticle $\Sigma(tt') = \Sigma(\tau)$	$E^0_{\mathbf{p}} \to \epsilon_{\mathbf{p}} = E^0_{\mathbf{p}} + \Delta_{\mathbf{p}}(\mathbf{R}T) + i\,2\gamma_{\mathbf{p}}(\mathbf{R}T)$ $U^0_{\mathbf{p}}(\mathbf{R} = 0, \omega, T) = \dfrac{\gamma_{\mathbf{p}}/\hbar}{(\omega - \frac{E^0_{\mathbf{p}}+\Delta_{\mathbf{p}}}{\hbar})^2 + (\frac{\gamma_{\mathbf{p}}}{\hbar})^2}$	(D) (E)
Homog.[2] field $\mathbf{A}(t)$	$\epsilon_{\mathbf{p}}\tau \to \int_{t'}^{t} d\tau\, \epsilon_{\tilde{\mathbf{p}}(\tau)}, \quad \tilde{\mathbf{p}}(\tau) = \mathbf{p} - \frac{e}{c}\mathbf{A}(\tau)$	(F)
Const. field[3] $\mathcal{E}_0$ and $\epsilon_{\mathbf{p}} = p^2/2m$	$U^0_p(tt') = \exp\left\{-\frac{i}{\hbar}\tilde{E}\tau\right\}, \quad \tilde{E} = \frac{p^2}{2m} + \frac{e^2\mathcal{E}_0^2\tau^2}{24c^2}$ $U^0_p(\omega) = \int_{-\infty}^{\infty} d\tau\, \exp\left[i(\omega - \tilde{E}/\hbar)\tau\right]$ **	(G) (H)
$\mathcal{E}(t) = \mathcal{E}_0 \cos\omega_0 t$ and $\epsilon_{\mathbf{p}} = p^2/2m$	$U^0_p(t, 0) = \sum_{n=-\infty}^{\infty} I_{2n}\left(\frac{\hat{E}}{2\hbar\omega_0}\right) e^{-\frac{i}{\hbar}\left(\frac{p^2}{2m} + \hat{E} - 2n\hbar\omega_0\right)t}$ $\hat{E} = \frac{e^2\mathcal{E}_0^2}{2c^2\omega_0^2}, \quad I_m$ - Bessel function	(I)
Relations:	$U^{0\pm}_p(tt') = \Theta[\pm\tau]\, U^0_p(tt') = i\hbar g^{\pm}_p(tt')$	

Table 11.2: Quasiparticle propagators $U(tt')$ in different approximations for the electromagnetic field **A** and the retarded selfenergy Σ^+. Abbreviations: $T = (t + t')/2$, $\tau = t - t'$, [1] E^0 - general single-particle energy dispersion, $\mathcal{U}$ - general spatial modulation, e.g. in solids, periodic part of Bloch function; [2] for extensions to inhomogeneous fields see [Iaf]; [3] see e.g. [Jau]; ** the real part is often expressed in terms of Airy functions [LL62, Jau]

in Wigner or momentum representation, respectively. In the case of a non-parabolic energy dispersion, it is often convenient to use a more general basis, e.g. in periodic systems, the Bloch basis. This will be discussed in detail below in Sec. 11.3. On the other hand, the basic representations considered in Ch. 2, are constructed on eigenstates of the carrier system. This is suitable for carriers in a weak external field too, since there, the eigenstates will be close to those of the field-free system. However, if the field is strong, this is no longer the case, and it will be convenient to use as a basis the eigenstates of the coupled particle–field system. Due to the possible time-dependence of the field, this basis evolves in time (e.g. "instantaneous Bloch basis" [Iaf]).[20]

11.3 Interaction of optical fields with multiband systems

We now return to the discussion of the Hamiltonian (11.7), now focusing on optical fields ($\phi^{ext} = 0$) where the single-particle Hamiltonian is given by Eq. (11.9). Considering moderate intensities, we may neglect the A^2 term and use the dipole approximation

$$H_1(t) = H_1^0(\mathbf{p}) - \frac{e_1}{m_1 c}\left(\mathbf{A} + \mathbf{A}^{ext}\right) \cdot \mathbf{p}_1 = H_1^0(\mathbf{p}) - \mathbf{d_1} \cdot \mathcal{E}_{tot}, \qquad (11.20)$$

where $\mathbf{d}_i = -e_i \mathbf{r}$ is the operator of the dipole momentum of particle i, and $\mathbf{A}$ obeys Maxwell's equations (11.2). The last part of Eq. (2.93) is obvious for monochromatic fields oscillating like $e^{-i\omega t}$, then the total electric field and total vector potential are related by $\mathcal{E}_{tot} = \frac{i\omega}{c}\mathbf{A}_{tot}$ and $\mathbf{p}_i = -im_i\omega\mathbf{r}$, see also [PKM93]. For semiconductors, a more general derivation which takes into account the band structure, is given in [BK95].

[20] Often it is convenient to use the propagators in a gauge invariant form U^{inv}, which is obtained from U^0 by [HM83b, Jau],

$$\begin{aligned} U^{inv}(\mathbf{p}, \omega, \mathbf{R}, T) &= \int d\mathbf{r} d\tau \, e^{i\Phi(\omega,\tau,T,\mathbf{p},\mathbf{r},\mathbf{R})} \, U^0(\mathbf{r}, \tau, \mathbf{R}, T) \\ \Phi(\omega, \tau, T, \mathbf{p}, \mathbf{r}, \mathbf{R}) &= \int_{-1/2}^{1/2} d\lambda \Big\{ \tau \left[\omega + \phi(\mathbf{R} + \lambda\mathbf{r}, T + \lambda\tau)\right] - \mathbf{r}\left[\mathbf{p} + \mathbf{A}(\mathbf{R} + \lambda\mathbf{r}, T + \lambda\tau)\right] \Big\}, \end{aligned}$$

and simplifies, e.g. for constant electrical fields $\mathcal{E}(\mathbf{r}, t) = \mathcal{E}_0$ to [Jau]

$$U^{inv}(\mathbf{p}, tt') = U^0(\mathbf{p} - \mathcal{E}_0 T, T + \frac{\tau}{2}, T - \frac{\tau}{2}).$$

For further discussion of the gauge problem in transport theory, we refer to [HJ96].

Transverse polarization. Within the dipole approximation, the field–matter interaction is completely described by a single quantity - the dipole moment. It determines the response of the system to the field and the macroscopic properties, such as absorption or transmission, i.e. the quantities measured in an optical experiment. The latter are determined by the dipole density, i.e. the density of the macroscopic polarization $\mathbf{P}$. This quantity is just the average of the dipole moment which is, according to Eq. (2.14), given by the one-particle density operator and, on the other hand, it is related to the induced current $\mathbf{j}$

$$\boxed{\mathbf{P}(t) = \frac{n}{\mathcal{V}}\mathrm{Tr}_1\ \mathbf{d}_1 F_1(t) = -\frac{n}{\mathcal{V}} e_1 \mathrm{Tr}_1\ \mathbf{r}_1 F_1(t) = \int\limits_{-\infty}^{t} dt'\,\mathbf{j}(t')} \tag{11.21}$$

$\mathbf{P}$ governs the dynamics of the electromagnetic field, which becomes obvious from Maxwell's equation (11.2) where we differentiate the first equation with respect to time and in the third act with $\nabla\times$,

$$\left\{\frac{1}{c^2}\frac{\partial^2}{\partial t^2} - \Delta_R\right\}\mathcal{E}_t(\mathbf{R},t) = -\frac{4\pi}{c^2}\frac{\partial^2}{\partial t^2}\mathbf{P}(\mathbf{R},t) - \frac{4\pi}{c^2}\frac{\partial}{\partial t}\mathbf{j}^{ext}(\mathbf{R},t). \tag{11.22}$$

Eqs. (11.21), (11.22) show clearly, how the particle properties influence the field dynamics. To compute the field evolution, we, therefore, need the one-particle density operator $F_1(t)$, which follows from the kinetic equation.

Properties of the polarization. The polarization may be space dependent due to inhomogeneities or a special geometry of the sample (then $\mathbf{d}$ depends on $\mathbf{R}$). Furthermore, the polarization and, more generally, the field–matter interaction, crucially depend on the microscopic properties of the system, such as existence of bound states, chemical composition or band structure. In general, there are three types of processes possible:

1. A free particle may absorb/emit a photon, thereby remaining free. This means, the field interacts only with the scattering part of the one–particle spectrum. In solids, this corresponds to intraband transitions.

2. Absorption of a photon leads to excitation of a bound particle into the continuum, i.e. to photoionization, (or to recombination in case of emission of a photon),

3. Absorption/emission of a photon leads to transitions of a particle in the discrete part of the spectrum, from one level (band) to another.

Correspondingly, the dipole moment and total polarization are the sum of three contributions, e.g. $P = P^{f\leftrightarrow f} + P^{f\leftrightarrow b} + P^{b\leftrightarrow b}$, where "f" and "b" denote a free and bound state, respectively. Each of these contributions is calculated according to Eq. (11.21), with the trace running over the corresponding states. To perform the trace operation one needs to introduce an appropriate set of basis functions which leads us to a generalized Bloch representation.

Bloch basis. Let us assume that the quantum–mechanical part of the problem has been solved, and the complete set of eigenstates $|b\rangle$ for the *field–free case* has been found.[21] In general, they consist of scattering and bound states $|\psi\rangle$ and $|\lambda\rangle$, respectively, with

$$\begin{aligned} \langle\psi(x)|\psi(x')\rangle &= \delta(x-x'), \\ \langle\lambda|\lambda'\rangle &= \delta_{\lambda\lambda'}, \\ \int dx\, |\psi(x)\rangle\, \langle\psi(x)| + \sum_\lambda |\lambda\rangle\, \langle\lambda| &= 1. \end{aligned} \tag{11.23}$$

Here, x and λ comprise all the relevant quantum numbers ($x = \mathbf{r}, s, \ldots$, whereas λ contains all quantum numbers, band indices etc. and may also depend on the coordinate).[22] In the Bloch basis, the dipole operator is represented by the dipole matrix, the elements of which are readily computed from the wave functions of the initial and final state, $|b_1(\mathbf{r}_1)\rangle$ and $|b_1'(\mathbf{r}_1')\rangle$ ($\langle b| = |b\rangle^*$),

$$\mathbf{d}^{b_1b_1'}(\mathbf{R}) = -e\int d\mathbf{r}\; \mathbf{r}\cdot \left|b_1\left(\mathbf{R}+\frac{\mathbf{r}}{2}\right)\right\rangle^* \cdot \left|b_1'\left(\mathbf{R}-\frac{\mathbf{r}}{2}\right)\right\rangle, \tag{11.24}$$

where we introduced center of mass and difference variables $R = (r_1+r_1')/2$ and $r = r_1 - r_1'$. Clearly, the dipole matrix element is determined by the overlap of the wave functions of the initial and final state (weighted by the radius vector). If this overlap is small, or the product of the wave functions has reflection symmetry, the transitions $b \to b'$ and $b' \to b$ are (dipole) forbidden. In many cases, only a limited number of transitions needs to be taken into account (selection rules).

From Eq. (11.24) it is clear that, in many cases, the dipole moment for free–free transitions will be close to zero which is due to the symmetry of

[21] We mention that the choice of a basis in terms of field–free eigen functions may not be suitable at high field intensities, when the field causes a strong modification of the one–particle spectrum. In that case it is more convenient to use the eigen states of the particle–field systems (Volkov states), [Kel64b, Fai73, Rei80].

[22] This subdivision is not always useful. In particular, in the region of the Mott density (pressure ionization) bound states merge into the continuum (where they continue to show up as resonances), and a clear distinction becomes meaningless.

the corresponding states. These transitions may be important only if this symmetry is broken, e.g. in case of strong inhomogeneity or as a result of external fields. Bound–free transitions usually require a substantial photon energy to overcome the ionization gap and are not relevant in the considered low-intensity optical regime. Both processes are more important at high field intensities which gives rise to multi–photon absorption or sub–threshold ionization, which we will not consider here. Consequently, for low intensity optical fields we expect the major contributions to arise from bound–bound transitions between different energy levels or bands, $\lambda \neq \lambda'$. As a result, $d^{\lambda\lambda'} \sim (1 - \delta_{\lambda\lambda'})$. The reduced Bloch basis consists of the bound state vectors $|\lambda x\rangle$, where λ labels all possible energy levels or energy bands, and we explicitly take into account the space dependence. In the low intensity regime where the contributions from the scattering states may be neglected, these vectors form a complete orthonormal basis with $\langle x\lambda|\lambda' x'\rangle = \delta_{\lambda\lambda'}\delta(x - x')$ and $\sum_\lambda \int dx \, |\lambda x\rangle \, \langle x\lambda| = 1$.

What is left now is to transform the single–particle density operator into the Bloch representation. As was discussed in detail in Sec. 2.3, this leads to a matrix with the elements $\langle b_1|F_1|b_1'\rangle = F^{b_1 b_1'}$, i.e. for interband transitions, we have to consider the reduced matrix $F^{\lambda_1\lambda_1'}(x_1 x_1')$. In the homogeneous case, we will use the momentum variable k instead of x and obtain the simpler matrix elements $nF_1 \to f^{\lambda_1\lambda_1'}(k_1)\,\delta(k_1 - k_1')$. In that case, the total polarization density (11.21) is given by

$$\mathbf{P}(t) = \frac{1}{\mathcal{V}} \sum_{\lambda\bar{\lambda}} \sum_{k} \mathbf{d}^{\lambda\bar{\lambda}}(k)\, f^{\lambda\bar{\lambda}}(k,t). \tag{11.25}$$

Since $d^{\lambda\lambda'} \sim 1 - \delta_{\lambda\lambda'}$, only band-off-diagonal matrix elements of density operator enter, which describe the probability of transitions between different bands.

11.4 Bloch representation of the first hierarchy equation

We already demonstrated how to transform the BBGKY-hierarchy for the reduced density operators $F_{1\ldots s}$ into the Bloch representation in Sec. 2.4.1. We will do the same now, but with the first hierarchy equation rewritten in terms of binary correlations g_{12}, which proved more convenient for the treatment of correlation effects, and which explicitly contain spin statistics effects. Thus we start from Eq. (3.19), which was derived in Ch. 3,

$$i\hbar\frac{\partial}{\partial t}F_1 - [\bar{H}_1, F_1] = n\mathrm{Tr}_2[V_{12}^{\pm}, g_{12}], \tag{11.26}$$

with the effective single–particle Hamiltonian

$$\bar{H}_1 = H_1 + H_1^{HF} = H_1 + n\mathrm{Tr}_2 V_{12}^{\pm} F_2, \tag{11.27}$$

where H_1 contains the external potential $\mathcal{U}_1$, which is still completely general. Compared to the previously derived equation for the Bloch matrix $f^{\lambda\lambda'}$, Eq. (2.96), we anticipate two modifications, which are related to the exchange terms in the Hartree-Fock Hamiltonian and in the collision integral on the r.h.s., Eqs. (11.27) and (11.26), respectively.

General case. Inhomogeneous systems. We begin with the general case of an inhomogeneous system. We first evaluate the Bloch-matrix elements of the (anti-)symmetrized potential $V_{12}^{\pm}$ and the single–particle Hamilton operator, in the coordinate representation, cf. Sec. 2.3.1,

$$\begin{aligned}
\langle x_1\lambda_1 x_2\lambda_2|V_{12}^{\pm}|x_2'\lambda_2' x_1'\lambda_1'\rangle &= \langle x_1\lambda_1 x_2\lambda_2|V_{12}|x_2'\lambda_2' x_1'\lambda_1'\rangle \\
&\pm \langle x_1\lambda_1 x_2\lambda_2|V_{12}|x_1'\lambda_1' x_2'\lambda_2'\rangle = \\
&= V(x_1 - x_2)\Big\{\delta_{\lambda_1\lambda_1'}\delta_{\lambda_2\lambda_2'}\delta(x_1 - x_1')\delta(x_2 - x_2') \\
&\pm \delta_{\lambda_1\lambda_2'}\delta_{\lambda_2\lambda_1'}\delta(x_1 - x_2')\delta(x_2 - x_1')\Big\}, \qquad (11.28)\\
\langle x_1\lambda_1|H_1|x_1'\lambda_1'\rangle &= \tilde{E}^{\lambda_1\lambda_1'}(x_1)\delta(x_1 - x_1'), \qquad (11.29)
\end{aligned}$$

where $\tilde{E}$ denotes the one particle energy eigenvalues in the presence of the external field. Notice that they are, in general, not diagonal in the band index. The field in H_1 is still very general, it may, for example, be a dipole field (11.20), a longitudinal field (11.8), or a combination of both.

Expanding Eq. (11.26) in terms of Bloch states $|x\lambda\rangle$ and using the results (11.28,11.29), we obtain the Bloch equation for an inhomogeneous system

$$\begin{aligned}
i\hbar\frac{\partial}{\partial t}f^{\lambda_1\lambda_1'}(x_1x_1',t) &- \sum_{\bar{\lambda}_1}\Big\{\tilde{E}^{\lambda_1\bar{\lambda}_1}(x_1)f^{\bar{\lambda}_1\lambda_1'}(x_1x_1',t) - f^{\lambda_1\bar{\lambda}_1}(x_1x_1',t)\tilde{E}^{\bar{\lambda}_1\lambda_1'}(x_1')\Big\} \\
&- \sum_{x_2\lambda_2}\{V(x_1 - x_2) - V(x_1' - x_2)\} \times \\
&\Big\{f^{\lambda_1\lambda_1'}(x_1x_1')f^{\lambda_2\lambda_2}(x_2x_2) \pm f^{\lambda_1\lambda_2}(x_1x_2)f^{\lambda_2\lambda_1'}(x_2x_1')\Big\} \qquad (11.30)\\
= \sum_{x_2\lambda_2}\{V(x_1 - x_2) - V(x_1' - x_2)\}&\Big\{g^{\lambda_1\lambda_1'}_{\lambda_2\lambda_2}(x_1x_2;x_1'x_2) \pm g^{\lambda_1\lambda_2}_{\lambda_2\lambda_1'}(x_1x_2;x_2x_1')\Big\}.
\end{aligned}$$

The second term on the l.h.s. contains the influence of the field on the single particle energies. Since the Bloch states $|x\lambda\rangle$ are, in general, not eigenstates of the particle field complex, this term is not diagonal in the band index. The sum contains, along with the diagonal elements ($\bar{\lambda}_1 = \lambda_1'$) which correspond to the field renormalized energies, also off-diagonal contributions which are related to transitions into other states. This becomes particularly clear within the dipole approximation. Then the one–particle energy in Eq. (11.30) is given by

$$\tilde{E}^{\lambda_1\lambda_1'}(x_1) = E^{\lambda_1}(x_1)\,\delta_{\lambda_1,\lambda_1'} - \mathbf{d}^{\lambda_1\lambda_1'}(x_1)\,(1-\delta_{\lambda_1,\lambda_1'})\,\mathcal{E}(x_1), \tag{11.31}$$

with $E^{\lambda_1}(x_1)$ being the energy eigenvalue of the field–free single–particle Hamiltonian. The last term on the l.h.s. of Eq. (11.30) is the mean field contribution which contains the classical mean field (Hartree field) and the exchange (Fock) terms, (terms with the $\pm$ sign). The r.h.s. contains the effect of binary correlations, i.e. the collision integral, again with direct and exchange contributions. Notice that the terms on the l.h.s. (except the first) are real quantities. This means, they only cause oscillations of $f^{\lambda_1\lambda_1'}$. In contrast, the collision term on the r.h.s. is complex and hence leads to relaxation of the band population or dephasing of the interband polarization. This will become more clear in the homogeneous case.

Eq. (11.30) is the most general form of the Bloch equations for a non-relativistic system. The index λ may refer simply to different particle species (then f would have only diagonal components $f^{\lambda_1\lambda_1} \to f^{\lambda_1}$). In a multi–level or multi–band system, also off–diagonal elements appear, corresponding to transitions from one band to another. An alternative form of the inhomogeneous Bloch equations (11.30) is obtained if instead of the coordinate representation the Wigner representation is used. The derivation is exactly the same as given in Sec. 2.3.2.

Homogeneous systems. In the case of spatial homogeneity, it is convenient to use the momentum representation, cf. Sec. 2.3.3. The Bloch-matrix elements of the (anti-)symmetrized potential $V_{12}^{\pm}$ and the single–particle Hamiltonian are now given by

$$\langle p_1\lambda_1 p_2\lambda_2|V_{12}^{\pm}|p_2'\lambda_2' p_1'\lambda_1'\rangle = \tag{11.32}$$
$$\left\{V(p_1-p_1')\delta_{\lambda_1\lambda_1'}\delta_{\lambda_2\lambda_2'} \pm V(p_1-p_2')\delta_{\lambda_1\lambda_2'}\delta_{\lambda_2\lambda_1'}\right\}\frac{\delta(p_1+p_2-p_1'-p_2')}{(2\pi\hbar)^3},$$

$$\langle p_1\lambda_1|H_1|p_1'\lambda_1'\rangle = \tilde{E}^{\lambda_1\lambda_1'}(p_1)\,\delta(p_1-p_1') =$$
$$\left\{E^{\lambda_1\lambda_1'}(p_1) - \mathbf{d}^{\lambda_1\lambda_1'}(p_1)\,(1-\delta_{\lambda_1\lambda_1'})\,\mathcal{E}(t)\right\}\delta(p_1-p_1'). \tag{11.33}$$

Again, $\tilde{E}$ denotes the energy eigenvalue in the presence of the field, and applies to the general situation, whereas the last line of Eq. (11.33) corresponds to the dipole approximation and the assumption of low field intensity which justifies the neglect of the A^2 contribution. Using Eqs. (11.32) and (11.33), we obtain from Eq. (11.26) the Bloch equations for a homogeneous system

$$
\begin{aligned}
&\left\{i\hbar\frac{\partial}{\partial t} - \left(E_{p_1}^{\lambda_1} - E_{p_1}^{\lambda_1'}\right)\right\} f^{\lambda_1\lambda_1'}(p_1,t) \\
&+\mathcal{E}(t) \sum_{\bar\lambda_1}{}' \left\{\mathbf{d}^{\lambda_1\bar\lambda_1}(p_1)\, f^{\bar\lambda_1\lambda_1'}(p_1,t) - f^{\lambda_1\bar\lambda_1}(p_1,t)\, \mathbf{d}^{\bar\lambda_1\lambda_1'}(p_1)\right\} \\
&\mp\sum_{\lambda_2}\int \frac{dp_2}{(2\pi\hbar)^3} V(p_2-p_1) \\
&\times\left\{f^{\lambda_1\lambda_2}(p_2,t)\, f^{\lambda_2\lambda_1'}(p_1,t) - f^{\lambda_1\lambda_2}(p_1,t)\, f^{\lambda_2\lambda_1'}(p_2,t)\right\} = I^{\lambda_1\lambda_1'}(p_1,t),
\end{aligned}
\tag{11.34}
$$

where in the second line $\sum'$ indicates that the band–diagonal elements of the dipole operator are missing (they are equal to zero). The last term on the l.h.s. corresponds to the Hartree–Fock contribution (only the exchange term is nonzero in the homogeneous case).

The Hartree–Fock term leads to a renormalization of the single–particle energy and the external field, so we can rewrite Eq. (11.35) as

$$
\begin{aligned}
&\left\{i\hbar\frac{\partial}{\partial t} - \left(\bar E_{p_1}^{\lambda_1} - \bar E_{p_1}^{\lambda_1'}\right)\right\} f^{\lambda_1\lambda_1'}(p_1,t) \\
&-\hbar\sum_{\bar\lambda_1}{}' \left\{\bar\Omega_{p_1}^{\lambda\bar\lambda_1}(t)\, f^{\bar\lambda_1\lambda_1'}(p_1,t) - f^{\lambda_1\bar\lambda_1}(p_1,t)\, \bar\Omega_{p_1}^{\bar\lambda_1\lambda_1'}(t)\right\} = I^{\lambda_1\lambda_1'}(p_1,t),
\end{aligned}
\tag{11.35}
$$

where we introduced effective single–particle energies $\bar E$ and effective Rabi energies $\hbar\bar\Omega$ given by

$$
\bar E_{p_1}^{\lambda_1}(t) = E_{p_1}^{\lambda_1} \pm \int \frac{dp_2}{(2\pi\hbar)^3} V(p_2-p_1)\, f^{\lambda_1\lambda_1}(p_2,t), \tag{11.36}
$$

$$
\hbar\bar\Omega^{\lambda_1\lambda_1'}(p_1,t) = -\mathbf{d}^{\lambda_1\lambda_1'}(p_1)\,\mathcal{E}(t) \pm \int \frac{dp_2}{(2\pi\hbar)^3} V(p_2-p_1)\, f^{\lambda_1\lambda_1'}(p_2,t). \tag{11.37}
$$

As one can see, the single–particle energy of band λ becomes renormalized by the exchange interaction of carriers in this band. At the same time, the field energy which drives the matrix element $f^{\lambda\lambda'}$ is modified by the exchange energy arising from all matrix elements $f^{\bar\lambda\lambda'}$ with $\bar\lambda \neq \lambda$. For the band–diagonal matrix elements $f^{\lambda\lambda}$, the single particle energy contributions $\bar E^{\lambda}$ cancel. For the off–diagonal elements they give rise to oscillatory behavior with a frequency corresponding to the energy difference $\bar E^{\lambda} - \bar E^{\lambda'}$. These oscillations are driven by the external field and are undamped as long as correlations are neglected.

The effect of correlations is contained in the collision integrals of the Bloch equations which are given by

$$I_{p_1}^{\lambda_1\lambda_1'} = \sum_{p_2\lambda_2\bar{p}_1\bar{p}_2} V_{p_1-\bar{p}_1}\delta_{p_1+p_2,\bar{p}_1+\bar{p}_2} \left\{g_{\lambda_2\lambda_2}^{\lambda_1\lambda_1'}(\bar{p}_1\bar{p}_2;p_1p_2) - g_{\lambda_2\lambda_2}^{\lambda_1\lambda_1'}(p_1p_2;\bar{p}_1\bar{p}_2)\right\}$$
$$\pm \sum_{p_2\lambda_2\bar{p}_1\bar{p}_2} V_{p_1-\bar{p}_2}\delta_{p_1+p_2,\bar{p}_1+\bar{p}_2} \left\{g_{\lambda_1\lambda_2}^{\lambda_2\lambda_1'}(\bar{p}_1\bar{p}_2;p_1p_2) - g_{\lambda_2\lambda_1'}^{\lambda_1\lambda_2}(p_1p_2;\bar{p}_1\bar{p}_2)\right\}, \quad (11.38)$$

where the term with the $\pm$ sign is the exchange scattering term. The collision integrals express the correlation (scattering) contributions to the Bloch equations in terms of Bloch matrix elements of the binary correlation operator. Notice that in contrast to the corresponding collision integral for the one–band case, Eq. (2.85), for $\lambda_1 \neq \lambda_1'$ not only the imaginary part of g_{12} appears, because the terms in parentheses are not pairwise adjoint, instead $[g_{\lambda_2\lambda_2}^{\lambda_1\lambda_1'}(\bar{p}_1\bar{p}_2;p_1p_2)]^* = g_{\lambda_2\lambda_2}^{\lambda_1'\lambda_1}(p_1p_2;\bar{p}_1\bar{p}_2)$. This has important consequences as it leads to additional contributions to the collision integral coming from the real part of g.

The result (11.38) is still exact, since it couples via the matrix elements of g_{12} to the rest of the hierarchy. Approximations are introduced by choosing an approximation for g_{12}, see e.g. Sec. 2.6. Notice further, that the field in Eqs. (11.36) is very general (the low intensity and dipole approximations have been used only to obtain formally simpler expressions, but they are not necessary for the current derivations). In particular, there has no assumption been made yet on the time dependence of the field.

Examples: Two–band case. Periodic fields. For illustration, we consider the simplest example of the Bloch equations, the *case of two bands* which allows to study all the basic features of multi–band systems. Using the notation of semiconductor optics, the energetically upper and lower band will be labeled "c" and "v", respectively (for conduction and valence band). Also, we will denote the matrix elements $f^{cc} \to f^c$, $f^{vv} \to f^v$ and $f^{cv} = [f^{vc}]^* = P$, which is usually called "interband polarization", see Fig. 11.1. Then, we obtain from Eqs. (11.36) the following system of equations,

$$\frac{\partial}{\partial t} f^c(p,t) + 2\mathrm{Im}[\bar{\Omega}^*(p,t)P(p,t)] = -\frac{i}{\hbar} I^c(p,t), \quad (11.39)$$

$$\frac{\partial}{\partial t} f^v(p,t) - 2\mathrm{Im}[\bar{\Omega}^*(p,t)P(p,t)] = -\frac{i}{\hbar} I^v(p,t), \quad (11.40)$$

$$\left(\frac{\partial}{\partial t} + \frac{i}{\hbar}\bar{E}_p^{cv}\right) P(p,t) + i\bar{\Omega}^*(p,t)\left[f^v(p,t) - f^c(p,t)\right] = -\frac{i}{\hbar} I^{cv}(p,t), \quad (11.41)$$

where we defined $\bar{\Omega} = \bar{\Omega}^{cv}$ and $\bar{E}_p^{cv} = \bar{E}^c(p) - \bar{E}^v(p)$. One clearly sees the different effects of the field on the band populations and interband transitions:

in the equations for the Wigner distributions of the conduction and valence band, the field acts as a source term which, at the same time, creates particles in the conduction band and annihilates them in the valence band and vice versa. On the other hand, the interband transition probability P is driven by the field also, but here the field causes an oscillatory behavior of P. The frequency of these oscillations increases with growing difference of the band populations (which is different for different momenta). It is zero when half of the carriers moved from the valence band to the conduction band. In case of population inversion, the phase changes its sign. Furthermore, $\bar{\Omega}$ depends on the intensity of the field. Thus with increasing field strength all these effects are enhanced: we have to expect increasing particle generation (density) in the conduction band and increasingly rapid oscillations of P.

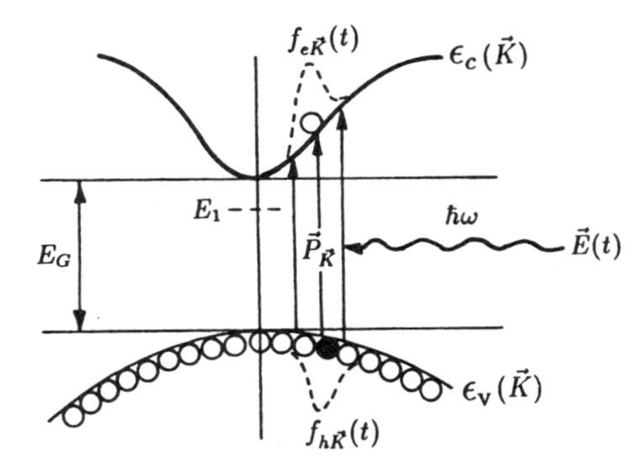

Fig. 11.1. Schematic picture of a two-band semiconductor. The total electromagnetic field lifts electrons across the band gap E_G from the valence (v) to the conduction (c) band, creating nonequilibrium electron and hole distributions f_e and f_h. For photon energies $\hbar\omega < E_G$, electrons may form bound states ("coherent" excitons) with a hole with the binding energy E_1. At short times, the field is spectrally broad (see Fig. 11.2), therefore, f_e, f_h and the interband polarization P are broad too.

If scattering effects (terms on the r.h.s.) are not taken into account, the band populations and interband polarization evolve in a dynamical and reversible way without dissipation. The distribution functions do not relax toward equilibrium and the polarization does not decay (dephase), i.e. it remains coherent. This corresponds to the (time–dependent) Hartree–Fock approximation or mean field approximations which has been extensively studied for one–component and two–component plasmas in Ch. 4. Nevertheless, the two–band case has a number of peculiarities: The most important one is the role of the *exchange (Fock) terms.* It is the interband exchange term (which is not present in a conventional plasma) in $\bar{\Omega}$ that gives rise to the formation of bound states (excitons) in the presence of an electro–magnetic field [ST86]. Instead of considering the physical picture of an electron being moved from the valence band to the conduction band (or back) by the field, it is often convenient to describe the same process as a creation of an electron in the conduction band and of a hole (i.e. a missing electron) in the valence band. The hole is a well–defined particle with positive charge and the Wigner distribution $f_h(k) = 1 - f^v(-k)$ (see Fig. 11.1.) which shows an attractive interaction on the electrons in the conduction band. In this picture it is not surprising that there may appear also bound states between electrons and holes [23], which are mediated by the field (their energy is denoted by E_1 in Fig. 11.2.). We will return to this effect of bound states in Sec. 12.5 where also numerical results will be given. For a detailed analysis of the Bloch equations in Hartree–Fock and the influence of duration and strength of the exciting field, the reader is referred to review articles and text books on semiconductor optics [HK93, PKM93, BK95, HJ96].

At this point let us consider the important case of a *periodic external field*, which, in general, can be written as a superposition of the base harmonic ω_0 and higher harmonics. Then, also the effective field $\mathcal{E}$ which is the solution of Maxwell's equation (11.22) can be written as a Fourier series,

$$\mathcal{E}(t) = \sum_{l=-\infty}^{\infty} \mathcal{E}_l(t)\, e^{-il\omega_0 t}, \tag{11.42}$$

where $\mathcal{E}_l(t)$ are slowly varying amplitudes of the l-th harmonic. Let us determine the response of the two–band system to the field (11.42). For this, we consider the Bloch equations (11.39–11.41). It is straightforward to look for solutions $f^{c,v}$ and P in form of a superposition of field harmonics. Expanding $f^{c,v}$, P and the collision terms I into a Fourier series analogous to (11.42), we find an algebraic system of equations for the amplitudes $f_l^{c,v}(p,t)$ and $P_l(p,t)$,

$$\left\{\frac{\partial}{\partial t} - il\omega_0\right\} f_l^{c,v} \pm 2 \sum_{\bar{l}=-\infty}^{\infty} \mathrm{Im}[\bar{\Omega}_{-\bar{l}} P_{l+\bar{l}}] = -\frac{i}{\hbar} I_l^{c,v}\,, \tag{11.43}$$

$$\left\{\frac{\partial}{\partial t} + \frac{i}{\hbar}\left(\bar{E}_p^{cv} - l\hbar\omega_0\right)\right\} P_l + i \sum_{\bar{l}=-\infty}^{\infty} \bar{\Omega}_{-\bar{l}} \left[f_{l+\bar{l}}^v - f_{l+\bar{l}}^c\right] = -\frac{i}{\hbar} I_l^{cv}\,, \tag{11.44}$$

where we used $\bar{\Omega}_l^*(p,t) = \bar{\Omega}_{-l}(p,t)$, and the $\pm$ sign refers to f^c and f^v, respectively. As one can see, there is a coupling between different harmonics of f and P. The number of harmonics excited depends on the effective (Hartree-Fock renormalized) field.

[23] Indeed, the Schrödinger equation for that attractive potential (Coulomb potential) is just the one for the hydrogen atom, which has bound state solutions. The main difference is the similar mass of electrons and holes (given by the similar curvature of the conduction and the valence band) and the large value of the background dielectric constant which results in binding energies being three orders of magnitude smaller than for the hydrogen atom, see e.g. [HK93].

Let us consider some examples.

1. *Weak harmonic field*: Only the first harmonic (and its conjugate) are excited, $\bar{\Omega}_l \sim \delta_{l,\pm 1}$. Then there are only two terms in the sums in Eqs. (11.43,11.44), and there appears only a coupling between $P_{l\pm 1}$ and f_l and between $f_{l\pm 1}$ and P_l,

$$\frac{\partial}{\partial t} f_0^{c,v} \pm \mathrm{Re}[\, dE_L^0\, (P_1 - P_{-1})\,] \;=\; -\frac{i}{\hbar} I_0^{c,v}\,, \tag{11.45}$$

$$\left\{\frac{\partial}{\partial t} + \frac{i}{\hbar}\left(\bar{E}_p^{cv} \mp \hbar\omega_0\right)\right\} P_{\pm 1} \mp \frac{dE_L^0}{2}\,[f_0^v - f_0^c] \;=\; -\frac{i}{\hbar} I_{\pm 1}^{cv}\,, \tag{11.46}$$

where only the lowest harmonics have been kept, and the field has been taken in the form (11.47). This is the typical situation for low–intensity laser pulses. In many cases, the electric field of the laser is of the form

$$E_L(t) = E_L^0(t)\,\sin\omega_0 t, \qquad E_L^0(t) = E_0\, e^{-\frac{(t-t_0)^2}{\tau_p^2}}\,, \tag{11.47}$$

where $E_L^0(t)$ is a slowly varying amplitude (envelope) of the field determining the pulse shape, its duration τ_p and intensity E_0 (see Fig. 11.2.).

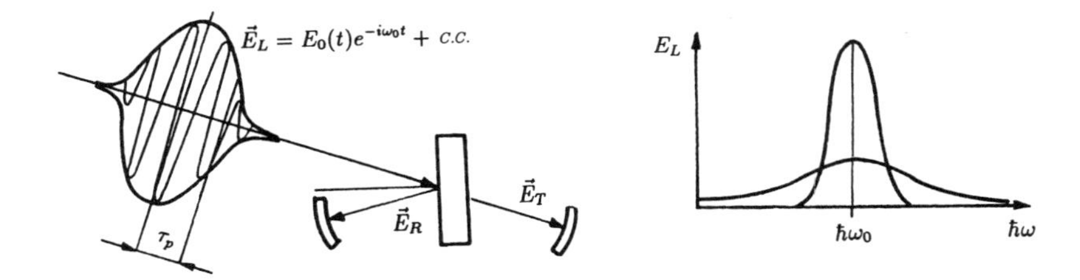

Fig. 11.2. Electric field of a short-pulse laser. For pulse duration τ_p less than $10fs$, the pulse contains only a few cycles. The energy spectrum of the pulse is broad. For observation times shorter than the pulse, it is even broader (flat curve). E_T and E_R denote respectively the field transmitted through and reflected from the material which is measured in an experiment. These fields are modified by the response of the material.

2. *Small excess energies. Rotating wave approximation*: Let us compare the different energy values. For example, under semiconductor conditions, typical kinetic energies of the electrons are in the range of several meV to several tens of meV, which is small compared to the optical photon energy $\hbar\omega_0 \sim 1eV$. If the distribution functions have initially only a time–independent component ($l = 0$, which is the case in most situations), under the action of the laser field (11.47), the polarization obtains two components corresponding to $l = \pm 1$, cf. Eq. (11.46). However, these components are spectrally far apart, and due to the comparably small particle energies, there will be almost no interaction between the two components. Thus, considering the evolution of P_1, we can safely neglect the effect of P_{-1}, i.e. of $\bar{\Omega}_{-1}$ and obtain from Eqs. (11.43,11.44)

$$\frac{\partial}{\partial t} f^{c,v} \pm 2\mathrm{Im}[\bar{\Omega}^*\, P] \;=\; -\frac{i}{\hbar} I^{c,v}\,, \tag{11.48}$$

$$\left\{\frac{\partial}{\partial t} + \frac{i}{\hbar} E_p^{cv}\right\} P + i\bar{\Omega}\,[f^v - f^c] \quad = \quad -\frac{i}{\hbar} I^{cv}\,, \tag{11.49}$$

where $f = f_0$, $I^{c,v} = I_0^{c,v}$, $P = P_1$, $I^{cv} = I_1^{cv}$, $\bar{\Omega} = \bar{\Omega}_1$ and $E_p^{cv} = \bar{E}_p^{cv} - \hbar\omega_0$.

3. *Strong fields*: As we mentioned, the Bloch equations which were derived above are valid for strong fields too. In that case, one has to recall that the total field which appears in the equations is not identical to the exciting external field, but is the solution of Maxwell's equations (11.22). Furthermore, even when the exciting field is monochromatic (cf. the previous example), the total field might be much more complex, because the Bloch equations may generate higher field harmonics, cf. Eqs. (11.43-11.44). The number of harmonics eventually excited depends on the field intensity and on the magnitude of the dipole moment at higher energies. On the other hand, at high intensities there may be of other absorption mechanisms activated, such as excitation of electrons into higher lying bands or photoionization. In that case, these effects have to be included into the model as well.

This analysis can be straightforwardly generalized to more complex situations. If there are several external fields present (e.g. several laser pulses), then one can expand all quantities in terms of harmonics of all fields. A more efficient way, however would be an expansion in terms of all higher harmonics and all possible frequency combinations (sum and differences). Moreover, the same analysis can be performed for spatially inhomogeneous systems too. Then one should use the coordinate representation (see above) and expand the quantities into a spatial Fourier series also, see e.g. [BK95].

11.5 *Bloch representation of the solution $g(t)$

In the previous section we derived the Bloch equations in a very general form, e.g. Eq. (11.30) and analyzed the interaction of the particles with the external field. This coupled particle–field dynamics is fully coherent (dissipation–less) as long as the terms on the r.h.s. of the Bloch equations (the collision terms) are neglected. In real systems, however, there exist always correlations among the particles which lead to scattering and equilibration of the carrier distributions. On the other hand, we have to expect that scattering events will destroy the coherence of the interband polarization P, i.e. cause its "dephasing". In this section we consider in detail the incoherent terms in the Bloch equations, which are related to the binary correlations g_{12}. The general effect of binary correlations has been studied in great detail above, cf. Chapters 6–10, and we will take advantage of these results in the following.

In the framework of the Bloch equations, the correlations enter in terms of Bloch matrix elements of g_{12}, cf. Eqs. (11.30) and (11.38), which we are now going to compute. This requires the following steps:

1. Choice of an approximations for the binary correlations and formal solution for $g_{12}(t)$, cf. Chs. 6–10.

2. Expression of the solution in terms of free single–particle (quasiparticle) propagators U_1 and dynamical quantities, such as the T-matrix (Ch. 8) or the two–time screened potential V_s, (Ch. 9),

3. Bloch representation of the propagators, and

4. Bloch representation of the dynamical quantities.

We begin the analysis with the simplest approximation for the binary correlations, the static second Born approximation, Chs. 6, 7. Then, for the Bloch representation of g_{12} it is sufficient to solve point 3, which also will serve as the basis for the treatment of more complex approximations below.

Born approximation. The equation of motion for the correlation operator in the static Born approximation is given by

$$i\hbar\frac{\partial}{\partial t}g_{12} - [\bar{H}_1 + \bar{H}_2, g_{12}] = Q_{12} = \hat{V}_{12}F_1F_2 - F_1F_2\hat{V}_{12}^{\dagger}, \tag{11.50}$$

with the effective single–particle Hamiltonian (11.27), $\bar{H}_1 = H_1 + H_1^{HF} + \Sigma_1^+$, where H_1 is given by Eq. (11.16) and contains the external potential $\mathcal{U}_1$. Eq. (11.50) has the formal solution (cf. Ch. 6),

$$g_{12}(t) = \frac{1}{i\hbar}\int_{t_0}^{t} d\bar{t}\, U_1(t\bar{t})\, U_2(t\bar{t})\, \bar{Q}_{12}(\bar{t})\, U_1^{\dagger}(t\bar{t})\, U_2^{\dagger}(t\bar{t}), \tag{11.51}$$

where $\bar{Q}$ includes initial correlations, $\bar{Q}_{12}(t) = Q_{12}(t) + i\hbar\delta(t-t_0)g^0$. We now transform this solution into the Bloch representation, in analogy to the transformation of the first hierarchy equation, Sec. 11.4. Using the short notation for the Bloch basis vectors, $|\lambda_1 x_1\rangle = |1\rangle$, the matrix elements of g_{12} become,

$$\begin{aligned}\langle 12|g(t)|2'1'\rangle &= \frac{1}{i\hbar}\sum_{\bar{1}\bar{2}\bar{1}'\bar{2}'}\int_{t_0}^{t} d\bar{t}\ \langle 1|U(t\bar{t})|\bar{1}\rangle\ \langle 2|U(t\bar{t})|\bar{2}\rangle \\ &\times\ \langle\bar{1}\bar{2}|\bar{Q}(\bar{t})|\bar{2}'\bar{1}'\rangle\ \langle\bar{1}'|U^{\dagger}(t\bar{t})|1'\rangle\ \langle\bar{2}'|U^{\dagger}(t\bar{t})|2'\rangle\,.\end{aligned} \tag{11.52}$$

Besides the Bloch matrix elements of the propagators (see below) we need the matrix elements of $\bar{Q}$, $\langle 12|\bar{Q}(t)|2'1'\rangle = \langle 12|Q(t)|2'1'\rangle + i\hbar\delta(t-t_0)\,\langle 12|g^0|2'1'\rangle$.

After straightforward calculations, we obtain[24]

$$
\begin{aligned}
\langle 12|Q(t)|2'1'\rangle &= \sum_{\bar\lambda_1\bar\lambda_2}\left\{\delta_{\lambda_1\bar\lambda_1}\,\delta_{\lambda_2\bar\lambda_2} \pm f_{p_1}^{\lambda_1\bar\lambda_1}\,\delta_{\lambda_2\bar\lambda_2} \pm \delta_{\lambda_1\bar\lambda_1}\,f_{p_2}^{\lambda_2\bar\lambda_2}\right\} \\
&\times \quad V(p_1 - p_1')\,\delta_{p_1+p_2,p_1'+p_2'}\,f_{p_1'}^{\bar\lambda_1\lambda_1'}\,f_{p_2'}^{\bar\lambda_2\lambda_2'} - \text{h.c.} \\
&= V(p_1 - p_1')\,\delta_{p_1+p_2,p_1'+p_2'}\sum_{\bar\lambda_1\bar\lambda_2}\Big\{ \qquad (11.56)\\
&\times \left(\delta_{\lambda_1\bar\lambda_1} \pm f_{p_1}^{\lambda_1\bar\lambda_1}\right)\left(\delta_{\lambda_2\bar\lambda_2} \pm f_{p_2}^{\lambda_2\bar\lambda_2}\right) f_{p_1'}^{\bar\lambda_1\lambda_1'}\,f_{p_2'}^{\bar\lambda_2\lambda_2'} - [f \leftrightarrow \delta \pm f]\Big\}.
\end{aligned}
$$

With the initial correlations g^0 assumed given, also their Bloch matrix elements are known. What is left now in the computation of the Bloch matrix elements of the pair correlation operator is to evaluate the matrix elements of the quasiparticle propagators.

Bloch representation of the quasiparticle propagators. While Q_{12} basically contains the effect of the statistics (the distributions), the dynamics and also field effects come into play via the propagators which obey the operator equation of motion (11.15). We now derive the Bloch representation of this equation. In the homogeneous case, the single–particle quantities $\bar H_1$ and U_1 are diagonal in the momentum variables, and, using the notation $\langle 1|U(tt')|1'\rangle = U_{p_1}^{\lambda_1\lambda_1'}\delta_{p_1p_1'}$, we obtain

$$
i\hbar\frac{\partial}{\partial t}U_{p_1}^{\lambda_1\lambda_1'}(tt') - \sum_{\bar\lambda_1}\bar H_{p_1}^{\lambda_1\bar\lambda_1}\,U_{p_1}^{\bar\lambda_1\lambda_1'}(tt') = 0, \qquad U_{p_1}^{\lambda_1\lambda_1'}(tt) = \delta_{\lambda_1\lambda_1'}. \quad (11.57)
$$

[24] We outline the main steps. The matrix elements of Q are

$$
\langle 12|Q(t)|2'1'\rangle = \sum_{\bar1\bar2}\langle 12|\hat V(t)|\bar2\bar1\rangle\ \langle\bar1|F(t)|1'\rangle\ \langle\bar2|F(t)|2'\rangle - \text{h.c.}, \qquad (11.53)
$$

and involve the matrix elements of the shielded potential $\hat V_{12} = (1 \pm nF_1 \pm nF_2)V_{12}$, for which we find

$$
\langle 12|\hat V(t)|2'1'\rangle = \sum_{\bar1\bar2}\{\delta_{1\bar1}\delta_{2\bar2} \pm \langle 1|nF(t)|\bar1\rangle\,\delta_{2\bar2} \pm \delta_{1\bar1}\,\langle 2|nF(t)|\bar2\rangle\}\,\langle\bar1\bar2|V|2'1'\rangle. \qquad (11.54)
$$

The Bloch matrix elements of the bare potential have been given above, Eqs. (11.28 and 11.32), so we restrict ourselves to the homogeneous case where the final result is

$$
\langle 12|\hat V|2'1'\rangle = \left\{\delta_{\lambda_1\lambda_1'}\,\delta_{\lambda_2\lambda_2'} \pm f_{p_1}^{\lambda_1\lambda_1'}\,\delta_{\lambda_2\lambda_2'} \pm \delta_{\lambda_1\lambda_1'}\,f_{p_2}^{\lambda_2\lambda_2'}\right\} V(p_1 - p_1')\,\delta_{p_1+p_2,p_1'+p_2'}. \qquad (11.55)
$$

Obviously, the Pauli blocking factors make the shielded potential $\hat V$ non–diagonal in the band indices, in contrast to the bare potential. With Eq. (11.55) inserted into Eq. (11.53), we obtain the result (11.57).

Notice that Eq. (11.57) is similar to the Bloch equations for the matrix elements $f^{\lambda_1\lambda_1'}$. The main difference is the selfenergy term in the Hamiltonian in Eq. (11.57)[25]. So all results of Sec. 11.4, can be used here too, including the general spatially inhomogeneous situation, Eq. (11.30). To keep the analysis more transparent, we will continue studying the homogeneous case, Eq. (11.57), and we will focus on dipole interaction again.

Following the derivations of Sec. 11.4, we rewrite Eq. (11.57) in analogy to Eq. (11.36),

$$\left\{ i\hbar\frac{\partial}{\partial t} - \bar{\epsilon}^{\lambda_1}_{p_1} \right\} U^{\lambda_1\lambda_1'}_{p_1}(tt') - \hbar \sum_{\bar{\lambda}_1}{}' \bar{\Omega}^{\lambda_1\bar{\lambda}_1}_{p_1}(t)\, U^{\bar{\lambda}_1\lambda_1'}_{p_1}(tt') = 0, \tag{11.58}$$

where, as before, the prime indicates that the band–diagonal elements have been excluded from the sum. For completeness, we give the equation for $U^\dagger$, the matrix elements of which are also needed to compute g_{12}, Eq. (11.52)[26]. Calculating the hermitean adjoint of Eq. (11.15), we find

$$i\hbar\frac{\partial}{\partial t} U^{\dagger\lambda_1\lambda_1'}_{p_1}(tt') + \bar{\epsilon}^{\lambda_1'}_{p_1} U^{\dagger\lambda_1\lambda_1'}_{p_1}(tt') + \hbar \sum_{\bar{\lambda}_1}{}' U^{\dagger\lambda_1\bar{\lambda}_1}_{p_1}(tt')\, \bar{\Omega}^{\dagger\bar{\lambda}_1\lambda_1'}_{p_1}(t') = 0. \tag{11.59}$$

In Eqs. (11.58, 11.59), we introduced effective single–particle energies $\bar{\epsilon}$, which are renormalized by selfenergy and Hartree–Fock terms, and effective Rabi energies $\hbar\bar{\Omega}$,

$$\begin{aligned} \bar{\epsilon}^{\lambda_1}_{p_1}(t) &= \epsilon^{\lambda_1}_{p_1}(t) \pm \int \frac{dp_2}{(2\pi\hbar)^3} V(p_2 - p_1)\, f^{\lambda_1\lambda_1}(p_2, t), && (11.60)\\ \epsilon^{\lambda_1}_{p_1}(t) &= E^{\lambda_1}_{p_1} + \Sigma^{\lambda_1\lambda_1}_{p_1}(t), && (11.61)\\ \hbar\bar{\Omega}^{\lambda_1\lambda_1'}_{p_1}(t) &= -\mathbf{d}^{\lambda_1\lambda_1'}(p_1)\, \mathcal{E}(t) \\ &\pm \int \frac{dp_2}{(2\pi\hbar)^3} V(p_2 - p_1)\, f^{\lambda_1\lambda_1'}(p_2, t) + \Sigma^{\lambda_1\lambda_1'}_{p_1}(t). && (11.62) \end{aligned}$$

Here, $\Sigma^{\lambda_1\lambda_1'}_{p_1}(t)\,\delta_{p_1p_1'} = \langle 1|\Sigma_1^+|1'\rangle$ are the Bloch matrix elements of the retarded single–particle selfenergy which will be determined below in Sec. 11.6. From Eqs. (11.60–11.62), we clearly see that the free single–particle energy as well as the "classical" transition energy (Rabi energy) are strongly modified by medium effects, i.e. by the presence of the surrounding carriers.

With Eq. (11.58) and the initial condition $U^{\lambda_1\lambda_1'}_{p_1}(tt) = \delta_{\lambda_1\lambda_1'}$ the Bloch matrix of the quasiparticle propagators is completely determined. With them,

[25] In fact, the equation for U is simpler: it is not a commutator equation, and it contains no collision integral.

[26] In fact, we could calculate them from $U^{\lambda\lambda'}_p$ using the relation $[U^\dagger(tt')]^{\lambda\lambda'}_p = [U^{\lambda'\lambda}_p]^*(tt')$.

the Bloch matrix elements of g_{12}, Eq. (11.52), and, furthermore, the collision integrals (11.38) in the Bloch equations, are fully defined. Thus, we have derived a selfconsistent system of equations, which fully describes the coupled matter–field interaction, including carrier–carrier scattering. Despite its complexity, it is well suited for numerical study of the time–dependent response of a many–particle system to an external field.

Analytical solutions for the quasiparticle propagators. So far, we made no explicit use of any particular form of the field. Now, we consider several special cases where explicit solutions of the propagator equation (11.58) can be found. In particular, we continue the analysis of the examples discussed in Sec. 11.4. Let us consider the structure of the system (11.58) more closely. In the case of N bands, we have a system of N^2 coupled equations. Due to the symmetry relation between $U^{\lambda\lambda'}$ and $U^{\lambda'\lambda}$, there are only $N(N+1)/2$ independent matrix elements. Moreover, Eqs. (11.58) couple only those elements which have the same second index. Thus, the system reduces to essentially simpler subsystems of dimension N.

All the basic properties are most easily uderstood in the *two–band case*, which we consider in the following. Denoting the bands again by "c" and "v", $\bar{\Omega}^{cv} \to \bar{\Omega}$, $U^{cc} \to U^c$, $U^{vv} \to U^v$, and suppressing the momentum index, we obtain from Eq. (11.58)

$$\begin{aligned} \left\{\frac{\partial}{\partial t} + \frac{i}{\hbar}\bar{\epsilon}^c(t)\right\} U^c(tt') + i\bar{\Omega}(t)\, U^{vc}(tt') &= 0, \\ \left\{\frac{\partial}{\partial t} + \frac{i}{\hbar}\bar{\epsilon}^v(t)\right\} U^{vc}(tt') + i\bar{\Omega}^*(t)\, U^c(tt') &= 0, \end{aligned} \tag{11.63}$$

and similar equations for the second pair of matrix elements U^v, U^{cv}. Notice that these equations have, in general, a very complicated time dependence and nonlinear structure, what is due to the selfenergy contributions in $\bar{\epsilon}^{c,v}$ and $\bar{\Omega}$, which depend on the propagators U also. Analytical results can be obtained within the *local approximation* for the selfenergy, i.e. $\Sigma^+(tt') \approx \Sigma^+(t-t')$, which leads to the local approximation for the propagators too[27].

We now perform a very instructive approximate solution of the system (11.63)[28]. For this we assume that $\bar{\epsilon}^{c,v}$ and $\bar{\Omega}$ do not explicitly depend on U although they may weakly depend on time (the precise condition will become clear below). Then, Eqs. (11.63) may be treated approximately as a linear system with weakly time-dependent coefficients, with the solution

$$U^c(t-t') \;=\; a_1^c(t)\, \exp\left\{\int_{t'}^t d\bar{t}\, r_1(\bar{t}),\right\} + a_2^c(t)\, \exp\left\{\int_{t'}^t d\bar{t}\, r_2(\bar{t})\right\}, \tag{11.64}$$

[27] For a detailed discussion of this approximation see Appendix D.

[28] This approach is similar to that of Galitski et al. [GGE69], see also [Kel95].

$$U^{vc}(t-t') = a_1^{vc}(t) \exp\left\{\int_{t'}^{t} d\bar{t}\, r_1(\bar{t})\right\} + a_2^{vc}(t) \exp\left\{\int_{t'}^{t} d\bar{t}\, r_2(\bar{t})\right\}, \quad (11.65)$$

which is applicable as long as $da_{1,2}/dt \ll r_{1,2}$. The eigenfrequencies of the system are given by

$$r_{1,2} = -\frac{i}{\hbar}\left\{\frac{\bar{\epsilon}^c + \bar{\epsilon}^v}{2} \pm \sqrt{\frac{(\bar{\epsilon}^c - \bar{\epsilon}^v)^2}{4} + \hbar^2\,|\bar{\Omega}|^2}\right\}. \quad (11.66)$$

With the initial conditions $U^c(0) = 1$ and $U^{vc}(0) = 0$, the coefficients $a_{1,2}$ are readily calculated to $a_1^c = -|\bar{\Omega}|^2/\Delta$, $a_2^c = (r_1 - A^c)(r_2 - A^v)/\Delta$ and $a_1^{vc} = -a_2^{vc} = i\bar{\Omega}^*(r_1 - A^c)/\Delta$, where we denoted $A^{c,v} = -i\bar{\epsilon}^{c,v}/\hbar$ and $\Delta = (r_1 - A^c)(r_2 - A^v) + |\bar{\Omega}|^2$. This solution describes the superposition of two modes which correspond to field modified energy bands. In the case of strong fields, these modes may essentially deviate from the field-free energies [Kel95].

Let us again consider some special cases. In the weak field limit, $\alpha = \hbar|\Omega|/(\bar{\epsilon}^c - \bar{\epsilon}^v) \ll 1$, and we may expand the square root to obtain the simpler solution

$$U^c(t-t') = \exp\left\{\int_{t'}^{t} d\bar{t}\, r_1(\bar{t})\right\} + O(\alpha^2), \quad (11.67)$$

$$U^{vc}(t-t') = \frac{\hbar\bar{\Omega}^*}{\bar{\epsilon}^c - \bar{\epsilon}^v}\left[\exp\left\{\int_{t'}^{t} d\bar{t}\, r_1(\bar{t})\right\} - \exp\left\{\int_{t'}^{t} d\bar{t}\, r_2(\bar{t})\right\}\right], \quad (11.68)$$

$$\text{with} \quad r_1 = -\frac{i}{\hbar}\left(\bar{\epsilon}^c + \alpha\,\hbar|\bar{\Omega}|\right), \quad r_2 = -\frac{i}{\hbar}\left(\bar{\epsilon}^v - \alpha\,\hbar|\bar{\Omega}|\right), \quad (11.69)$$

and U^v follows from U^c by substituting $r_1 \longrightarrow r_2$. At vanishing field intensities, $\alpha \to 0$, U^c and U^v become the free quasiparticle propagators of the respective energy band. They still contain selfenergy effects and are damped. U^{vc} vanishes, as in this limit no carrier transitions between the two bands are possible. The effect of the field is two-fold: first, it causes a shift of the quasiparticle energies which is quadratic in the field, and second, it leads to an increasing amplitude of the interband propagator U^{vc}. The latter oscillates with both quasiparticle energies, and the dominating low energy spectral component results from the beating of both, i.e. from oscillations with the renormalized gap energy $\bar{\epsilon}^c - \bar{\epsilon}^v$.

So far we did not specify the external field and its time dependence. In case of a *harmonic field*, e.g. $\bar{\Omega}(t) = \bar{\Omega}_0 \sin\omega_0 t$, $\bar{\Omega}_0 = const$, we may perform the time integral in Eqs. (11.67, 11.68). Assuming weak time dependence of $\bar{\epsilon}^{c,v}$ and $\bar{\Omega}_0$, we obtain

$$\int_0^t d\bar{t}\,\bar{\Omega}^2(\bar{t}) \approx \frac{\bar{\Omega}_0^2}{2}\,t - \frac{\bar{\Omega}_0^2}{4\omega_0}\,\sin 2\omega_0 t, \quad (11.70)$$

and with the identity (A.4), we may express the quasiparticle and interband propagators in terms of Bessel functions of order n,

$$U^{c,v}(t) = \sum_{n=-\infty}^{\infty} I_n\left[\frac{1}{4}\alpha\beta\right] e^{-\frac{i}{\hbar}\left(\bar{\epsilon}^{c,v} \pm \alpha\hbar\bar{\Omega}_0 \pm 2n\hbar\omega_0\right)t}, \quad (11.71)$$

$$U^{vc}(t) = \frac{\hbar\bar{\Omega}_0^*\,\sin\omega_0 t}{\bar{\epsilon}^c - \bar{\epsilon}^v}\sum_{n=-\infty}^{\infty} I_n\left[\frac{1}{4}\alpha\beta\right] \times \left\{e^{-\frac{i}{\hbar}\left(\bar{\epsilon}^c + \alpha\hbar\bar{\Omega}_0 + 2n\hbar\omega_0\right)t} - e^{-\frac{i}{\hbar}\left(\bar{\epsilon}^v - \alpha\hbar\bar{\Omega}_0 - 2n\hbar\omega_0\right)t}\right\}, \quad (11.72)$$

where $\beta = \bar{\Omega}_0/\omega_0$, and α was defined above. The propagators are given as a sum of field harmonics (even harmonics only). The oscillations are modulated by the field renormalized quasiparticle energy $\bar{\epsilon}^{c,v} \pm \alpha\hbar\bar{\Omega}_0$. α and β are the two relevant parameters which relate the intensity of the field respectively to its central frequency (photon energy) and the characteristic energy scale of the material - in this case the band gap energy (level separation). The number of harmonics which contribute to the propagator, depends on the product $\alpha \cdot \beta$. The behavior of the system is readily understood from the properties of the Bessel functions. For vanishing field intensity, the argument vanishes and all higher harmonics disappear, $I_n(\alpha\beta/4) \longrightarrow \delta_{n,0}$. With increasing intensity, the zeroth order harmonics decays while the higher harmonics grow successively larger.

Let us now consider the modifications arising from a *finite duration* of the field. A rough estimate is obtained if the pulse is modeled by a rectangular envelope, with $\Omega_0(t) = \Omega_0$ for $0 \le t \le T$ and zero otherwise. Obviously, for $t \le T$ all results remain the same, while for $t > T$ we have to replace $t \to T$ in the field contribution, i.e. the term in parentheses in the exponent of Eq. (11.71) becomes $\bar{\epsilon}^{c,v}t \pm \left(\alpha\hbar\bar{\Omega}_0 + 2n\hbar\omega_0\right) T$. The same analysis can be performed for smooth pulse shapes. E.g. for the pulse (11.47), the integral (11.70) yields ($t_0 = 0$, for the definition of the error function, see App. A)

$$\frac{1}{8}\sqrt{\frac{\pi}{2}}\,\bar{\Omega}_0^2\,\tau_p\left\{2\,\mathrm{erf}(2^{1/2}t/\tau_p) - e^{-\frac{1}{2}\omega_0^2\tau_p^2}\left[\mathrm{erf}\left(\frac{i\omega_0\tau_p^2 + 2t}{2^{1/2}\tau_p}\right) - \mathrm{erf}\left(\frac{i\omega_0\tau_p^2 - 2t}{2^{1/2}\tau_p}\right)\right]\right\}$$

The analysis of this expression is straightforward, and we will not continue it here. The same analysis can be applied to other pulse shapes or to superpositions of several pulses.

We mention that another analytical approach would be to perform an expansion of the propagator equations (11.63) in terms of the field harmonics and the application of the rotating wave approximation. This idea was discussed in detail in Sec. 11.4, and we will not repeat it here.

11.6 *Correlation operator, non–Markovian collision integral and selfenergy in an electromagnetic field

Bloch matrix of the binary correlations. With the results for the quasiparticle and interband propagators from Sec. 11.5, we are now ready to write down the explicit result for the Bloch matrix elements of the correlation operator in an electromagnetic field. Inserting the result for the inhomogeneity Q, Eq. (11.57), into Eq. (11.52), we obtain (homogeneous case)

$$g^{\lambda_1\lambda_1'}_{\lambda_2\lambda_2'}(p_1p_2;p_1'p_2',t) = \sum_{\bar{\lambda}_1\bar{\lambda}_2\bar{\lambda}_1'\bar{\lambda}_2'} U^{\lambda_1\bar{\lambda}_1}_{p_1}(tt_0)\,U^{\lambda_2\bar{\lambda}_2}_{p_2}(tt_0)\,g^{\bar{\lambda}_1\bar{\lambda}_1'}_{\bar{\lambda}_2\bar{\lambda}_2'}(p_1p_2;p_1'p_2',t_0)$$
$$\times\, U^{*\bar{\lambda}_1'\lambda_1'}_{p_1'}(tt_0)\,U^{*\bar{\lambda}_2'\lambda_2'}_{p_2'}(tt_0)$$

$$+\frac{1}{i\hbar}V(p_1-p_1')\,\delta_{p_1+p_2,p_1'+p_2'}\sum_{\bar\lambda_1\bar\lambda_2}\sum_{\bar\lambda_1\bar\lambda_2\bar\lambda_1'\bar\lambda_2'}\int_{t_0}^{t}d\bar t$$
$$\times\, U_{p_1}^{\lambda_1\bar\lambda_1}(t\bar t)\,U_{p_2}^{\lambda_2\bar\lambda_2}(t\bar t)\,U_{p_1'}^{*\bar\lambda_1'\lambda_1'}(t\bar t)\,U_{p_2'}^{*\bar\lambda_2'\lambda_2'}(t\bar t)$$
$$\times\left\{\left(\delta_{\bar\lambda_1\tilde\lambda_1}\pm f_{p_1}^{\bar\lambda_1\tilde\lambda_1}\right)\left(\delta_{\bar\lambda_2\tilde\lambda_2}\pm f_{p_2}^{\bar\lambda_2\tilde\lambda_2}\right)f_{p_1'}^{\tilde\lambda_1\bar\lambda_1'}f_{p_2'}^{\tilde\lambda_2\bar\lambda_2'}-[f\leftrightarrow\delta\pm f]\right\}\Big|_{\bar t}. \quad (11.73)$$

The first term is due to initial correlations in the system.[29] The second term describes the correlation buildup with the typical retardation structure (the distributions enter at earlier times $\bar t\le t$). Notice that there are, in general, 6 band index sums. Of these, four are summations involving the quasiparticle and the interband propagators. This is important for strong excitation fields. As we saw in Sec. 11.5, the relative magnitude of the interband components is proportional to the field intensity, e.g. Eq. (11.68). Thus, in the weak field limit, these components may be neglected, leaving us with only two sums over $\tilde\lambda_1,\tilde\lambda_2$.

Let us consider the field effects in this expression. Each propagator contains the field, cf. e.g. Eq. (11.67 and 11.68), but at a different momentum value. So, for example, for a weak harmonic field (long pulse), the four propagators (neglecting the interband terms) contributing to g_{vv}^{cc} (two band case) give rise to the factor

$$\exp\left\{-\frac{i}{\hbar}\left(\tilde\epsilon_{p_1}^{c}+\tilde\epsilon_{p_2}^{v}-\tilde\epsilon_{p_1'}^{c*}-\tilde\epsilon_{p_2'}^{v*}\right)(t-\bar t)\right\}\sum_{n_1n_2n_1'n_2'}\times \quad (11.74)$$
$$I_{n_1}\left[\frac{\alpha_{p_1}\beta_{p_1}}{4}\right]I_{n_2}\left[\frac{\alpha_{p_2}\beta_{p_2}}{4}\right]I_{n_1'}\left[\frac{\alpha_{p_1'}^{*}\beta_{p_1'}}{4}\right]I_{n_1'}\left[\frac{\alpha_{p_2'}^{*}\beta_{p_2'}}{4}\right]e^{-2i\left(n_1-n_1'-n_2+n_2'\right)\omega_0(t-\bar t)},$$

with the field renormalized band energies $\tilde\epsilon_p^{c,v}=\bar\epsilon_p^{c,v}\pm\dfrac{\hbar^2|\bar\Omega_p^0|^2}{2(\bar\epsilon_p^c-\bar\epsilon_p^v)}$. In the weak field limit, the quasiparticle damping is not modified by the field, $\mathrm{Im}\tilde\epsilon\approx\mathrm{Im}\bar\epsilon=\mathrm{Im}\Sigma^+$.

Eq. (11.73) is a very general result for the binary correlation operator and valid for quite general fields. The specifics of the latter is accounted for in the propagators, as was discussed in detail in the Section before. This expression can be straightforwardly generalized to include longitudinal fields also. For this, we would replace all momenta by the kinematic momentum $p\to p+e\int\mathcal{E}_l^{ext}(t)dt$, (cf. Sec. 11.2), which further modifies the quasiparticle energies.

Non-Markovian collision integrals of the Bloch equations. Using the result for the Bloch matrix elements of g_{12}, Eq. (11.73), we can now

[29]In many cases $g(t_0)$ is zero, e.g. in solid state systems, if the initial time is chosen sufficiently long before the pulse. On the other hand, in pre–excited systems, if t_0 is chosen after the first pulse and before a test pulse, the system already contains carriers which are correlated, and g_0 will affect the dynamics.

immediately complete the Bloch equations (11.36) by computing the terms on the r.h.s. Indeed, what is left in order to calculate the *collision integrals* is to insert the result (11.73) into Eq. (11.38). This yields two collision integrals, one due to initial correlations and the other due to the correlation buildup. Both contain direct and exchange terms. This is straightforward, so there is no need to rewrite the lengthy expressions.

Bloch representation of the correlation selfenergy. As was mentioned above, the quasiparticle and interband propagators are not propagators of free particles (here in an external field), but they contain many–body effects. These are collective and correlation effects which are contained in the Hartree–Fock and correlation selfenergy, respectively, leading to a renormalization of the single–particle Hamiltonian $H_1 \to \bar{H}_1$, given by Eq. (11.16). These effects enter the Bloch equations via the Bloch matrix elements of $\bar{H}_1$: the band–diagonal elements contribute to the quasiparticle energies $\bar{\epsilon}$ and the off–diagonal elements to the effective Rabi energy $\bar{\Omega}$, cf. Eqs. (11.60,11.62). Thus, to explicitly calculate $\bar{\epsilon}$ and $\bar{\Omega}$, we need the *Bloch matrix elements of the retarded selfenergy*, $\Sigma_p^{\lambda\lambda'}$. The derivation follows the same lines as for the Bloch matrix elements of the correlation operator. We first transform the operator expression of the selfenergy into the Bloch representation and then calculate all Bloch matrix elements needed.

The general form of the operator of the retarded selfenergy was found to be $(t \geq t')$

$$\Sigma_1(tt') = n\mathrm{Tr}_3\, V_{13}^{\pm}\, h_{13}(tt'), \tag{11.75}$$

where, in second Born approximation,

$$\begin{aligned} h_{13}(tt') &= \frac{1}{i\hbar} U_1(tt')\, U_3(tt')\, R_{13}(t')\, U_3^{\dagger}(tt'), &(11.76)\\ R_{13}(t) &= F_1^{>}(t)F_3^{>}(t)V_{13}F_3^{<} \mp F_1^{<}(t)F_3^{<}(t)V_{13}F_3^{>}, &(11.77)\end{aligned}$$

cf. Sec. 7.4. In Eq. (11.77), we omitted the initial correlation term as well as the higher order contributions, which are treated in complete analogy. The Bloch representation of Σ_1 is readily calculated with the final result

$$\begin{aligned} \Sigma_{p_1p_1'}^{\lambda_1\lambda_1'}(tt') = \Sigma_{p_1}^{\lambda_1\lambda_1'}(tt')\,\delta_{p_1p_1'} &= \sum_{p_3\bar{p}_1\bar{p}_3}\sum_{\lambda_3}\delta_{p_1+p_3,\bar{p}_1+\bar{p}_3}\times &(11.78)\\ \Big\{V_{p_1-\bar{p}_1}\, h_{\lambda_3\lambda_3}^{\lambda_1\lambda_1'}(\bar{p}_1\bar{p}_3;p_1'p_3,tt') &\pm V_{p_1-\bar{p}_3}\, h_{\lambda_1\lambda_3}^{\lambda_3\lambda_1'}(\bar{p}_1\bar{p}_3;p_1'p_3,tt')\Big\}, \\ h_{\lambda_3\lambda_3'}^{\lambda_1\lambda_1'}(p_1p_3;p_1'p_3',tt') &= \frac{1}{i\hbar}\sum_{\bar{\lambda}_1\bar{\lambda}_3\bar{\lambda}_3'} U_{p_1}^{\lambda_1\bar{\lambda}_1}(tt')\, U_{p_3}^{\lambda_3\bar{\lambda}_3}(tt') \\ &\times R_{\bar{\lambda}_3\bar{\lambda}_3'}^{\bar{\lambda}_1\lambda_1'}(p_1p_3;p_1'p_3',t')\, U_{p_3'}^{*\bar{\lambda}_3'\lambda_3'}(tt'),\end{aligned}$$

$$R^{\lambda_1\lambda_1'}_{\lambda_3\lambda_3'}(p_1p_3;p_1'p_3',t) \;=\; V_{p_1-p_1'}\,\delta_{p_1+p_3,p_1'+p_3'}\sum_{\bar\lambda_3}$$
$$\Big\{\Big(\delta_{\lambda_1\lambda_1'}\pm f^{\lambda_1\lambda_1'}_{p_1}\Big)\Big(\delta_{\lambda_3\bar\lambda_3}\pm f^{\lambda_3\bar\lambda_3}_{p_3}\Big)\,f^{\bar\lambda_3\lambda_3'}_{p_3'} \;\mp\; f^{\lambda_1\lambda_1'}_{p_1}f^{\lambda_3\bar\lambda_3}_{p_3}\Big(\delta_{\bar\lambda_3\lambda_3'}\pm f^{\bar\lambda_3\lambda_3'}_{p_3}\Big)\Big\}\Big|_t.$$

As one expects, the structure of $\Sigma^{\lambda\lambda'}$ is similar to that of the collision integral $I^{\lambda\lambda'}$ of the Bloch equations. It consists of a direct and an exchange term. The kernel h is similar to the binary correlation matrix (11.73), with the difference that the matrix Q is replaced by R and the fourth propagator $U^*_{p_1'}$ and the time integration are missing. The dynamical properties enter again via the single–particle propagators, the Bloch matrix of which has been discussed in detail above. Thus, via the propagators, the selfenergy contains field effects too, in full consistency with the collision integral. This explicit result for the retarded selfenergy closes the system of equations for the generalized non–Markovian Bloch equations.

11.7 *Non-Markovian Bloch equations beyond the static Born approximation

Until now, we considered the Bloch equations with non–Markovian carrier–carrier collision integrals in static second Born approximation. In Chs. 8–10, we derived quantum kinetic equations for the Wigner distribution function which include correlation effects (incoherent effects) which are beyond the Born approximation, including strong coupling effects (Ch. 8) and dynamical screening (Ch. 9). We now briefly discuss, how to incorporate these improved approximations into the generalized Bloch equations.

As we have seen in Chs. 8 and 9, there exist two equivalent representations of the collision integral. The first uses certain complex dynamical propagators U_{12} - the two–particle propagator, in case of the ladder approximation, and the dielectric propagator, in case of the RPA. The alternative form follows if these propagators are expressed by the simpler single–particle (quasiparticle) propagators U_1 and certain dynamical quantities: the T-matrix or the screened potential, respectively. Using the latter representation, the collision integrals in the Bloch equations follow straightforwardly from the results obtained before for the one-band case. With the Bloch matrix of the quasiparticle propagators known, cf. Sec. 11.5, what is left now is to derive the Bloch representation of the dynamically screened potential and the T-matrix.

Bloch representation of the non–Markovian Balescu–Lenard collision integral. We now consider the collision integral which takes into account dynamical screening. This integral contains terms of the form (cf. Eq. (9.40)

$$U^{0\pm}\Phi^{\gtrless}V_s^{\pm}V_s^{\mp}\Phi^{\gtrless}U^{0\mp}, \tag{11.79}$$

where we omit the time and momentum arguments and integrations. In Eq. (11.79), $\Phi^{\gtrless}$ denote combinations of distribution functions which were introduced in Eq. (9.40), and $U^{0\pm}$ are related to the quasiparticle propagators $U^{\pm}$ by $U^{0\pm}(k,k',tt') = U^{\pm}(k,tt')U^{\mp}(k',t't)$. We already know how to handle these terms in the collision integrals of the Bloch equations. The only new problem beyond the static Born approximation is, therefore, to find the Bloch representation of the screened potential $V_s(tt')$. The central equation for V_s is the Dyson equation (9.33), with the nonequilibrium polarization function, $\Pi(q,tt')$, defined by Eq. (9.27). One readily confirms that the screened potential is diagonal in the band index, as is the bare Coulomb potential. The same holds for the retarded polarization. On the other hand, Π now contains sums over the bands which have their origin in the propagator U^0. Further, the trace in Π, being performed in the Bloch representation, contains now, besides the momentum integral, a band sum too,

$$\Pi(q,tt') = \frac{1}{i\hbar}\sum_{\lambda_1\lambda_1'}\int\frac{dp_1}{(2\pi\hbar)^3}U_{p_1}^{\lambda_1\lambda_1'}(tt')[U_{p_1+q}^{\lambda_1\lambda_1'}(tt')]^*\left\{f_{p_1+q}^{\lambda_1'\lambda_1} - f_{p_1}^{\lambda_1\lambda_1'}\right\}|_{t'}. \tag{11.80}$$

With this result, it is straightforward to write down the complete collision integral with dynamical screening effects included, which can be found, e.g. in [HJ96].

Collision terms in the strong coupling case. We now consider the *Bloch representation of the retarded and advanced T–matrix*, $T^{\pm}(tt')$, which determine the collision integral in ladder approximation, Eq. (8.62). This complicated integral has the same structure as discussed above: it contains the quasiparticle propagators, the T-matrix and the distribution functions. The latter appear in the combination familiar from scattering rates, similar as for the Born approximation (cf. function Q in Sec. 11.5). To obtain the band index picture of this expression in the multi–band case, we consider the first term under the momentum and time integrals in Eq. (8.62), suppressing the momentum indices. It has the structure (integration over repeating time arguments is implied)

$$I_1(t) \sim T_{12}^{+}(tt_1)U_{12}^{0+}(t_1t_2)T_{12}^{-}(t_2t_3)U_{12}^{0-}(t_3t)Q_{12}(t_2t_3), \tag{11.81}$$

where $U_{12}^{0\pm} = U_1^{0\pm}U_2^{0\pm}$, and Q contains the distribution functions as in Eq. (11.53), (notice the two time arguments of Q in Eq. (8.62)). In the multiband case, the collision integral I is replaced by a matrix $I^{\lambda_1\lambda_1'}$ as are all operators in it. There is a summation over all intermediate band indices and an additional sum over λ_2 resulting from the trace over 2 in the collision term

$$\begin{aligned} I^{\lambda_1\lambda_1'}(t) \quad &\sim \quad \sum_{\lambda_2}\sum_{\lambda_3\lambda_4}\sum_{\lambda_5\lambda_6}\sum_{\lambda_7\lambda_8}\sum_{\lambda_9\lambda_{10}} T^{+\lambda_1\lambda_3}_{\lambda_2\lambda_4}(tt_1)U^{0+\lambda_3\lambda_5}_{\lambda_4\lambda_6}(t_1t_2) \\ &\times \quad T^{-\lambda_5\lambda_7}_{\lambda_6\lambda_8}(t_2t_3)U^{0-\lambda_7\lambda_9}_{\lambda_8\lambda_{10}}(t_3t)Q^{+\lambda_9\lambda_1'}_{\lambda_{10}\lambda_2}(t_2t_3). \end{aligned} \tag{11.82}$$

The central equation for $T^{\pm}$ is the Lippmann–Schwinger equation (8.37), which has the Bloch representation

$$\begin{aligned} T^{+\lambda_1\lambda_1'}_{\lambda_2\lambda_2'}(tt') &= V_{12}\,\delta(t-t')\,\delta_{\lambda_1,\lambda_1'}\,\delta_{\lambda_2,\lambda_2'} \\ &- \frac{i}{\hbar}V_{12}\sum_{\bar\lambda_1\bar\lambda_2}\int_{-\infty}^{+\infty} d\bar t\, G^{0+\lambda_1\bar\lambda_1}_{\lambda_2\bar\lambda_2}(t\bar t)\,T^{+\bar\lambda_1\lambda_1'}_{\bar\lambda_2\lambda_2'}(\bar t t'), \end{aligned} \tag{11.83}$$

$$\begin{aligned} T^{-\lambda_1\lambda_1'}_{\lambda_2\lambda_2'}(tt') &= V_{12}\,\delta(t-t')\,\delta_{\lambda_1,\lambda_1'}\,\delta_{\lambda_2,\lambda_2'} \\ &- \frac{i}{\hbar}\sum_{\bar\lambda_1\bar\lambda_2}\int_{-\infty}^{+\infty} d\bar t\, T^{-\lambda_1\bar\lambda_1}_{\lambda_2\bar\lambda_2}(t\bar t)\,G^{0-\bar\lambda_1\lambda_1'}_{\bar\lambda_2\lambda_2'}(\bar t t')\,V_{12}. \end{aligned} \tag{11.84}$$

The propagators $G^{0\pm}$ are given by Eq. (8.27) and have the matrix form

$$\begin{aligned} G^{0\pm\lambda_1\lambda_1'}_{\lambda_2\lambda_2'}(tt') &= \pm\Theta[\pm(t-t')]\sum_{\bar\lambda_1\bar\lambda_2}\Big\{U^{0+}_{\lambda_1\bar\lambda_1}(tt')\,U^{0+}_{\lambda_2\bar\lambda_2}(tt')N^{\bar\lambda_1\lambda_1'}_{\bar\lambda_2\lambda_2'}(t') \\ &+ \quad N^{\lambda_1\bar\lambda_1}_{\lambda_1\bar\lambda_1}(t)U^{0-}_{\bar\lambda_1\lambda_1'}(tt')U^{0-}_{\bar\lambda_1\lambda_1'}(tt')\Big\}, \end{aligned} \tag{11.85}$$

where $N^{\lambda_1\lambda_1'}_{\bar\lambda_2\lambda_2'} = \delta_{\lambda_1,\lambda_1'}\delta_{\lambda_2,\lambda_2'} - \delta_{\lambda_1,\lambda_1'}f^{\lambda_2\lambda_2'} - \delta_{\lambda_2,\lambda_2'}f^{\lambda_1\lambda_1'}$.

This formidable collision integral which contains four expressions of the form (11.82) appearing under three time and three momentum integrals, is certainly not (yet) feasible for numerical analysis. However, its structure is easy to understand, and it is a good starting point for the derivation of simpler approximations. The matrix $T^{\pm\lambda_1\lambda_1'}_{\lambda_2\lambda_2'}$ describes the *strong interaction between two electrons* either in the same band ($\lambda_1 = \lambda_1' = \lambda_2 = \lambda_2'$) or in different bands ($\lambda_1 = \lambda_1' \neq \lambda_2 = \lambda_2'$), but also between electrons that (one or both) undergo interband transitions (e.g. $\lambda_1 \neq \lambda_1'$). Again, we see that the field effects have been separated, they enter via the quasiparticle propagators for which we already obtained expressions above. In the weak field limit, the band–off–diagonal matrix elements vanish, $U_{\lambda\lambda'} \sim \delta_{\lambda,\lambda'}$ what significantly reduces the number of summations.

Markovian T-matrix scattering integrals. From this collision integral we can derive simpler expressions using the various approximation schemes discussed in detail in Ch. 8, in particular, the retardation expansion. Here, we restrict ourselves to the zeroth order term which yields the Markovian collision integral, if the Boltzmann limit is performed, i.e. the initial time $t_0 \to -\infty$, and initial correlations are neglected, $\lim_{t_0 \to -\infty} g(t_0) = 0$. In this limit, the retardation in the distributions is neglected, i.e. all distributions in Q depend on the current (non–retarded) time t.[30] Furthermore, we consider the weak field, where the propagators are band–diagonal, and the field can be neglected in $U^{0\pm}$. Also, the local approximation for $T^{\pm}$ and $U^{0\pm}$ is used and selfenergy is neglected, with the result for the propagators $U_{12}^{0\pm}(\tau) = \Theta[\pm\tau] \exp[-iE_{12}\tau/\hbar]$. After these drastic simplifications, the complete collision integral [i.e. all four terms of type (11.82)] turns into [the momentum arguments are the same as in Eq. (8.62)]

$$\begin{aligned} I^{\lambda_1\lambda_1'}(t) &= \frac{2}{\hbar} \sum_{\lambda_2} \sum_{\lambda_3\lambda_4} \sum_{\lambda_5\lambda_6} \sum_{\lambda_7\lambda_8} \sum_{p_2\bar{p}_1\bar{p}_2} \left\{ \bar{f}^{>\lambda_5\lambda_7}_{\lambda_6\lambda_8} f^{<\lambda_7\lambda_1'}_{\lambda_8\lambda_2} - \bar{f}^{<\lambda_5\lambda_7}_{\lambda_6\lambda_8} f^{>\lambda_7\lambda_1'}_{\lambda_8\lambda_2} \right\} \\ &\times \left\{ T^{+\lambda_1\lambda_3}_{\lambda_2\lambda_4}(\bar{E}^{+\lambda_3}_{\lambda_4}) T^{-\lambda_3\lambda_5}_{\lambda_4\lambda_6}(-E^{-\lambda_5}_{\lambda_6})\, \delta\left(\bar{E}^{\lambda_3}_{\lambda_4} - E^{\lambda_5}_{\lambda_6}\right) \right. \\ &\left. - T^{+\lambda_1\lambda_3}_{\lambda_2\lambda_4}(-E^{+\lambda_1}_{\lambda_2}) T^{-\lambda_3\lambda_5}_{\lambda_4\lambda_6}(\bar{E}^{-\lambda_3}_{\lambda_4})\, \delta\left(\bar{E}^{\lambda_3}_{\lambda_4} - E^{\lambda_1}_{\lambda_2}\right) \right\}, \qquad (11.86) \end{aligned}$$

where the energy arguments of the retarded/advanced T-matrix are $E^{\pm} = E \pm i\varepsilon$ and $E^{\lambda_1}_{\lambda_2} = E_{\lambda_1} + E_{\lambda_1}$, and further $f^{\gtrless\lambda_1\lambda_3}_{\lambda_2\lambda_4} = f^{\gtrless\lambda_1\lambda_3} f^{\gtrless\lambda_2\lambda_4}$. This is the ladder approximation for the carrier–carrier scattering term in the Bloch equations. This result is more complicated than in the one–band case, cf. Eq. (8.67). The previous form is recovered only for the terms in the band sums whith $\lambda_5 = \lambda_1$ and $\lambda_6 = \lambda_2$. Then, the two T-matrices are adjoint and yield $|T(E^+)|^2$. In similar fashion one can derive higher retardation approximations (cf. Ch. 8) and also include the effect of stronger fields into the propagators.

Concluding remarks. We have seen that our results of the previous Chapters are straightforwardly extended to systems in an external field. The main modification of the previous kinetic equations is that the propagators are replaced by generalized field-dependent propagators. They are determined from an effective Schrödinger (or Dyson) equation which includes the field, solutions of which have been extensively studied in the literature.

[30] Moreover, bound states (i.e. incoherent excitons) are excluded, unless one takes into account the additional collision term considered at the end of Ch. 8.

Chapter 12

*Green's Functions Approach to Field–Matter Dynamics

In this Chapter, we give an introduction to an alternative approach to the interaction of particles with electro-magnetic fields in nonequilibrium - the method of nonequilibrium Green's functions. This is the second of the theoretical approaches listed in Sec. 1.3.3 which are able to describe ultrafast relaxation and, certainly, the most powerful one. The Green's functions formalism has been covered in many excellent text books, e.g. [KB89, AGD62, FW71, KKER86], its extension to nonequilibrium, following the concepts of Kadanoff and Baym [KB89] and Keldysh [Kel64a] is the subject of recent monographs [HJ96, KKS98]. The Green's functions approach to particle–field interaction was developed in classical papers by Dubois and others [DuB67, Kor66], including its generalization to the relativistic case [DuB67, BD72, Bec81].

Our goal here is to outline the basic ideas and results of the method of two-time correlation functions. The discussion will be pragmatic, focusing on a formulation of the results which is suitable for *direct numerical solution*. This, naturally, leads to the *Kadanoff-Baym equations*, for which first numerical results could be achieved in recent years, and even more remarkable results will be possible in the near future. Furthermore, we consider the Green's functions derivation of quantum kinetic equations for the Wigner function and compare them to our density operator results, Sec. 12.6. For readers interested in the derivation of the results, we provide the main ideas in footnotes, for further details as well as for applications, references to the special literature are given.

Section 12.1 gives an overview on the relativistic quantum statistical treatment of field and matter, which allows for the most general and, at the same time, most consistent approach to the problem. The non-relativistic limit is considered in the sections thereafter.

12.1 Basic concepts of relativistic quantum electrodynamics

For many charged carrier-field interaction phenomena, a relativistic treatment is necessary. This includes the behavior of highly energetic carriers in a variety of astrophysical situations or in cosmology. Furthermore, relativistic behavior is observed in numerous laboratory experiments in high energy physics or solid state and plasma physics, where particles gain energy respectively in accelerators, or from particle beams or high intensity lasers. Understanding of relativistic behavior is indispensable for numerous fundamental theoretical problems, but also for practical applications which are only emerging, ranging from medicine and material sciences to inertial confinement fusion. But even for many-particle systems which are far from relativistic velocities, the approach of relativistic QED provides the most complete and consistent, and, at the same time, the conceptually most simple description. Many derivations are more clear in this picture as is the nature of many approximations, and taking the nonrelativistic limit in the final results is usually a comparably easy task.

The Green's functions approach to relativistic charged particle–photon interaction is due to Dubois and Bezzerides [DuB67, BD72] and will be the natural basis for our discussion. Nevertheless, we mention impressive early work on relativistic kinetic theory by Belyaev and Budker [BB57], Klimontovich [Kli60a, Kli60b], Silin [Sil62] and also Akhiezer/Berestezki and Bjorken/Drell, see their classical text books [AB59, BD64] and references therein, which are the basis for the modern concepts.

Definitions. 4-dimensional notation. We will use standard 4-vector notation with the convention, $a^\mu = (a_0, \mathbf{a})$, and $a_\mu = (a_0, -\mathbf{a}) = g_{\mu\nu}a^\nu$, where $g^{\mu\nu} = g_{\mu\nu}$ is the metric tensor, with the diagonal components $1, -1, -1, -1$ e.g. [LL80b, BD64]. A scalar product is then $ab = a_\mu b^\mu = a_0 b_0 - \mathbf{ab}$, where summation over repeated indices is implied. In particular, we will write for the coordinates $x^\mu = (ct, \mathbf{r})$, momentum $p^\mu = (E/c, \mathbf{p})$, charge current $j^\mu = (c\rho, \mathbf{j})$ and the vector potential $A^\mu = (c\phi, \mathbf{A})$, where ρ and ϕ denote the charge density and the scalar potential, respectively. Differential operators are given by $\partial_\mu = \partial/\partial x^\mu$, and the wave operator is $\Box = \partial^\mu \partial_\mu = \frac{1}{c^2}\partial/\partial t^2 - \Delta$.

Electromagnetic field operators. The electromagnetic field is represented by the fluctuating field operator (4-vector) $\hat{A}^\mu(1)$, where $1 = x_1^\mu$, and obeys Maxwell's equations

$$\mathcal{D}^\mu_\nu(1)\hat{A}^\nu(1) = \frac{4\pi}{c}\left\{\hat{j}^\mu(1) + j^{\mu\, ext}(1)\right\}, \tag{12.1}$$

Lorentz (general)	**Coulomb** (transverse†)	**Feynman** (4-transverse)
$\begin{pmatrix} -\Delta & \partial_0^1 & \partial_0^2 & \partial_0^3 \\ \partial_1^0 & \Box_1^T & \partial_1^2 & \partial_1^3 \\ \partial_2^0 & \partial_2^1 & \Box_3^T & \partial_2^3 \\ \partial_3^0 & \partial_3^1 & \partial_3^2 & \Box_3^T \end{pmatrix}$	$\begin{pmatrix} -\Delta & 0 & 0 & 0 \\ 0 & \Box_1^T & \partial_1^2 & \partial_1^3 \\ 0 & \partial_2^1 & \Box_3^T & \partial_2^3 \\ 0 & \partial_3^1 & \partial_3^2 & \Box_3^T \end{pmatrix}$	$\begin{pmatrix} \Box & 0 & 0 & 0 \\ 0 & \Box & 0 & 0 \\ 0 & 0 & \Box & 0 \\ 0 & 0 & 0 & \Box \end{pmatrix}$
–	$\mathrm{div}\mathbf{A} = 0$	$\partial_\mu A^\mu = 0$
Lorentz inv.	not inv.	not inv.
general media	isotropic media	vacuum

Table 12.1: Common gauges and their properties. First line: Explicit expressions for the matrix Maxwell operator $\mathcal{D}^\mu_\nu$, second line: analytic gauge condition, third line: Lorentz invariance (in the general case of a plasma), fourth line: application. $\partial_i^j = \partial_i\partial^j$; $\Box_i^T = \Box + \partial_i^i$. In Fourier space $\partial_i\partial^j \to -k_i k^j$. $i,j = 1,2,3$, $\mu,\nu = 0,1,2,3$; † Maxwell's equations contain only the transverse current j^T, Eq. (12.61).

where $\mathcal{D}^\mu_\nu$ is a general 4×4 matrix operator.[1] The field is created by induced and external currents (and charges) given by $\hat{j}^\mu$, and $j^{\mu\, ext}$, respectively (the first is an operator, whereas the latter is a c-number). While $j^{\mu\, ext}$ is due to externally controlled sources (see also Sec. 12.3), $\hat{j}^\mu$ is generated by the carriers in the system and related to the particle operators via Eq. (12.4).

[1]Recall that $\hat{A}^\mu$ generates the electromagnetic field tensor $\hat{F}^{\mu\nu} = \partial^\nu\hat{A}^\mu - \partial^\mu\hat{A}^\nu$, which yields the Lorentz-invariant form of Maxwell's equations

$$\partial_\nu \hat{F}^{\mu\nu} = \frac{4\pi}{c}\left\{\hat{j}^\mu + j^{\mu\, ext}\right\}. \tag{12.2}$$

This form does not change under (gauge) transformations $A_\mu \to A_\mu + ic/ep_\mu\chi$. On the other hand, the *explicit form* of the operator on the l.h.s. of Eq. (12.1) does depend on the gauge, several common forms are listed in table 12.1. The Lorentz form follows from Eq. (12.2) without additional conditions. The Coulomb gauge is the most suitable one for isotropic media, where longitudinal and transverse components decouple, the Feynman gauge is the appropriate choice for vacuum.

Relativistic particle operators. Dirac equation. The requirement of Lorentz invariance leads to the representation of particles by a pair of operators $\hat{\psi}, \hat{\zeta}$, which form the particle propagator $\hat{\Psi} = (\hat{\psi}, \hat{\zeta})$. For particles with spin, both $\hat{\psi}$ and $\hat{\zeta}$, become spinors, i.e. they have $2s+1$ components, where s is the spin projection. Thus, the propagator $\hat{\Psi}$ of fermions (the Dirac field, $s = 1/2$) has four components. In the case of an applied electromagnetic field given by $\hat{A}$, the spinor $\hat{\Psi}$ obeys the Dirac equation which includes the total momentum $p - e/cA$,[2]

$$\left\{\gamma^\mu \left(\hat{p}_\mu - \frac{e}{c}\hat{A}_\mu\right) - mc\right\}\hat{\Psi} = 0, \qquad \hat{\bar{\Psi}}\left\{\gamma^\mu \left(\hat{p}_\mu + \frac{e}{c}\hat{A}_\mu\right) + mc\right\} = 0, \quad (12.3)$$

where e, and m are the particle charge and mass, $\gamma^\mu = (\gamma^0, \vec{\gamma})$ is the 4-vector of the Dirac matrices[3] and $\hat{\bar{\Psi}}$ is the adjoint operator defined by $\hat{\bar{\Psi}} = \hat{\Psi}^\dagger\gamma^0$. The 4-vector of the particle momentum operator is given by (coordinate representation) $\hat{p}^\mu = i\hbar\partial^\mu = i\hbar(\frac{1}{c}\partial_t, -\nabla)$, and the charge current operator is the generalization of the standard quantum mechanical expression

$$\hat{j}^\mu = ec\hat{\bar{\Psi}}\gamma_\mu\hat{\Psi}. \quad (12.4)$$

It obeys the charge conservation equation $\partial_\mu \hat{j}^\mu = 0$ and also the positive definiteness of the number density operator $\hat{n} = \hat{j}^0/e = \hat{\bar{\Psi}}\gamma^0\hat{\Psi} \geq 0$.

Thus, we have written down the coupled relativistic equations of motion for particle and field propagators. The carrier dynamics is influenced by the electromagnetic field via the operator $\hat{A}$ in the Dirac equation, whereas the field dynamics is determined by the particle current (12.4) which acts as a source term in Maxwell's equations (12.1).

---

[2]This combination arises from the requirement that the Dirac equation remains invariant under gauge transformations. With the transformation $\Psi \to e^{i\chi}\Psi$, in the Dirac equation appears an additional term $i\gamma^\mu\hat{p}_\mu\chi(x)$ which is exactly compensated if, instead of the free momentum, $p - e/cA$ is used, with A having the gauge transform $A_\mu \to A_\mu + ic/ep_\mu\chi$. This is fulfilled for the vector potential. In other words, the electromagnetic field is the gauge field of the Dirac equation for fermions. This basic idea is generalized by modern gauge field theories to application to particle physics (QCD).

[3]Recall the definition of the Dirac matrices [LL80b]:

$$\gamma^0 = \begin{pmatrix} 0 & \mathbf{1} \\ \mathbf{1} & 0 \end{pmatrix}, \quad \vec{\gamma} = \begin{pmatrix} 0 & -\vec{\sigma} \\ \vec{\sigma} & 0 \end{pmatrix}, \quad \text{where } \vec{\sigma} = (\sigma_x, \sigma_y, \sigma_z) \text{ are the Pauli matrices}$$

$$\sigma_x = \begin{pmatrix} 0 & 1 \\ 1 & 0 \end{pmatrix}, \quad \sigma_y = \begin{pmatrix} 0 & -i \\ i & 0 \end{pmatrix}, \quad \sigma_z = \begin{pmatrix} 1 & 0 \\ 0 & -1 \end{pmatrix}, \quad \text{and} \quad \mathbf{1} = \begin{pmatrix} 1 & 0 \\ 0 & 1 \end{pmatrix}.$$

Statistical description in nonequilibrium. To gain information on physical observables, the operator equations (12.1), (12.3) have to be transformed into equations for suitably chosen averages. As usual, averages are computed as the trace over the (equilibrium or nonequilibrium) density operator, for example $A(1) \equiv \langle \hat{A}(1) \rangle = \mathrm{Tr}\{\rho \hat{A}(1)\}$. In equilibrium (in the absence of external fields and sources) the expectation values of the induced 4-current vanishes, $\langle \hat{j}^\mu \rangle \equiv j^\mu = 0$. With it, also the average total field is zero, $\langle \hat{A}^\mu \rangle = A^\mu = 0$. Nonzero average fields are induced only by perturbing external source j^{ext}. In that case, the time development of all observables can be expressed via the (forward or backward) time evolution operators $S_\pm$ according to

$$S_\pm(tt_0) = \mathrm{T}_\pm \exp\{\mp \frac{i}{\hbar} \int_{t_0}^{t} d\bar{1}\, \frac{1}{c} j^{ext}_{\pm\nu}(\bar{1})\, \hat{A}^\nu(\bar{1})\}, \tag{12.5}$$

and $T_+(T_-)$ orders all operators with greater time arguments left (right) from those with smaller ones. The basic quantities for a quantum statistical description of carriers and fields are Green's functions, which are defined as averages of products of (two or more) field operators (see below). It turns out that the results of equilibrium Green's functions theory can be directly extended to nonequilibrium, if the time arguments of the field operators are formally considered to belong to a *double-time contour* $\mathcal{C}$ (Keldysh contour) [Sch61, Kel64a]. Here, we briefly summarize the main results which are independent of the relativistic behavior, for details, see e.g. [Dan84a, HJ96, BK95].

1. The time contour $\mathcal{C}$ runs from the initial moment t_0 "forward" to the current time t (on the "upper" branch, $t = t_+$) and "back" to t_0 (on the "lower" branch, $t = t_-$). Each quantity is represented by a pair of quantities, which have time arguments belonging to the upper and lower branch, respectively. For example, the external current is represented by $j^{ext}_\pm$, cf. Eq. (12.5). At the end of the calculations, the "physical limit" is taken, i.e. in all results, $j^{ext}_+ \to j^{ext}_- \to j^{ext}(t)$, where $-\infty < t < \infty$.

2. A two-operator product on $\mathcal{C}$ is denoted by $C(\underline{1}, \underline{2}) = A(\underline{1})B(\underline{2})$ and describes in fact a 2×2 matrix, the elements corresponding to the four possibilities for t_1 and t_2 to belong to t_+ or t_-, respectively.

3. Integrals over matrix variables are understood as

$$\int d\underline{1} B(\underline{1}) = \int_{-\infty}^{\infty} d1\, [B_+(1) + B_-(1)]\,. \tag{12.6}$$

4. Averages of two-operator products are related to the evolution operator by

$$\langle C(\underline{1},\underline{2})\rangle = \eta_2 \frac{\langle T_c S_c A(\underline{1}) B(\underline{2})\rangle}{\langle S_c\rangle}, \tag{12.7}$$

where $S_c = S_- S^+$, $\eta_2 = \pm 1$, depending whether $t_2 = t_+, t_-$, and T_c is the contour-ordering operator[4]. In the physical limit, $S_c \to 1$.

5. Averages of products of N operators which include the field operator A can be obtained by functional differentiation with respect to j^{ext} of $(N-1)$-operator averages, due to the structure of $S_\pm$, Eq. (12.5). For example, for an arbitrary operator $\hat{B} = B + \delta\hat{B}$ and $\langle\hat{B}\rangle \equiv B$, we have

$$\langle \delta\hat{A}^\mu(\underline{1})\delta\hat{B}(\underline{2})\rangle = \langle \hat{A}^\mu(\underline{1})\hat{B}(\underline{2})\rangle - A^\mu(\underline{1})B(\underline{2}) = -\frac{i}{\hbar c}\eta_1 \frac{\delta B(\underline{2})}{\delta j^{ext}_\mu(\underline{1})}, \tag{12.8}$$

which holds for all elements of the matrix at once and which is easily generalized to more complicated operator products.

6. The "canonical"[5] notation for the elements of an arbitrary Keldysh matrix C relating them, in the physical limit, to the familiar correlation functions ($c^>$, $c^<$) and retarded/advanced functions $c^\pm$ is [DuB67]

$$\begin{pmatrix} C_{++} & C_{+-} \\ C_{-+} & C_{--} \end{pmatrix} = \begin{pmatrix} c^+ + c^< & -c^< \\ c^> & c^+ - c^> \end{pmatrix}, \tag{12.9}$$

where $c^\pm$ are defined as (the singular part is missing in many cases)

$$c^\pm(1,2) = c_0\delta(t_1 - t_2) \pm \Theta\left[\pm(t_1 - t_2)\right]\left\{c^>(1,2) - c^<(1,2)\right\}, \tag{12.10}$$

and they obey the relation (which also defines the spectral function $\hat{c}$)

$$c^>(1,2) - c^<(1,2) = c^+(1,2) - c^-(1,2) \sim \frac{1}{i}\,\hat{c}(1,2). \tag{12.11}$$

While in equilibrium, one of these functions is sufficient to describe the system, in nonequilibrium two functions are independent. Nevertheless, the matrix notation which includes all four elements is convenient in the formal derivations.

[4] $T_c = T_\pm$, if both t_1 and t_2 belong to t_+ or t_-, in the mixed case, it orders operators with the "-" index left from those with a "+" index. In the case of Fermion operators, each permutation adds a minus sign to the whole expression. For two operators, this is accounted for by the factor η_2.

[5] Here, we follow Dubois' notation, however, also other definitions appear in the literature. Also, we will use capitals for the quantities on the contour and small letters for the physical quantities.

Green's functions for photons and carriers. The central quantities for the statistical description of fields and particles are the Green's functions which are defined as *averages of products of field operator fluctuations.*[6] We define the photon Green's function and the one-particle carrier Green's function as the corresponding two-operator product, which can also be obtained by functional differentiation according to Eq. (12.8)[7], (recall that $A \equiv \langle \hat{A} \rangle$)

$$\boxed{\begin{aligned} \frac{4\pi}{c} D^{\mu\nu}(\underline{1}, \underline{1}') &= \frac{\delta A^\mu(\underline{1})}{\delta j_\nu^{ext}(\underline{1}')} = -\frac{i}{\hbar c}\eta' \left\{ \langle \hat{A}^\mu(\underline{1}) \hat{A}^\nu(\underline{1}') \rangle - A^\mu(\underline{1}) A^\nu(\underline{1}') \right\} \\ G(\underline{1}, \underline{1}') &= -\frac{i}{\hbar}\eta_1' \, \langle \hat{\Psi}(\underline{1}) \hat{\overline{\Psi}}(\underline{1}') \rangle \end{aligned}} \tag{12.12}$$

Notice that in the carrier case, the product of one–operator averages is missing because $\langle \hat{\Psi} \rangle = \langle \hat{\overline{\Psi}} \rangle = 0$.[8] We underline that Eqs. (12.12) define very complex quantities. In the definition of the photon Green's functions, $\mu, \nu = 0, 1, 2, 3$, so $D^{\mu\nu}$ fully contains longitudinal and transverse electromagnetic fields, as well as their coupling.[9] On the other hand, recall that the operators in the definition of G are 4-spinors. Finally, each element of $D^{\mu\nu}$ and G is itself a 2×2 matrix in the time indices.

Equations of motion for the Green's functions. The equation for $D^{\mu\nu}$ follows directly from Eq. (12.1) after differentiation with respect to j_ν^{ext},

$$\mathcal{D}^\mu_\nu(\underline{1}) D^{\nu\lambda}(\underline{1}, \underline{1}') = \delta^{\mu\lambda}(\underline{1} - \underline{1}') + \frac{\delta j^\mu(\underline{1})}{\delta j_\lambda^{ext}(\underline{1}')}, \tag{12.13}$$

where the first term on the r.h.s. comes from differentiation of $j^{ext\,\mu}$ in Eq. (12.1). Transformation of the second term using the chain rule yields [10],

$$\mathcal{D}^\mu_\nu(\underline{1}) D^{\nu\lambda}(\underline{1}, \underline{1}') - \int d\underline{2}\, \Pi^\mu_\nu(\underline{1}, \underline{2}) D^{\nu\lambda}(\underline{2}, \underline{1}') = \delta^{\mu\lambda}(\underline{1} - \underline{1}'), \tag{12.14}$$

[6]The classical analogue appears in *Klimontovich's phase space density technique* [Kli57], where he also considers correlations of particle and field fluctuations $\langle \delta N \delta N \rangle$, $\langle \delta E \delta E \rangle$ and cross correlations $\langle \delta N \delta E \rangle$, for which he derives equations of motion, e.g. [Kli75], see also Sec. 1.3.3.

[7]A functional derivative for the carriers can be formally defined too by introduction of a fictitious external perturbation which is linear in $\hat{\Psi}$ and $\hat{\overline{\Psi}}$ which vanishes after differentiation.

[8]An exception are "anomalous" situations, such as superconductivity or Bose condensation.

[9]We emphasize that only for isotropic media a separation of longitudinal and transverse components is possible. Only in that case, with the use of the Coulomb gauge, the (00) component contains the longitudinal part, while the (ij) components $(i, j = 1, 2, 3)$ corresponds to the transverse part, cf. table 12.1.

[10]I.e. we write $\frac{\delta j^\mu(\underline{1})}{\delta j_\lambda^{ext}(\underline{1}')} = \int d\underline{2}\, \frac{\delta j^\mu(\underline{1})}{\delta A^\nu(\underline{2})} \frac{\delta A^\nu(\underline{2})}{\delta j_\lambda^{ext}(\underline{1}')}$, where summation over ν is implied.

where we defined the **photon selfenergy** (polarization matrix),

$$\boxed{\Pi^{\mu}_{\nu}(\underline{1},\underline{2}) = \frac{4\pi}{c}\frac{\delta j^{\mu}(\underline{1})}{\delta A^{\nu}(\underline{2})}} \tag{12.15}$$

which contains all effects of QED–vacuum polarization and plasma polarization, including collective excitations, instabilities and screening.

In similar manner, we derive the *equation of motion for the carrier Green's function.* Multiplying Eq. (12.3) by the adjoint spinor $\overline{\Psi}\cdot(-i\eta_1'/\hbar)$ and averaging the resulting equation, we obtain

$$\left\{\gamma_\mu p^{\mu}_{\underline{1}} - mc\right\} G(\underline{1},\underline{1}') + \frac{ie}{c\hbar}\eta_1'\gamma_\mu \langle \hat{A}^{\mu}(\underline{1})\hat{\Psi}(\underline{1})\hat{\overline{\Psi}}(\underline{1}')\rangle \;=\; 0. \tag{12.16}$$

This equation for G is not closed, since it involves a three–operator average. One possible solution is to derive an equation for this quantity, which, in turn, couples to 4-operator averages, giving rise to an infinite hierarchy of equations, which, as in the case of the BBGKY-hierarchy, can be decoupled by introducing suitable approximations. However, a more convenient alternative procedure is to express $\langle \hat{A}^{\mu}_{\underline{1}}\hat{\Psi}(\underline{1})\hat{\overline{\Psi}}(\underline{1}')\rangle$ by a functional derivative of the particle Green's function using Eq. (12.8)

$$\begin{aligned}\langle \hat{A}^{\mu}(\underline{1})\hat{\Psi}(\underline{1})\hat{\overline{\Psi}}(\underline{1}')\rangle - A^{\mu}(\underline{1})\,\langle\hat{\Psi}(\underline{1})\hat{\overline{\Psi}}(\underline{1}')\rangle \;&=\; i\hbar\eta_1'\,\langle \delta\hat{A}^{\mu}(\underline{1})\delta G(\underline{1},\underline{1}')\rangle \\ &=\; \frac{1}{c}\eta_1\eta_1'\,\frac{\delta G(\underline{1},\underline{1}')}{\delta j^{ext}_{\mu}(\underline{1})}.\end{aligned} \tag{12.17}$$

The second term on the l.h.s. is the Hartree (mean field) term and can be combined with the particle momentum to yield the operator of the kinematic momentum $(p - e/cA)$ times G. The derivative on the right gives rise to the particle selfenergy Σ, which includes only correlation contributions. Furthermore, a contribution from initial particle–field correlations appears,[11]

$$\left\{\gamma_\mu\left(p^{\mu}_{\underline{1}} - \frac{e}{c}A^{\mu}(\underline{1})\right) - mc\right\} G(\underline{1},\underline{1}') - \int d\underline{2}\,\Sigma(\underline{1},\underline{2})G(\underline{2},\underline{1}')$$

[11] We outline the main steps of the derivation: First, the functional derivative in Eq. (12.17) is transformed with the chain rule and definition (12.12) to

$$\frac{\delta G(\underline{1},\underline{1}')}{\delta j^{ext}_{\mu}(\underline{1})} = \int d\underline{2}\,\frac{\delta G(\underline{1},\underline{1}')}{\delta A^{\nu}(\underline{2})}\frac{4\pi}{c}D^{\nu\mu}(\underline{21}). \tag{12.18}$$

For the further derivation, we introduce the inverse propagators G^{-1} and $\tilde{G}^{-1}$ by

$$\int d\underline{2}\,G^{-1}(\underline{1},\underline{2})G(\underline{2},\underline{1}') = \int d\underline{2}\,G(\underline{1},\underline{2})\tilde{G}^{-1}(\underline{2},\underline{1}') = \delta(\underline{1}-\underline{1}'), \tag{12.19}$$

where it can be shown that, except for the initial moment $t_1 = t_1' = t_0$, $G^{-1} = \tilde{G}^{-1}$.

$$-\gamma_\mu \int d\underline{2}\, C_\nu(\underline{1},\underline{1}',\underline{2}) \frac{4\pi}{c} D^{\nu\mu}(\underline{2},\underline{1}) = \delta(\underline{1}-\underline{1}'), \tag{12.21}$$

$$\Sigma(\underline{1},\underline{2}) = -iec\hbar\eta_1\gamma_\mu \int d\underline{3}d\underline{4}\, G(\underline{1},\underline{3}) \frac{\delta G^{-1}(\underline{3},\underline{2})}{\delta A^\nu(\underline{4})} \frac{4\pi}{c} D^{\nu\mu}(\underline{4},\underline{1}). \tag{12.22}$$

One readily confirms that C_ν defines the initial value ($t_1 \to t_1' \to t_0$) of the joint "field-matter fluctuation" Eq. (12.17), which in general is nonzero,[12]

$$\langle \hat{A}^\mu(t_0)\hat{\Psi}(t_0)\hat{\bar{\Psi}}(t_0)\rangle - A^\mu(t_0)\,\langle \hat{\Psi}(t_0)\hat{\bar{\Psi}}(t_0)\rangle = \frac{4\pi\eta_1\eta_1'}{c^2} \int d\underline{2}\, C_\nu(t_0 t_0 \underline{2}) D^{\nu\mu}(\underline{2}t_0).$$

With Eqs. (12.14) and (12.21), we have obtained a coupled system of equations for the photon and carrier Green's functions. Higher order correlations are formally eliminated by introduction of the particle and photon selfenergies Σ, Π, which also contain all coupling effects between carriers and electromagnetic field.

Equations (12.14) and (12.21) may be inverted using Eq. (12.19) and yield for the inverse matrix propagators (for $t, t' > t_0$),

$$D^{-1\mu}_{\ \ \nu}(\underline{1},\underline{1}') = D^{-1\mu}_{0\ \nu}(\underline{1},\underline{1}') - \Pi^\mu_{\ \nu}(\underline{1},\underline{1}'), \tag{12.23}$$

$$G^{-1}(\underline{1},\underline{1}') = G_0^{-1}(\underline{1},\underline{1}') - \Sigma(\underline{1},\underline{1}'), \tag{12.24}$$

where D_0^{-1} and G_0^{-1} are the interaction–free inverse Green's functions, which are given by

$$D^{-1\mu}_{0\ \nu}(\underline{1},\underline{1}') = \mathcal{D}^\mu_{\ \nu}(\underline{1})\delta(\underline{1}-\underline{1}'), \tag{12.25}$$

$$G_0^{-1}(\underline{1},\underline{1}') = \left\{\gamma_\mu\left(p^\mu_{\underline{1}} - \frac{e}{c}A^\mu_{\underline{1}}\right) - mc\right\}\delta(\underline{1}-\underline{1}'). \tag{12.26}$$

Integral equations for the Green's functions. Eqs. (12.14) and (12.21) contain certain freedom due to the gauge dependence of the electromagnetic field. It is, therefore, advantageous to rewrite them in the form of

---

Differentiation of the left part of Eq. (12.19) with respect to A^ν yields

$$\frac{\delta G(\underline{1},\underline{1}')}{\delta A^\nu(\underline{2})} = -\int d\underline{3}d\underline{4}\, G(\underline{1}\underline{3}) \frac{\delta G^{-1}(\underline{3},\underline{4})}{\delta A^\nu(\underline{2})} G(\underline{4},\underline{1}') + C_\nu(\underline{1},\underline{1}',\underline{2}), \tag{12.20}$$

which is verified by inserting it back into Eq. (12.19). It is important to realize that Eq. (12.20) is defined only up to an arbitrary function $C_\nu(\underline{1},\underline{1}',\underline{2})$ which vanishes in Eq. (12.19) upon action of G^{-1}. The analysis shows that (1) C is related to initial correlations, see below; (2) $\int d\bar{\underline{1}} G^{-1}(\underline{1},\bar{\underline{1}})C_\nu(\bar{\underline{1}},\underline{1}',\underline{2}) = \int d\bar{\underline{1}} C_\nu(\underline{1},\bar{\underline{1}},\underline{2})G^{-1}(\bar{\underline{1}},\underline{1}') = \int d\bar{\underline{2}} G^{-1}(\underline{2},\bar{\underline{2}})C_\nu(\bar{\underline{1}},\underline{1}',\underline{2}) = \int d\bar{\underline{2}} C_\nu(\underline{1},\bar{\underline{1}},\underline{2})G^{-1}(\bar{\underline{2}},\underline{2}') = 0$; and (3) C is the same for all Keldysh matrix components. More details are given in [Sem96, KSB].

[12] This is especially transparent in the nonrelativistic limit, see Sec. 12.4.

integral equations which turn out to be of *gauge-invariant structure* (although the explicit expressions may vary),

$$D^{\mu\nu}(\underline{1},\underline{1}') = D_0^{\mu\nu}(\underline{1},\underline{1}') + \int d\underline{2}d\underline{3}\, D^{\mu}_{0\lambda}(\underline{1},\underline{2})\Pi^{\lambda\gamma}(\underline{2},\underline{3})D_{\gamma}^{\ \nu}(\underline{3},\underline{1}'), \quad (12.27)$$

$$G(\underline{1},\underline{1}') = G_0(\underline{1},\underline{1}') + \int d\underline{2}d\underline{3}\, G_0(\underline{1},\underline{2})\Sigma(\underline{2},\underline{3})G(\underline{3},\underline{1}'). \quad (12.28)$$

The crucial point is [BD72] that, in this form, the whole gauge problem is completely determined by the gauge properties of the free photon propagator D_0 which is defined by the well-known matrix $\mathcal{D}^{\mu}_{\nu}$ (see table 12.1).

Vertex function. The strength of quantum electrodynamics is that it treats the particle–electromagnetic field system as one whole complex. Nevertheless, so far, the dynamics of particles and field is governed by two different quantities, the selfenergies Σ and Π. This is conceptually not satisfactory and moreover problematic if one thinks about introducing simplifications to the exact equations: a fundamental requirement for any approximation, obviously, has to be full consistency in the treatment of particles and field. However, inspection of the photon and particle selfenergies, Eqs. (12.15) and (12.22) shows that these quantities can be written in completely symmetric way if they are expressed in terms of the *vertex function* Γ,[13]

$$\Pi^{\mu}_{\nu}(\underline{1},\underline{2}) = \eta_1\gamma^{\mu}\int d\underline{3}d\underline{4}\, G(\underline{1},\underline{3})\,\Gamma_{\nu}(\underline{3},\underline{4};\underline{2})\,G(\underline{4},\underline{1}), \quad (12.29)$$

$$\Sigma(\underline{1},\underline{2}) = \eta_1\gamma_{\mu}\int d\underline{3}d\underline{4}\, G(\underline{1},\underline{3})\,\Gamma_{\nu}(\underline{3},\underline{2};\underline{4})\,D^{\nu\mu}(\underline{4},\underline{1}), \quad (12.30)$$

$$\Gamma_{\nu}(\underline{1},\underline{1}';\underline{2}) = -\frac{4\pi}{c}ie\hbar\frac{\delta G^{-1}(\underline{1},\underline{1}')}{\delta A^{\nu}(\underline{2})}. \quad (12.31)$$

Notice that Π defined by Eq. (12.15) implicitly contains the initial correlation term C, which, for symmetry reasons, we do not include in Eq. (12.29). Thus, the previous definition is substituted by Eq. (12.29) according to $\Pi^{\mu}_{\nu}|_{(12.15)} \longrightarrow \Pi^{\mu}_{\nu}|_{(12.29)} + \eta_1\gamma^{\mu}C_{\nu}(\underline{11},\underline{2})$. The C term will appear explicitly in the equation of motion for D below.

To come to a closed system of equations, we now need to find an equation for Γ. It is readily derived from functional differentiation of the explicit form

[13] Expression (12.30) follows immediately from the definition of Σ, Eq. (12.22). To obtain (12.29) from the definition (12.15), one first has to express the current variation δj^{μ} in terms of δG using Eq. (12.4). Finally, δG is expressed via δG^{-1} by means of the identity (12.20), which again gives rise to the initial correlation contribution C_{ν}.

of the inverse Green's function, Eq. (12.24), with respect to A^ν,[14]

$$\Gamma_\nu(\underline{1},\underline{1}';\underline{2}) = i\frac{4\pi e^2}{c}\hbar\eta_1\gamma_\mu\,\delta^\mu_\nu\delta(\underline{1}-\underline{1}')\delta(\underline{1}-\underline{2}) + \int d\underline{3}d\underline{4}d\underline{5}d\underline{6} \times \frac{\delta\Sigma(\underline{1},\underline{1}')}{\delta G(\underline{3},\underline{4})}\left\{G(\underline{3},\underline{5})\,\Gamma_\nu(\underline{5},\underline{6};\underline{2})\,G(\underline{6},\underline{4}) + C_\nu(\underline{34};\underline{2})\right\}. \quad (12.32)$$

Thus, we have succeeded in writing the particle and photon selfenergies in a completely symmetric form which involves one fundamental quantity Γ common to both, particles and photons. Moreover, with Eq. (12.32), we have on hand a convenient tool for systematic derivation of approximations. In deriving approximations for $\delta\Sigma/\delta G$ one can generate approximations for Γ and, with it simultaneously, approximations for Σ and Π which will be fully consistent with each other. We will derive various important approximations later in this Chapter.

To conclude the general derivations, we rewrite the equations of motion for the Green's functions using the definitions (12.25), (12.26) and (12.29), where integrations over the intermediate variable $\underline{2}$ is implied,

$$\left\{D_{0\,\nu}^{-1\mu}(\underline{1},\underline{2}) - \Pi^\mu_\nu(\underline{1},\underline{2}) - \eta_1\gamma^\mu C_\nu(\underline{1},\underline{1},\underline{2})\right\} D^{\nu\lambda}(\underline{2},\underline{1}') = \delta^{\mu\lambda}(\underline{1}-\underline{1}'), \quad (12.33)$$

$$\left\{G_0^{-1}(\underline{1},\underline{2}) - \Sigma(\underline{1},\underline{2})\right\} G(\underline{2},\underline{1}') - \gamma_\mu C_\nu(\underline{1},\underline{1}',\underline{2})\frac{4\pi}{c}D^{\nu\mu}(\underline{2},\underline{1}) = \delta(\underline{1}-\underline{1}'). \quad (12.34)$$

12.2 Relativistic Kadanoff–Baym equations for particles and photons

We now consider the physical limit of the equations for the Green's functions, which yields four equations for the components of the Keldysh matrix. First, let us rewrite the integral equations (12.27,12.28) separately for the components of the Keldysh matrices which are defined already on the real time axis, and apply the notation introduced above in Eq. (12.9) to D, G, D_0, G_0, Π and Σ,

$$d^{\mu\nu\gtrless}(1,1') = \int d2d3\, d^{\mu+}_\lambda(1,2)\,\pi^{\lambda\gamma\gtrless}(2,3)\,d^{\nu-}_\gamma(3,1'), \quad (12.35)$$

$$g^\gtrless(1,1') = \int d2d3\, g^+(1,2)\,\sigma^\gtrless(2,3)\,g^-(3,1'), \quad (12.36)$$

[14] In Eq. (12.32), the first term comes from the differentiation of G_0^{-1}, whereas the second comes from applying the chain rule to $\delta\Sigma/\delta A \sim \delta\Sigma/\delta G \cdot \delta G/\delta A$. Finally, the variation of G is transformed into a variation of G^{-1} according to Eq. (12.20).

$$\begin{aligned} d^{\mu\nu\pm}(1,1') &= d_0^{\mu\nu\pm}(1,1') + \int d2d3\, d_{0\,\lambda}^{\mu\pm}(1,2)\, \pi^{\lambda\pm}_{\gamma}(2,3)\, d^{\gamma\nu\pm}(3,1'), \quad (12.37) \\ g^{\pm}(1,1') &= g_0^{\pm}(1,1') + \int d2d3\, g_0^{\pm}(1,2)\, \sigma^{\pm}(2,3)\, g^{\pm}(3,1'). \quad (12.38) \end{aligned}$$

The relations (12.35,12.36) are generalized fluctuation-dissipation (or optical) theorems. They couple fluctuations of photons ($d^>$, $d^<$) and particles or anti-particles ($g^>$, $g^<$) to all possible dissipative processes, which are contained in the generalized scattering rates $\pi^>, \pi^<$ and $\sigma^>, \sigma^<$. In particular, $\pi^>, \pi^<$ include emission and absorption of longitudinal plasma excitations and electromagnetic field oscillations, and $\sigma^>, \sigma^<$ all particle-field interactions, including particle-particle interaction (via longitudinal fields), pair creation and annihilation etc.

The nonequilibrium behavior of the particle–photon system is fully determined if two of the four Green's functions are known. Among the possible equivalent choices, we will prefer to work with the *correlation functions* $g^>, g^<, d^>, d^<$, which is motivated by the fact that these quantities are closest to physical observables, including distribution functions, as well as by numerical experience (see below). We will derive a closed system of differential equations for these functions, which is a generalization of the well-known Kadanoff–Baym equations. To this end, we consider Eqs. (12.14, 12.21) for the off-diagonal components of the Keldysh matrix (12.9). For these components, the delta functions are absent, and the only problem left is to evaluate the integrals involving the selfenergies Π and Σ, respectively. After straightforward calculations,we obtain the **relativistic Kadanoff–Baym equations**

$$\boxed{\begin{gathered} \mathcal{D}^{\mu}_{\nu}(1) d^{\nu\lambda\gtrless}(11') = \int_{t_0}^{t_1} d\bar{1}\, [\pi^{\mu>}_{\nu}(1\bar{1}) - \pi^{\mu<}_{\nu}(1\bar{1})]\, d^{\nu\lambda\gtrless}(\bar{1}1') \\ - \int_{t_0}^{t_1'} d\bar{1} \pi^{\mu\gtrless}_{\nu}(1\bar{1}) [d^{\nu\lambda>}(\bar{1}1') - d^{\nu\lambda<}(\bar{1}1')] + I^{\mu\lambda\gtrless}_{ICd}(11') \\ \left\{\gamma_\mu \left(p_1^\mu - \frac{e}{c} A^\mu(1)\right) - mc\right\} g^{\gtrless}(11') = I^{\gtrless}_{ICg}(11') + \\ \int_{t_0}^{t_1} d\bar{1}\, [\sigma^>(1\bar{1}) - \sigma^<(1\bar{1})]\, g^{\gtrless}(\bar{1}1') - \int_{t_0}^{t_1'} d\bar{1} \sigma^{\gtrless}(1\bar{1})\, [g^>(\bar{1}1') - g^<(\bar{1}1')] \end{gathered}} \quad (12.39)$$

where the integrals I_{IC} arise from the initial correlation C and die out after a few collisions (cf. the discussion in Ch. 5). These equations have to be supplemented with initial conditions for the Green's functions and the initial correlations, $g^{\gtrless}(t_0t_0), d^{\gtrless}(t_0t_0)$ and $C(t_0t_0t_0)$. Furthermore, the adjoint equations are necessary.[15] The Kadanoff–Baym equations (12.39) have a very

[15] We briefly outline the derivation of Eqs. (12.39). The ">" component of an integral

general structure which remains the same also in the non-relativistic limit. As we will see in Sec. 12.3, the only changes appear in the actual form of the inverse Green's functions and in the specific matrix structure of the Green's functions.

Approximations for the selfenergies

We now consider important approximations for the selfenergies. As we have seen above, the level of approximation is completely determined by the vertex function Γ. With Γ fixed, the selfenergies Σ and Π are determined by Eqs. (12.30) and (12.29), respectively. Then, the evolution of the Green's functions G and D is given by the coupled Kadanoff–Baym equations or, equivalently, by the integral equations (12.28), (12.27). The latter are, combined with the integral equation for Γ (12.32), of advantage for the derivation of explicit approximations for the selfenergies. The main idea is to use an iterative scheme where one starts with the zeroth order for the vertex function $\Gamma^{(0)}$, which inserted into Eq. (12.32) on the r.h.s., $\Gamma \to \Gamma^{(0)}$, yields $\Gamma^{(1)}$ on the l.h.s. and so on. Before giving explicit examples, we mention that such a procedure can be carried out in various ways, depending on the choice of the smallness parameter in the expansion. In particular, (I) one can use the interaction in the carrier equation, i.e. Σ compared to G_0^{-1} as small parameter - this leads to an expansion *in terms of free Green's functions* G_0, or (II), one can iterate the equation for Γ, using only *full Green's functions* G in all expressions. These expansions are conveniently performed for all 4 components of the Keldysh matrix at once, taking the physical limit at the end. Also, we will not consider

---

over the product or two Keldysh matrices A and B, $C(\underline{1},\underline{1}') = \int d\underline{\bar{1}} A(\underline{1},\underline{\bar{1}})B(\underline{\bar{1}},\underline{1}')$ follows from elementary matrix multiplication, using the notation (12.9) and the rule (12.6),

$$c^>(11') = \int d\bar{1}\,\{a^>(1\bar{1})\,[b^+(\bar{1}1') + b^<(\bar{1}1')] + [a^+(1\bar{1}) - a^>(1\bar{1})]\,b^>(\bar{1}1')\}\,. \qquad (12.40)$$

Finally, Eq. (12.10) allows to eliminate the "±" functions and also specifies the limits of the time integration, leading to the result which is easily generalized to the $c^<$ component,

$$c^{\gtrless}(11') = \int_{t_0}^{t_1} d\bar{1}\,[a^>(1\bar{1}) - a^<(1\bar{1})]\,b^{\gtrless}(\bar{1}1') - \int_{t_0}^{t_1'} d\bar{1} a^{\gtrless}(1\bar{1})\,[b^>(\bar{1}1') - b^<(\bar{1}1')]\,. \qquad (12.41)$$

In particular, on the time diagonal we have

$$c^{\gtrless}(tt) = \int_{t_0}^{t} d\bar{1}\,[a^>(1\bar{1})b^<(\bar{1}1') - a^<(1\bar{1})b^>(\bar{1}1')]\,. \qquad (12.42)$$

the initial correlation contribution to Γ [KSB].

Zeroth order. Relativistic Vlasov equation. The trivial starting point of all expansions is the complete neglect of correlations, i.e. of Σ and Π. This means, in the relativistic Kadanoff–Baym equations (12.39), all terms on the r.h.s. (collision integrals) are neglected. As in the nonrelativistic case (cf. Sec. 2.6), this yields the mean-field approximation of the Vlasov type (Sec. 4). This approximation describes the collisionless motion of the relativistic carriers in the mean electromagnetic field which includes external fields, as well as induced longitudinal (Coulomb) and transverse fields which evolve according to Maxwell's equations. Obviously, in this approximation the two-time picture does not yield additional information, and one can simply consider the equation for the Wigner function, i.e. take $g^{\gtrless}$ on the time diagonal, $t_1 = t_1'$ (see Sec. 12.4).

I. **Expansion in terms of G_0 and D.**

First order: Neglecting the integral term in Eq. (12.32), and also Σ in G^{-1}, Eq. (12.24) we obtain the relativistically generalized *random phase approximation (RPA)* for the particle and photon selfenergies,

$$\Sigma^{(1)}(\underline{1},\underline{2}) = i\eta_1 \frac{4\pi e^2}{c} \hbar \gamma_\mu \gamma_\nu G_0(\underline{1},\underline{2}) D^{\nu\mu}(\underline{2},\underline{1}), \tag{12.43}$$

$$\Pi^{\mu(1)}_{\nu}(\underline{1},\underline{2}) = -i\eta_1 \frac{4\pi e^2}{c} \hbar \mathrm{Tr}\left\{\gamma^\mu \gamma_\nu G_0(\underline{1},\underline{2}) G_0(\underline{2},\underline{1})\right\}. \tag{12.44}$$

It is convenient to represent these results graphically in terms of Feynman diagrams (leaving out the prefactors)[16]:

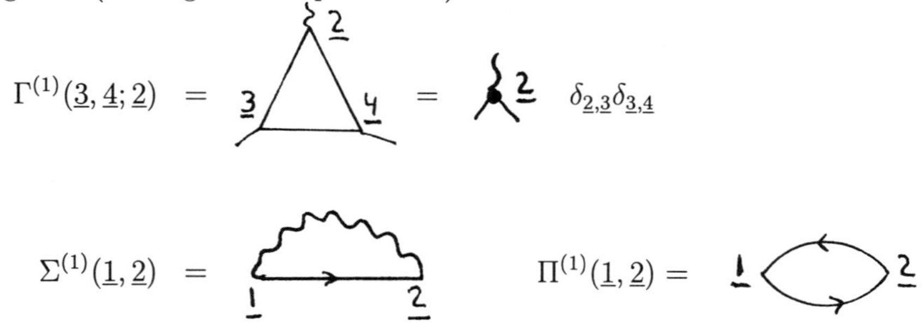

$\Gamma^{(1)}(\underline{3},\underline{4};\underline{2}) =$ [diagram: triangle with vertices 3, 4, 2] $=$ [diagram: vertex 2] $\delta_{\underline{2},\underline{3}}\delta_{\underline{3},\underline{4}}$

$\Sigma^{(1)}(\underline{1},\underline{2}) =$ [diagram: line from 1 to 2 with wavy line] $\qquad \Pi^{(1)}(\underline{1},\underline{2}) =$ [diagram: loop between 1 and 2]

Fig. 12.1. Feynman diagrams for the first order result for the vertex function and the selfenergies. Thin straight lines denote G_0 and wavy lines D.

[16] The rules are the following [BD72]: (1) For each particle-field vertex (thick dot), a factor $\sqrt{-e^2/\hbar c}\gamma_\mu$ is assigned, for each photon line a factor $4\pi\hbar\eta_\nu D^{\mu\nu}$ and for each particle line of species "s" a factor $\hbar G_s$ (or $\hbar G_{0s}$). (2) Over all internal indices (numbers) integration over space and time and summation over the Keldysh branches is implied. (3) For each closed Fermion loop a factor -1 arises. Readers interested in details are referred to Ref. [Mat76].

With the result for $\Sigma^{(1)}$ inserted in Eq. (12.28), we obtain the first order contribution to the particle Green's function,

$$G^{(1)} = \int d\underline{3}d\underline{4}G_0(\underline{13})G_0(\underline{34})D(\underline{43})G_0(\underline{42}). \tag{12.45}$$

Using the canonical representation (12.9), we can now derive the corresponding expressions for the greater/less components of the selfenergy matrices, which are defined on the real time axis,

$$\sigma^{(1)\gtrless}(12) = i\frac{4\pi e^2}{c}\hbar\gamma_\mu\gamma_\nu g_0^{\gtrless}(12)d^{\nu\mu\gtrless}(21), \tag{12.46}$$

$$\pi_\nu^{\mu(1)\gtrless}(12) = -i\frac{4\pi e^2}{c}\hbar\mathrm{Tr}\left\{\gamma^\mu\gamma_\nu g_0^{\gtrless}(12)g_0^{\lessgtr}(21)\right\}. \tag{12.47}$$

We mention that *exchange effects (Fock terms)* which are not contained in the relativistic Vlasov approximation above, appear in the RPA. Indeed, using for D the longitudinal Coulomb field only and taking the free (unscreened) limit, yields σ^{HF} from Eq. (12.46).

Second order: The second order for Γ may be calculated by differentiating $\Sigma^{(1)}$, Eq. (12.43) with respect to A. This yields three terms for $\Gamma^{(2)}$ which are shown in Fig. 12.2. Two arise from $\delta D^{(1)}/\delta A$ (which leads to $\delta\Pi^{(1)}/\delta A$, diagrams (c) and (d)) and one from $\delta G^{(1)}/\delta A$ (diagram (a)).

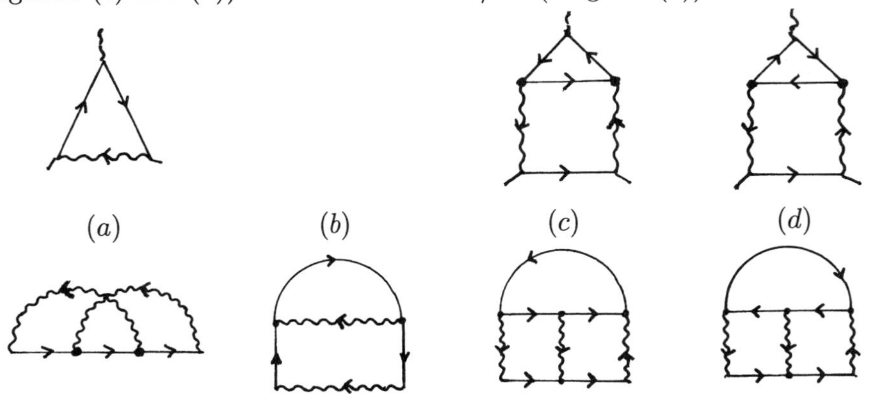

Fig. 12.2. Feynman diagrams for the second order results $\Gamma^{(2)}$ (first line) and $\Sigma^{(2)}$ (second line). The selfenergy diagram (b) arises from $\Sigma^{(1)}$ where $G_0 \to G^{(1)}$.

With the results for $\Gamma^{(2)}$, the selfenergies in second order follow from Eqs. (12.30) and (12.29). Furthermore, additional contributions appear from iterating the first order results (12.44) and (12.43), i.e. by replacing $G_0 \to G^{(1)}$.

This yields one contribution to $\Sigma^{(2)}$ ((b) in Fig. 12.2.) and two to $\Pi^{(2)}$ ((b1) and (b2) in Fig. 12.3).

The second order terms describe a large variety of physical effects. Diagrams (a) are the exchange correction to the RPA. Diagrams (b) are the first of a series leading to one full G function in the RPA expression. Diagram (c) is the first of a ladder-type series where the ladders are screened by full photon lines, and (d) corresponds to higher order polarization effects (electron-hole excitations). So all diagrams known from nonrelativistic quantum statistics can be generated from these expansion. In addition, they include all relativistic many-body effects, such as pair creation/annihilation, Compton scattering or Bremsstrahlung, for more details see Ref. [BD72].

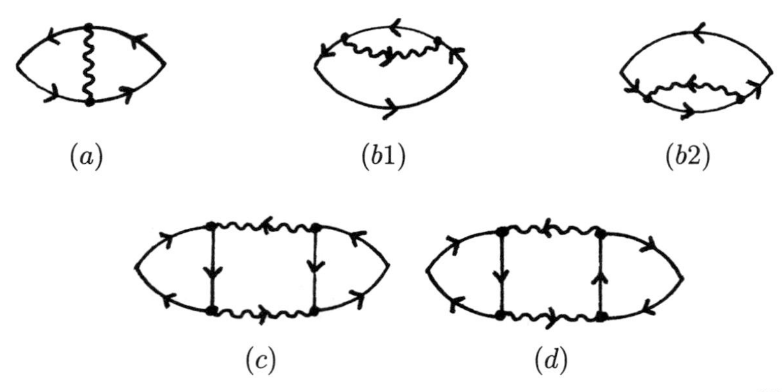

Fig. 12.3. Feynman diagrams for the second order polarization $\Pi^{(2)}$. The two diagrams (b1) and (b2) arise from $\Pi^{(1)}$ where one G_0 is replaced by $G^{(1)}$.

II. **Expansion in terms of G and D.** This expansion differs from the former, because for the particle Green's functions always full G's are taken. As a result, several diagrams do not appear or appear in other orders.[17]

First order: Neglect of the integral term in Eq. (12.32), leads to the relativistically generalized *selfconsistent random phase approximation* for the particle and photon selfenergies. The result is formally the same as above, but differs from Eqs. (12.46), (12.47) by the appearance of the full G's (see first line of Fig. 12.4),

$$\Sigma^{(1)}(\underline{1},\underline{2}) = i\eta_1 \frac{4\pi e^2}{c}\hbar\gamma_\mu\gamma_\nu G(\underline{1},\underline{2})D^{\nu\mu}(\underline{2},\underline{1}), \tag{12.48}$$

$$\Pi^{\mu(1)}_{\nu}(\underline{1},\underline{2}) = -i\eta_1 \frac{4\pi e^2}{c}\hbar\mathrm{Tr}\left\{\gamma^\mu\gamma_\nu G(\underline{1},\underline{2})G(\underline{2},\underline{1})\right\}. \tag{12.49}$$

[17] In the nonrelativistic limit, the two types of expansions have been compared e.g. in [KB89].

Second order: The second order is obtained from inserting $\Gamma^{(1)}$ for Γ in the integral term of Eq. (12.32). The results are summarized in Fig. 12.4. (second line).

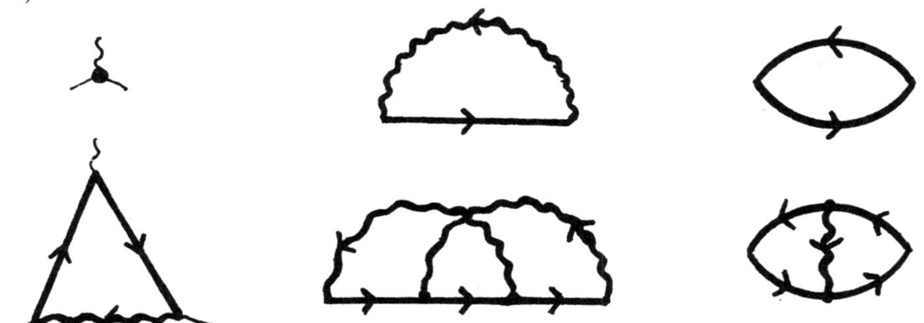

Fig. 12.4. Feynman diagrams for the first and second order (first and second line, respectively) of the vertex function Γ (left) and the selfenergies Σ (center) and Π (right). A thick straight line denotes a *full* G, a wavy line D.

We see that in second order we recover only one of the four selfenergy diagrams of the former expansion. One has to go to higher orders to identify further types of diagrams, what is straightforward. Obviously, the number of diagrams rapidly grows, especially in the first expansion. Therefore, it is important to choose subgroups of diagrams (such as "bubbles" or "ladders") which are appropriate to describe the phenomena of particular interest. In conclusion we mention that the first expansion is useful for systematic perturbation theory in the strength of the interaction, whereas the second one, due to its self-consistency, is very well suited for numerical solutions of the Kadanoff-Baym equations.

Adiabatic approximation for the electromagnetic field. The Kadanoff–Baym equations (12.39) allow for the most general statistical description of the dynamics of particles and photons. However, the question arises, how these equations are related to the more familiar *kinetic equations for charged particles alone*, such as the Landau or Balescu-Lenard equation (Markovian or non-Markovian, see Chs. 7–9) which are successfully used in plasma or solid state theory and other fields. The answer is that in many situations, the typical time and spatial scales of the relaxation of photons is much shorter than that of the particles. Then, we may expect a two stage process: a first stage where the photons reach their quasi stationary distribution while the particles remain close to their initial state, and, second, the stage of almost independent particle kinetics which contains only stationary photon properties. The first stage is then approximately described by the photon Kadanoff–Baym equations (first equation of (12.39)) where in the selfenergies $\pi^>$ and $\pi^<$ one has to replace $g^{\gtrless}(11') \to g^{\gtrless}(t_0t_0)$.[18] On the other hand, the second stage is described by the particle Kadanoff–Baym equations (second equation of (12.39), using the (quasi-)stationary values for the photon Green's functions $d_{st}^{>}$ and $d_{st}^{<}$ in the selfenergies $\sigma^{>}$, $\sigma^{<}$. This approximation is called *adiabatic approximation* because the photons are assumed to adjust to the evolution of the particles immediately, always remaining in the stationary state, i.e. d_{st} is still time and space dependent, but only via the particles, $d_{st}(11') = d_{st}(\{g^{\gtrless}(11')\})$.[19] d_{st} are the stationary solutions of the first equation (12.39) which lead to vanishing of the collision terms on the r.h.s. A sufficient condition for vanishing of the collision integrals of the time diagonal equation ($t_1 = t_1'$) is, according to Eq. (12.42), that the integrand equals zero, i.e.

$$\pi_{\nu}^{\mu>}(1\bar{1})d^{\nu\lambda<}(\bar{1}1) - \pi_{\nu}^{\mu<}(1\bar{1})d^{\nu\lambda>}(\bar{1}1) = 0, \tag{12.50}$$

for all values $\bar{1}, \mu, \lambda$. This corresponds to a state of local detailed balance between absorption and emission of photons, including electromagnetic waves and longitudinal excitations (plasmons). If condition (12.50) is fulfilled, the photon Green's functions are stationary, and the particles evolve according to their own dynamics. Of course, the detailed balance (12.50) is not applicable to all situations. A well known exception are unstable collective modes in a plasma, where plasmon excitation may be strongly enhanced. We mention that there exist various approaches for an approximate treatment of the photon kinetics beyond the adiabatic limit, such as gradient expansions [GH98] or inclusion of resonant plasmon contributions [BD72].

12.3 Nonrelativistic Kadanoff–Baym equations

In this Section, we will derive the nonrelativistic limit of the Green's functions results obtained so far. We will consider a system of fermions (allowing different species) in an electromagnetic field. The extension of these results to multiband systems will be given in section 12.5.

Nonrelativistic limit. Pauli equation. To obtain the familiar nonrelativistic wave equation, it is useful to rewrite the Dirac equation (12.3) as a Schrödinger-type equation by identifying the relativistic particle Hamiltonian H^{REL}, see e.g. [LL80b]. Returning to the three–dimensional vectors $\mathbf{p}$ and $\mathbf{A}$, and generalizing to different species of fermions ($s = 1, \ldots, M$) we have

$$i\hbar\frac{\partial\hat{\Psi}_s}{\partial t} = H_s^{REL}\hat{\Psi}_s, \tag{12.51}$$

[18]If the system is correlated initially, the g's are needed also for other time arguments.

[19]This is analogous, to Bogolyubov's functional hypothesis (see Ch. 5).

$$H_s^{REL} = c\alpha\left(\hat{\mathbf{p}} - \frac{e_s}{c}\hat{\mathbf{A}}\right) + \beta m_s c^2 + e_s\hat{\phi}, \tag{12.52}$$

where the 4×4 matrices α and β are related to the Dirac matrices (see above) by $\alpha = \gamma^0\vec{\gamma}/c$ and $\beta = \gamma^0$. The total relativistic energy (12.52) contains the rest mass contribution which adds a pure phase factor to the wave function. For the nonrelativistic limit, this term is eliminated by means of the substitution $\hat{\Psi}_s \to \exp\{-im_s c^2 t/\hbar\}\hat{\Psi}_s$. Finally, taking the limit $v/c \ll 1$, the 4-spinor $\hat{\Psi}_s$ retains only two independent components, it reduces to a 2-spinor $\hat{\psi}_s$ (for fermions, with the components corresponding to the spin projections $+1/2$ and $-1/2$, respectively), and the Dirac equation (12.51) transforms into the Pauli equation with the Pauli Hamiltonian H^P,[20]

$$i\hbar\frac{\partial\hat{\psi}_s}{\partial t} = \left[\frac{1}{2m_s}\left(\hat{\mathbf{p}} - \frac{e_s}{c}\hat{\mathbf{A}}\right)^2 + e_s\hat{\phi} - \vec{\mu}_s\cdot\nabla\times\hat{\mathbf{A}}\right]\hat{\psi}_s = H_s^P\,\hat{\psi}_s, \tag{12.55}$$

where the last term in brackets is the magnetic dipole energy of the particle due to its magnetic moment $\vec{\mu}_s = \frac{e_s\hbar}{2m_s c}\vec{\sigma}$ in the magnetic field $\nabla\times\hat{\mathbf{A}}$. This term is important for the description of a variety of magnetic field phenomena, including spin-orbit coupling, which, however, are beyond the scope of this book, and we will neglect this term below. Note that the formal simplification of the Pauli equation over the Dirac equation - the field operator is now only a 2-spinor - has the important trade-off that the total momentum appears now squared.

With this result, we now readily derive the nonrelativistic limit of the equations of motion for the Green's functions. To do this, we summarize the modifications arising in this limit (in first order in v/c):

1. The equations for the electromagnetic field operators do not change. As a result, the equation for the photon Green's function remains the same.

[20] We outline the main steps: First, we recall that the 4-spinor $\hat{\Psi}_s$ is composed of two bi-spinors $(\hat{\psi}_s, \hat{\zeta}_s)$, which may be chosen in a way that $\zeta_s/\psi_s \sim v/c$. Then, Eq. (12.51) is equivalent to the system

$$\left(i\hbar\frac{\partial}{\partial t} - e_s\hat{\phi}\right)\hat{\psi}_s = c\vec{\sigma}\left(\hat{\mathbf{p}} - \frac{e_s}{c}\hat{\mathbf{A}}\right)\hat{\zeta}_s, \tag{12.53}$$

$$\left(i\hbar\frac{\partial}{\partial t} - e_s\hat{\phi} + 2m_s c^2\right)\zeta_s = c\vec{\sigma}\left(\hat{\mathbf{p}} - \frac{e_s}{c}\hat{\mathbf{A}}\right)\hat{\psi}_s. \tag{12.54}$$

If the kinetic and potential energy of the particles is small compared to the rest energy, the first two terms on the l.h.s. of Eq. (12.54) may be neglected, and the equation can be solved for ζ. Inserting the result into Eq. (12.53), yields, after some algebra (which can be found e.g. in [LL80b]), the Pauli equation (12.55).

2. The Dirac Hamiltonian (12.52) of the particles is replaced by the Pauli Hamiltonian (12.55).

3. The field operators of the particles become 2-spinors instead of 4-spinors.

4. Consequently, the particle Green's function becomes a 2×2 matrix, instead of a 4×4 matrix. Accordingly, the equation of motion for G is reduced from a 4-vector to a 2-vector form.

5. The Keldysh matrix structure does not change.

With these simple transformations, one can immediately write down the nonrelativistic equations of motion for the photon and particle Green's functions. These equations differ slightly from the equations which are commonly derived in nonrelativistic theories:

i) Usually one does not use 2-spinors for the particles, but scalar field operators. The transition is trivial and will be briefly explained below.

ii) In most cases, isotropy is assumed. Then, with the *Coulomb gauge* (see table 12.1) the electromagnetic field splits conveniently into a longitudinal and a transverse part. Correspondingly, the photon Green's function decomposes into a longitudinal (D^{00}) component ("plasmon" Green's function) and a transverse 3×3 matrix function D^{ij}.

Below, we will use these conventions.

Before we proceed, let us briefly discuss the introduction of *scalar field operators* for the particles. The Dirac and the Pauli equation are written for the spinors $\hat{\Psi}_s$ and $\hat{\psi}_s$, respectively, (i.e. vectors, the components of which represent all possible spin projections). Alternatively, one can use a different representation for the particle field, where the basic quantities are the individual *spinor components*. In nonrelativistic Green's functions theory, usually these spinor components are called "field operators", denoted by $\psi_a(1)$, where the index "a" labels, in addition to the species, also the spin projections (i.e., $a = 1, \dots, 2 \cdot M$). We will follow this convention in the reminder of this Chapter. Furthermore, we use the notation $1 = \mathbf{r}_1, t_1$ (coordinate representation).

The field operators for the particles obey the following commutation rules, which guarantee that the symmetry postulate for bosons/fermions (upper/lower sign) is fulfilled, [21]

$$\psi_a(1)\psi_b(2) \mp \psi_b(2)\psi_a(1) \quad = \quad \psi_a^\dagger(1)\psi_b^\dagger(2) \mp \psi_b^\dagger(2)\psi_a^\dagger(1) = 0,$$

[21] This is one possible definition for the field operators. On the other hand, these commutation rules are derived straightforwardly from the spinor representation of the field based on the requirement for the total energy to be positive defined, e.g. [LL80b].

$$\psi_a(1)\psi_b^\dagger(2) \mp \psi_b^\dagger(2)\psi_a(1) \quad = \quad \delta(1-2)\,\delta_{a,b}. \tag{12.56}$$

Green's functions for field and carriers. Proceeding as explained above, we now rewrite the definitions of the Green's functions (12.12), defining G in terms of scalar field operators. The photon Green's function is split into a (00) and a (ij) component, $(i,j=1,2,3)$[22] where we use the fact that in an isotropic system no cross terms appear.

$$\boxed{\begin{aligned}
\frac{4\pi}{c}D_{ik}(\underline{1}\underline{1}') &= -\frac{\delta A_i(\underline{1})}{\delta j_k^{ext}(\underline{1}')} = -\frac{i}{\hbar}\eta_1'\left\{\langle \hat{A}_i(\underline{1})\hat{A}_k(\underline{1}')\rangle - A_i(\underline{1})A_k(\underline{1}')\right\}\\
\frac{4\pi}{c}D_{00}(\underline{1}\underline{1}') &= -\frac{\delta \phi(\underline{1})}{\delta \rho^{ext}(\underline{1}')} = V(\underline{1}-\underline{1}') - \frac{i}{\hbar}\eta_1'\left\{\langle \hat{\phi}(\underline{1})\hat{\phi}(\underline{1}')\rangle - \phi(\underline{1})\phi(\underline{1}')\right\}\\
G_a(\underline{1}\underline{1}') &= -\frac{i}{\hbar}\eta_1'\,\langle \psi_a(\underline{1})\psi_a^\dagger(\underline{1}')\rangle
\end{aligned}} \tag{12.57}$$

The definition for G is formally unchanged, but one has to recall that now G is defined with scalar field operators, in contrast to the 4-vectors (spinors) which appeared in the relativistic definition (12.12). Also, we recall that statistical averages are denoted by $A_i \equiv \langle \hat{A}_i\rangle$ and $\phi \equiv \langle \hat{\phi}\rangle$ and so on.[23] As before, all Green's functions are defined on the contour $\mathcal{C}$, i.e. they are 2×2 matrices.

Equations of motion for the Green's functions. The differential and integral equations for the particle and photon Green's functions follow directly from their relativistic counterparts, Eqs. (12.21) and (12.14), where the transverse (photon) and longitudinal (plasmon) components of the latter read ($D \equiv D_{00}$ and $\Pi \equiv \Pi_{00}$),

$$\Box^2_{\underline{1}} D_{ik}(\underline{1}\underline{1}') \quad = \quad \delta^T_{ik}(\underline{1}-\underline{1}') + \int d\bar{\underline{1}}\,\Pi_{i\bar{k}}(\underline{1}\bar{\underline{1}})D_{\bar{k}k}(\bar{\underline{1}}\underline{1}'), \tag{12.58}$$

$$\nabla^2_{\underline{1}} D(\underline{1}\underline{1}') \quad = \quad \delta(\underline{1}-\underline{1}') + \int d\bar{\underline{1}}\,\Pi(\underline{1}\bar{\underline{1}})D(\bar{\underline{1}}\underline{1}'). \tag{12.59}$$

In Eqs. (12.58, 12.59) we used the transverse and longitudinal components of the relativistic photon selfenergy (polarization) tensor (12.15), which read explicitly

$$\Pi_{ik}(\underline{1}\underline{1}') = \frac{4\pi}{c}\frac{\delta j_i^T(\underline{1})}{\delta A_k(\underline{1}')}, \qquad \Pi(\underline{1}\underline{1}') = \frac{4\pi}{c}\frac{\delta \rho(\underline{1})}{\delta \phi(\underline{1}')}, \tag{12.60}$$

[22] In the 3-vector notation, there is no distinction between upper and lower indices necessary, so we will use subscripts. Still summation over repeated indices is implied.

[23] The Coulomb potential in the plasmon Green's function arises from the functional differentiation of ρ^{ext} in the Maxwell equation for the static potential, i.e. the "0" component of the 4-vector equation (12.1) in Coulomb gauge, $\Delta\phi = -4\pi(\rho + \rho^{ext})$.

where we defined the transverse projection of the current j^T in terms of a "transverse delta function" δ^T [DuB67],

$$j_i^T(\underline{1}) = \int d^3\bar{r}_1 \delta_{ij}^T(\mathbf{r}_1 - \bar{\mathbf{r}}_1) j_i(\underline{1}), \tag{12.61}$$

$$\delta_{ij}^T(\mathbf{r}_1 - \bar{\mathbf{r}}_1) = \delta_{ij}\delta(\mathbf{r}_1 - \bar{\mathbf{r}}_1) - \frac{1}{4\pi}\nabla_i\nabla_j \frac{1}{|\mathbf{r}_1 - \bar{\mathbf{r}}_1|}. \tag{12.62}$$

The photon and plasmon Green's functions completely determine the behavior of the electromagnetic field and its fluctuations. The longitudinal component yields the selfconsistent (dynamically screened) Coulomb potential, i.e. the screened interaction between the charge carriers. We notice that, in this approximation, longitudinal and transverse photon Green's functions do not couple.

Now we obtain the equation of motion for the particle Green's function as the nonrelativistic limit of the relativistic expression,

$$\left\{ i\hbar \frac{\partial}{\partial t_{\underline{1}}} - \frac{1}{2m_a}\left(\mathbf{p}_{\underline{1}} - \frac{e_a}{c}\mathbf{A}(\underline{1})\right)^2 - e_a\phi(\underline{1}) \right\} G_a(\underline{1}\underline{1}') - \int d\bar{\underline{1}} \Sigma_a(\underline{1}\bar{\underline{1}}) G_a(\bar{\underline{1}}\underline{1}')$$
$$- \int d\underline{2} \left\{ C_{a0}(\underline{1}, \underline{1}', \underline{2}) \frac{4\pi}{c} D(\underline{2}, \underline{1}) + C_{ai}(\underline{1}, \underline{1}', \underline{2}) \frac{4\pi}{c} D_{ij}(\underline{2}, \underline{1}) \right\} = \delta(\underline{1} - \underline{1}'), \tag{12.63}$$

where the particle selfenergy $\Sigma = \Sigma^L + \Sigma^T$ is the nonrelativistic limit of Eq. (12.22), and the A^2 contribution to Σ is neglected,

$$\Sigma_a^L(\underline{1}, \underline{2}) = -ie_a\eta_1'\hbar \int d\underline{3}d\underline{4}\, G_a(\underline{1}, \underline{3}) \Gamma_{a0}(\underline{3}, \underline{2}; \underline{4}) \frac{4\pi}{c} D(\underline{4}, \underline{1}), \tag{12.64}$$

$$\Sigma_a^T(\underline{1}, \underline{2}) = -ie_a\eta_1'\hbar p_{ai}(\underline{1}) \int d\underline{3}d\underline{4}\, G_a(\underline{1}, \underline{3}) \Gamma_{ai}(\underline{3}, \underline{2}; \underline{4}) \frac{4\pi}{c} D_{ik}(\underline{4}, \underline{1}). \tag{12.65}$$

Here, we introduced the longitudinal and transverse components of the vertex function, which are defined in analogy to Eq. (12.31),

$$\Gamma_{a0}(\underline{1}, \underline{1}'; \underline{2}) = -\frac{4\pi}{c} ie_a\hbar \frac{\delta G_a^{-1}(\underline{1}, \underline{1}')}{\delta\phi(\underline{2})}, \quad \Gamma_{ai}(\underline{1}, \underline{1}'; \underline{2}) = -\frac{4\pi}{c} ie_a\hbar \frac{\delta G_a^{-1}(\underline{1}, \underline{1}')}{\delta A_i(\underline{2})},$$

and which again allow us to rewrite the photon selfenergies in a form symmetric with the particle selfenergy

$$\Pi(\underline{1}, \underline{2}) = \eta_1 \sum_b \int d\underline{3}d\underline{4}\, G_b(\underline{1}, \underline{3})\, \Gamma_{b0}(\underline{3}, \underline{4}; \underline{2})\, G_b(\underline{4}, \underline{1}), \tag{12.66}$$

$$\Pi_{ik}(\underline{1}, \underline{2}) = \eta_1 \sum_b p_{bi}(\underline{1}) \int d\underline{3}d\underline{4}\, G_b(\underline{1}, \underline{3})\, \Gamma_{bk}(\underline{3}, \underline{4}; \underline{2})\, G_b(\underline{4}, \underline{1}). \tag{12.67}$$

As before, we excluded the initial correlation term from the definitions (12.66), (12.67) and restore it in the Kadanoff-Baym equations below, see the discussion after Eq. (12.29). Again, the system of equations is closed by integral equations for the vertex functions, cf. Eq. (12.32),

$$\begin{aligned}
\Gamma_{a0}(\underline{1},\underline{1}';\underline{2}) &= i\frac{4\pi e_a^2}{c}\hbar\eta_1\delta(\underline{1}-\underline{1}')\delta(\underline{1}-\underline{2}) + \int d\underline{3}d\underline{4}d\underline{5}d\underline{6}\,\times \qquad (12.68)\\
&\quad \frac{\delta\Sigma_{ab}(\underline{1},\underline{1}')}{\delta G_b(\underline{3},\underline{4})}\left\{G_b(\underline{3},\underline{5})\,\Gamma_{b0}(\underline{5},\underline{6};\underline{2})\,G_b(\underline{6},\underline{4}) + C_{b0}(\underline{3},\underline{4};\underline{2})\right\},\\
\Gamma_{ai}(\underline{1},\underline{1}';\underline{2}) &= i\frac{4\pi e_a^2}{c}\hbar\eta_1\delta(\underline{1}-\underline{1}')\delta(\underline{1}-\underline{2}) + \int d\underline{3}d\underline{4}d\underline{5}d\underline{6}\,\times \qquad (12.69)\\
&\quad \frac{\delta\Sigma_{ab}(\underline{1},\underline{1}')}{\delta G_b(\underline{3},\underline{4})}\left\{G_b(\underline{3},\underline{5})\,\Gamma_{bi}(\underline{5},\underline{6};\underline{2})\,G_b(\underline{6},\underline{4}) + C_{bi}(\underline{3},\underline{4};\underline{2})\right\}.
\end{aligned}$$

This system of equations of motion for the particle, plasmon and photon Green's functions completely describes the dynamics of the complex many-particle system. It is very well suited for systematic derivation of approximations or iterative solution, as was explained in the previous section.

To obtain information on physical quantities and for numerical investigations, it is again useful, to derive from the equations on the Keldysh contour, Eq. (12.63), (12.58) and (12.59) equations for the correlation functions of the electromagnetic field and the particles. Repeating the derivation of Eq. (12.39), we obtain the **nonrelativistic Kadanoff–Baym equations for particles and photons**,

$$\boxed{\begin{aligned}
\Delta_1 d^{\gtrless}(11') &= \int_{t_0}^{t_1} d\bar{1}\,[\pi^{>}(1\bar{1}) - \pi^{<}(1\bar{1})]\,d^{\gtrless}(\bar{1}1')\\
-\int_{t_0}^{t_1'} d\bar{1}\pi^{\gtrless}(1\bar{1})[d^{>}(\bar{1}1') &- d^{<}(\bar{1}1')] + I^{\gtrless}_{\mathrm{IC}\,0}\\
\Box_1 d^{\gtrless}_{ij}(11') &= \int_{t_0}^{t_1} d\bar{1}\,[\pi^{>}_{ik}(1\bar{1}) - \pi^{<}_{ik}(1\bar{1})]\,d^{\gtrless}_{kj}(\bar{1}1')\\
-\int_{t_0}^{t_1'} d\bar{1}\pi^{\gtrless}_{ik}(1\bar{1})[d^{>}_{kj}(\bar{1}1') &- d^{<}_{kj}(\bar{1}1')] + I^{\gtrless}_{\mathrm{IC\,ij}}\\
\left\{i\hbar\frac{\partial}{\partial t_1} - \frac{1}{2m_a}\left(\mathbf{p}_1 - \frac{e_a}{c}\mathbf{A}_1\right)^2 \right. &- \left. e_a\phi_1\right\}g^{\gtrless}_a(11') = I^{\gtrless}_{a\mathrm{IC}}(11') +\\
\int_{t_0}^{t_1} d\bar{1}\,[\sigma^{>}_{ab}(1\bar{1}) - \sigma^{<}_{ab}(1\bar{1})]\,g^{\gtrless}_b(\bar{1}1') &- \int_{t_0}^{t_1'} d\bar{1}\sigma^{\gtrless}_{ab}(1\bar{1})\,[g^{>}_b(\bar{1}1') - g^{<}_b(\bar{1}1')]
\end{aligned}}$$

(12.70)

The coupled evolution of the particle and photon Green's functions is completely determined by the selfenergies and by the initial correlations. Approximations for the selfenergy are straightforwardly derived from the relativistic results of Sec. 12.2. We give some common nonrelativistic approximations below in Sec. 12.4. There we will also consider the initial correlation integral more in detail.

With these results we conclude the general discussion of coupled Kadanoff-Baym equations for carriers and photons. These equations are, in principle, well suited for numerical analysis of correlated many-particle systems, including short-time phenomena such as the build up of correlations. Of course, such an analysis has to start from the simplest case, for which the properties of the equations are known in detail.

12.4 Particle Kadanoff-Baym equations. Properties and approximations. Numerical results

In this Section we consider situations where the particle dynamics is decoupled from that of the photons and discuss the properties of the resulting equations. As discussed in Sec. 12.2, if the photon relaxation is fast compared to that of the particles, on can use the *adiabatic limit.* Then, in the particle equations, the photon and plasmon Green's functions are replaced by their quasistationary limits d^{st} and d^{st}_{ij}, and the particle dynamics is defined by the last equation of (12.70) alone, where the selfenergy σ contains d^{st} and d^{st}_{ij}. To further simplify the analysis, we will consider the one-component case

$$\left(i\hbar\frac{\partial}{\partial t_1} - \frac{p_1^2}{2m}\right) g^{\gtrless}(11') - \int d\bar{r}_1\, \sigma^{\mathrm{HF}}(1\bar{1})g^{\gtrless}(\bar{1}1') = I^{\gtrless}_{\mathrm{IC}}(11') +$$

$$\int_{t_0}^{t} d\bar{t}_1 \left\{\sigma^{>}(1\bar{1}) - \sigma^{<}(1\bar{1})\right\} g^{\gtrless}(\bar{1}1') - \int_{t_0}^{t'} d\bar{t}\sigma^{\gtrless}(1\bar{1}) \left\{g^{>}(\bar{1}1') - g^{<}(\bar{1}1')\right\}, \quad (12.71)$$

where the mean field and exchange (Fock) term are contained in the Hartree-Fock selfenergy σ^{HF}, see below.[24] As before, Eq. (12.71) has to be supplemented by the adjoint equation and initial conditions for $g^{>}$ and $g^{<}$.

Initial correlations. Notice that in Eq. (12.71) t_0 is a *finite* initial time, and I_{IC} contains arbitrary initial binary correlations c_{12}

$$I^{\gtrless}_{\mathrm{IC}}(11') = \int d2 V(r_1 - r_2) \int d\bar{r}_1 d\bar{r}_2 d\tilde{r}_1 d\tilde{r}_2 \times$$

$$g^{+}_{12}(12; \bar{r}_1 t_0 \bar{r}_2 t_0)\, c_{12}(\bar{r}_1 t_0 \bar{r}_2 t_0; \tilde{r}_1 t_0 \tilde{r}_2 t_0)\, g^{-}(\tilde{r}_1 t_0; 1') g^{-}(\tilde{r}_2 t_0; 2). \quad (12.72)$$

[24] Here and below, σ denotes, as usually, the selfenergy *beyond Hartree-Fock.*

This additional collision integral is derived from $I_{\rm IC}$ in Eqs. (12.70). Here it is given for the special case of longitudinal fields, therefore, it is coupled to the Coulomb potential $V = e\phi$ only, and not to D_{ij}. In contrast to Eqs. (12.70) where C is related to particle-field initial correlations, here c_{12} is a *two-particle* initial correlation. Notice that the initial correlations are evolved with one full two-particle propagator g^+_{12} and a free one (i.e. two factorized one-particle propagators) $g^-_1 g^-_2$.[25] The result on the time diagonal is obtained from the difference of the equations for $g^>$ and $g^<$. In this case, also the product $g^-_1 g^-_2$ is replaced by g^-_{12} as it follows from the density operator formalism, see Ch. 8. Details of the derivation are given in [Sem96, KSB].

Approximations for the selfenergy

Approximations for the selfenergy are readily obtained from the more general relativistic expressions of Sec. 12.2 taking into account the above mentioned rules for the nonrelativistic limit.

Collisionless Approximation. This approximation remains the same as in the relativistic case. It follows from taking in the last equation (12.70) only the terms on the l.h.s. Again, with correlations being neglected, this approximation is equivalent to the (one-time) quantum Vlasov equation for the Wigner function, cf. Ch. 4. The particle kinetics is coupled to that of the (mean) electromagnetic field which obeys Maxwell's equations.

Hartree–Fock Selfenergy. This approximation contains, beyond the previous one, exchange corrections in the mean-field selfenergy. If we restrict ourselves to longitudinal fields, the direct and exchange selfenergy are called "Hartree–Fock selfenergy" which is given by

$$\sigma^{HF}(11') = i\hbar V(11')g^<(11') \pm i\hbar\delta(1-1')\int d\bar{1}V(1-\bar{1})g^<(\bar{1}\bar{1}). \quad (12.73)$$

The exchange term (second term) follows from correlation contributions, as discussed in Sec. 12.2.

[25] We mention that the introduction of *initial correlations* in the Kadanoff-Baym equations is essentially more complicated than in the density operator technique (cf. Sec. 6), due to the two-time structure of the former. This question has been discussed by many authors before, e.g. [Fuj65, Hal75, Dan84a], see also the recent text books [HJ96, ZMR96]. Apparently, the most satisfactory treatment is due to Danielewicz [Dan84a], who gives two formulations. One is based on the deformation of the Keldysh contour to imaginary times, but this is applicable only to ground state or equilibrium initial conditions. His second derivation is more general and uses a generalization of Wick's theorem, and the result agrees with ours. An important requirement noted by Danielewicz is that the initial correlation terms must on the time diagonal coincide with the respective density operator expressions.

Dynamically screened second Born approximation (RPA). The standard random phase approximation for the selfenergy follows if in the particle–photon Kadanoff–Baym equations (12.70) the transverse (photon) part can be neglected. Then we are left with coupled equations for the particle and plasmon Green's functions. We obtain from the relativistic Eqs. (12.46) and (12.47), introducing the common notation for the plasmon Green's function $d \to V_s$ (two-time dynamically screened potential),

$$\sigma_a^{\gtrless}(kt_1t_2) = i\hbar \sum_{k'} V_s^{\gtrless}(k'-k, t_1t_2)\, g_a^{\gtrless}(k't_1t_2), \tag{12.74}$$

$$V_s^{\gtrless}(t_1t_2) = \int_{t_0}^{t_1} d\bar{t}_1 \int_{t_0}^{t_2} d\bar{t}_2 V_s^{+}(t_1\bar{t}_1)\, \pi^{\gtrless}(\bar{t}_1\bar{t}_2) V_s^{-}(\bar{t}_2, t_2), \tag{12.75}$$

$$V_s^{\pm}(t_1t_2) = V\delta(t_1-t_2) + V\int_{t_1}^{t_2} d\bar{t}\, \pi^{\pm}(t_1\bar{t})\, V_s^{\pm}(\bar{t}, t_2), \tag{12.76}$$

$$\pi^{\pm}(t_1t_2) = \pm\Theta\Big(\pm(t_1-t_2)\Big)\, \{\pi^{>}(t_1t_2) - \pi^{<}(t_1t_2)\}, \tag{12.77}$$

$$\pi^{\gtrless}(q, t_1t_2) = -i\hbar \sum_{k'\,b} g_b^{\gtrless}(k'+q, t_1t_2)\, g_b^{\lessgtr}(k', t_2t_1). \tag{12.78}$$

Here, V is the bare Coulomb potential and the momentum arguments in V, V_s, and Π have been omitted in Eqs. (12.75 – 12.77)[26]. This system of equations for the RPA selfenergies has been solved numerically in Ref. [MHH$^+$95] using a simple relaxation time approximation for the carrier kinetics as an input for Eq. (12.78). In Ref. [BKK$^+$], these equations were solved using Green's functions from Kadanoff–Baym calculations with static screening (see below) as an input. Fully selfconsistent KB calculations in RPA have not been carried out yet, although they seem to come within reach.

Statically screened second Born approximation. Due to the quite complex structure of the random phase approximation for the carrier selfenergy, for numerical purposes further simplifications are desirable. The simplest approximation is obtained if the dynamical potential is replaced by a quasi-static one, i.e. in Eq. (12.76), $V_s^{\pm}(t_1t_2) \to V(t)\delta(t_1-t_2)$. This yields the quasi-static second Born approximation for the selfenergies

$$\begin{aligned}\sigma_a^{\gtrless}(kt_1t_2) &= i\hbar \sum_{k'} V_s(k'-k, t_1)\, V_s(k'-k, t_2) \\ &\times\ \pi^{\gtrless}(k-k', t_1t_2)\, g_a^{\gtrless}(k't_1t_2),\end{aligned} \tag{12.79}$$

where the polarization function was defined in Eq. (12.78). The expressions for the selfenergy diverge for the bare Coulomb potential V due to its long

[26]Notice that the equations for $V_s^{\pm}$ and π fully agree with our density operator result which was derived in Ch. 9.

range, but with screening effects taken into account, $\sigma^>$ and $\sigma^<$ are finite. So, one has to use a statically screened potential $V_s(t)$ instead of V in Eq. (12.79). The simplest approximation for the screened Coulomb potential is the static limit of the RPA-screened potential $V_s^{\pm}$, which may weakly depend on time (on the macroscopic time only).

Exchange Scattering Selfenergy contributions. The exchange corrections to the selfenergies (12.74) and (12.79) follow straightforwardly from the relativistic result, cf. Sec. 12.2 (Fig. 12.4).

Ladder (T-matrix) approximation. The selfenergy diagrams of ladder structure have been identified in Sec. 12.2, Fig. 12.2. They can be expressed by the T-matrices, (e.g. [KKER86, Dan90])[27]

$$\sigma_a^{\gtrless}(k_1t_1t_2) \;=\; i\hbar\sum_{k_2b}\langle k_1k_2|T_{ab}^{\gtrless}(t_1t_2)|k_2k_1\rangle\; g_b^{\lessgtr}(k_2t_1t_2), \qquad (12.80)$$

which obey the optical theorem

$$\begin{aligned}\langle k_1k_2|T_{ab}^{\gtrless}(t_1t_2)|k_2k_1\rangle &= \int_{t_0}^{t_1}d\bar t_1\int_{t_0}^{t_2}d\bar t_2\;\langle k_1k_2|T_{ab}^{+}(t_1\bar t_1)|k_2k_1\rangle\\ &\times\langle k_1k_2|\mathcal{G}_{ab}^{0\gtrless}(\bar t_1\bar t_2)|k_2k_1\rangle\;\langle k_1k_2|T_{ab}^{-}(\bar t_2,t_2)|k_2k_1\rangle\,, \qquad (12.81)\end{aligned}$$

where the retarded and advanced T-matrices are the solutions of the Lippmann–Schwinger equation

$$T_{ab}^{\pm}(t_1t_2) \;=\; V\delta(t_1-t_2)+V\int_{t_1}^{t_2}d\bar t\,\mathcal{G}_{ab}^{0\pm}(t_1\bar t)\,T_{ab}^{\pm}(\bar t,t_2), \qquad (12.82)$$

and $\mathcal{G}_{ab}^{0}$ denotes the free two-particle propagator,

$$\begin{aligned}\mathcal{G}_{ab}^{0\gtrless}(k_1k_2t_1t_2) &= g_a^{\gtrless}(k_1t_1t_2)\,g_b^{\gtrless}(k_2t_1t_2),\\ \mathcal{G}_{ab}^{0\pm}(k_1k_2t_1t_2) &= \pm\Theta\big(\pm(t_1-t_2)\big)\{\mathcal{G}_{ab}^{0>}(k_1k_2t_1t_2)-\mathcal{G}_{ab}^{0<}(k_1k_2t_1t_2)\}.\end{aligned}$$

Properties of the Kadanoff-Baym equations. The Kadanoff-Baym equations have a number of important properties, which make their analysis and solution very attractive compared to other approaches of nonequilibrium statistics:

i) $g^{\gtrless}$ contain the complete *statistical* (single-particle) information: The equal time limit of $g^<$ defines the Wigner distributions f_a,

$$f_a(kt) \;=\; \pm i\hbar g_a^{<}(ktt). \qquad (12.83)$$

[27] These equations have been derived in Ch. 8 using the density operator formalism.

ii) Due to the dependence on two times, the functions $g^{\gtrless}$ contain also spectral information (information about the correlations), which is determined from the spectral function,

$$\begin{aligned} A_a(kt_1t_2) &= i\hbar[g_a^>(kt_1t_2) - g_a^<(kt_1t_2)] \\ &= i\hbar[g_a^+(kt_1t_2) - g_a^-(kt_1t_2)], \end{aligned} \tag{12.84}$$

and this also allows one to compute *two–particle quantities*, such as the potential energy, from $g^<$:

$$\begin{aligned} \langle V\rangle(t) &= \frac{1}{4}\mathcal{V}\hbar\sum_b\int\frac{d\mathbf{p}}{(2\pi\hbar)^3}\left\{\left(i\hbar\frac{\partial}{\partial t} - i\hbar\frac{\partial}{\partial t'}\right) - \frac{p^2}{m_\mu}\right\} \\ &\times\ (\mp i)\, g_b^<(\mathbf{p},t,t')|_{t=t'}. \end{aligned} \tag{12.85}$$

iii) Conservation of total energy: Total (kinetic + potential) energy of the particles is readily computed within the Kadanoff–Baym approach, cf. Eq. (12.85),

$$\begin{aligned} \langle H\rangle(t) &= \frac{1}{4}\mathcal{V}\hbar\sum_b\int\frac{d\mathbf{p}}{(2\pi\hbar)^3}\left\{\left(i\hbar\frac{\partial}{\partial t} - i\hbar\frac{\partial}{\partial t'}\right) + \frac{p^2}{m_b}\right\} \\ &\times\ (\mp i)\, g_b^<(\mathbf{p},t,t')|_{t=t'}. \end{aligned} \tag{12.86}$$

Kadanoff and Baym have shown that this quantity is conserved for the quite relaxed condition on the two–particle Green's function: $g(12;1'2') = g(21;2'1')$ [BK61, KB89]. This condition if fulfilled for the exact two–particle Green's function, but also for all practically relevant approximations for the selfenergy [28] and is easily generalized to the inhomogeneous case as well as to external fields.

iv) Both $g's$ have the following symmetry properties which follow from Eqs. (12.56):

$$g_a^{\gtrless}(kt_1t_2) = -[g_a^{\gtrless}(kt_2t_1)]^*, \tag{12.87}$$

and they are related to each other on the time diagonal by

$$g_a^>(ktt) = \frac{1}{i\hbar} + g_a^<(ktt). \tag{12.88}$$

[28] Notice the similarity of this condition with the conservation criterion for the density operators, Sec. 2.2.2.

The corresponding symmetry relation for the retarded and advanced functions is

$$g_a^+(kt_1t_2) = [g_a^-(kt_2t_1)]^*. \tag{12.89}$$

An important consequence of Eq. (12.97) is the property of the spectral function $A_a(kt_1t_1) = 1$.

Numerical results

These properties, together with their fully selfconsistent structure make the Kadanoff-Baym equations very attractive for numerical investigations[29] (we used the Born approximation, for details of the numerical solution, see Appendix F). In particular, they allow for a straightforward treatment of correlations. We illustrate this on the example of an uncorrelated electron gas, which relaxes beginning from an initial Wigner distribution, in Figs. 12.5 and 12.6 below.

Correlated equilibrium distribution. In Fig. 12.5., we started with a Fermi distribution. Obviously, a conventional Boltzmann-type kinetic equation would show no relaxation at all. But the Kadanoff-Baym equations yield a relaxation towards a *correlated equilibrium distribution.*

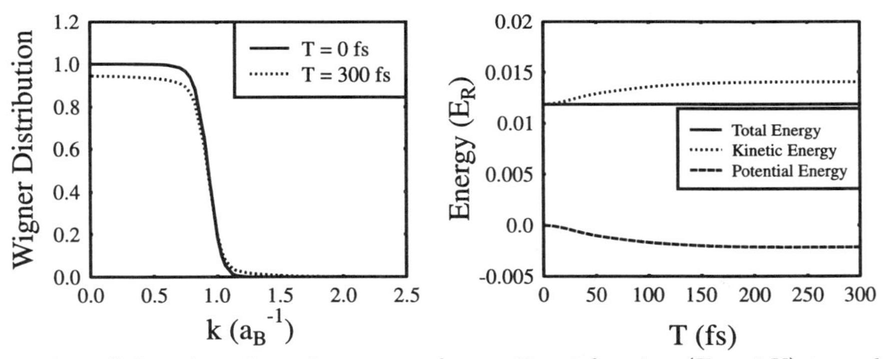

Fig. 12.5: Relaxation of an electron gas from a Fermi function ($T = 3K$) towards a correlated equilibrium distribution (**left figure**, dotted line). Parameters are for electrons in a bulk semiconductor (GaAs). The **right figure** shows the buildup of correlation energy and the corresponding kinetic energy increase [BKSK].

An example of Kadanoff-Baym relaxation of a nonequilibrium initial distribution was shown in Sec. 7.3.1, cf. Fig. 7.1., where we compared different approximations of quantum kinetic theory.

---

[29] First numerical solutions were reported in nuclear matter by Danielewicz [Dan84b] which were truly amazing for that time, Greiner et al. [GWR94] and Köhler, e.g. [Köh95]. First semiconductor applications are due to Schäfer [Sch96], see below.

Kinetic energy relaxation. The right part of Fig. 12.5. demonstrates that indeed total (kinetic + potential) energy is conserved during the relaxation. However, in contrast to Markovian kinetic equations, kinetic energy changes, due to the buildup of correlations in the system. This behavior is in full agreement with the results from non-Markovian kinetic equations for the Wigner function, which were discussed in detail in Ch. 7. Interestingly, there are situations where the relaxation of kinetic (and potential) energy is more complex. Especially at low temperatures, one observes, at early times, a non-monotonic relaxation [SBK]. This effect is shown in Fig. 12.6[30]. Notice that the correlation energy starts with zero only in the case of zero initial correlations. Furthermore, for a correlated initial state, there may be the opposite trend in the time dependence as was shown in Sec. 7.3.2, cf. Fig. 7.3.

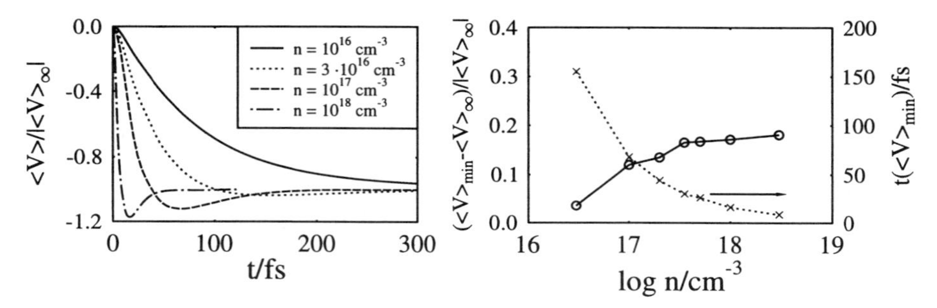

Fig. 12.6 left: Potential energy relaxation normalized to the stationary value, for different densities at $T = 10K$ (same parameters as in Fig. 12.5.). **right fig.**: Depth of the minimum normalized to the stationary value of $\langle V \rangle$ (full line) and time when the minimum is reached (dotted line, right scale), [SBK].

Spectral function. We can study correlation effects not only through the evolution of correlation energy, but more in detail from the spectral function A which contains the complete dynamical information. To understand its behavior, we first consider the *case of free particles.* The ideal spectral function is readily obtained from the Kadanoff-Baym equations for a free particle (no selfenergy). Denoting $\tau = t - t'$ and $T = (t + t')/2$, one finds with the conditions on the diagonal following from Eqs. and (12.83) and (12.88)

$$g^<_{\text{free}}(p, T, \tau) \;=\; \mp \frac{i}{\hbar} f(p) e^{-\frac{i}{\hbar}E(p)\tau}, \tag{12.90}$$

$$g^>_{\text{free}}(p, T, \tau) \;=\; -\frac{i}{\hbar}[1 \pm f(p)] e^{-\frac{i}{\hbar}E(p)\tau}; \tag{12.91}$$

[30]Molecular Dynamics simulations indicate that for strongly coupled systems the relaxation may even be oscillatory, see e.g. [ZTR]. This has not been observed in Kadanoff-Baym calculations due to the strong selfconsistent damping (selfenergy).

$$A_{\text{free}}(p,T,\tau) \quad = \quad e^{-\frac{i}{\hbar}E(p)\tau}, \tag{12.92}$$

where f is time-independent. This means, the spectral function of a free function represents a harmonic oscillation. On the other hand, *correlation effects* disturb this form by modifying the frequency (energy shift) and introducing damping of the wave. Thus, medium effects lead to a finite "lifetime" of the particles, for illustration, see Fig. 12.12. below. In Fig. 12.7., the calculated nonequilibrium spectral function corresponding to the conditions of Fig. 12.5. is shown and compared to the free case.

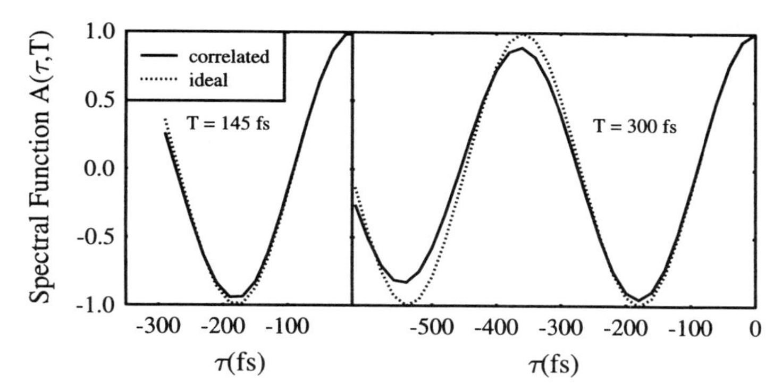

Fig. 12.7: Real part of the nonequilibrium spectral function for a weakly correlated electron gas compared to an ideal one (dotted lines). The spectral function is drawn across the diagonal in the time plane, τ is the distance from the diagonal. In the numerical solution, with increasing T, also the length of the τ-interval grows, [BKSK].

12.5 Interband Kadanoff-Baym equations. Numerical results

We now consider the extension of the Kadanoff–Baym equations to multi-band (solid state or nuclear matter) systems where we will be primarily interested in the response of these systems to an external optical excitation, such as a short-pulse laser. This leads to the Bloch representation of the kinetic equations which has been discussed in detail in Chapter 11. We consider electrons in a solid which may be on different energy levels/bands μ. The definition of the Green's functions (12.57) on the Keldysh contour is naturally generalized according to

$$G_{\mu_1\mu_2}(\underline{1}\underline{1}') = -\frac{i}{\hbar}\eta_1' \langle \psi_{\mu_1}(\underline{1})\psi^\dagger_{\mu_2}(\underline{1}')\rangle \tag{12.93}$$

and analogously for the physical ("$\gtrless$" and "$\pm$") components and for the selfenergies. The discussion of the properties of the Kadanoff–Baym equations, Sec. 12.4, is straightforwardly extended to the multi-band case. The equal time limit of $g^<$ defines the Wigner distributions f and the transition probabilities (interband polarizations) P,

$$\begin{aligned} f_\mu(kt) &= \pm i\hbar g^<_{\mu\mu}(ktt), \\ P_{\mu_1\mu_2}(kt) &= \pm i\hbar g^<_{\mu_1\mu_2}(ktt), \quad \mu_1 \neq \mu_2, \end{aligned} \tag{12.94}$$

which in the case of a two-band system with particles ("e" in the conduction band "c") and holes ("h" in the valence band "v") are usually defined as

$$\begin{aligned} f_e(kt) &= \pm i\hbar g^<_{cc}(ktt); \quad f_h(kt) = \mp i\hbar g^<_{vv}(-ktt), \\ P(kt) &= \pm i\hbar g^<_{cv}(ktt). \end{aligned} \tag{12.95}$$

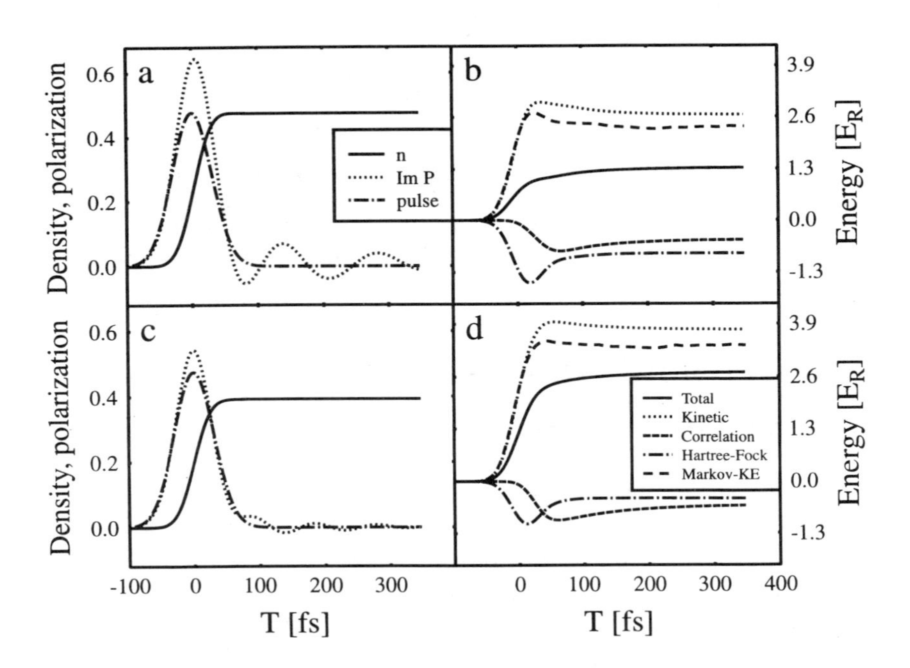

Fig. 12.8: **a, c**: Time evolution of density n and interband polarization P. **b, d**: Various energy contributions (summed over conduction and valence band). Interband Kadanoff-Baym calculations are performed for a GaAs single quantum well excited with a $50fs$ pulse with an excess laser energy of $10meV$ (a,b) and $25meV$ (c,d) above E_G, respectively. n and P are in units of a_B^{-2} (3D exciton Bohr radius), $E_R = 4.2meV$, from Ref. [KBBK98].

The dynamical information is contained in the spectral function matrix,

$$\begin{aligned} A_{\mu_1\mu_2}(kt_1t_2) &= i\hbar[g^>_{\mu_1\mu_2}(kt_1t_2) - g^<_{\mu_1\mu_2}(kt_1t_2)] \\ &= i\hbar[g^+_{\mu_1\mu_2}(kt_1t_2) - g^-_{\mu_1\mu_2}(kt_1t_2)], \end{aligned} \tag{12.96}$$

for which numerical results are shown below in Fig. 12.14. The symmetry properties of the $g's$ are in the multi-band case:

$$g^{\gtrless}_{\mu_1\mu_2}(kt_1t_2) = -[g^{\gtrless}_{\mu_2\mu_1}(kt_2t_1)]^\dagger; \quad g^+_{\mu_1\mu_2}(kt_1t_2) = [g^-_{\mu_2\mu_1}(kt_2t_1)]^\dagger, \tag{12.97}$$

and on the time diagonal we have the relation

$$g^>_{\mu_1\mu_2}(ktt) = \frac{1}{i\hbar} + g^<_{\mu_1\mu_2}(ktt). \tag{12.98}$$

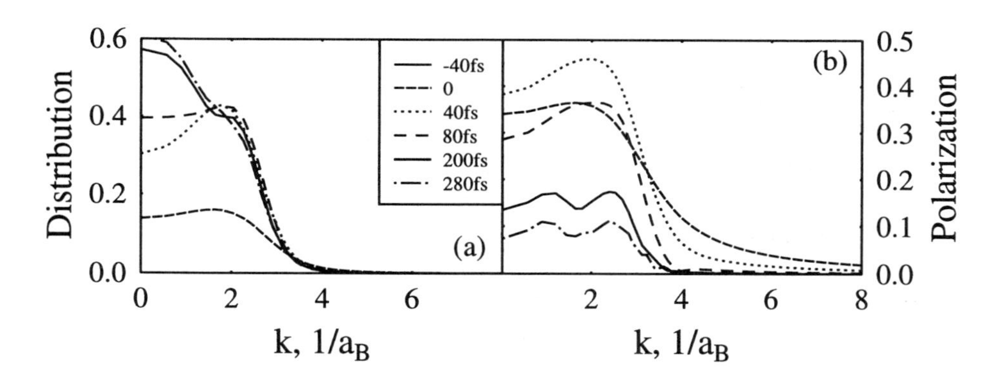

Fig. 12.9: **a**: Time evolution of the electron distribution function and, **b** of the interband polarization, [KBBK].

The equations of motion for the correlation functions in a multi-band system follow straightforwardly from Eq. (12.71), with two modifications: First, we have to account for the influence of the lattice on the electron behavior and second, we restore the transverse electromagnetic field. The first point is treated by including a periodic lattice potential in the particle Hamiltonian, cf. Ch. 11. Then, instead of a basis of free particle states, the appropriate basis is the Bloch basis which leads to generalized eigenvalues of the one-particle Hamiltonian, $p_\mu^2/2m_\mu \rightarrow \epsilon_\mu(p)$. The electromagnetic field is treated in the dipole approximation (see Ch. 11). Furthermore, we may neglect initial correlations if we consider, as an initial state of the system, the un-excited material

($g_{12} = 0$), otherwise, the corresponding term in Eq. (12.71) has to be included. In the spatially homogeneous case, the equations of motion are

$$\begin{aligned} \{i\hbar\frac{\partial}{\partial t_1} - \epsilon_{\mu_1}\} g^{\gtrless}_{\mu_1\mu_2}(t_1t_2) - \sum_{\bar\mu} \hbar\Omega_{\mu_1\bar\mu}(t_1) g^{\gtrless}_{\bar\mu\mu_2}(t_1t_2) &= I^{\gtrless}_{\mu_1\mu_2}(t_1t_2), \quad (12.99) \\ \{-i\hbar\frac{\partial}{\partial t_2} - \epsilon_{\mu_2}\} g^{\gtrless}_{\mu_1\mu_2}(t_1t_2) - \sum_{\bar\mu} g^{\gtrless}_{\mu_1\bar\mu}(t_1t_2) \hbar\Omega_{\bar\mu\mu_2}(t_2) &= -I^{\gtrless}_{\mu_2\mu_1}(t_2t_1), \end{aligned}$$

where ϵ_μ is the one-particle energy for band (component) "μ", and the momentum index "k" in the $g's$ and I's has been suppressed. In Eq. (12.100) we introduced an effective electric field term (renormalized Rabi energy)

$$\hbar\Omega_{\mu_1\mu_2}(t) = -\mathbf{d}_{\mu_1\mu_2}\mathcal{E}(t)(1 - \delta_{\mu_1\mu_2}) + i\hbar\sum_{k'} g^{<}_{\mu_1\mu_2}(k'tt)V_{\mu_1\mu_2}(k - k'), \quad (12.100)$$

where $V_{\mu_1\mu_2}$ is the bare interaction potential between particles in band μ_1 and μ_2, respectively. Notice that $\mathcal{E}$ carries some memory of the original coupled equations for carriers and photons: it is the total field (external + induced), which obeys Maxwell's equations. The conventional dipole energy is modified by mean field (Hartree-Fock) effects which, due to homogeneity, contribute only to the equations for the off-diagonal Green's functions ($\mu_1 \neq \mu_2$). Under the action of the laser, electrons are excited from the valence band into the conduction band what creates nonequilibrium electron and hole distributions in the two bands. We have discussed these processes in detail in Ch. 11, see e.g. the illustrative Fig. 11. 1.

Here, we concentrate on the peculiarities of the Green's functions treatment which are connected with the treatment of correlations which arise from the collision integrals in the two-time Bloch equations,

$$\begin{aligned} I^{\gtrless}_{\mu_1\mu_2}(t_1t_2) &= \sum_{\bar\mu}\int_{t_0}^{t_1} d\bar t\, [\sigma^{>}_{\mu_1\bar\mu}(t_1\bar t) - \sigma^{<}_{\mu_1\bar\mu}(t_1\bar t)]\, g^{\gtrless}_{\mu_1\bar\mu}(\bar t t_2) \\ &- \sum_{\bar\mu}\int_{t_0}^{t_2} d\bar t\, \sigma^{\gtrless}_{\mu_1\bar\mu}(t_1\bar t)\, [g^{>}_{\mu_1\bar\mu}(\bar t t_2) - g^{<}_{\mu_1\bar\mu}(\bar t t_2)]. \quad (12.101) \end{aligned}$$

If no external field $\mathcal{E}$ is applied, no interband polarization is generated and the equations reduce to coupled equations for the diagonal Green's functions, $\mu_1 = \mu_2$.

The approximations for the selfenergy discussed in Sec. 12.4 are naturally generalized to multi-band systems. As an example, we give the result for the **Statically screened second Born approximation:**

$$\begin{aligned} \sigma^{\gtrless}_{\mu_1\mu_2}(kt_1t_2) &= i\hbar\sum_{k'} V_{s\,\mu_1\mu_2}(k' - k, t_1)\, V_{s\,\mu_1\mu_2}(k' - k, t_2) \\ &\times\ \pi^{\gtrless}(k - k', t_1t_2)\, g^{\gtrless}_{\mu_1\mu_2}(k't_1t_2), \quad (12.102) \end{aligned}$$

where the polarization functions are now defined as

$$\pi^{\gtrless}(q, t_1 t_2) = -i\hbar \sum_{k'\lambda\mu} g_{\mu\lambda}^{\gtrless}(k' + q, t_1 t_2)\, g_{\lambda\mu}^{\lessgtr}(k', t_2 t_1), \tag{12.103}$$

and $V_s(t)$ is the statically screened Coulomb potential, which is (as above) the static limit of the RPA-screened potential.

We mention that the interband Kadanoff-Baym equations have been intensively analyzed in recent years in semiconductor optics, e.g. [ST86, Hen88, HH88, TH93, BK95], for a text book discussion see [HJ96].

Numerical results for ultrafast relaxation in fs-pulse excited semiconductors

Ultrafast relaxation of optically excited electron-hole plasmas in semiconductors has attracted great interest in recent years. With modern fs-lasers the temporal resolution of experiments has improved greatly, e.g. [Be95, CAH+96], see also [Sha89]. This was accompanied by numerical investigations of carrier-phonon scattering [Zim92, SKM94, SKM95, Be95] and carrier-carrier scattering [SBK92, BSP+92, Col93, RHK94], for a more complete overview see [HJ96]. Only recently *numerical solutions of the interband Kadanoff-Baym equations* became possible. First remarkable results for four-wave mixing experiments were performed by Schäfer, [Sch96] where also earlier references are given. Recently, further investigations have been published which were concerned with memory effects [BKS+96], momentum orientation relaxation [BKB97] and correlation effects [KBBK98]. For numerical details, see Appendix F.

Some typical results are summarized in Figs. 12.8-12.11. Fig. 12.8. shows the time evolution of macroscopic quantities during and after excitation with a $50fs$ laser pulse for two different photon energies. Fig. 12.9. shows the corresponding buildup and relaxation of the electron distribution in the conduction band and the formation and decay of the interband polarization (probability density of valence-conduction-band transitions). One sees clearly that, at early times, the excitation is energetically very broad (cf. also Fig. 11.1). This underlines our discussion of Ch. 1 that, due to Heisenberg's principle, sharp "initial" distributions cannot be generated by short excitations.

The polarization is an important quantity which can directly be compared to experiments. For example, Fig. 12.10 shows the Fourier transform, $\mathrm{Im}P(\omega)$, which is directly related to the absorption of the light by the semiconductor, see e.g. [HK93, PKM93]. Another measurable quantity is the electric field transmitted or reflected by the material. Fig. 12.11. shows the field reflected from a quantum well. Interestingly, quantum kinetic calculations allow to compute the full complex field, including its phase, which can be measured by modern optical techniques.[31] We have to mention that in these figures, correlation effects are weak, because of the rather high particle kinetic energy ("temperature"), cf. Ch. 1, Fig. 1.1.

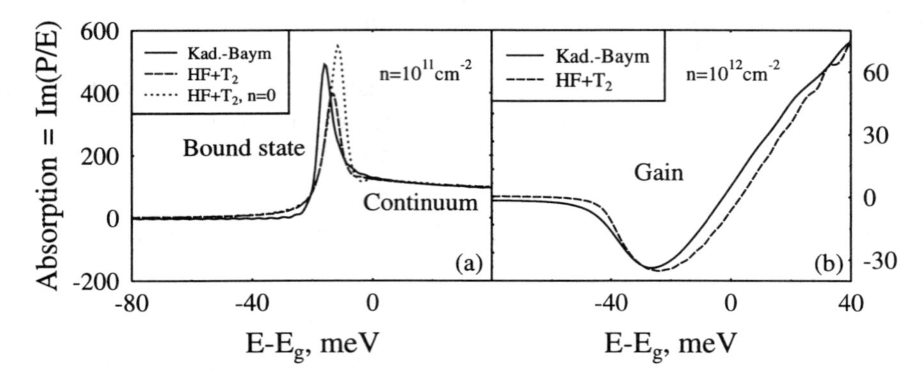

Fig. 12.10: Spectra of a 2D quantum well for two different electron densities, showing pure absorption **(a)** and gain (i.e. electron transitions back to the valence band, **(b)**. The Left figure clearly shows the exciton peak and the continuum of scattering states which, due to damping, are not separated here. Kadanoff-Baym results are compared to a Hartree-Fock calculation with a phenomenological damping rate (corresponding to a dephasing time $T_2 = 200fs$), and to the linear case (zero density). [KBBK].

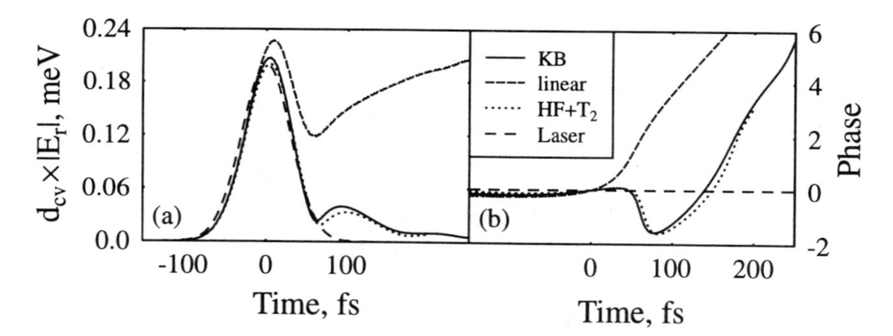

Fig. 12.11: Magnitude **(a)** and phase **(b)** of the electric field reflected from a quantum well at short times. Same approximations as in Fig. 12.10. The magnitude of the laser field has been reduced by a factor of 25. $n = 0.34 \times 10^{12} cm^{-2}$, d_{cv} is the dipole moment. Correlation effects are increasing with time [KBBK].

[31] In particular by the so-called FROG (frequency resolved optical gating) method developed in [TK93]. Interesting first results for optical diagnostics of semiconductors were obtained by Stolz and co-workers, see e.g. [NBK+98].

12.6 Kinetic equations for one–time functions. Comparison to the density operators

To derive kinetic equations for the single–time functions (Wigner distributions f_μ and interband polarization $P_{\mu_1\mu_2}$), we consider the Kadanoff–Baym equations (12.100) in the limit of equal times $t_1 = t_2$.[32] As a result, on the l.h.s. of Eqs. (12.100), all $g's$ are replaced by Wigner functions and interband polarizations, respectively. On the r.h.s., the collision integrals simplify to ($t_1 = t_2 = T$):

$$I^{\gtrless}_{\mu_1\mu_2}(T) = \sum_{\bar{\mu}} \int_{t_0}^{T} d\bar{t}\, [\sigma^{>}_{\mu_1\bar{\mu}}(T\bar{t})\, g^{<}_{\bar{\mu}\mu_2}(\bar{t}T) - \sigma^{<}_{\mu_1\bar{\mu}}(T\bar{t})\, g^{>}_{\bar{\mu}\mu_2}(\bar{t}T)], \quad (12.104)$$

where, obviously, the $g's$ cannot be eliminated (in terms of f and P), because also time-off-diagonal values are needed. To obtain closed equations for f and P, it is necessary to express ("reconstruct") the two–time quantities $g^>$ and $g^<$ in terms of their values on the time diagonal (i.e. by single–time functions). This can be done exactly (see below), however, the result is too complicated for practical use. Therefore, there have been proposed various approximation schemes to solve the reconstruction problem.

Equilibrium. Kadanoff-Baym ansatz. In the stationary case, all system properties are independent of the macroscopic time $T = (t_1 + t_2)/2$, and only the dependence on the difference time $\tau = t_1 - t_2$ remains, $g(t_1t_2) \longrightarrow g(\tau)$, see Fig. 12.6. In this case, it is convenient to Fourier transform with respect to τ and to use frequency dependent functions. The Kadanoff–Baym ansatz is defined as [KB89]

$$g^{\gtrless}(k\omega) = f^{\gtrless}(k\omega)\, A(k\omega), \quad (12.105)$$

where we denoted $f^< = f$ and $f^> = 1 \pm f$. The Kadanoff–Baym ansatz (12.105) allows to eliminate the $g's$ from the collision integrals (12.104) in lowest order. This ansatz has also been applied to nonequilibrium situations, where there remains a $T-$dependence of all quantities. Of course, this is correct only if the dependence on T is much weaker than that on τ, which justifies a separation of "macroscopic" (T) and "microscopic" (τ) scales. However, this ansatz fails to describe ultrafast processes, since it completely neglects retardation effects. This leads to the necessity of generalizations. For extensions to relativistic systems, see e.g. [GWR94].

---

[32]It follows by subtracting the equations for $g^>$ and $g^<$.

Exact reconstruction in nonequilibrium. From the Kadanoff-Baym equations one can derive the following expression for $g^>$ and $g^<$ in terms of single–time functions, which, for the one–band case, reads [LvV86],

$$g^{\gtrless}(t_1t_2) = \begin{cases} -i\hbar\, g^+(t_1t_2)\, f^{\gtrless}(t_2) + \int_{t_2}^{t_1} d\bar{t}_1 \int_{-\infty}^{t_2} d\bar{t}_2 \times \\ g^+(t_1\bar{t}_1)\left\{\sigma^+(\bar{t}_1\bar{t}_2)\, g^{\gtrless}(\bar{t}_2t_2) + \sigma^{\gtrless}(\bar{t}_1\bar{t}_2)\, g^-(\bar{t}_2t_2)\right\}, & t_1 \geq t_2, \\ \\ i\hbar\, f^{\gtrless}(t_1)\, g^-(t_1t_2) + \int_{t_1}^{t_2} d\bar{t}_1 \int_{-\infty}^{t_1} d\bar{t}_2 \times \\ \left\{g^{\gtrless}(t_1\bar{t}_1)\, \sigma^-(\bar{t}_1\bar{t}_2) + g^{\gtrless}(t_1\bar{t}_1)\, \sigma^-(\bar{t}_1\bar{t}_2)\right\} g^-(\bar{t}_2t_2), & t_1 < t_2. \end{cases}$$

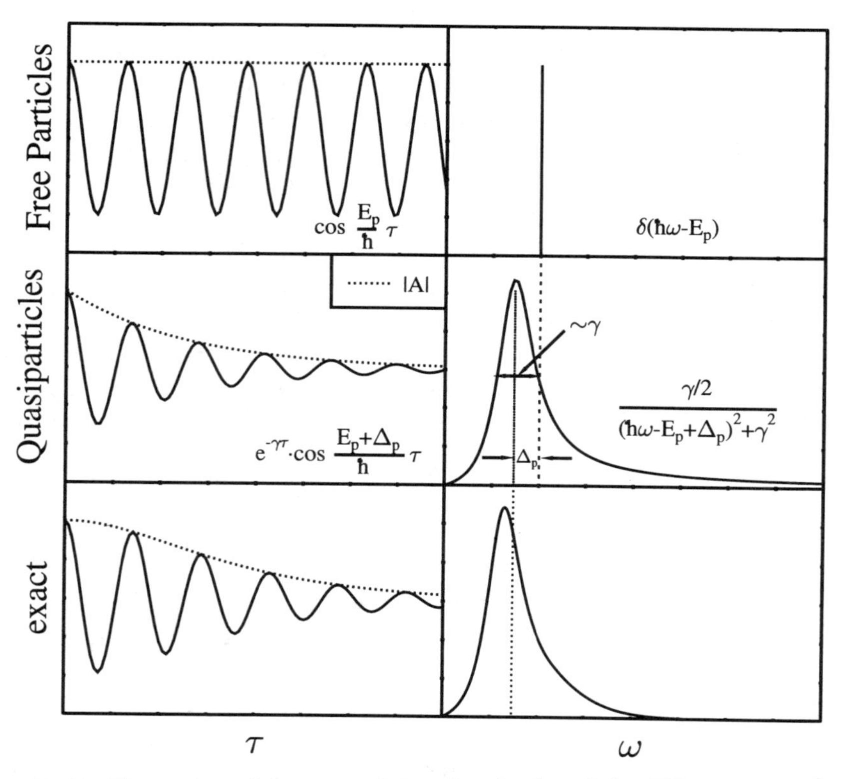

Fig. 12.12: Illustration of the spectral function (*real part*) for different approximations. **Left**: vs. microscopic time, **right**: vs. frequency (energy), (courtesy of D. Semkat).

This is a general result which is fully equivalent to the Kadanoff–Baym equations. The first terms on the r.h.s. contain the Wigner functions, but also the two–time retarded and advanced Green's functions. However, in the integral terms the functions $g^>$ and $g^<$ appear again. So far no explicit solution for the two–time functions is known, so Eq. (12.106) below can be used to construct approximations, e.g. for an iteration procedure. In most treatments, only the first term on the r.h.s., i.e. Eq. (12.106) is used. For further discussions, see [LvV86], extensions to the case of external fields and a gauge–invariant form are summarized in [HJ96].

Generalized Kadanoff-Baym ansatz (GKBA). Taking only the first term in the exact reconstruction formula above leads to the generalized Kadanoff–Baym ansatz of Lipavský, Špička and Velický [LvV86],

$$g^{\gtrless}(t_1t_2) = -i\hbar \left\{ g^+(t_1t_2)\, f^{\gtrless}(t_2) - f^{\gtrless}(t_1)\, g^-(t_1t_2) \right\}. \tag{12.106}$$

This expression is, of course, exact on the time diagonal, cf. Eq. (12.95). For $t_1 \geq t_2$ only the first term contributes and for $t_1 < t_2$, the second. This is an important generalization of the ansatz of Kadanoff and Baym (12.105) to nonequilibrium. In particular, it accounts for retardation effects and it has the correct causal structure, containing the distributions only at earlier time. The equilibrium ansatz (12.105) is recovered from Eq. (12.106) if the time retardation is neglected by approximating the distributions by their values at the macroscopic (middle) time $T = (t_1 + t_2)/2$. The generalization to multi–component systems is straightforward,

$$g_\mu^{\gtrless}(t_1t_2) = -i\hbar \left\{ g_\mu^+(t_1t_2)\, f_\mu^{\gtrless}(t_2) - f_\mu^{\gtrless}(t_1)\, g_\mu^-(t_1t_2) \right\}, \tag{12.107}$$

which is also exact on the time diagonal. Furthermore, the generalization to the multi–band case has been proposed in Ref. [SSH+94]

$$g_{\mu_1\mu_2}^{\gtrless}(t_1t_2) = -i\hbar \sum_{\bar\mu} \left\{ g_{\mu_1\bar\mu}^+(t_1t_2)\, f_{\bar\mu\mu_2}^{\gtrless}(t_2) - f_{\mu_1\bar\mu}^{\gtrless}(t_1)\, g_{\bar\mu\mu_2}^-(t_1t_2) \right\}. \tag{12.108}$$

The problem of spectral function in the GKBA. Relation to the Density operator approach. We have discussed in Chs. 6-10 how to derive generalized quantum kinetic equations for the Wigner function from the BBGKY-hierarchy. These results have to agree with the equations obtained from nonequilibrium Green's functions, what provides an important consistency test for both theories.

Recently, the generalized Kadanoff-Baym ansatz has been extensively used by several authors to derive non-Markovian kinetic equations for the Wigner function, e.g. [Kuz91, HE92, MR94, BKKS96] and non-Markovian generalized Bloch equations, e.g. [TH93, BK95]. The GKBA is very efficient to derive these equations in general form. However, as soon as one approaches any application, it immediately causes headaches. The reason is that, although Eqs. (12.106-12.108) do express $g^>$ and $g^<$ in terms of single-time functions, they still involve other two-time functions, the retarded and advance Green's functions (or, equivalently, the spectral function). Unfortunately, this theory does not yet provide a practicable approach for a consistent determination of the spectral function to be used in the GKBA. The simplest choice was to use free spectral functions. But this immediately lead to non-Markovian kinetic equations with *infinite memory depth*, the problems of which were discussed in Chs. 6 and 7. The first improvement was to add some phenomenological small damping, and indeed results improved greatly. Attempts to introduce damped spectral functions systematically, lead directly to the *quasiparticle approximation* or *local approximation*, where the free oscillation is modified by a correlation induced frequency shift Δ and damping γ, for illustration, see Fig. 12.12. We have computed these quantities selfconsistently as a function of momentum and time in Ch. 7, see Fig. 7.5. However, no matter how accurate and selfconsistent these calculations of *Lorentzian spectral functions* are, they do not conserve total energy, nor do they yield the correct asymptotic distribution [BKS+96, HB96]. The reason is their nonzero slope at $\tau = 0$ which translates into a long high-frequency tail which is not observed in the Kadanoff-Baym calculations (second and third lines of Fig. 12.12, respectively).

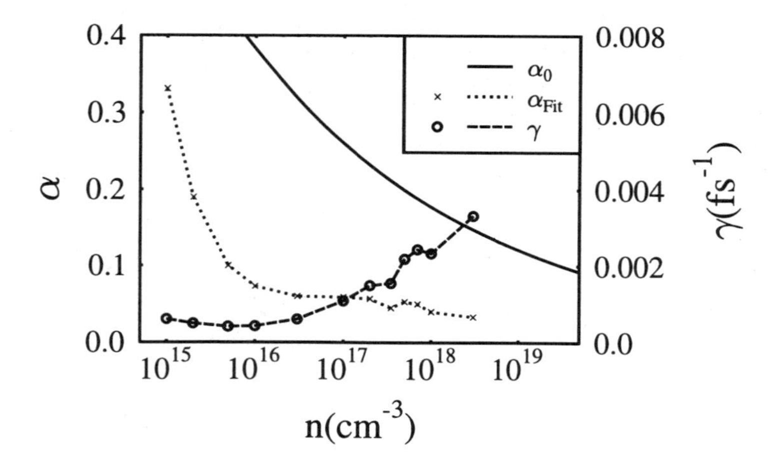

Fig. 12.13: Parameter γ in the $1/\cosh^{\alpha}(\omega_{pl}\tau)$ approximation for the spectral function fitted to Kadanoff-Baym carrier-carrier scattering calculations. $\alpha = \gamma/\omega_{pl}$ and α_0 is its analytical high-density limit [BSHB]. The wiggles indicate that more complex trial functions may be needed.

On the other hand, we have seen in Chs. 7-9, that the GKBA follows straightforwardly from the density operator formalism. Moreover, we have seen that the original results for the spectral functions are *nonlocal*. This suggests to use the GKBA but with improved *non-Lorentzian* spectral functions [HB96, BHG98, BSHB]. Haug and Banyai proposed a spectral function which interpolates between a Lorentzian at large τ and zero derivative at $\tau = 0$, $A(\tau) \sim 1/\cosh^{\alpha}(\omega_{pl}\tau)$ [HB96]. One can now use spectral functions from Kadanoff-Baym calculations and try to determine γ as a fit parameter to the numerical result. Fig. 12.13 shows results for the density dependence of γ in a bulk semiconductor (GaAs) at a temperature of $10K$.

The question of the appropriate spectral functions in the multi-band case is even more complex. Here, one can use Kadanoff-Baym calculations also to test various approximation as shown in Fig. 12.14. [KBBK98].

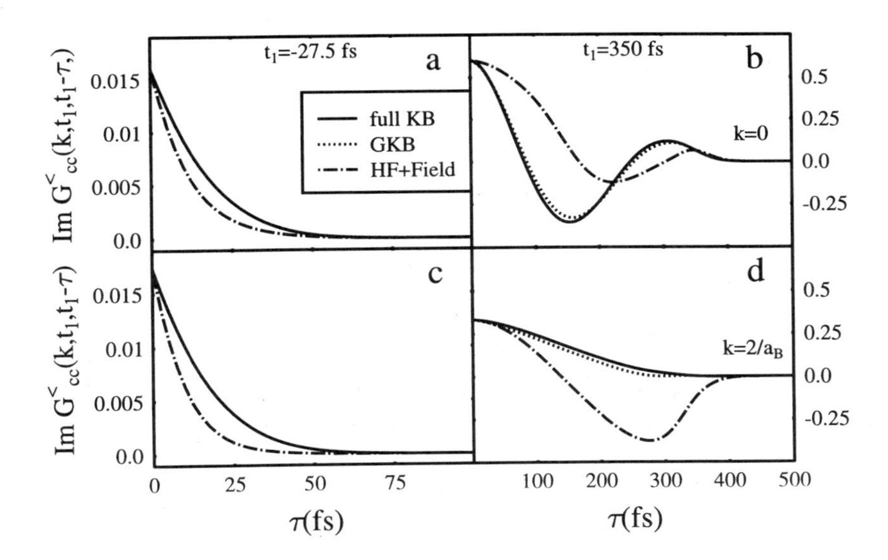

Fig. 12.14: Imaginary part of $g^{<}_{cc}(t, t-\tau)$ vs. distance τ from time diagonal for two values of t and two different momenta (figs. a,b vs. c,d). Curves show the full interband Kadanoff-Baym calculation, GKBA with the exact spectral function taken from the former and the free GKBA (Hartree-Fock spectral functions with field). Parameters as in Fig. 12.8, pulse duration $50fs$, [KBBK98].

We mention that there is another direction of research to go beyond free spectral functions, which is related to the so-called *extended quasiparticle concept.* This focuses on the long-time behavior of correlated many-particle systems. There one uses the usual quasiparticle approximation and derives correlation corrections to it, see e.g. [KKER86, KM93, vL94, BKKS96].

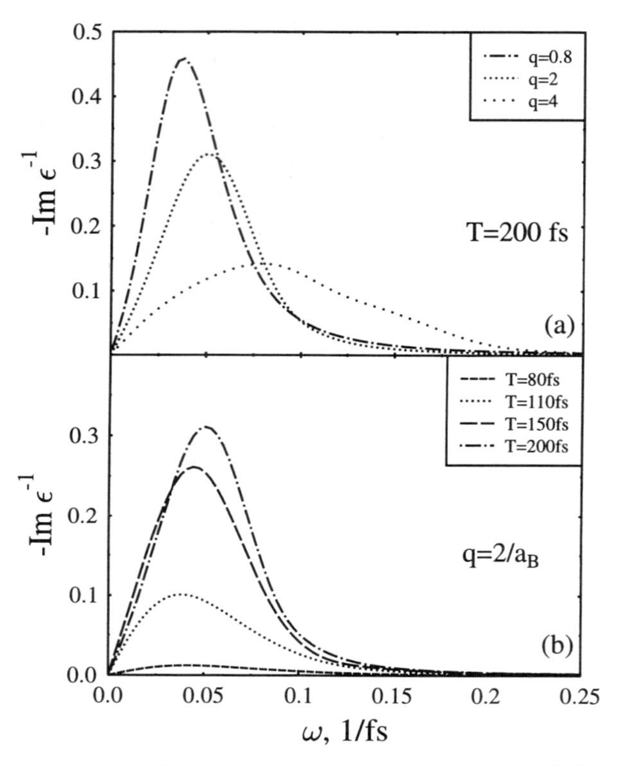

Fig. 12.15: Fs-buildup of the nonequilibrium plasmon spectral function for a quasi-2D electron gas in a quantum well excited by a 50 fs laser pulse. **a**: for different momenta and **b** as a function of time [BKK+].

12.7 Outlook. Advanced calculations

The investigation of improved spectral functions is a field of ongoing interest, e.g. [BHG98, BSHB]. On the other hand, all these problems do of course not appear, if full two-time Kadanoff-Baym equations are solved. We have seen above that this is now possible, at least for the selfenergies in static Born approximation. Moreover, it may be expected that also full RPA calculations are now within reach. The calculation of the two-time screened potential and of the RPA selfenergies in nonequilibrium is already possible [MHH+95, BKK+].

Some first results are shown below.

Femtosecond screening buildup. RPA results. The buildup of screening was discussed in many places in this book, within various theoretical concepts: quasi-classical kinetic equations for carriers and plasmons (Sec. 4.7) or the generalized polarization approximation to the BBGKY-hierarchy (Sec. 9.2). Finally, in this Chapter we made contact with the most general approach, the quantum electrodynamics for plasmas. The resulting coupled relativistic Kadanoff-Baym equations for carriers and photons are the appropriate frame to study all relevant phenomena of particle-photon interaction in correlated plasmas. Yet this is out of reach.

However, one can attempt to solve the simpler problem of carrier dynamics under the influence of the two-time screened potential $V_s(tt')$, i.e. in RPA. On the other hand, a correct description of these ultrafast processes requires to *include the generation process* of the plasma into the calculation as is done in the interband Kadanoff-Baym equations. The main problem for Kadanoff-Baym calculations with RPA selfenergies is the selfconsistent calculation of the carrier dynamics, which has not been possible so far. On the other hand, it is relatively easy to first solve the Kadanoff-Baym equations with quasi-static screening and then use the results to compute the longitudinal polarization Π. Using these time-dependent values of Π, one can compute the corresponding screened potential by using Dyson's equation for $V_s^+(tt')$ or $\epsilon^{-1}(tt')$.

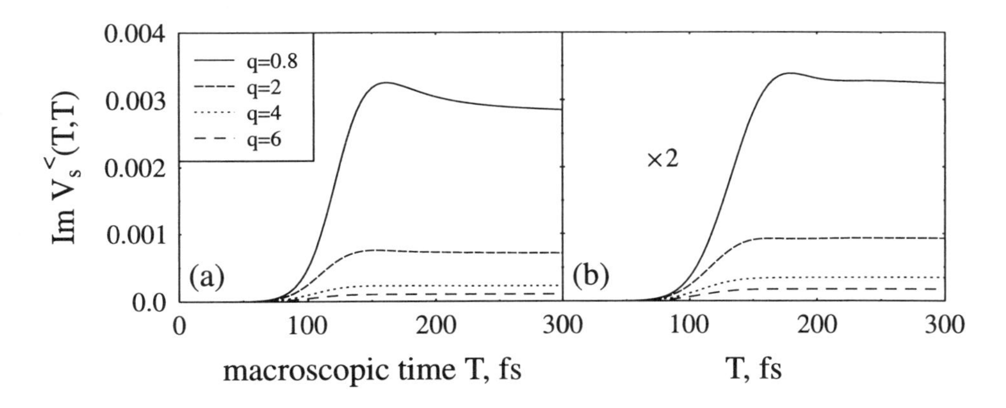

Fig. 12.16: Fs-buildup of the nonequilibrium plasmon distribution for a quasi-2D electron gas in a quantum well excited by a 50 fs laser pulse of different photon energy. **(a)**: Laser energy 25 meV and **(b)**: 10 meV above band gap, [BKK+].

Figs. 12.15 and 12.16. show first results of such calculations where the *interband Kadanoff-Baym* equations were solved. A 50 fs laser pulse excites electrons from the valence band to the conduction band which are generated unscreened. Simultaneously with the relaxation of the nonequilibrium distribution (see Fig. 12.9) the buildup of the screening cloud takes place. Fig. 12.15.b shows the buildup of the plasmon spectral function -Imϵ^{-1} which typically has peaks on the plasmon energies. One sees that these peaks are initially very broad (due to the uncertainty principle) until they obtain some stationary width after about the inverse plasma frequency. Fig. 12.15.a shows the wavevector dispersion which is typically strong in 2D, cf. Sec. 4.4.2. Notice that all peaks are rather broad. This is a characteristic feature of the Kadanoff-Baym approach where the damping of single-particle states is computed fully selfconsistently. Finally, Fig. 12.16. shows the screened potential along the time diagonal. In analogy to the carrier correlation functions, this is nothing but the Wigner function of the plasmons. One clearly sees the growth of the plasmon density[33] and its dependence on the wavenumber and the energy of the exciting photon.

Using the broader perspective of QED, these calculations yield the coupled evolution of the carrier and photon Green's functions, selfconsistently including the excitation by external transverse photons. The results contain the complete information on the single particle and correlation properties of the carrier-photon system. This opens amazing opportunities for quantum many-body theory of correlated plasmas in nonequilibrium, including ultrafast relaxation phenomena. Along these lines, the extension to relativistic carriers, full inclusion of transverse photons, consideration of bosons or more complex nuclear interactions seems within reach in the near future.

[33]We do not observe oscillations of $V_s^<$ which is, again, due to the strong selfconsistent damping. Earlier observations [MHH+95] were based on calculations with the free (undamped) GKBA.

Chapter 13

Quantum Kinetics vs. Molecular Dynamics

The preceding Chapters were devoted to the quantum kinetic approach to nonequilibrium many–particle systems. We have shown that the conventional description of relaxation processes which is based upon irreversible kinetic equations of the Boltzmann type, has serious limitations. In particular, it is valid only on times larger than the correlation time τ_{cor}, it does not include initial correlations, and conserves only kinetic energy instead of total energy. We showed in detail how one can derive generalized quantum kinetic equations which overcome these shortcomings. However, due to the complex form of these equations, so far only the Born approximation, i.e. the limit of small coupling parameters $\Gamma \ll 1$, where Γ is the ratio of potential and kinetic energy in the system (see Ch. 1), was feasible for systematic numerical studies.

In this Chapter, we consider a completely different approach to nonequilibrium many–particle systems - Molecular Dynamics simulations (MD). As we will see this approach is not limited to small coupling parameters, but it has other limitations. Therefore, a comparison, and eventually, a combination of both concepts could proof very useful in the future.

13.1 Classical Molecular Dynamics

The time evolution of a system of N interacting classical particles is completely determined by the fundamental equations of classical mechanics – N coupled Newton's equations,

$$m_i \frac{d^2\mathbf{r}_i}{dt^2} = \sum_{j \neq i} \mathbf{f}(|\mathbf{r}_i - \mathbf{r}_j|) + \mathbf{F}_i^{ext}(t), \quad i = 1 \ldots N, \tag{13.1}$$

which contain all binary interaction forces and also external forces[1]. The index "i" labels all particles and may include different species. Equations (13.1) have to be supplemented with the initial conditions[2]

$$\mathbf{r}_i(t_0) = \mathbf{r}_i^0, \qquad \dot{\mathbf{r}}_i(t_0) = \mathbf{v}_i^0, \tag{13.2}$$

which give rise to a well–defined mechanical initial value problem. The properties of the system (13.1,13.2) are well known: the equations are time reversible and obey a number of conservation laws (details depending on the external forces). In particular, for vanishing external forces which we will be concerned with, the system conserves total energy, i.e. the sum of the kinetic energies of all particles plus the sum of all binary interaction energies

$$E(t) = \sum_{i=1}^{N} \frac{p_i^2(t)}{2m_i} + \sum_{i<j} V_{ij}(t) = E(t_0). \tag{13.3}$$

Of course, this does not mean that kinetic energy remains constant. During the evolution the particle momenta and actual interaction forces change and, as a result, also kinetic and potential energy may change, but only in such a way that Eq. (13.3) is fulfilled. Obviously, the change of kinetic and potential energy will also depend on the initial state of the system,

$$\sum_{i=1}^{N} \frac{p_i^2(t)}{2m_i} - \sum_{i=1}^{N} \frac{p_i^2(t_0)}{2m_i} = -\left\{ \sum_{i<j} V_{ij}(t) - \sum_{i<j} V_{ij}(t_0) \right\}. \tag{13.4}$$

Physical observables. Averages and fluctuations. To obtain a result for a macroscopic quantity A (such as total energy) which is not influenced by an arbitrary choice of the initial conditions, one performs multiple (L) runs with varying initial conditions and, at the end, averages over all these realizations (l),[3]

$$\langle A \rangle (t) = \frac{1}{L} \sum_{l=1}^{L} A(t)|_l. \tag{13.5}$$

While for any one realization the evolution of $A^{(l)}(t)$ may show rather strong fluctuations, the ensemble averaged quantity (13.5) behaves, more or less,

[1] For an extension to relativistic situations one has to substitute $m \to m/\sqrt{1-(v/c)^2}$.

[2] For practical purposes of numerical solutions also boundary conditions have to be given, for details we refer to Ref. [Zwi94].

[3] In order to reproduce the correct ensemble average, the initial conditions are to be chosen according to the appropriate density operator at the initial time.

smoothly, if L is chosen sufficiently large. On the other hand, the fluctuations in any given realization may be reduced by averaging along the trajectory,

$$A(t)|_l \longrightarrow \overline{A}(t)|_{l,T} = \frac{1}{T}\int_{t-\frac{T}{2}}^{t+\frac{T}{2}} d\bar{t}\; A(\bar{t})|_l. \tag{13.6}$$

By increasing the particle number N of the simulations, the fluctuations decrease for each realization also, and the interval T for the time average (13.6) may be reduced, thus improving the time resolution of the simulation.

A major advantage of the MD–approach is, that it easily yields fluctuations and their time evolution. For example, in addition to simple one-particle averages of the form (13.5), it allows to calculate more complex quantities, such as correlation functions which may depend on multiple times, e.g.[4]

$$\overline{A(t_1)B(t_2)}|_{l,T} = \frac{1}{T}\int_{t_1}^{t_1+T} d\bar{t}\, A(\bar{t})|_l B(\bar{t}+\Delta t)|_l, \tag{13.7}$$

where $\Delta t = t_2 - t_1$ and $t_1 < t_2 < t_1 + T$. Averaging (13.7) again over multiple realization yields an ensemble averaged product, in this case a two–time correlation function $\langle A(t_1)B(t_2)\rangle$. Further, one can subtract the product of averages $\langle A(t_1)\rangle\, \langle B(t_2)\rangle$ resulting in a correlation function of the fluctuations of A and B, which is an extremely useful quantity and allows to compute many macroscopic quantities, such as auto–correlation functions and transport coefficients (e.g. the diffusion coefficient). These techniques are well developed for *classical* systems and have been applied in a wide range of fields (for a recent overview, see [HFPH95]), including systems with short range [AT87] as well as long range interaction.

Specifics of MD for Coulomb systems. The specifics of binary interactions in Coulomb systems leads to certain difficulties for MD simulations [Han73, HM78, HMV79, RT90, Zwi94]. These problems are related to both the long range and the short range behavior[5]. The first arises from the *long range* of the bare Coulomb interaction which leads to the necessity to include large spatial extensions. This problem is overcome by using certain ways of periodic continuation of the real system (periodic boundary conditions, Ewald summation etc.), [Han73] and corrections for the "lost potential", i.e. for the potential range which is left outside the box. On the other hand, also *close encounters* of two particle are problematic, due to the divergence of the bare

[4]The integration limits can be chosen more symmetrically with respect to t_1 and t_2.

[5]Of course, these are the same problems as we have encountered in the statistical treatment of Coulomb systems before. The solution is well known: the long–range problem is cured by screening, while the short–range one is "automatically" corrected in a quantum treatment.

interaction. This problem becomes more serious in dense systems, where the interparticle distance decreases.[6]

Attractive interactions. The problems on short distances become even more serious in a two–component (e.g. electron-ion/hole) plasma where, due to attractive interaction, close approaches are enhanced, including a possible collapse of pairs of oppositely charged particles. While these effects, which are related to strong collisions and bound states, are solved successfully in quantum statistics (we have discussed them in detail in Ch. 8), they pose a real challenge to the Molecular Dynamics. The traditional way out is to introduce an additional repulsive potential which dominates on short distances and accounts for the fact that two (classical) particles cannot penetrate each other. On the other hand, this excludes the formation of bound states, and also the question of a selfconsistent computation of this attractive potential remains open. Clearly, a satisfactory solution can only be given based on a quantum mechanical generalization of the whole MD concept.

13.2 *Quantum Molecular Dynamics

Problems which require a quantum approach. Let us summarize the problems which cannot be solved *in principle* by classical MD:

1. Close encounters of particles with the same charge sign (see above);

2. Close encounters of oppositely charged particles and their "collapse", identification of scattering events which lead to bound states; ionization/recombination and excitation/de-excitation of bound states;

3. Account of the wave character of highly energetic particles (De Broglie wave);

4. Incorporation of Heisenberg's uncertainty principle, i.e. lower limit for the resolution of the particle momentum (energy) in small space (time) intervals;

5. Incorporation of the spin statistics theorem, which requires the N-particle probability (N-particle wave function, $N \geq 2$) to be (anti-) symmetrized in the case of bosons (fermions), see Ch. 3;

[6]More precisely, most problems should be expected in *strongly coupled systems* where $\Gamma \gg 1$, since there the kinetic energy of the particles is comparably small, reducing their chance to escape these scattering events.

6. As a consequence, incorporation of the Pauli principle (Pauli blocking) and of exchange scattering effects;

7. Treatment of the interaction of particles with quantum fields (e.g. photons), see Ch. 12.

Quantum mechanical equations of motion. For quantum systems with finite degeneracy parameter, $\chi = n\Lambda^3/(2s+1) \geq 1$ (n is the density, s the spin projection and $\Lambda = h/\sqrt{2\pi m k_B T}$ the thermal De Broglie wave length), the mechanical treatment has to be based on the N-particle Schrödinger equation[7] (non-relativistic case),

$$i\hbar \frac{d}{dt}\Psi_{1\ldots N}(t) = H_{1\ldots N}\Psi_{1\ldots N}(t), \tag{13.8}$$

where the $N-$particle Hamilton operator has been defined in Eq. (2.4). Similarly as in the classical case, the mechanical equation (13.8) defines an initial value problem together with the initial condition $\Psi_{1\ldots N}(t_0) = \Psi^0_{1\ldots N}$ which fully determines the time evolution of the N-particle system. The solution of Eq. (13.8) yields the correct generalization of the classical trajectories $\{r_i(t), v_i(t)\}$ of each particle to the quantum case in terms of a wave function which is extended in space and momentum[8]. There are two major complications arising from the quantum picture: First, in an interacting system, there is no decoupling into separate wave functions of individual particles, the system is described by one collective function $\Psi_{1\ldots N}(t)$. Secondly, as a consequence of the spin statistics theorem (points 5. and 6. above), the solution of Eq. (13.8) has to be properly (anti-)symmetrized, cf. Ch. 3.

Unfortunately, both problems can currently be solved exactly only for small particle numbers. Therefore, there has been great activity over the last years to find alternative descriptions of the quantum mechanical problem, such as Feynman's path integrals [FH65], Wigner function methods or variational approaches, which promise to be easier to handle numerically. There have already been achieved remarkable results, see e.g. Ref. [KSHB95], some of which we briefly discuss in the following.

Quasi-classical approach. Effective potentials[9] This method uses the classical MD concept, accounting for quantum effects by introduction of

[7] Typically, degeneracy effects are important only for light particles, such as electrons or holes. In a plasma, ions and neutral particles can, in most cases, be treated classically.

[8] In general, the state of one particle is given by a superposition of different wave functions from a complete set (basis).

[9] Here, we follow the discussion given by Ebeling and co-workers, cf. [OSE97] and references therein.

certain semi-quantum interaction potentials. Space-dependent effective pair potentials have been derived from the Slater sum already by Kelbg [Kel63], who obtained for different particles a, b[10]

$$V_{ab}^K(r) = \frac{e_a e_b}{r\epsilon_b}\left\{1 - \exp\left[-x_{ab}^2\right] + \sqrt{\pi}\, x_{ab}\left[1 - \mathrm{erf}(x_{ab})\right]\right\}, \quad (13.9)$$

$$\text{where} \quad x_{ab} = \frac{r}{\lambda_{ab}}; \quad \lambda_{ab} = \hbar/\sqrt{2\pi m_{ab} kT}; \quad m_{ab} = \frac{m_a m_b}{m_a + m_b}, \quad (13.10)$$

which is finite at $r = 0$ as a result of quantum effects.[11] Quasiclassical MD simulations for electron-proton plasmas with similar potentials have been performed e.g. by Norman et al. [NV79], Hansen et al. [HM81] and Furukawa et al. [FN90].

The concept of pseudo-potentials was elaborated further by Kirschbaum et al., Dorso et al. and Ortner et al. e.g. [KW80, DDR87, OSE97], who phenomenologically included phase space filling effects (Pauli blocking). In Ref. [OSE97], the following Hamiltonian for a semiclassical electron gas was used,

$$H_{1\dots N} = \sum_{i=1}^{N} \frac{p_i^2}{2m} + \sum_{i<j}^{N} V_p\left(\frac{r_{ij}}{r_0}, \frac{p_{ij}}{p_0}\right) + \sum_{i<j}^{N} e^2 F\left(\frac{r_{ij}}{r_0}, \frac{p_{ij}}{p_0}\right), \quad (13.11)$$

$$\text{with } V_p(r,p) = V_0 \exp\left[-\left(\frac{r^2}{r_0^2} + \frac{p^2}{p_0^2}\right)\right]; \quad F(r,p) = \frac{1}{r}\mathrm{erf}\left(\frac{1}{\sqrt{2}}\frac{\mathrm{r}}{\mathrm{r}_0}\right), \quad (13.12)$$

where the potential V_p leads to a repulsion, if two particles come close to each other in phase space [DDR87], what simulates the Pauli principle. F is the Coulomb potential averaged over the two-particle wave function.[12] Despite its simplicity, this approach yields results for the thermodynamic quantities, such as the mean energy, which are in good agreement with Monte Carlo simulations [EM97], and also reproduces collective plasmon spectra of a correlated electron gas fairly well [OSE97]. On the other hand, the concept of effective potentials is not applicable to nonequilibrium situations.

Wave packet molecular dynamics. The idea of this approach is to replace classical point particles by localized wave packets accounting for the wave character of quantum particles, which evolve in time according to classical

[10] This result is of first order in e^2. Higher order terms as well as the equal particle result, which contains exchange are also known [Kel63, EHK67], see also [Bar71, DL72, KKER86].

[11] For the definition of the error function erf, see App. A.

[12] In Ref. [OSE97], a Gaussian wave packet was assumed which leads to the minimal product $r_0 p_0 = \hbar$. For the remaining parameters they used, based on thermodynamic considerations $V_0 = p_0^2/m = kT_Q$, with T_Q being the "quantum temperature" $kT_Q = kT(kT/E_F)^{3/2} I_{3/2}(\mu/kT)$, where I is defined in App. A.

equations of motion.[13] This approach has been developed to model nuclear collisions by Feldmaier et al. [Fel90, FBS95] and has been successfully applied to Coulomb systems by Barnes et al., Klakow et al., Ebeling et al. and others [BNNT93, KTR94, EM97]. The main idea is to parametrize the N-particle wave function of the electrons in terms of single-particle wave functions (wave packets) of simple analytical form, such as a Gaussian, e.g. [KTR94, KMRT]

$$\Psi_{1\ldots N}(x_1,\ldots,x_N;t) \approx \mathcal{A}^{\pm}\left\{\prod_i \phi_i(x_i;t)\right\}, \tag{13.13}$$

$$\phi_{\vec{q}}(x) = \left(\frac{3}{2\pi\beta}\right)^{3/4} e^{-\left(\frac{3}{4}\beta+i\frac{p_\beta}{\hbar^2}\right)(\mathbf{x}-\mathbf{r})^2-i\frac{\mathbf{p}}{\hbar}(\mathbf{x}-\mathbf{r})}, \tag{13.14}$$

where $\mathcal{A}^{\pm}$ denotes the (anti-)symmetrization operator. The index $\vec{q}$ comprises the variational parameters $(p_\beta, \beta, \mathbf{p}, \mathbf{r})$, where $\mathbf{p}$ and $\mathbf{r}$ are the momentum and position of the classical particle which is associated with the wave packet, and β is the square of the packet width and p_β the corresponding conjugate variable. These four parameters evolve in time and with them the wave packet.[14] With the wave packet picture, the Schrödinger equation is transformed into classical Hamilton's equations (their characteristics) which can be solved using classical techniques. Klakow et al. have analyzed this approach in detail[15] and have applied it to a variety of complex nonequilibrium phenomena in dense plasmas, including partial ionization, Mott effect, plasma phase transition and stopping power. On the other hand, there have been reported consistency problems, e.g. in the calculation of equilibrium quantities, such as the mean energy [EM97].

Wigner function methods. The idea is to start from a classical description and then find its quantum analogue which is treated by some approximation. A straightforward concept is to use a Wigner function description and then go to the classical limit with the first (or possibly higher) quantum correction included (i.e. terms of the order $\hbar$, cf. Sec. 2.3.2). This was done e.g. by Aichelen et al. [Aic91].

Filinov et al. have developed another interesting approach based on the Wigner representation [FMK95, Fil96]. Their starting point are binary time correlation functions $C_{AB}(t)$ for some operators A and B which are averaged

[13]Formally, this leads to an extension of the phase space by the introduction of additional "quantum" parameters which are used to simulate the underlying quantum dynamics.

[14]At each time step, the parameters are obtained from a variational principle, for details cf. [KTR94].

[15]Although the ansatz (13.13) has the form of a Hartree-Fock factorization, the choice of the variational parameters allows, in principle, to incorporate correlation effects. It has been shown that this approach is exact for free quantum particles and those in a harmonic potential.

with a canonical density operator. They introduce the Wigner transform of $C_{AB}(t)$ and are able to reduce its evolution to the simpler dynamics of a spectral density, for which they derive systematic iteration schemes. This method has been successfully applied to transport properties of strongly correlated systems in the linear response regime, including conductivity, tunneling probabilities of disordered systems etc.

Path integral molecular dynamics. This approach uses the formulation of the quantum mechanical equations of motion in terms of Feynman integrals [FH65]. Various groups are developing these methods, in analogy to Path integral Monte Carlo methods, cf. the works of Ceperley et al., e.g. [PBCM94], and there have been numerous reports about its application to accurate computation of equilibrium properties of quantum systems, including complex materials, e.g. [LD92, JD94, DJWC96].

Other methods. There are several other techniques, which cannot be discussed here in detail. We mention the method of *Car and Parinello* which uses a hybrid technique for multicomponent systems, where the electrons are treated quantum mechanically but the heavy particles are considered classically, [CP85, HNCM93, KH95]. Furthermore, there are so-called *orbital free molecular dynamics* techniques, which are simplified versions of the Car–Parinello method aiming at partially ionized systems, equation of state etc.[16]

Common to all methods is the particular difficulty arising in the treatment of fermion systems due to the Pauli principle and the wave function antisymmetrization. While this is relatively easy for small particle numbers leading to a limited number of possible permutations, it becomes a big problem for growing particle numbers. An approximate solution to this problem is to include only pair permutations, although the limits of this approximations are not exactly known.

Yet these methods are only emerging. To verify and improve them, extensive experimental data is needed. In cases which are not yet feasible for experiments, other tests and comparisons are required, including known analytical limits or independent numerical techniques. As we will show, one important possibility is to use the results of quantum kinetics [BSK97]. The latter allows one to test not only equilibrium results, but also true dynamical quantities, including characteristic times of relaxation processes. On the other hand, statistical physics uses a completely different approach to correlated many–particle systems, and it is not trivial to come to a quantitative comparison. To solve this principal problem, we will consider here only clas-

[16] The idea is to simplify the quantum treatment of the electron component (neglecting orbitals in the free energy functional of the electrons) by using low temperature expansions together with pseudo–potentials, [ZCP92, CPZ92, PCZ].

sical MD results. Generalizations to quantum molecular dynamics should be straightforward.

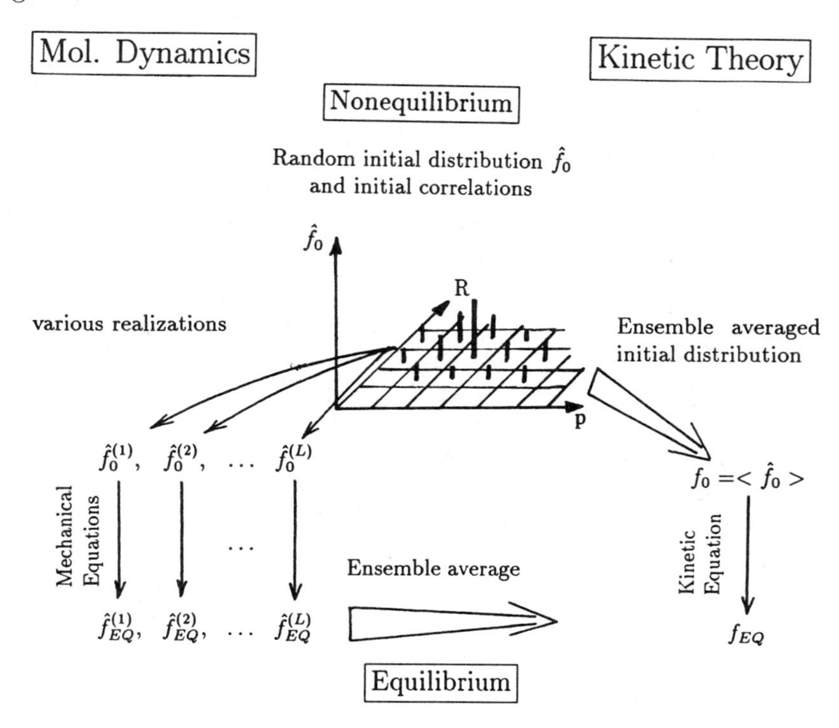

Fig. 13.1. Comparison of the concepts of (classical) molecular dynamics and kinetic theory. The nonequilibrium initial state of an N-particle system given by the random initial positions and momenta of the particles gives rise to a random distribution $\hat{f}_0(R,p)$. In contrast to *kinetic theory* which uses, from the beginning, *averaged distributions* and studies their relaxation (described by kinetic equations), *MD uses the microscopic distributions* taken for different realizations (microstates) of the same macro-state (given e.g. by fixed particle number and energy of the system). From the solution of the mechanical equations for each realization random final distributions are obtained. The ensemble average $f(t)$ agrees at all times with the statistical result.

13.3 Comparison of the concepts of Molecular Dynamics and Quantum Kinetic Theory

At first sight, there is a principal contradiction between the mechanical and the statistical description of macroscopic many–particle systems. It seems, reversible dynamical equations contradict to thermodynamic processes, relaxation and dissipation.[17] However, it is now well established that (reversible) mechanical dynamics is able to reproduce (irreversible) relaxation results of kinetic theory, on sufficiently large time scales. The macroscopic properties and equilibrium states generated from MD simulations are in excellent agreement with the corresponding statistical results. Of course, strictly speaking, there cannot be a stationary (final) state of a mechanical dynamics, defined by Eqs. (13.1) or (13.8). The simple solution is that, in contrast to kinetic theory, the mechanical evolution may, in principle, leave the stationary state again, but only after a long time (Poincaré revival time) which increases rapidly with the particle number and, in typical simulations, is far beyond the time of observation.

Nonequilibrium comparisons. Despite numerous comparisons of equilibrium molecular dynamics results with thermodynamic predictions, only recently a first comparison of the short-time behavior ($t_0 \leq t < \tau_{cor}$) was performed. This is not trivial, since one first has to clarify what kind of kinetic theory is equivalent to the MD–approach in this regime. It is the purpose of this section to answer this question. We compare the underlying concepts of

both approaches and show that only kinetic equations which are defined as an initial value problem may be equivalent to MD. The best candidate for this task are found to be the quantum kinetic equations of Kadanoff and Baym [KB89], in the generalized form which allows for arbitrary initial correlations, see Ch. 12. Let us consider in detail the concepts of the two approaches.

1. Both, MD and kinetic theory, are based on the fundamental equations of motion of classical or quantum mechanics (see above).

2. Both approaches differ only in the treatment of these equations: While MD works with microstates, kinetic theory uses ensemble averaged quantities, such as the N–particle density operator $\rho_{1\ldots N}$. The time evolution of this averaged quantity is the von Neumann equation (2.12). If it is supplemented with an initial condition $\rho_{1\ldots N}(t = t_0) = \rho^0_{1\ldots N}$, it is fully equivalent to the N–particle Schrödinger equation: both are mechanical equations which are time reversible and conserve energy (cf. Ch. 2).

3. Thus, both approaches invoke ensemble averages, only in different places: In the kinetic case, the ensemble average is performed *before* solving the equations of motion, while in case of MD, it is performed *afterwards*, on the solutions (realizations) of the microscopic equations, cf. Eq. (13.5). This comparison is shown schematically in Fig. 13.1.

[17] In fact, this was the subject of a continuous argument between L. Boltzmann and E. Zermelo a century ago, see for example Refs. [Bol96, Zer96].

4. Therefore, the theories must lead to the *same results for all ensemble averages* (observables), at each point of the time evolution.

5. The reduction of the exact $N-$particle problem is, in the MD–case, performed by simply taking only a small piece $N_{MD} \ll N$ of the system, and including *all interactions*, while the kinetic approach takes *all particles* but includes only a subset of the interactions.

The main conclusion of points 1.-4. is that, in fact, agreement between both concepts is possible also for *nonequilibrium* processes. If the relaxation starts from the same initial state, the two concepts must give the same result for ensemble averages for all times. This is a principal conclusion which is true for the case that both approaches are solved exactly. Of course, this alone would be a pure academic statement. For practical application it is more important to know, if this agreement can be achieved also for approximations which are feasible numerically. Again, this is not trivial since the approximation schemes of the two methods (cf. point 5.) are completely different. Thereby the MD scheme (at least the classical one) is straightforward: accuracy is improved by increasing the number of particles in the simulation as well as the number of realizations. More difficult is to come to an agreement from the side of statistical theory.

Choice of the statistical theory. Thus, we have to find formulations of kinetic theory which

a) are given as an initial value problem and allow for arbitrary initial correlations,

b) preserve the conservation properties of the full $N-$particle system, cf. point (2.) and, furthermore,

c) are straightforwardly and in a systematic way, extendable to all situations of interest, including in particular arbitrary values of the coupling and the degeneracy parameters Γ and χ.

As we have discussed in Ch. 12, the most general equations for the time evolution of quantum ensemble averages are generalized Kadanoff-Baym equations which, in the nonrelativistic case, are given by Eq. (12.71) As it was discussed in Ch. 12, the Kadanoff-Baym equations (12.71) have a number of remarkable properties: They trivially include quantum effects (arbitrary degeneracy). There are no consistency problems - the internal structure of approximations is completely determined by the approximation for the selfenergies $\Sigma^{\gtrless}$ alone.

Moreover, the conservation properties of Eqs. (12.71) depend on simple well-known symmetry properties of the selfenergies [KB89] which are easy to satisfy. Due to this consistency, the equations yield the correct asymptotic state of a correlated many–particle system and also the correct behavior on short times and are suitable to describe the buildup of correlations. So we may conclude that in the generalized form (12.71), the Kadanoff-Baym equations satisfy the above requirements a) – c). In this form, statistical mechanics and molecular dynamics are equivalent.[18]

13.4 Numerical comparisons of Molecular Dynamics and Quantum Kinetics

Choice of physical quantities. Let us discuss now what kind of comparisons between the two approaches are possible in nonequilibrium. As was pointed out (point 4.), comparable quantities are ensemble averages. Quantities which are sensitive to the short–time dynamics we are concerned with, have been found to be the kinetic and the correlation energy, see Ch. 7, [BK96, BKS+96, ZTR]. They measure the decay of initial correlations and the build-up of correlations due to scattering processes very well. Typically, at the initial stage, $t_0 \leq t \leq \tau_{cor}$, potential energy changes until it saturates around $t \sim \tau_{cor}$, see Ch. 7. For larger times, potential and kinetic energy remain approximately constant, indicating that the kinetic regime has been reached, see Ch. 5. The *absolute amount* of potential energy change is a measure for the strength of the interaction (and for the initial correlations), while the *time scale* of this change yields an estimate for the nonequilibrium correlation time τ_{cor}, and thus important quantitative information on the relevant relaxation mechanisms in the system. Hence, these quantities are very sensitive to the physical processes on short time scales and sensitive to the various physical approximations too. Furthermore, kinetic and potential energy have a clear definition in both approaches which is independent of the particular method of solution or approximations. Therefore, we will use the *time evolution of kinetic energy* for comparison below.

Choice of system and parameters. The next question is the appropriate choice of the system to study and of its parameters. Interestingly, the advantages of one method overlap with the weak points of the other: Quantum effects, which are problematic in the MD–approach, are trivially included in the Kadanoff–Baym equations. Vice versa, strong coupling effects, which are

[18]This does not rule out other types of kinetic equations, which, however, will be less general.

difficult to handle in the latter, are, no problem for the former, see Fig. 13.2.

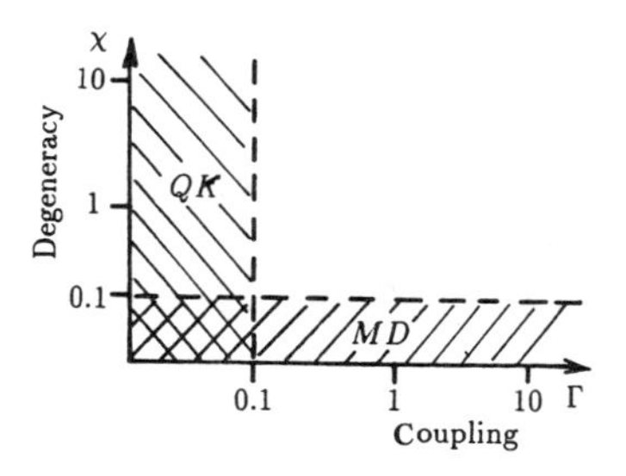

Fig. 13.2. Areas of applicability of classical MD and of Quantum kinetic theory ("QK", in Born approximation). The region of overlap is chosen for the numerical comparison.

This creates a great potential for the description of correlated quantum many–body systems over the whole range of parameters. So it is a challenging task to find ways for a systematic comparison and, eventually some combination of both methods in the future. There are three major problems to solve:

I. Identification of situations where both methods are expected to give qualitative and quantitative agreement and to find examples which are simple enough to be studied by MD and kinetic theory. Obviously, this will have to be *weakly coupled and weakly degenerate systems.*

II. Extension of the analysis into parameter regions where only one method is reliable:

1. Increasing the coupling parameter, at small degeneracy: here the classical MD should be sufficiently accurate. On the other hand, this region is very difficult to handle in kinetic treatments (see Ch. 8), where often additional approximations are unavoidable. Having MD results to compare with, may suggest new approximation schemes, which could be more efficient compared to conventional ones which are based on theoretical criteria, such as perturbation expansions or diagrammatic considerations, (cf. Sec. 2.6).
2. Increasing the degeneracy parameter at small coupling: here the quantum kinetic theory is expected to be accurate, whereas classical MD fails. The numerous concepts of quantum MD (Sec. 13.2) can be systematically tested on quantum kinetic results, eventually yielding conclusive information about the adequateness of one assumption or another in a given situation.

This should allow to systematically extend the region of validity of both, MD and quantum kinetics to higher degeneracy and stronger coupling, respectively.

III. Having reliable results for several limiting cases, one can try to design interpolating concepts for the whole parameter plane. For example, one could think of improved nonequilibrium effective potentials for MD simulations or better cross sections for collision integrals in kinetic equations.

Numerical parameters. Here we focus on the first point of this program. Before any comparison can be made, we have to identify situations where quantitative agreement of both approaches is expected. The simplest system to consider is a *homogeneous electron gas without initial correlations.* Furthermore, since the parameter ranges where both methods give reliable results overlap in the corner of small coupling and weak degeneracy (Fig. 13.2), we choose $\Gamma = 0.1$ and $\chi = 0.1$. For the comparison, we use results of Zwicknagel et al. [ZTR], who performed classical MD–simulations for these parameters. They used periodic boundary conditions with $N_{MD} = 500$ particles. The MD–simulation started from a micro-state with initial coordinates and velocities, $\mathbf{r}_i(t_0), \mathbf{v}_i(t_0)$, which were distributed homogeneous in coordinate space and isotropic in momentum space, according to a given initial one–particle distribution $f(\mathbf{r}, \mathbf{v}, t_0) = f(v, t_0) \sim \exp[-64(v-1)^2]$. Since the initial state was uncorrelated, the potential energy was $\langle V \rangle(t_0) \approx 0$.[19] The starting value of the coupling parameter was $\Gamma_0 = 0.1$, which was calculated using the initial kinetic energy instead of temperature $\Gamma_0 = (4\pi n/3)^{1/3} e^2/(kT_0)$, with $3/2kT_0 = E_{kin}(t_0)$.

The Kadanoff-Baym equations were solved as discussed in App. F. For the weak-coupling/weak degeneracy limit, we are concerned with, the selfenergy was calculated in second order Born approximation, according to Eq. (12.79). For the matrix element of the interaction potential $V(q)$, we used the static long-wavelength limit of the screened potential in random phase approximation, where the inverse screening length $\kappa(t)$ was calculated selfconsistently from the current nonequilibrium distribution $f(p,t)$, (3D system)

$$V(q,t) = \frac{4\pi e^2/\epsilon_r}{q^2 + \kappa^2(t)}, \qquad \kappa^2(t) = \frac{4m}{\pi} \int_0^\infty dp\, f(p,t). \tag{13.15}$$

For our calculations, we chose typical semiconductor parameters (bulk GaAs), $\epsilon_r = 13.99, m/m_r = 1.284$ and took the same initial distribution as above. Here we face a typical problem of such comparisons: Since the kinetic approach explicitly involves approximations of statistical theory, it contains additional parameters (which do not appear in the MD approach) - here an additional length parameter - the screening length κ^{-1}. As a result, there is some additional freedom which allows to reproduce the MD-conditions by different

[19] In fact, this is one of the complicated points for simulations.

distributions (e.g. different peak positions in units of κ). Our choice was dictated by the requirement of weak degeneracy, $\chi = 0.1$ and negligible Pauli blocking effects.[20]

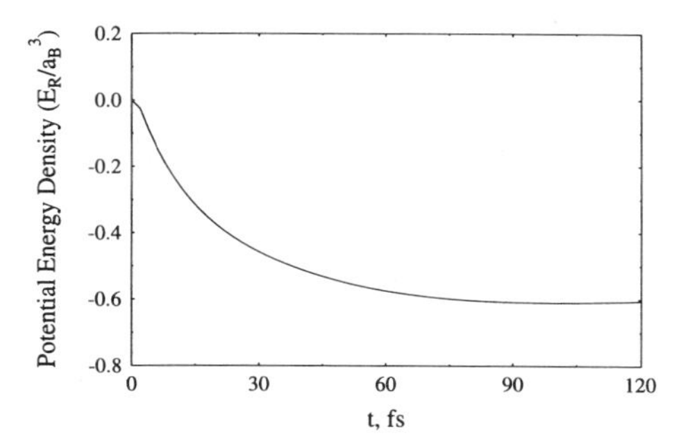

Fig. 13.3. Evolution of potential energy according to the KB calculations

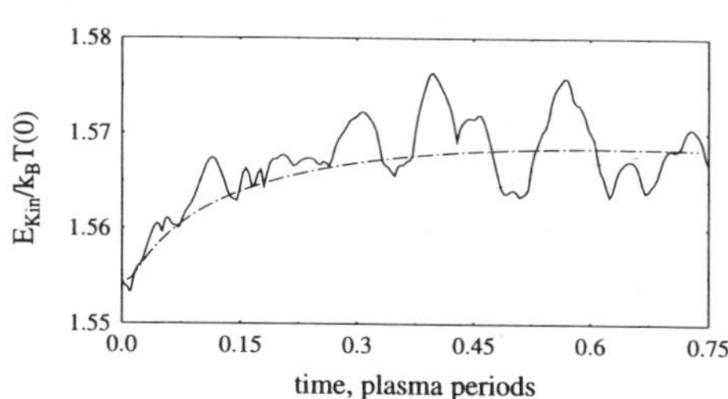

Fig. 13.4. Comparison of KB and MD results for the kinetic energy relaxation.

Numerical results. In Fig. 13.3, the Kadanoff-Baym result for the evolution of potential energy density during the initial period is given, cf. Eq. (12.85). As one can see, the potential energy change saturates after about $120fs$. One readily verifies that this time is close to the plasma period $\tau_p = \sqrt{\frac{\pi m \epsilon_r}{ne^2}} \approx 160fs$, which is the correlation time in this system. Due to energy conservation, at the same time kinetic energy increases, see Fig. 13.4. To compare the Kadanoff-Baym calculation with the MD–data of Zwicknagel et al., we had to rescale our results for the kinetic energy in their units (units of kT_0). The results of both calculations are shown in Fig. 13.4. Notice that the MD–data in Ref. [ZTR] was given for a single realization and, therefore, contains fluctuations around the ensemble average, which are not present in the Kadanoff-Baym run. Nevertheless, the agreement between both calculations, in particular with respect to the time scale of the evolution is rather good.

13.5 Discussion and outlook

In fact, the agreement of the absolute values of the mean kinetic energy between the two calculations raises questions. The reason is that the kinetic theory result is obtained with a statically screened Coulomb potential, for which the correlation energy approaches the value $-N\kappa e^2/4$ (which agrees with the value in Fig. 13.3. In contrast, the simulations are (ideally) done with the bare

[20] This requires some care, since the conventional definition of degeneracy (cf. the definition of χ above) applies to equilibrium systems only.

Coulomb interaction. Only during the relaxation, carriers start to rearrange themselves which "automatically" gives rise to dynamical screening. Therefore, the correlation energy should approach the dynamical (RPA) result, i.e. $-N\kappa e^2/2$, cf. Sec. 9.3.2.[21] This suggests that in the simulations, effectively screening was introduced (obviously, a bare Coulomb potential would lead to a divergency).

Thus, the comparison between MD and kinetic theory has already proved useful. Clearly, further improvements of the simulation of Coulomb systems (especially for weak coupling) are necessary, in particular to extract reliable results for *time-dependent quantities on the initial stage.* This is a challenging and rather new task [still the vast majority of MD-simulations aim at equilibrium quantities, where (a) the results are averaged over a sufficiently long period and (b) some initial equilibration is used where all transients are neglected][22]. The goal should be a well tested and commonly accepted approach to the simulation of the initial relaxation.

The next step of the program which was outlined above, is to extend the comparison to larger values of the coupling and degeneracy parameters. Kadanoff–Baym calculations for coupling parameters larger than one have to use T-matrix cross sections in some suitable approximation, which is essentially more complicated than the Born approximation. First nonequilibrium results were presented in Ch. 8). Of course, future comparisons can also use other quantities, including higher moments of the distribution function or two-time correlation functions. Developments in Statistical Physics, such as strong-coupling approximations for the Kadanoff-Baym equations and also quantum molecular dynamics concepts will benefit from the proposed comparison. Advances to novel approximations in one field can be checked quantitatively on well–established results of the other. This comparison will eventually improve the theoretical understanding and numerical modeling of strongly correlated quantum many–particle systems, including their relaxation behavior on long and short time scales.

---

[21] This has been pointed out also in Ref. [MvL].

[22] We mention that results analogous to those of Ref. [ZTR] have been obtained also in Ref. [PG96] and also for ionic mixtures (Kalugin et al., priv. comm. 1997).

Chapter 14

Conclusion

An old field is returning into the focus of science and technology: the interaction of matter with electromagnetic radiation. Due to recent advances in femtosecond laser technology, even small laboratories are now aquiring the ability to directly study the behavior of atoms and charged particles under a very broad range of excitation conditions. The observed phenomena include coherent effects, collective plasma oscillations, transport processes and optical response up to relativistic energies. Not only the more traditional questions of quasi–stationary behavior can be studied, but also transient processes on the scale of a few femtoseconds.

In this book the attempt was made to summarize the first results in the new field of *ultrafast relaxation processes in many-particle systems.* It was shown that these phenomena which are of interest in various fields, including semiconductors, dense plasmas and nuclear matter, cannot be described using traditional kinetic theory. This is due to fundamental limitations of the latter: conventional kinetic equations do not properly account for correlation effects (they conserve kinetic instead of total energy and describe the relaxation towards ideal equilibrium distributions), and they are applicable only to times larger than the correlation time (they completely neglect initial correlations and do not describe the buildup of correlations). Therefore, it was necessary to derive generalized quantum kinetic equations. We have derived *non-Markovian generalizations of the Landau, Balescu-Lenard and Boltzmann equation*, which fully include memory (retardation) effects, collisional energy broadening, initial correlations and selfenergy. In addition, the latter two equations describe, respectively, dynamical screening and strong coupling (T-matrix effects) in nonequilibrium.

We have seen that there exist two alternative statistical approaches - the method of *reduced density operators* and *nonequilibrium Green's functions* techniques. Both agree in the resulting equations for the Wigner distribution function, but are very different in the treatment of particular problems. The density operator method is conceptually very simple (Chs. 2, 3). It trivially includes initial correlations and works well for simple approximations of many-body theory (e.g. perturbation theory) for which the formal solution for the binary correlation operator is known. On the other hand, Green's functions methods (Ch. 12) are advantageous for the description of complex phenomena such as energy renormalization (selfenergy), dynamical screening and bound states, (but we have shown in Chs. 7–10 that these phenomena can also be derived from the BBGKY-hierarchy). Furthermore, Green's functions take advantage of powerful systematic (diagrammatic) approximation schemes and are readily extended to the description of fluctuations. Therefore, a development of both approaches appears to be important in order to combine the strengths of each and to have independent tests and comparisons.

Finally, in Ch. 13 we have discussed a third approach to relaxation processes far from equilibrium - Molecular Dynamics. In contrast to statistical theories, Molecular Dynamics is based on a mechanical approach to many-body systems and works well for classical systems at arbitrary coupling, and its proper generalization to quantum systems is currently actively debated. We have compared the underlying concepts of the dynamical and the statistical approaches in detail and showed that the results of the two have to agree for ensemble averages. Interestingly, strengths and weaknesses of both complement each other. This suggests to perform detailed numerical comparisons of both (we presented first results) and eventually to attempt a combination of the two methods in the future.

Besides discussing different theoretical approaches to ultrafast relaxation phenomena, we also presented numerical results. Although non-Markovian quantum kinetic equations are essentially more difficult to solve than conventional kinetic equations, there have recently been developed efficient and robust numerical techniques a survey of which was provided in Appendix F. It is now also possible to solve the *Kadanoff-Baym equations* numerically, including their generalizations to multi-component and multiband systems and the selfconsistent coupling to external electromagnetic fields. Due to their internal consistency and conservation properties, the Kadanoff-Baym equations appear to be the most promising method for the investigation of ultrafast relaxation phenomena in correlated systems. With continuously increasing computational capabilities, it will be possible in the near future to solve the Kadanoff-Baym equations with full RPA selfenergies, coupled Kadanoff-Baym equations for carriers, photons and plasmons as well as the relativistic Kadanoff-Baym equations.

Appendix A

Used Mathematical Formulas

In this Appendix we briefly list, in alphabetical order, mathematical functions and definitions and provide formulas for them which are being used in this book.

Basis. An orthonormal basis is given by a set of Hilbert state vectors $|\psi^{(1)}\rangle$, $|\psi^{(2)}\rangle \ldots |\psi^{(M)}\rangle$ which obey the properties

$$\langle\psi^{(k)}|\psi^{(l)}\rangle = \delta_{k,l}, \tag{A.1}$$

$$\sum_{k=1}^{M} |\psi^{(k)}\rangle\langle\psi^{(k)}| = 1, \tag{A.2}$$

where the Kronecker symbol $\delta_{k,l}$ is defined below. In the case of a continuous system, $\sum \to \int$.

Bessel function or 1st kind. One definition is

$$I_m(z) = \left(\frac{z}{2}\right)^m \sum_{k=0}^{\infty} \frac{(-1)^k}{k!\,\Gamma(m+k+1)} \left(\frac{z}{2}\right)^{2k}, \quad |\arg z| < \pi, \tag{A.3}$$

where Γ is the gamma function. For further properties, see ref. [KK68].

Harmonics expansion of periodic functions. Exponentials of a periodic function can be expressed as a series of harmonics according to

$$e^{\pm ia\sin\omega_0 t} = \sum_{n=-\infty}^{\infty} I_n(a)\, e^{\pm in\omega_0 t}, \tag{A.4}$$

where I_n is the Bessel function of order n (see above).

Bose/Fermi function. The Bose and Fermi functions are given by

$$f_{B/F}(p) = \frac{1}{e^{\beta(\frac{p^2}{2m}-\mu)} \mp 1}, \tag{A.5}$$

where β and μ are the inverse temperature and chemical potential, respectively.

Broadened delta function. We define

$$\delta^{\varepsilon}(x) \equiv \frac{1}{\pi}\frac{\varepsilon}{x^2+\varepsilon^2}, \tag{A.6}$$

which has the properties

$$\lim_{\varepsilon\to 0}\delta^{\varepsilon}(x) = \delta(x), \qquad \text{and} \quad \lim_{\varepsilon\to 0}\frac{d}{d\varepsilon}\delta^{\varepsilon}(x) = \frac{\mathcal{P}'}{x}, \tag{A.7}$$

where $\mathcal{P}$ and $\delta(x)$ denote the principal value and the delta function being defined below.

Cauchy formulas. For a function f which is analytic in some region which includes a contour C, the following formulas hold

$$\int_C f(\zeta)d\zeta = 0, \tag{A.8}$$

$$\frac{n!}{2\pi i}\int_C \frac{f(\zeta)}{(\zeta-z)^{n+1}}d\zeta = f^{(n)}(z), \tag{A.9}$$

where $f^{(n)}$ denotes the n-th derivative, $f^{(0)} \equiv f$.

Commutator. The commutator of two operators A and B is defined as

$$[A,B] = AB - BA. \tag{A.10}$$

Its classical limit is given by $[A,B] \longrightarrow \frac{1}{i\hbar}\{A,B\}$, with the Poisson brackets being defined below.

Delta function. Dirac's delta distribution is defined as $(a < b)$ [KK68]

$$\int_a^b f(y)\delta(y-X)dy = \begin{cases} 0 & , \quad X < a \\ \frac{1}{2}f(X+0) & , \quad X = a \\ \frac{1}{2}f(X-0) & , \quad X = b \\ \frac{1}{2}\left[f(X+0)+f(X-0)\right] & , \quad a < X < b \end{cases} \tag{A.11}$$

where $f(x)$ is an arbitrary function which is finite around X. In particular, $\int_{-\infty}^{\infty} dx\, \delta(x) = 1$. We mention the following properties of the delta function $\delta(-x) = \delta(x)$; $x\delta(x) = 0$; $\int_{-\infty}^{\infty} dx\, \delta(x-a)\delta(x-b) = \delta(a-b)$, and also

$$\delta[f(x)] = \sum_a \frac{\delta(x)}{|f'(x_a)|}, \quad \text{with } f(x_a) = 0, \tag{A.12}$$

$$
\begin{aligned}
\delta^{(n)}(x) &= (-1)^n\, n!\, \frac{\delta(x)}{x^n}, \qquad n = 0, 1, 2, \ldots, &\text{(A.13)}\\
\delta(\mathbf{p}) &= \frac{1}{(2\pi\hbar)^3} \int d\mathbf{r} e^{-\frac{i}{\hbar}\mathbf{pr}}, &\text{(A.14)}\\
\delta(0) &= \frac{\mathcal{V}}{(2\pi\hbar)^3}. &\text{(A.15)}
\end{aligned}
$$

Dirac identity. see Plemlj formula below.
Error function. The error function is defined as

$$
\operatorname{erf}(x) = \frac{2}{\sqrt{\pi}} \int_0^x d\bar{x}\, e^{-\bar{x}^2}. \tag{A.16}
$$

Fermi integral. The Fermi integral of order ν is defined as [KKER86]

$$
I_\nu(y) = \sum_{l=1}^{\infty} (-1)^{l+1} \frac{e^{ly}}{l^{\nu+1}}; \quad \nu > -1. \tag{A.17}
$$

Heaviside step function. The step function is defined as

$$
\Theta(x) = \begin{cases} 1 \;, & x \ge 0 \\ 0 \;, & x < 0 \end{cases} \tag{A.18}
$$

and is related to the delta function (see above) by $\frac{d}{dx}\Theta(x) = \delta(x)$.

Kronecker delta symbol. The discrete analogon to the delta function (see above) is the Kronecker symbol,

$$
\delta_{ab} = \begin{cases} 1 \;, & a = b \\ 0 \;, & a \neq b \end{cases} \tag{A.19}
$$

Permutation operator. The binary permutation operator changes the order of two given operator indices, e.g.

$$
P_{12} A_{312} = A_{321}, \tag{A.20}
$$

and has the following properties,

$$
\begin{aligned}
P_{ij}^2 &= 1,\\
\mathrm{Tr}_j P_{ij} &= 1,\\
P_{ij} A_{ij} P_{ij} &= A_{ji},\\
P_{ij} B_{ij} &= B_{ij} P_{ij}, \quad \text{if} \quad B_{ij} = B_{ji}. &\text{(A.21)}
\end{aligned}
$$

Plemlj formula (Dirac identity). The formula

$$\lim_{\varepsilon \to +0} \frac{1}{x \pm i\epsilon} = \mathcal{P}\frac{1}{x} \mp i\pi\delta(x), \tag{A.22}$$

may be used as the definition of the principal value and the delta function.

Poisson brackets. For two s-particle operators the definition is

$$\{A_{1\ldots s}, B_{1\ldots s}\} = \sum_{i=1}^{s} \left(\nabla_{p_i} A_{1\ldots s} \nabla_{R_i} B_{1\ldots s} - \nabla_{p_i} B_{1\ldots s} \nabla_{R_i} A_{1\ldots s} \right) \tag{A.23}$$

Principal value $\mathcal{P}$. See Plemlj formula above.

Time ordering operator. The T-operator regroups a product of operators which all depend on a scalar parameter (e.g. time) into a sequence of operators with monotonically increasing arguments,

$$T\, A(t_1)B(t_2) = \begin{cases} A(t_1)B(t_2) & , \quad t_1 \leq t_2 \\ B(t_2)A(t_1) & , \quad t_1 > t_2 \end{cases} \tag{A.24}$$

More generally, the parameter t may run along a curve in higher dimensional space, then the ordering will be with respect to the distance from some chosen point on the curve (contour ordering).

Trace. The trace of an operator A which is represented by an $N \times N$ matrix $A(x_1 \ldots x_N; x_1' \ldots x_N')$ is defined as the sum of the diagonal elements and yields a c-number

$$\mathrm{Tr}_{1\ldots N} A_{1\ldots N} = \sum_{x_1 \ldots x_N} A(x_1 \ldots x_N; x_1 \ldots x_N) = C. \tag{A.25}$$

More generally, the *partial trace* over variables $s+1 \ldots N$ yields an $s \times s$ matrix

$$\mathrm{Tr}_{s+1\ldots N} A = \sum_{x_{s+1} \ldots x_N} A(x_1 \ldots x_N; x_1' \ldots x_s', x_{s+1} \ldots x_N) = C(x_1 \ldots x_s; x_1' \ldots x_s').$$

An important property is the *(cyclic) invariance of the trace* over operators the arguments of which are affected by the trace, for example

$$\mathrm{Tr}_1 A_1 B_{12} C_{123} = \mathrm{Tr}_1 B_{12} C_{123} A_1, \tag{A.26}$$

whereas the order of B and C cannot be altered due to their dependence on variable 2.

Appendix B

Wigner Representation of the BBGKY-Hierarchy

We consider the BBGKY-hierarchy in coordinate representation, Eq. (2.44) which involves matrices of the type $F_{1\ldots s}(r_1', \ldots r_s'; r_1'', \ldots r_s'', t)$. Introducing center of mass and relative coordinates, R_i and r_i for each particle, according to $r_i' = R_i + r_i/2$ and $r_i'' = R_i - r_i/2$, or, vice versa, $R_i = (r_i' + r_i'')/2$ and $r_i = r_i' - r_i''$, the above matrix transforms into

$$\begin{aligned} & F_{1\ldots s}(r_1', \ldots r_s'; r_1'', \ldots r_s'', t) \\ = \; & F_{1\ldots s}(R_1 + \frac{r_1}{2} \ldots R_s + \frac{r_s}{2}; R_s - \frac{r_s}{2} \ldots R_s - \frac{r_s}{2}, t) \\ = \; & \tilde{F}_{1\ldots s}(R_1, r_1 \ldots R_s, r_s, t) = \frac{1}{n^s} \tilde{f}_{1\ldots s}(R_1, r_1 \ldots R_s, r_s, t), \end{aligned} \tag{B.1}$$

we can rewrite the hierarchy equations suppressing the spin variables:

$$\begin{aligned} & \left\{ i\hbar \frac{\partial}{\partial t} - H_{1\ldots k}(R_1 + \frac{r_1}{2}, \ldots, R_k + \frac{r_k}{2}) + H_{1\ldots k}(R_1 - \frac{r_1}{2}, \ldots, R_k - \frac{r_k}{2}) \right\} \\ \times \; & \tilde{f}_{1\ldots k}(R_1, r_1 \ldots R_k, r_k, t) \\ = \; & \sum_{i=1}^{k} \int dR_{k+1} \left\{ V(R_i - R_{k+1} + \frac{r_i}{2}) - V(R_i - R_{k+1} - \frac{r_i}{2}) \right\} \\ \times \; & \tilde{f}_{k+1}(R_1, r_1 \ldots R_{k+1}, 0, t). \end{aligned} \tag{B.2}$$

The kinetic energy terms and the binary interaction potentials V_{ij} in the Hamiltonians transform to ($r_{ij} = r_i - r_j$; $R_{ij} = R_i - R_j$)

$$\begin{aligned} V(r_i' - r_j') - V(r_i'') - V(r_j'') &= V(R_{ij} + \frac{r_{ij}}{2}) - V(R_{ij} - \frac{r_{ij}}{2}) \\ -\frac{\hbar^2}{2\,m_i}\left(\nabla^2_{r_i'} - \nabla^2_{r_i''}\right) &= -2\frac{\hbar^2}{2\,m_i} \nabla_{R_i} \nabla_{r_i}. \end{aligned}$$

We now introduce the *Wigner transformation* with respect to the relative coordinates $r_1, \ldots r_s$ and the inverse transform according to

$$\begin{aligned} f_{1\ldots s}(R_1, p_1 \ldots R_s, p_s, t) &= \int \frac{dr_1}{(2\pi\hbar)^3} \cdots \frac{dr_s}{(2\pi\hbar)^3} \exp\{-i\,(p_1 r_1 + \ldots p_s r_s)/\hbar\} \\ &\times\ \tilde{f}_{1\ldots s}(R_1, r_1 \ldots R_s, r_s, t) \qquad \text{(B.3)} \\ \tilde{f}_{1\ldots s}(R_1, r_1 \ldots R_s, r_s, t) &= \int dp_1 \ldots dp_s \exp\{i\,(p_1 r_1 + \ldots p_s r_s)/\hbar\}, \\ &\times\ f_{1\ldots s}(R_1, p_1 \ldots R_s, p_s, t). \qquad \text{(B.4)} \end{aligned}$$

The Wigner transform (B.3) of Eq. (B.2) is

$$\left\{ i\hbar \frac{\partial}{\partial t} + i\hbar \sum_{i=1}^{k} \frac{p_i}{m_i} \nabla_{R_i} \right\} f(R_1 p_1 \ldots R_k p_k) - \sum_{1 \le i < j \le k} V_k^{(ij)} - \sum_i U_k^{(i)} = \sum_{i=1}^{k} F_{k+1}^{(i)},$$

where the only non-trivial terms are those containing the external potential or the interaction potential which are denoted U_k^i, $V_k^{(ij)}$ and $F_{k+1}^{(i}$, respectively. We consider $V_k^{(ij)}$ more in detail,

$$\begin{aligned} V_k^{(ij)} &= \int \frac{dr_1}{(2\pi\hbar)^3} \cdots \frac{dr_k}{(2\pi\hbar)^3} \exp\{-\frac{i}{\hbar}\,(p_1 r_1 + \ldots p_k r_k)\} \\ &\times\ \left\{ V(R_{ij} + \frac{r_{ij}}{2}) - V(R_{ij} - \frac{r_{ij}}{2}) \right\} \tilde{f}_{1\ldots k}(R_1 r_1 \ldots R_k r_k), \end{aligned}$$

which, after using for $\tilde{f}$ Eq. (B.4), transforms to

$$\begin{aligned} &\int \frac{dr_1}{(2\pi\hbar)^3} \cdots \frac{dr_k}{(2\pi\hbar)^3} d\bar{p}_1 \ldots d\bar{p}_k \exp\{-\frac{i}{\hbar}\,(p_1 r_1 + \ldots p_k r_k - \bar{p}_1 r_1 - \bar{p}_k r_k)\} \\ \times\ &\left\{ V(R_{ij} + \frac{r_{ij}}{2}) - V(R_{ij} - \frac{r_{ij}}{2}) \right\} f_{1\ldots k}(R_1 \bar{p}_1 \ldots R_k \bar{p}_k). \end{aligned}$$

The integrals over all coordinates and momenta, except those with the indices i and j can be carried out according to (G is an arbitrary continuous function of momentum, see the properties of the delta function, cf. App. A)

$$\int \frac{dr_a}{(2\pi\hbar)^3} d\bar{p}_a \exp\{-i\,(p_a - \bar{p}_a)\, r_a/\hbar\} G(\bar{p}_a) = G(p_a), \qquad \text{(B.5)}$$

whereas the argument of the remaining exponential is conveniently rearranged as $p_i r_i + p_j r_j - \bar{p}_i r_i - \bar{p}_j r_j = (p_i - \bar{p}_i)(r_i - r_j) + (p_i - \bar{p}_i + p_j - \bar{p}_j) r_j$. The second term yields $\delta(p_i - \bar{p}_i + p_j - \bar{p}_j)$, and we finally obtain

$$\begin{aligned} V_k^{(ij)} &= \int \frac{dr_{ij}}{(2\pi\hbar)^3} d\bar{p}_i \exp\{-i\,(p_i - \bar{p}_i)\, r_{ij}/\hbar\} \\ &\times\ \left\{ V(R_{ij} + \frac{r_{ij}}{2}) - V(R_{ij} - \frac{r_{ij}}{2}) \right\} \\ &\times\ f_{1\ldots k}(R_1, p_1, \ldots, R_i, \bar{p}_i, \ldots, R_j, p_i - \bar{p}_i + p_j, \ldots, R_k, p_k). \qquad \text{(B.6)} \end{aligned}$$

In similar way, we transform $U_k^{(i)}$ and $F_{k+1}^{(i)}$, with the final result

$$\begin{aligned}
U_k^{(i)} &= \int \frac{dr_i}{(2\pi\hbar)^3}\, d\bar{p}_i\, \exp\{-i\,(p_i - \bar{p}_i)\, r_i/\hbar\} \\
&\times \left\{U(R_i + \frac{r_i}{2}) - U(R_i - \frac{r_i}{2})\right\} \\
&\times f_{1\ldots k}(R_1, p_1, \ldots, R_{i-1}, p_{i-1} R_i, \bar{p}_i, R_{i+1}, p_{i+1}, \ldots, R_k, p_k), \qquad \text{(B.7)} \\
F_{k+1}^{(i)} &= \int \frac{dr_i}{(2\pi\hbar)^3}\, d\bar{p}_i dR_{k+1}\, dp_{k+1}\, \exp\{-i\,(p_i - \bar{p}_i)\, r_i/\hbar\} \\
&\times \left\{V(R_{i,k+1} + \frac{r_i}{2}) - V(R_{i,k+1} - \frac{r_i}{2})\right\} \\
&\times f_{1\ldots k+1}(R_1, p_1, \ldots, R_i, \bar{p}_i, \ldots, R_k, p_k, R_{k+1}, p_{k+1}). \qquad \text{(B.8)}
\end{aligned}$$

Classical limit and quantum corrections

To obtain the classical limit, all quantities have to be expanded in terms of the microscopic coordinates r_a with the subsequent limit $r_a \to 0$. In particular, the binary interaction potentials V_{ij} and external potentials U_i, have the expansions

$$\begin{aligned}
V(R_{ij} \pm \frac{r_{ij}}{2}) &= \sum_{l=0}^{\infty} \frac{(\pm 1)^l}{l!} \frac{\partial^{(l)}}{\partial R_i^l} V(R_{ij}) \left(\frac{r_{ij}}{2}\right)^l, \qquad \text{(B.9)} \\
U(R_i \pm \frac{r_i}{2}) &= \sum_{l=0}^{\infty} \frac{(\pm 1)^l}{l!} \frac{\partial^{(l)}}{\partial R_i^l} U(R_i) \left(\frac{r_i}{2}\right)^l. \qquad \text{(B.10)}
\end{aligned}$$

For the differences of potentials appearing in the hierarchy equations only the odd powers contribute:

$$\begin{aligned}
V(R_{ij} + \frac{r_{ij}}{2}) - V(R_{ij} - \frac{r_{ij}}{2}) &= \sum_{l=0}^{\infty} \frac{2}{(2l+1)!} \frac{\partial^{(2l+1)}}{\partial R_i^{2l+1}} V(R_{ij}) \left(\frac{r_{ij}}{2}\right)^{2l+1}, \\
U(R_i + \frac{r_i}{2}) - U(R_i - \frac{r_i}{2}) &= \sum_{l=0}^{\infty} \frac{2}{(2l+1)!} \frac{\partial^{(2l+1)}}{\partial R_i^{2l+1}} U(R_i) \left(\frac{r_i}{2}\right)^{2l+1}.
\end{aligned}$$

We now make use of the identity

$$\begin{aligned}
&\int \frac{dr_i}{(2\pi\hbar)^3}\, d\bar{p}_i\, r_i^l \exp\{-i\,(p_i - \bar{p}_i)\, r_i/\hbar\}\, G(\bar{p}_i) \qquad \text{(B.11)} \\
&= \int \frac{dr_i}{(2\pi\hbar)^3}\, d\bar{p}_i\, (-\frac{\hbar}{i})^l \frac{\partial^{(l)}}{\partial p^l} \exp\{-i\,(p_i - \bar{p}_i)\, r_i/\hbar\}\, G(\bar{p}_i) = (i\,\hbar)^l \frac{\partial^{(l)}}{\partial p^l} G(p_i),
\end{aligned}$$

where the last line follows from the property (B.5). Using this result, we can write down the classical form of the BBGKY-hierarchy (terms with $l = 0$)

with quantum corrections of all orders (in $l > 0$), with the following results for $V_k^{(ij)}$, $U_k^{(i)}$ and $F_{k+1}^{(i)}$,

$$\begin{aligned}
V_k^{(ij)} &= \sum_{l=0}^{\infty} \frac{(i\,\hbar)^{(2\,l+1)}}{2^{2l}(2l+1)!} \frac{\partial^{(2\,l+1)}}{\partial R_i^{2\,l+1}} V(R_{ij}) \frac{\partial^{(2\,l+1)}}{\partial p_i^{2\,l+1}} f_{1\ldots k}(R_1, p_1, \ldots, R_k, p_k), \\
U_k^{(i)} &= \sum_{l=0}^{\infty} \frac{(i\,\hbar)^{(2\,l+1)}}{2^{2l}(2l+1)!} \frac{\partial^{(2\,l+1)}}{\partial R_i^{2\,l+1}} U(R_i) \frac{\partial^{(2\,l+1)}}{\partial p_i^{2\,l+1}} f_{1\ldots k}(R_1, p_1, \ldots, R_k, p_k), \\
F_{k+1}^{(i)} &= \int dR_{k+1} dp_{k+1} \sum_{l=0}^{\infty} \frac{(i\,\hbar)^{(2\,l+1)}}{2^{2l}(2l+1)!} \frac{\partial^{(2\,l+1)}}{\partial R_i^{2\,l+1}} V(R_{ij}) \\
&\times \frac{\partial^{(2\,l+1)}}{\partial p_i^{2\,l+1}} f_{1\ldots k+1}(R_1, p_1, \ldots, R_{k+1}, p_{k+1}).
\end{aligned} \tag{B.12}$$

This is still an exact result, equivalent to the full hierarchy equations (provided, the Taylor expansions (B.9), (B.10) converge). The classical limit is given by the lowest order term of the expansions ($l = 0$, whereas the the second order term ($l = 1$) gives the first quantum corrections.

Appendix C

Equations of Motion for Binary and Ternary Correlations

Derivation of the equation for g_{12}, Eq. (2.99)

Inserting the cluster expansions for F_{12} and F_{123} into the second hierarchy equation (2.99), we obtain

$$\begin{aligned}
& i\hbar\frac{\partial}{\partial t}(F_1F_2+g_{12}) - [H^0_{12}, F_1F_2+g_{12})] - [V_{12}, g_{12}] \qquad \text{(C.1)}\\
& = [V_{12}, F_1F_2] + n\mathrm{Tr}_3[V_{13}+V_{23}, F_1F_2F_3] + n\mathrm{Tr}_3[V_{13}+V_{23}, F_3g_{12}] + \\
& n\mathrm{Tr}_3[V_{13}+V_{23}, F_1g_{23}] + n\mathrm{Tr}_3[V_{13}+V_{23}, F_2g_{13}] + n\mathrm{Tr}_3[V_{13}+V_{23}, g_{123}].
\end{aligned}$$

The time derivative of the one-particle operators can be eliminated using the first hierarchy equation, which can be transformed to

$$i\hbar\frac{\partial}{\partial t}F_1F_2 - [\bar{H}_1+\bar{H}_2, F_1F_2] = F_2n\mathrm{Tr}_3[V_{13}, g_{13}] + F_1n\mathrm{Tr}_3[V_{23}, g_{23}]. \qquad \text{(C.2)}$$

We further use the following relations

$$\begin{aligned}
n\mathrm{Tr}_3[V_{13}+V_{23}, F_1F_2F_3] &= [U_1^H+U_2^H, F_1F_2], \qquad \text{(C.3)}\\
n\mathrm{Tr}_3[V_{13}+V_{23}, F_3g_{12}] &= [U_1^H+U_2^H, g_{12}],
\end{aligned}$$

with the effective (Hartree) potential $U_1^H = n\mathrm{Tr}_3V_{13}F_3$. These terms, together with $[H^0_{12}, F_1F_2]$ and $[H^0_{12}, g_{12}]$, yield, respectively, $[\bar{H}^0_{12}, F_1F_2]$ and $[\bar{H}^0_{12}, g_{12}]$. After cancellation of the terms $[\bar{H}^0_{12}, F_1F_2]$, $n\mathrm{Tr}_3[V_{13}, F_2g_{13}]$ and $n\mathrm{Tr}_3[V_{23}, F_1g_{23}]$ with the corresponding terms from Eq. (C.2), we arrive at the final equation

$$\begin{aligned}
& i\hbar\frac{\partial}{\partial t}g_{12} - [\bar{H}^0_{12}+V_{12}, g_{12}] = [V_{12}, F_1F_2] + \\
n \quad & \mathrm{Tr}_3\Big\{[V_{13}, F_1g_{23}] + [V_{23}, F_2g_{13}] + [V_{13}+V_{23}, g_{123}]\Big\}.
\end{aligned}$$

Derivation of the equation for g_{123}, Eq. (2.99)

Using the Ursell-Mayer expansions for F_{123} and F_{1234}, Eqs. (2.98), the third hierarchy equation can be written as

$$\begin{aligned} & i\hbar\frac{\partial}{\partial t}(I+II+III+IV+V)-[H_{123},I+II+III+IV+V] \\ = & \; n\mathrm{Tr}_4[V_{14}+V_{24}+V_{34},A+B+C+D+E], \qquad \text{(C.4)} \\ I = & \; F_1F_2F_3, \quad II=F_1g_{23}, \quad III=F_2g_{13}, \quad IV=F_3g_{12}, \quad V=g_{123}, \\ A = & \; F_1F_2F_3F_4, \quad B=F_1g_{234}+F_2g_{134}+F_3g_{124}+F_4g_{123}, \\ C = & \; g_{12}g_{34}+g_{13}g_{24}+g_{14}g_{23}, \\ D = & \; F_1F_2g_{34}+F_1F_3g_{24}+F_1F_4g_{23}+F_2F_4g_{13}+F_3F_4g_{12}, \quad E=g_{1234}. \end{aligned}$$

To eliminate the time derivative of F_1 (term I), we use the first hierarchy equation, that can be written as

$$\begin{aligned} & i\hbar\frac{\partial}{\partial t}F_1F_2F_3-[\bar{H}_1+\bar{H}_2+\bar{H}_3,F_1F_2F_3] \\ & =F_1F_2n\mathrm{Tr}_4[V_{34},g_{34}]+F_1F_3n\mathrm{Tr}_4[V_{24},g_{24}]+F_2F_3n\mathrm{Tr}_4[V_{14},g_{14}]. \qquad \text{(C.5)} \end{aligned}$$

For the terms II-IV, we need the 1st and 2nd hierarchy equations in the form

$$\begin{aligned} & i\hbar g_{23}\frac{\partial}{\partial t}F_1-g_{23}[\bar{H}_1,F_1]=g_{23}n\mathrm{Tr}_4[V_{14},g_{14}], \qquad \text{(C.6)} \\ & i\hbar F_1\frac{\partial}{\partial t}g_{23}-F_1[\bar{H}_{23},g_{23}]-F_1[V_{23},F_2F_3]=F_1n\mathrm{Tr}_4[V_{24},F_2g_{34}] \\ & +F_1n\mathrm{Tr}_4[V_{34},F_3g_{24}]+F_1n\mathrm{Tr}_4[V_{24}+V_{34},g_{234}], \qquad \text{(C.7)} \end{aligned}$$

and the corresponding permutations. We now notice that all terms on the r.h.s. of Eqs. (C.5, C.6 and C.7) cancel (the r.h.s. of Eq. (C.5) with terms in D, the r.h.s. of Eq. (C.6) with terms in C and the r.h.s. of Eq. (C.7) with terms in B and D). The remaining terms in A-D can be transformed into

$$\begin{aligned} A & \rightarrow [U_1^H+U_2^H+U_3^H,F_1F_2F_3], \\ B & \rightarrow [U_1^H+U_2^H+U_3^H,g_{123}] \\ & + n\mathrm{Tr}_4[V_{14},F_1g_{234}]+n\mathrm{Tr}_4[V_{24},F_2g_{134}]+n\mathrm{Tr}_4[V_{34},F_3g_{124}], \\ C & \rightarrow n\mathrm{Tr}_4[V_{14}+V_{24},g_{12}g_{34}]+n\mathrm{Tr}_4[V_{14}+V_{34},g_{13}g_{24}], \\ & + n\mathrm{Tr}_4[V_{24}+V_{34},g_{14}g_{23}] \\ D & \rightarrow [U_1^H+U_2^H+U_3^H,F_1g_{23}+F_2g_{13}+F_3g_{12}]. \qquad \text{(C.8)} \end{aligned}$$

Terms A and D are exactly compensated by the effective potentials occurring in Eqs. (C.5,C.6, and C.7), and the effective potential in B allows to renormalize the Hamiltonian in the equation for g_{123} to $\bar{H}_{123}=H_{123}+U_1^H+U_2^H+$

U_3^H. Finally, we have to take into account that the potential contributions in $[H_{123}, F_1F_2F_3 + F_1g_{23} + F_2g_{13} + F_3g_{12}]$ are only partially compensated by Eqs. (C.5, C.6, and C.7), giving rise to the inhomogeneity in the equation for g_{123}. Collecting the remaining terms from B and C and also E, we obtain the desired exact result

$$i\hbar\frac{\partial}{\partial t}g_{123} - [\bar{H}_{123}, g_{123}] = [V_{12} + V_{13} + V_{23}, F_1F_2F_3] + \\ [V_{13} + V_{23}, F_3g_{12}] + [V_{12} + V_{23}, F_2g_{13}] + [V_{12} + V_{13}, F_1g_{23}] + \\ n\mathrm{Tr}_4[V_{14} + V_{24}, g_{12}g_{34}] + n\mathrm{Tr}_4[V_{14} + V_{34}, g_{13}g_{24}] + n\mathrm{Tr}_4[V_{24} + V_{34}, g_{14}g_{23}] + \\ n\mathrm{Tr}_4[V_{14}, F_1g_{234}] + n\mathrm{Tr}_4[V_{24}, F_2g_{134}] + n\mathrm{Tr}_4[V_{24}, F_3g_{124}] + \\ n\mathrm{Tr}_4[V_{14} + V_{24} + V_{34}, g_{1234}]. \tag{C.9}$$

(Anti-)Symmetrization of the equation for g_{12}

We now derive Eq. (3.22). Using the equation for F_{12} (2.20), where F_{12} and F_{123} are decomposed using the (anti-)symmetrized version of the Ursell-Mayer expansion, Eqs. (2.98,3.8), yields[1]

$$\begin{aligned} & i\hbar\frac{\partial}{\partial t}\Big(F_1F_2 + g_{12}\Big)\Lambda^{\pm}_{12} - [H_{12}, F_1F_2 + g_{12}]\Lambda^{\pm}_{12} \\ = & \Big\{n\mathrm{Tr}_3[V_{13} + V_{23}, F_1F_2F_3] + n\mathrm{Tr}_3[V_{13} + V_{23}, F_1g_{23}] \\ + & n\mathrm{Tr}_3[V_{13} + V_{23}, F_2g_{13}] + n\mathrm{Tr}_3[V_{13} + V_{23}, F_3g_{12}] \\ + & n\mathrm{Tr}_3[V_{13} + V_{23}, g_{123}]\Big\}\Lambda^{\pm}_{123}. \end{aligned} \tag{C.10}$$

We first notice that, due to the factorization property of $\Lambda^{\pm}_{123}$, Eq. (3.6), an overall factor $\Lambda^{\pm}_{12}$ can be can-celled, what results in the equation

$$\begin{aligned} & i\hbar\frac{\partial}{\partial t}\Big(F_1F_2 + g_{12}\Big) - [H_{12}, F_1F_2 + g_{12}] \\ = & \Big\{n\mathrm{Tr}_3[V_{13} + V_{23}, F_1F_2F_3] + n\mathrm{Tr}_3[V_{13} + V_{23}, F_1g_{23}] \\ + & n\mathrm{Tr}_3[V_{13} + V_{23}, F_2g_{13}] + n\mathrm{Tr}_3[V_{13} + V_{23}, F_3g_{12}] \\ + & n\mathrm{Tr}_3[V_{13} + V_{23}, g_{123}]\Big\}(1 + \epsilon P_{13} + \epsilon P_{23}). \end{aligned} \tag{C.11}$$

Eliminating the derivatives of the one-particle operators using Eq.(3.20), we obtain

$$\begin{aligned} & i\hbar\frac{\partial}{\partial t}g_{12} - [H_{12}, g_{12}] - [V_{12}, F_1F_2] = \Big\{ - n\mathrm{Tr}_3[V_{13}, F_2F_{13}\Lambda^{\pm}_{13}] \\ + & n\mathrm{Tr}_3[V_{13}, F_{123}](1 + \epsilon P_{13} + \epsilon P_{23}) + (1 \longleftrightarrow 2)\Big\}. \end{aligned} \tag{C.12}$$

---

[1] For the properties of the permutation operator, see App. A.

Inserting for F_{123} the cluster expansion, Eq. (2.98), the first term on the r.h.s. of Eq. (C.12) cancels,

$$i\hbar\frac{\partial}{\partial t}g_{12} - [H_{12}, g_{12}] - [V_{12}, F_1F_2] = \Big\{ - n\mathrm{Tr}_3[V_{13}, F_2F_{13}P_{23}] + \\ n\mathrm{Tr}_3[V_{13}, g_{23}F_1 + g_{12}F_3]\,\Lambda^{\pm}_{123} + n\mathrm{Tr}_3[V_{13}, g_{123}]\,\Lambda^{\pm}_{123} + (1 \longleftrightarrow 2)\Big\}. \quad \text{(C.13)}$$

We now notice that the first term on the r.h.s. can be decomposed into terms of the form of Eqs. (3.11), (3.16), giving rise to Pauli blocking in the Hamiltonian and in the last term on the l.h.s. ($V_{12} \to \hat{V}_{12} = (1 + n\epsilon F_1 + n\epsilon F_2)V_{12})$).

Now we transform the second term on the r.h.s. of Eq. (C.13). We first separate the Hartree-Fock term, Eq. (3.9), and then transform the remaining terms which yield the polarization contributions, [DKBB97]

$$\begin{aligned} n\mathrm{Tr}_3V_{13}(g_{23}F_1 + g_{12}F_3)(1 + \epsilon P_{13} + \epsilon P_{23}) &= \\ n\mathrm{Tr}_3V_{13}(g_{12}F_3 - g_{23}F_1P_{13}) + n\mathrm{Tr}_3V_{13}(g_{23}F_1\Lambda^{\pm}_{23} + \epsilon g_{12}F_3(P_{13} + P_{23})) &= \\ U_1^{HF}g_{12} + n\mathrm{Tr}_3V_{13}(g_{23}F_1\Lambda^{\pm}_{23} + \epsilon P_{13}P_{13}g_{12}F_3P_{13}(1 + P_{13}P_{23})), & \quad \text{(C.14)} \end{aligned}$$

where the second and the fourth properties (3.7) of the permutation operators have been used. Consider now $(1 + P_{13}P_{23}) = \Lambda^{\pm}_{23} + (1 - \epsilon P_{13})P_{23} = \Lambda^{\pm}_{23} + P_{23}(1 - \epsilon P_{12})$. If we recall now that an overall factor $\Lambda^{\pm}_{12}$ has been can-celled, the term containing $1 - \epsilon P_{12}$ vanishes exactly. Thus only the factor $\Lambda^{\pm}_{23}$ remains in the polarization term, and we obtain the final result of Eq. (3.22).

(Anti-)Symmetrization of the equation for g_{123}

We now derive Eq. (3.24). Using the equation for F_{123}, (2.21), and decomposing F_{123} and F_{1234} by means of the (anti-)symmetrized version of the Ursell-Mayer expansion, Eqs. (2.98), (3.8), yields

$$\begin{aligned} & i\hbar\frac{\partial}{\partial t}\Big(F_1F_2F_3 + g_{12}F_3 + \ldots + g_{123}\Big)\Lambda^{\pm}_{123} \qquad \text{(C.15)} \\ & \quad -[H_{123}, F_1F_2F_3 + g_{12}F_3 + \ldots + g_{123}]\Lambda^{\pm}_{123} \\ = & \ \Big\{n\mathrm{Tr}_4[V_{14} + V_{24} + V_{34}, F_1F_2F_3F_4] \\ + & \ n\mathrm{Tr}_4[V_{14} + V_{24} + V_{34}, F_1F_2g_{34}] + \ldots + n\mathrm{Tr}_4[V_{14} + V_{24} + V_{34}, g_{12}g_{34}] + \ldots \\ + & \ n\mathrm{Tr}_4[V_{14} + V_{24} + V_{34}, F_1g_{234}] + \ldots + n\mathrm{Tr}_4[V_{14} + V_{24} + V_{34}, g_{1234}]\Big\}\Lambda^{\pm}_{1234}, \end{aligned}$$

where "..." denotes all permutations of the preceding term. Due to the factorization property of $\Lambda^{\pm}_{1234}$, Eq. (3.6), an overall factor $\Lambda^{\pm}_{123}$ can be can-celled. This leaves on the r.h.s. of Eq. (C.15) a factor $(1 + \epsilon P_{14} + \epsilon P_{24} + \epsilon P_{34})$. The derivatives of the one-particle and two-particle operators are eliminated using Eqs. (3.20) and (3.22) and their permutations in the form

$$i\hbar\frac{\partial}{\partial t}F_1F_2F_3 - [H_1+H_2+H_3, F_1F_2F_3] = F_2F_3 n\mathrm{Tr}_4[V_{14}, F_{14}]\Lambda^{\pm}_{14} + \mathcal{P}(123),$$
$$g_{23}i\hbar\frac{\partial}{\partial t}F_1 - g_{23}[H_1, F_1] = g_{23}n\mathrm{Tr}_4[V_{14}, F_{14}]\Lambda^{\pm}_{14},$$
$$F_1 i\hbar\frac{\partial}{\partial t}g_{23} - F_1[H_{23}, g_{23}] - F_1[V_{23}, F_2F_3] = F_1\Big\{ - n\mathrm{Tr}_4[V_{24}, F_3F_{24}\Lambda^{\pm}_{24}] +$$
$$n\mathrm{Tr}_4[V_{24}, F_2F_3F_4 + F_2g_{34} + F_3g_{24} + F_4g_{23} + g_{234}](1+\epsilon P_{24}+\epsilon P_{34}) + \mathcal{P}(12)\Big\},$$

where $\mathcal{P}(abc)$ denotes all permutations of particles a, b, c in all terms in the parentheses. With these relations, the one and two-particle operators F_a and g_{ab} on the l.h.s. of Eq. (C.15) can be eliminated, except certain commutators of these operators with the binary potentials V_{ab}, which can be regarded as inhomogeneity I_{123}. Thus we obtain

$$i\hbar\frac{\partial}{\partial t}g_{123} - [H_{123}, g_{123}] = I_{123} + n\mathrm{Tr}_4 A_{1234}, \tag{C.16}$$
$$\text{with} \quad I_{123} = V_{12}F_1F_2F_3 + (V_{12}+V_{13})F_1g_{23} + \mathcal{P}(123) - \text{h.c.} \tag{C.17}$$

We now examine the terms contributing to A_{1234} under the trace. Consider first the products $F_1F_2F_3F_4$. For example, we have

$$n\mathrm{Tr}_4 V_{14}F_1F_2F_3F_4(1+\epsilon P_{14}+\epsilon P_{24}+\epsilon P_{34} - \Lambda^{\pm}_{14}) =$$
$$n\epsilon\mathrm{Tr}_4 V_{14}F_1F_2F_3F_4(P_{24}+P_{34}) = n\epsilon(F_2V_{12}+F_3V_{13})F_1F_2F_3, \tag{C.18}$$

where we used Eq. (3.11). Taking all terms of the form (C.18) we obtain the Pauli blocking corrections to the product of free one-particle operators in I_{123}, i.e. everywhere we may replace $V_{ab} \to \hat{V}_{ab}$. Next, consider the three-particle correlation operators in A_{1234}. These terms are treated exactly like in the case of the second hierarchy equation (Appendix C) the two-particle correlation operators under the trace over "3". These terms give rise to the Hartree–Fock and the Pauli blocking terms in the three-particle ladders, e.g.

$$n\mathrm{Tr}_4 V_{14}(F_4g_{123} + \epsilon F_1 g_{234}P_{14}) = U_1^{HF} g_{123}, \tag{C.19}$$
$$n\epsilon\mathrm{Tr}_4 V_{24}F_1g_{234}P_{14} = n\epsilon F_1 V_{12}g_{123}. \tag{C.20}$$

All these terms together allow to replace on the l.h.s. of Eq. (C.16) in the Hamiltonian $H_a \to \bar{H}_a = H_a + U_a^{HF}$ and $V_{ab} \to \hat{V}_{ab}$. The remaining terms involving g_{abc} yield the three–particle polarizations including exchange polarization contributions (directly generalizing the results of Appendix C),

$$n\mathrm{Tr}_4[V^{\pm}_{14}, F_1]g_{234}(1+\epsilon P_{24}+\epsilon P_{34}) + \mathcal{P}(123). \tag{C.21}$$

Next we consider the contributions of products $g_{ab}g_{cd}$ to A_{1234}:

$$\begin{aligned}\Big\{n\mathrm{Tr}_4[V_{14}, g_{12}g_{34} + g_{13}g_{24} + g_{14}g_{23}(1+\epsilon P_{14}+\epsilon P_{24}+\epsilon P_{34})\\ -n\mathrm{Tr}_4[V_{14}, g_{14}\Lambda^{\pm}_{14}]g_{23}\Big\} + \mathcal{P}(123) =\\ \Big\{n\mathrm{Tr}_4[V_{14}+V_{24}, g_{12}g_{34}](1+\epsilon P_{14}+\epsilon P_{24}+\epsilon P_{34})\\ +n\epsilon\mathrm{Tr}_4[V_{14}, g_{14}]g_{23}(P_{24}+P_{34})\Big\} + \mathcal{P}(123).\end{aligned} \qquad \text{(C.22)}$$

The first line on the r.h.s. corresponds to generalized polarization terms, whereas the last line can be transformed further according to

$$\begin{aligned} n\epsilon\mathrm{Tr}_4 V_{14}g_{14}g_{23}P_{24} &= n\epsilon g_{23}V_{12}g_{12},\\ n\epsilon\mathrm{Tr}_4 P_{14}g_{14}g_{23}V_{14} &= n\epsilon g_{12}V_{12}g_{23}.\end{aligned} \qquad \text{(C.23)}$$

These terms add to the inhomogeneity I_{123}, where for each term of the form $V_{13}F_3g_{12}$ there appears one new term of the type $-n\epsilon g_{13}V_{13}g_{12}$.

Finally, we consider those terms in A_{1234} that contain $F_aF_bg_{cd}$. From the six possible index combinations (all indices different) three are of the form $A = F_aF_bg_{c4}$ and three of the form $B = F_4F_ag_{bc}$. E.g. the term $A = F_1F_2g_{34}$ after cancellation gives rise to

$$\begin{aligned} & n\epsilon\mathrm{Tr}_4(V_{14}AP_{24} + V_{24}AP_{14}) - \mathrm{h.c.}\\ = \; & n\epsilon(F_2V_{12}F_1g_{23} + F_1V_{12}F_2g_{13}) - \mathrm{h.c.}\end{aligned} \qquad \text{(C.24)}$$

These are Pauli blocking corrections for the inhomogeneity (I_{123}) contributions F_1g_{23} and F_2g_{13}, respectively. Consider now one of the three terms "B", e.g. $B = F_3F_4g_{12}$, which yields the result

$$\begin{aligned} n\epsilon\mathrm{Tr}_4(V_{14}+V_{24})BP_{34} + V_{34}B(P_{14}+P_{24}) - \mathrm{h.c.} =\\ n\epsilon F_3(V_{13}+V_{23})F_3g_{12} + n\epsilon V_{34}B(P_{14}+P_{24}) - \mathrm{h.c.}\end{aligned} \qquad \text{(C.25)}$$

The first two terms on the r.h.s. are again Pauli blocking corrections to the inhomogeneity terms F_3g_{12}. Combined with the corresponding terms in Eq. (C.24) and permutations, this leads to the substitution $V_{ab} \to \hat{V}_{ab}$ in I_{123} in all terms containing F_ag_{bc}. The last two terms in Eq. (C.25) and their adjoint are transformed according to

$$\begin{aligned} n\epsilon\mathrm{Tr}_4 V_{34}F_3F_4g_{12}P_{14} &= n\epsilon g_{12}V_{13}F_3F_1\\ n\epsilon\mathrm{Tr}_4 P_{14}F_3F_4g_{12}V_{34} &= n\epsilon F_1F_3V_{13}g_{12}.\end{aligned} \qquad \text{(C.26)}$$

These terms add to the inhomogeneity I_{123}, where for each term of the form $V_{13}F_3g_{12}$ there appears one new term of the type $-n\epsilon F_1F_3V_{13}g_{12}$.

Collecting all terms together yields the (anti-)symmetrized third hierarchy equation, Eq. (3.24).

Appendix D

Properties of the Free Propagators U^0 and $U^{0\pm}$

In this appendix we consider the properties of the free $s-$particle propagators which are used to construct the solution for the correlation operators $g_{1...s}$. In general, they depend on two times $U^0_{1...s} = U^0_{1...s}(t,t')$ and are not hermitean. We distinguish between the propagator $U^0_{1...s}$ and the retarded and advanced propagators $U^{0\pm}_{1...s}$, which are related by

$$\begin{aligned} U^{0\pm}_{1...s}(t,t') &= \Theta[\pm(t-t')]\, U^0_{1...s}(t,t'), && \text{(D.1)} \\ \text{or} \quad U^0_{1...s}(t,t') &= U^{0\,+}_{1...s}(t,t') + U^{0\,-}_{1...s}(t,t'), && \text{(D.2)} \end{aligned}$$

where $\Theta(\tau)$ is the Heaviside step function which equals one for $\tau > 0$ and zero for $\tau < 0$, and we further have

$$\lim_{t\to t'+0} U^+_{1...s}(t,t') = 1, \qquad \lim_{t\to t'-0} U^-_{1...s}(t,t') = 1. \tag{D.3}$$

Properties of the propagators $U_{1...s}$

The propagator $U^0_{1...s}$ is defined by the following equation of motion:

$$\left(i\hbar\frac{\partial}{\partial t} - H^{0\,\mathrm{eff}}_{1...s}(t) \right) U^0_{1...s}(t,t') = 0, \quad U^0_{1...s}(t,t) = 1, \tag{D.4}$$

where we introduced the effective free s–particle Hamiltonian which, in general, contains one-particle selfenergy corrections (see below)

$$H^{0\,\mathrm{eff}}_{1...s}(t) = \sum_{i=1}^{s} H^{\mathrm{eff}}_i(t). \tag{D.5}$$

Due to the additivity of the Hamiltonian, $U^0_{1\ldots s}$ factorizes,

$$\left(i\hbar\frac{\partial}{\partial t} - H_1^{\rm eff}(t)\right) U_1(t,t') = 0, \quad U_1(t,t) = 1. \tag{D.6}$$

Eq. (D.6) can be solved formally with the result

$$U_1(t,t') = \mathrm{T}\, e^{-\frac{i}{\hbar}\int\limits_{t'}^{t} d\bar t H_1^{\rm eff}(\bar t)}, \tag{D.7}$$

where T is the time-ordering operator, and $H_1^{\rm eff}$ is defined below.

Properties of the propagators $U^{\pm}_{1\ldots s}$

The equations of motion for the retarded and advanced propagators,

$$\left(i\hbar\frac{\partial}{\partial t} - H^{0\,\rm eff}_{1\ldots s}(t)\right) U^{+}_{1\ldots s}(t,t') = i\hbar\,\delta(t-t'), \tag{D.8}$$

which again factorizes,

$$\left(i\hbar\frac{\partial}{\partial t} - H_1^{\rm eff}(t)\right) U_1^{+}(t,t') = i\hbar\,\delta(t-t'). \tag{D.9}$$

The equations for the advanced propagator follow from the adjoint of Eqs. (D.8) and (D.9) with the help of the symmetry relation

$$[U^{0\,\pm}_{1\ldots s}(t,t')]^\dagger = U^{0\,\mp}_{1\ldots s}(t',t). \tag{D.10}$$

Taking the hermitean adjoint of Eqs. (D.8) and (D.9), we obtain

$$i\hbar\frac{\partial}{\partial t} U^{-}_{1\ldots s}(t',t) + U^{-}_{1\ldots s}(t',t)\, H^{0\,\rm eff\,\dagger}_{1\ldots s}(t) = i\hbar\,\delta(t-t'), \tag{D.11}$$

$$i\hbar\frac{\partial}{\partial t} U^{-}_{1}(t',t) + U^{-}_{1}(t',t)\, H^{\rm eff\,\dagger}_{1}(t) = i\hbar\,\delta(t-t'). \tag{D.12}$$

The one–particle Hamiltonian, is defined as

$$H_1^{\rm eff}(t)\, U_1^{+}(t,t') = \left\{H_1 + \mathcal{U}_1(t) + \Sigma_1^{HF}(t)\right\} U_1^{+}(t,t') + \int d\bar t\, \Sigma_1^{+}(t,\bar t)\, U_1^{+}(\bar t,t').$$

$$U_1^{-}(t',t)\, H_1^{\rm eff\,\dagger}(t) = U_1^{-}(t',t)\left\{H_1 + \mathcal{U}_1(t) + \Sigma_1^{HF}(t)\right\} + \int d\bar t\, U_1^{-}(t',\bar t)\, \Sigma_1^{-}(\bar t,t).$$

The equations of motion for $U^{\pm}$ are equations of the Dyson type of Green's functions theory, and $U_1^{\pm}$ are related to the retarded and advanced Green's functions by

$$U_1^{+}(t,t') = i\hbar\, G^R(t,t'); \qquad U_1^{-}(t,t') = i\hbar\, G^A(t,t'). \tag{D.13}$$

This allows, to make use of the results of the theory of nonequilibrium Green's functions [KB89, KKER86].

Properties of $U_1^{\pm}$. Local approximation

We give a brief discussion of some important properties of $U_1^{\pm}$, starting from the coordinate representation of Eq. (D.9). For this purpose it is useful to introduce the microscopic and macroscopic variables by $r = r_1 - r_1'$, $\tau = t - t'$ and $R = \frac{1}{2}(r_1 + r_1')$, $T = \frac{1}{2}(t + t')$, respectively. The Fourier transform with respect to the microscopic variables is then defined by

$$U_1^{\pm}(p\omega, RT) = \int dr d\tau e^{-\frac{i}{\hbar}pr + i\omega\tau} U_1^{\pm}(r\tau, RT)\,.$$

The analytic properties of the propagators are well-known. First, the propagators may be continued analytically into the complex ω–plane. The analytic continuation of $U_1{}^{\pm}$ may be written as a Cauchy–type integral (cf. App. A)

$$U^{\pm}{}_1(pz, RT) = i \int \frac{d\bar{\omega}}{2\pi} \frac{A(p\bar{\omega}, RT)}{z - \bar{\omega}}, \tag{D.14}$$

with the spectral function

$$A(p\omega, RT) = U_1^{+}(p\omega, RT) - U_1^{-}(p\omega, RT)\,. \tag{D.15}$$

$U_1^{\pm}$ are analytic in the upper/lower half plane and may be continued into the lower (upper) half plane by

$$U_1^{\pm}(pz, RT) = U_1^{\mp}(pz, RT) \pm A(pz, RT)\,.$$

The propagators can be determined explicitly, if the local approximation is applied to Eq. (D.9), i.e. if all quantities depend only on the difference variables. Then, the solution of Eq. (D.9) is given by

$$U_1^{\pm}(p\omega, RT) = \frac{1}{\hbar\omega - \frac{p^2}{2m} - \Sigma_1^{\pm}(p\omega, RT) \pm i\varepsilon}\,. \tag{D.16}$$

Using the result of Eq. (D.16) and Eq. (D.15), we obtain for the spectral function

$$A(p\omega, RT) = \frac{\gamma_1(p\omega, RT)}{\left[\hbar\omega - \frac{p^2}{2m} - \mathrm{Re}\Sigma_1^{+}(p\omega, RT)\right]^2 + \left[\frac{1}{2}\gamma_1(p\omega, RT)\right]^2}. \tag{D.17}$$

with $\gamma_1 = -2\mathrm{Im}\Sigma_1^{+}$. This is a rather general result. In order to demonstrate the meaning of the spectral function, Eq. (D.17), we consider a further simplification: We calculate $Re\Sigma_1^{+}$ and γ_1 substituting the argument $\hbar\omega \to p^2/2m$. This leads to a Lorentz shape of the spectral function

$$A(p\omega RT) = \frac{\gamma_1(pRT)}{\left[\hbar\omega - E_1(pRT)\right]^2 + \left[\frac{1}{2}\gamma_1(pRT)\right]^2}. \tag{D.18}$$

This result for the spectral function yields the following expression for the single–particle propagators

$$U_1^{\pm}(\tau) = \Theta(\pm\tau)e^{-\frac{i}{\hbar}(E_1 \mp i\gamma_1)\tau}. \tag{D.19}$$

Though Eq. (D.19) has the familiar exponential form, there are some important differences in comparison to the propagator of a free particle. We have effective one–particle energies given by

$$E_1(pRT) = \frac{p^2}{2m} + \mathrm{Re}\Sigma_1^+(p\omega RT)\Big|_{\hbar\omega = E_1(pRT)}, \tag{D.20}$$

and damping of the one-particle states given by the imaginary part of the self energy, γ_1. Therefore Eq. (D.19) describes the propagator of damped quasi-particles.

Properties of the free propagators. The free propagators follow from the quasiparticle propagators (D.19) if the selfenergy corrections are neglected, $E_1 \to E_1^0$. The Fourier transform of $U_{1\ldots s}^{0\pm}$ is a special case of Eq. (D.16),

$$U_{1\ldots s}^{0\pm}(\omega, E) = \int_{-\infty}^{\infty} d\tau\, \Theta[\pm\tau]e^{-iE\tau/\hbar}e^{i\omega\tau} = \frac{-i\hbar}{\hbar\omega - E \pm i\delta}, \tag{D.21}$$

where $E = \sum_i^s E_i^0$. Very useful expressions follow for the sum and difference of the propagators, by applying the Dirac identity (cf. Appendix A),

$$U_{1\ldots s}^{0+}(\omega, E) + U_{1\ldots s}^{0-}(\omega, E) = -2\pi i\hbar\delta\left(\hbar\omega - E\right), \tag{D.22}$$

$$U_{1\ldots s}^{0+}(\omega, E) - U_{1\ldots s}^{0-}(\omega, E) = -2\hbar\frac{\mathcal{P}}{\hbar\omega - E}. \tag{D.23}$$

Appendix E

Retardation Expansion

I. Integral $B^0(\omega, t-t_0) = \int_0^{t-t_0} d\tau \; \cos[\omega\tau]\, \Phi(t-\tau)$

We calculate B^0 by expanding the integrand into a series in powers of τ, where Φ and its time derivatives are assumed to exist and to be continuous. After expanding Φ around $\tau = 0$, with

$$\Phi(t-\tau) = \Phi(t) - \tau\Phi'(t) + \ldots + \frac{(-1)^n}{n!}\tau^n \Phi^{(n)}(t) + \ldots,$$

the result of the τ–integration in B^0 can be written in the form

$$B^0(\omega, t-t_0) = \sum_{m=0}^{\infty} B^0_m(\omega, t-t_0) = \sum_{m=0}^{\infty} \frac{(-1)^m}{m!} D^0_m(\omega, t-t_0)\Phi^{(m)}(t-t_0). \quad \text{(E.1)}$$

Denoting $x = \omega(t-t_0)$, the coefficients are given as

$$\begin{aligned}
D^0_m(\omega, x) &= R_m(\omega, x) + T_m(\omega, x)\cos x + W_m(\omega, x)\sin x, \\
R_m(\omega, x) &= \begin{cases} 0, & m \text{ even}, \\ \frac{(-1)^{\frac{m+1}{2}} m!}{\omega^{m+1}}, & m \text{ odd}, \end{cases} \\
T_m(\omega, x) &= \frac{m!}{\omega^{m+1}} \sum_{s=0}^{s_{max}} (-1)^s \frac{x^{m-2s-1}}{(m-2s-1)!}, \quad s_{max} = \begin{cases} \frac{m-1}{2}, & m \text{ odd}, \\ \frac{m}{2} - 1, & m \text{ even}, \end{cases} \\
W_m(\omega, x) &= \frac{m!}{\omega^{m+1}} \sum_{s=0}^{s_{max}} (-1)^s \frac{x^{m-2s}}{(m-2s)!}, \quad s_{max} = \begin{cases} \frac{m}{2}, & m \text{ even}, \\ \frac{m-1}{2}, & m \text{ odd}. \end{cases}
\end{aligned}$$

From Eq. (E.2), the expansion terms up to the second order are

$$D^0_0(\omega, x) = \frac{\sin x}{\omega}; \quad D^0_1(\omega, x) = \frac{1}{\omega^2}\left(-1 + \cos x + x\sin x\right), \quad \text{(E.2)}$$

$$D^0_2(\omega, x) = \frac{2}{\omega^3}\left[x\cos x + \left(\frac{x^2}{2} - 1\right)\sin x\right]. \quad \text{(E.3)}$$

The expansion terms have the following properties:

1. All D_m^0 and B_m^0 may be expressed by the zeroth order terms:

$$D_m^0(\omega,t) = \begin{cases} (-1)^{\frac{m-1}{2}} \frac{d^m}{d\omega^m} C_0^0(\omega,t), & m \text{ odd}, \\ (-1)^{\frac{m}{2}} \frac{d^m}{d\omega^m} D_0^0(\omega,t), & m \text{ even}, \end{cases}$$

$$B_m^0(\omega,t) = \frac{\Phi^{(m)}(t)}{m!} \begin{cases} (-1)^{\frac{3m-1}{2}} \frac{d^m}{d\omega^m} C_0^0(\omega,t), & m \text{ odd}, \\ (-1)^{\frac{3m}{2}} \frac{d^m}{d\omega^m} D_0^0(\omega,t), & m \text{ even}, \end{cases}$$

   where C_0^0 is defined below.

2. The limit $t \to t_0$ of Eqs. (E.2) is $\lim_{t\to t_0} D_n(\omega, t-t_0) = 0$, and the leading terms at short times are

$$D_0^0(\omega, t-t_0) \approx t - t_0 + O\left[(t-t_0)^3\right], \tag{E.4}$$

$$D_1^0(\omega, t-t_0) \approx \frac{1}{2}(t-t_0)^2 + O\left[(t-t_0)^3\right]. \tag{E.5}$$

3. In the Markov limit $(t-t_0) \to \infty$, we have the asymptotic

$$D_m(\omega) \longrightarrow D_m^{0M}(\omega) = \begin{cases} (-1)^{\frac{3m-1}{2}} \frac{d^m}{d\omega^m} \mathcal{P}\left(\frac{1}{\omega}\right), & m \text{ odd}, \\ 0, & m \text{ even}. \end{cases}$$

   Notice, that for the even orders, the differentiation has to be done before the Markov limit. The asymptotic results for the expressions (E.2) are:

$$D_0^{0M}(\omega) = 0; \quad D_1^{0M}(\omega,) = -\mathcal{P}\left(\frac{1}{\omega^2}\right); \quad D_2^{0M}(\omega) = 0. \tag{E.6}$$

II. Integral $A^0(\omega, t-t_0) = \int_0^{t-t_0} d\tau \, \sin[\omega\tau] \, \Phi(t-\tau)$

Analogously as above, we obtain for A^0,

$$A^0(\omega, t-t_0) = \sum_{m=0}^{\infty} A_m^0(\omega, t-t_0) = \sum_{m=0}^{\infty} \frac{(-1)^m}{m!} C_m^0(\omega, t-t_0)\, \Phi^{(m)}(t-t_0), \tag{E.7}$$

where the following relations hold,

$$C_m^0(\omega,x) = X_m(\omega,x) + Y_m(\omega,x)\cos x + Z_m(\omega,x)\sin x, \tag{E.8}$$

$$X_m(\omega,x) = \begin{cases} \frac{(-1)^{\frac{m}{2}} m!}{\omega^m}, & m \text{ even}, \\ 0, & m \text{ odd}, \end{cases}$$

$$Y_m(\omega,x) = -W_m(\omega,x); \quad Z_m(\omega,x) = T_m(\omega,x).$$

From Eq. (E.8), the expansion terms up to the second order are

$$C_0^0(\omega,x) = \frac{1-\cos x}{\omega}; \quad C_1^0(\omega,x) = \frac{1}{\omega^2}(-x\cos x + \sin x), \tag{E.9}$$

$$C_2^0(\omega,x) = \frac{2}{\omega^3}\left[\left(-\frac{x^2}{2}+1\right)\cos x + x\sin x\right]. \tag{E.10}$$

The expansion terms have the following properties:

1. The C_m^0 and A_m^0 may be expressed by the zeroth order terms:

$$C_m^0(\omega,t) = \begin{cases} (-1)^{\frac{m}{2}} \frac{d^m}{d\omega^m} C_0^0(\omega,t), & m \text{ even}, \\ (-1)^{\frac{m+1}{2}} \frac{d^m}{d\omega^m} D_0^0(\omega,t), & m \text{ odd}, \end{cases}$$

$$A_m^0(\omega,t) = \frac{\Phi^{(m)}(t)}{m!} \begin{cases} (-1)^{\frac{3m}{2}} \frac{d^m}{d\omega^m} C_0^0(\omega,t), & m \text{ even}, \\ (-1)^{\frac{3m+1}{2}} \frac{d^m}{d\omega^m} D_0^0(\omega,t), & m \text{ odd}. \end{cases}$$

2. The limit $t \to t_0$ of Eqs. (E.9) is $\lim_{t\to t_0} C_n(\omega, t-t_0) = 0$, and the leading terms at short times are

$$C_0^0(\omega,t-t_0) \approx \frac{\omega}{2}(t-t_0)^2 + O\left[(t-t_0)^3\right], \tag{E.11}$$

$$C_1^0(\omega,t-t_0) \approx O\left[(t-t_0)^3\right]. \tag{E.12}$$

3. In the Markov limit, $t - t_0 \to \infty$, we have the asymptotics

$$C_m^0(\omega) \longrightarrow C_m^{0M}(\omega) = \begin{cases} (-1)^{\frac{3m}{2}} \frac{d^m}{d\omega^m} \mathcal{P}\left(\frac{1}{\omega}\right), & m \text{ even}, \\ 0, & m \text{ odd}. \end{cases}$$

The asymptotic results for the expressions (E.9), (E.10) are:

$$C_0^{0M}(\omega) = \mathcal{P}\left(\frac{1}{\omega}\right); \quad C_1^{0M}(\omega) = 0; \quad C_2^{0M}(\omega) = -2\mathcal{P}\left(\frac{1}{\omega^3}\right). \tag{E.13}$$

III. Integral $B(\omega,\Gamma,t-t_0) = \int_0^{t-t_0} d\tau\, \cos[\omega\tau]\, e^{-\Gamma\tau}\, \Phi(t-\tau)$

After expanding Φ around $\tau = 0$, the $\tau-$integration in B can be carried out,

$$B(\omega,\Gamma,t-t_0) = \sum_{m=0}^{\infty} B_m(\omega,\Gamma,t-t_0) = \sum_{m=0}^{\infty} \frac{(-1)^m}{m!} D_m(\omega,\Gamma,t-t_0)\Phi^{(m)}(t-t_0). \tag{E.14}$$

Introducing the variables x and also $y = \Gamma t$ and $z^2 = \omega^2 + \Gamma^2$, the weight functions up to the second retardation order are

$$\begin{aligned}
D_0(\omega,\Gamma,t) &= \frac{\Gamma}{z^2} + \frac{e^{-y}}{z^2}\left(\omega \sin x - \Gamma \cos x\right), \\
D_1(\omega,\Gamma,t) &= -\frac{\omega^2-\Gamma^2}{z^4} + \frac{e^{-y}}{z^4}\left[(2\Gamma\omega + z^2 x)\sin x + \left(\omega^2 - \Gamma^2 - z^2 y\right)\cos x\right], \\
D_2(\omega,\Gamma,t) &= -\frac{2\Gamma}{z^6}\left(3\omega^2 - \Gamma^2\right) \\
&+ \frac{e^{-y}}{z^6}\left\{\left[-2\omega\left(\omega^2 - 3\Gamma^2\right) + 4\Gamma\omega z^2 x + z^4 \tau x\right]\sin x\right. \\
&+ \left.\left[2\Gamma\left(3\omega^2 - \Gamma^2\right) + 4\left(\omega^4 - \Gamma^4\right)\tau - 2z^4 y\tau\right]\cos x\right\}.
\end{aligned}$$

The expansion terms have the following properties:

1. All D_m and B_m may be expressed by the zeroth order terms:

$$\begin{aligned}
D_m(\omega,\Gamma,t) &= \begin{cases} (-1)^{\frac{m-1}{2}} \frac{\partial^m}{\partial\omega^m} C_0(\omega,\Gamma,t), & m \text{ odd}, \\ (-1)^{\frac{m}{2}} \frac{\partial^m}{\partial\omega^m} D_0(\omega,\Gamma,t), & m \text{ even}, \end{cases} \\
&= (-1)^m \frac{\partial^m}{\partial\Gamma^m} D_0(\omega,\Gamma,t),
\end{aligned} \tag{E.15}$$

$$\begin{aligned}
B_m(\omega,\Gamma,t) &= \frac{\Phi^{(m)}(t)}{m!} \begin{cases} (-1)^{\frac{3m-1}{2}} \frac{\partial^m}{\partial\omega^m} C_0(\omega,\Gamma,t), & m \text{ odd}, \\ (-1)^{\frac{3m}{2}} \frac{\partial^m}{\partial\omega^m} D_0(\omega,\Gamma,t), & m \text{ even}, \end{cases} \\
&= \frac{\Phi^{(m)}(t)}{m!} \frac{\partial^m}{\partial\Gamma^m} D_0(\omega,\Gamma,t),
\end{aligned} \tag{E.16}$$

   where C_0 is defined below.

2. The limit $t \to t_0$ of Eqs. (E.15) is $\lim_{t\to t_0} D_n(\omega,\Gamma,t-t_0) = 0$, where the dominating terms at short times are

$$\begin{aligned}
D_0(\omega,\Gamma,t-t_0) &\approx t - t_0 + \frac{(t-t_0)^2}{2}\frac{\omega^2 - 2\omega\Gamma - \Gamma^2}{\omega^2+\Gamma^2}\Gamma + O\left[(t-t_0)^3\right], \\
D_1(\omega,\Gamma,t-t_0) &\approx \frac{(t-t_0)^2}{2} + O\left[(t-t_0)^3\right].
\end{aligned}$$

3. In the Markov limit $(t - t_0) \to \infty$, the asymptotic are

$$D_m(\omega,\Gamma) \longrightarrow D_m^M(\omega,\Gamma) = \begin{cases} (-1)^{\frac{m-1}{2}} \frac{\partial^m}{\partial\omega^m} C_0^M(\omega,\Gamma), & m \text{ odd}, \\ (-1)^{\frac{m}{2}} \frac{\partial^m}{\partial\omega^m} D_0^M(\omega,\Gamma), & m \text{ even}, \end{cases}$$

$$= (-1)^m \frac{\partial^m}{\partial \Gamma^m} D_0^M(\omega, \Gamma).$$

The asymptotic results for the expressions (E.15) are:

$$D_0^M(\omega,\Gamma) = \frac{\Gamma}{z^2}; \quad D_1^M(\omega,\Gamma) = -\frac{\omega^2-\Gamma^2}{z^4}, \tag{E.17}$$

$$D_2^M(\omega,\Gamma) = -\frac{2\Gamma}{z^6}\left(3\omega^2-\Gamma^2\right). \tag{E.18}$$

4. $\lim_{\Gamma\to 0} D_n(\omega,\Gamma,t-t_0) = D_n^0(\omega,t-t_0)$, cf. Eqs. (E.2), (E.15).

IV. Integral $A(\omega,\Gamma,t-t_0) = \int_0^{t-t_0} d\tau\, \sin[\omega\tau]\, e^{-\Gamma\tau}\, \Phi(t-\tau)$

After expanding Φ around $\tau = 0$, the result of the τ–integration in A can be written as

$$A(\omega,\Gamma,t-t_0) = \sum_{m=0}^{\infty} A_m(\omega,\Gamma,t-t_0) = \sum_{m=0}^{\infty} \frac{(-1)^m}{m!} C_m(\omega,\Gamma,t-t_0)\Phi^{(m)}(t-t_0). \tag{E.19}$$

With the variables x, y and z^2 defined above, we have explicitly,

$$C_0(\omega,\Gamma,t) = \frac{\omega}{z^2} + \frac{e^{-y}}{z^2}\left\{-\Gamma\sin x - \omega\cos x\right\}, \tag{E.20}$$

$$\begin{aligned} C_1(\omega,\Gamma,t) &= \frac{2\Gamma\omega}{z^4} \\ &+ \frac{e^{-y}}{z^4}\left\{\left[\omega^2-\Gamma^2-z^2y\right]\sin x - \left[2\omega\Gamma + z^2x\right]\cos x\right\}, \end{aligned} \tag{E.21}$$

$$\begin{aligned} C_2(\omega,\Gamma,t) &= -\frac{2\omega}{z^6}\left(\omega^2-3\Gamma^2\right) \\ &+ \frac{e^{-y}}{z^6}\left\{\left[2\Gamma\left(3\omega^2-\Gamma^2\right) + 2\left(\omega^4-\Gamma^4\right)\tau - z^4y\tau\right]\sin x\right. \\ &+ \left.\left[2\omega\left(\omega^2-2\Gamma^2\right) - 4z^2\omega y - z^4xy\right]\cos x\right\}. \end{aligned} \tag{E.22}$$

The expansion terms have the following properties:

1. The C_m and A_m may be expressed by the zeroth order terms:

$$C_m(\omega,\Gamma,t) = \begin{cases} (-1)^{\frac{m}{2}} \frac{\partial^m}{\partial\omega^m} C_0(\omega,\Gamma,t), & m \text{ even}, \\ (-1)^{\frac{m+1}{2}} \frac{\partial^m}{\partial\omega^m} D_0(\omega,\Gamma,t), & m \text{ odd}, \end{cases}$$

$$= (-1)^m \frac{\partial^m}{\partial \Gamma^m} C_0(\omega, \Gamma, t), \tag{E.23}$$

$$A_m(\omega, \Gamma, t) = \frac{\Phi^{(m)}(t)}{m!} \begin{cases} (-1)^{\frac{3m}{2}} \frac{\partial^m}{\partial \omega^m} C_0(\omega, \Gamma, t), & m \text{ even}, \\ (-1)^{\frac{3m+1}{2}} \frac{\partial^m}{\partial \omega^m} D_0(\omega, \Gamma, t), & m \text{ odd}, \end{cases}$$

$$= \frac{\Phi^{(m)}(t)}{m!} \frac{\partial^m}{\partial \Gamma^m} C_0(\omega, \Gamma, t). \tag{E.24}$$

2. The limit $t \to t_0$ of Eqs. (E.20) $\lim_{t \to t_0} C_n(\omega, \Gamma, t - t_0) = 0$, with the leading terms at short times

$$C_0(\omega, \Gamma, t - t_0) \approx \frac{\omega}{2}(t - t_0)^2 + \frac{\omega \Gamma^2}{\omega^2 + \Gamma^2} + O\left[(t - t_0)^3\right], \tag{E.25}$$

$$C_1(\omega, \Gamma, t - t_0) \approx O\left[(t - t_0)^3\right]. \tag{E.26}$$

3. In the Markov limit, $t - t_0 \to \infty$, the asymptotic are

$$C_m(\omega, \Gamma) \longrightarrow C_m^M(\omega, \Gamma) = \begin{cases} (-1)^{\frac{m}{2}} \frac{\partial^m}{\partial \omega^m} C_0(\omega, \Gamma), & m \text{ even}, \\ (-1)^{\frac{m+1}{2}} \frac{\partial^m}{\partial \omega^m} D_0(\omega, \Gamma), & m \text{ odd}. \end{cases}$$

The asymptotic results for the expressions (E.20) - (E.22) are:

$$C_0^M(\omega, \Gamma) = \frac{\omega}{z^2}; \quad C_1^M(\omega, \Gamma) = \frac{2\Gamma\omega}{z^4}; \quad C_2^M(\omega, \Gamma) = -\frac{2\omega}{z^6}\left(\omega^2 - 3\Gamma^2\right).$$

4. $\lim_{\Gamma \to 0} C_n(\omega, \Gamma, t - t_0) = C_n^0(\omega, t - t_0)$, cf. Eqs. (E.9), (E.20).

Appendix F

Numerical Solution of Quantum Kinetic Equations

In this Appendix, we briefly outline the schemes for the solution of various quantum kinetic equations,[1] (1) for the Wigner distribution f_a and (2) for the Kadanoff–Baym equations for the two-time correlation functions $g^{\gtrless}$. These equations are of integro-differential structure,

$$\mathcal{D}_{ab}(\mathbf{R}, \mathbf{p}, t) f_b(\mathbf{R}, \mathbf{p}, t) = I_a(\mathbf{R}, \mathbf{p}, t; \{f\}), \tag{F.1}$$

in the first case (a, b are component indices), and

$$\mathcal{D}^{\gtrless}_{ab}(\mathbf{R}, \mathbf{p}, t) g^{\gtrless}_b(\mathbf{R}\mathbf{p}t; \mathbf{R}'\mathbf{p}'t') = I^{\gtrless}_a(\mathbf{R}\mathbf{p}t; \mathbf{R}'\mathbf{p}'t'; \{g^{\gtrless}\}), \tag{F.2}$$

$$\mathcal{D}^{\gtrless}_{ab}(\mathbf{R}', \mathbf{p}', t') g^{\gtrless}_b(\mathbf{R}\mathbf{p}t; \mathbf{R}'\mathbf{p}'t') = I^{\gtrless}_a(\mathbf{R}\mathbf{p}t; \mathbf{R}'\mathbf{p}'t'; \{g^{\gtrless}\}), \tag{F.3}$$

in the second. They contain complicated integral terms (collision integrals) I or $I^{\gtrless}$, which in a nonlinear way depend on f or $g^{\gtrless}$ and which involve multiple momentum or space integrations and, in the non-Markovian case, also time integrations. The efficient evaluation of the collision integrals is, therefore, crucial for the performance of the whole numerical scheme. On the other hand, these equations describe the evolution of functions in time, space or/and momentum space, containing on the l.h.s. first or second order differential operators $\mathcal{D}$ which require a stable and efficient solution algorithm. Eqs. (F.1) and (F.2, F.3) are supplemented with appropriate initial conditions on f and $g^{\gtrless}$, respectively.

[1] We limit ourselves to direct integration methods. For other techniques, such as Monte Carlo methods, see e.g. Refs. [RHK94, HJ96] and references therein.

Discretization and time stepping

Discretization. The simplest and most reliable approach to solve the above equations numerically, is to introduce suitable grids in t, R and p space, i.e. the continuous variables are replaced by discrete ones, $t \to t_i, i = 1 \dots N$, $R \to r_j, j = j_{min} \dots j_{max}$ and $p \to p_k, k = k_{min} \dots k_{max}$. The discretization of R and p depends on the dimensionality and the symmetry of the system. In the general case, r_j and p_k are three-dimensional vector indices. There exist special techniques how to choose suitable grids, techniques which use multiple grids or grids which change during the evolution (adaptive grids) etc. The latter are useful if the character of the evolution is well known in advance. For codes which should be applicable to a very broad range of problems, the most robust approach is to use equidistant grids.[2] Physical quantities which depend on the continuous variables $\mathbf{R}, \mathbf{p}, t$ etc., such as the distributions f, are replaced by their values on the grid points, i.e. by a matrix, $f(\mathbf{R}, \mathbf{p}, t) \to f_{jki}$. If values of f between grid points are needed, one can use standard interpolation techniques.

Accuracy. Grid spacing and minimal/maximal values (grid boundaries) are dictated by a balance of (i) stability requirements, (ii) needed accuracy, and (iii) efficiency goals. Typically, the physical problem is characterized by a number of conservation laws on certain macroscopic variables $B(\mathbf{R}, \mathbf{p}, t)$, such as particle number, mean momentum, mean energy etc. These quantities are computed by integration over the distribution functions, which translates into summations over f_{jki}, for example,

$$\langle B \rangle (t) = \int d\mathbf{R} d\mathbf{p} W(\mathbf{R}\mathbf{p}) B(\mathbf{R}\mathbf{p}t) f(\mathbf{R}\mathbf{p}t) \approx \sum_{j=j_{min}}^{j_{max}} \sum_{k=k_{min}}^{k_{max}} W_{jk} B_{jki} f_{jki}, \qquad \text{(F.4)}$$

(where W is some weight function), see below. Obviously, the accuracy to which the sum is evaluated, depends crucially on the grid spacing. Furthermore, it must be assured that the summation covers that part of the parameter space completely where the integrand in Eq. (F.4) is nonzero.

Numerical treatment of integro-differential equations. Eqs. (F.1) and (F.2, F.3) contain the unknown functions under a differential operator (l.h.s.) and under the collision integrals. This makes it, usually, impossible to solve explicitly for f or $g^{\gtrless}$. Fortunately, this poses no problem for the numerics. To propagate the solution one step in time and in $\mathbf{p}$ or $\mathbf{R}$ space, it is always possible to take for the functions under the integral on the r.h.s. the

[2] or piecewise equidistant grids, where particular important parameter ranges are covered with a denser grid. Moreover, there exist hierarchical schemes, where processes with different typical scales are described with different grids.

(known) values from the previous step.[3]

Time stepping in one-time equations. We illustrate the treatment of the differential operator on the l.h.s. of Eq (F.2) on the simplest case, where $\mathcal{D}_{ab} = \delta_{ab}\, d/dt$. Then the discretization transforms the equation into a system of $J*K$ coupled first order ordinary differential equations (ODE) for the matrix elements f_{jk}, for which there exist numerous solution schemes, such as Runge-Kutta or predictor-corrector methods, see e.g. [PTVF92]. Rewriting f_{jk} as a vector $\mathbf{f}$, the time stepping can formally be written as for a single ODE. The intuitive scheme to advance $\mathbf{f}$ from t_i to $t_{i+1} = t_i + h$ is $\mathbf{f}_{i+1} = \mathbf{f}_i + h\mathbf{I}(t_i, \mathbf{f}_i)$ and has an error $O(h^2)$. The error can be reduced by a special choice of substeps. For example, a simple and reliable scheme is the *4th order Runge-Kutta formula* which involves four evaluation of the right hand side of the kinetic equation and has an error $O(h^5)$, and thus, works stable with a bigger time step:

$$\begin{aligned}
\mathbf{k}_1 &= h\mathbf{I}(t_i, \mathbf{f}_i); \quad \mathbf{k}_2 = h\mathbf{I}(t_i + h/2, \mathbf{f}_i + \mathbf{k}_1/2); \\
\mathbf{k}_3 &= h\mathbf{I}(t_i + h/2, \mathbf{f}_i + \mathbf{k}_2/2); \quad \mathbf{k}_4 = h\mathbf{I}(t_i + h, \mathbf{f}_i + \mathbf{k}_3); \\
\mathbf{f}_{i+1} &= \mathbf{f}_i + \frac{\mathbf{k}_1}{6} + \frac{\mathbf{k}_2}{3} + \frac{\mathbf{k}_3}{3} + \frac{\mathbf{k}_4}{6} + O(h^5). \qquad \text{(F.5)}
\end{aligned}$$

Time stepping in two-time equations. To solve the Kadanoff-Baym equations, one needs to advance two functions $g^>$ and $g^<$ simultaneously[4] in two time directions along t and t', starting from the initial point (t_0, t_0). There are two relations which are crucial for the solution:

$$g^{\gtrless}_{\mu_1\mu_2}(t, t') = -[g^{\gtrless}_{\mu_2\mu_1}(t', t)]^*; \qquad g^{>}_{\mu_1\mu_2}(t, t) = \left[\frac{1}{i\hbar} + g^{<}_{\mu_1\mu_2}(t, t)\right], \qquad \text{(F.6)}$$

which is valid for each momentum value. The first shows that, in fact, each of the two functions needs to be known only in half of the $t - t'$ plane (and on the diagonal). On the other hand, the second relates the two functions to one another on the diagonal. With these relations, one can directly apply the schemes from the one-time case discussed above to the simultaneous evolution in t and t' direction also. Both are "synchronized" on the diagonal by the second condition (F.6).

[3] This is only a matter of a sufficiently small time (or R, p) step). Typically, the collision terms change, as a function of R, p, t less rapid than the terms on the l.h.s.

[4] In nonequilibrium, the knowledge of two Green's functions is needed (see Ch. 12). Of the many choices, the functions $g^>$ and $g^<$ are numerically the most convenient pair.

Evaluation of Markovian quantum collision integrals

In the Markov limit, the collision integral is related to the in ("<") and out (">") scattering rates by

$$I_a(\mathbf{p},t) = \sum_b I_{ab}(\mathbf{p},t) = \Sigma_a^<(\mathbf{p},t) f_a^>(\mathbf{p},t) - \Sigma_a^>(\mathbf{p},t) f_a^<(\mathbf{p},t), \qquad \text{(F.7)}$$

where $f_a^< = f_a$ and $f_a^> = 1 - f_a$. It is useful to introduce the abbreviations $E_{ab} = E_a + E_b$, where $E_a = p^2/2m_a$, $\overline{E}_a = \overline{p}^2/2m_a$, and also

$$\begin{aligned} \Phi_{ab}(\overline{p},\overline{p}';p,p',t) &= f_a(\overline{p},t) f_b(\overline{p}',t)\,[1 - f_a(p,t)]\,[1 - f_b(p',t)] \\ &- f_a(p,t) f_b(p',t)\,[1 - f_a(\overline{p},t)]\,[1 - f_b(\overline{p}',t)]\,. \end{aligned} \qquad \text{(F.8)}$$

Further, we denote by V_{ab} the Fourier transform of the Debye (Yukawa) potential $V_{ab}(q,t) = 4\pi e_a e_b/[q^2 + \kappa^2(t)]$, where the potential range κ^{-1} is allowed to slowly vary in time.

Numerical Integration. The critical part in quantum scattering integral are the multiple momentum integrations. The discretization is trivially performed,

$$\int dp F(p) \longrightarrow \sum_{k=k_{min}}^{k_{max}} W_k F_k, \qquad \text{(F.9)}$$

by introduction of suitable weights W_k. In the simplest case of the trapezoidal rule, $W_{k_{min}} = W_{k_{max}} = 1/2$ and $W_k = 1$ elsewhere. There exists a great variety of weights optimized for special types of functions F. For a flexible code, however, simple weights are a good choice.

Markovian quantum Landau collision integral

The Landau collision integral (static Born approximation), Ch. 7, is of second order in the interaction potential,

$$\begin{aligned} I_{ab}^{ML\pm}(\mathbf{p}t) &= \int \frac{d\mathbf{p}'}{(2\pi)^3}\frac{d\overline{\mathbf{p}}}{(2\pi)^3}\frac{d\overline{\mathbf{p}}'}{(2\pi)^3}\, V_{ab}(|\mathbf{p}-\overline{\mathbf{p}}|)\,[V_{ab}(|\mathbf{p}-\overline{\mathbf{p}}|) \mp V_{ab}(|\mathbf{p}-\overline{\mathbf{p}}'|)] \\ &\times\; 2\pi\delta(E_{ab} - \overline{E}_{ab})(2\pi)^3\delta(\overline{\mathbf{p}} + \overline{\mathbf{p}}' - \mathbf{p} - \mathbf{p}')\Phi_{ab}(\overline{\mathbf{p}},\overline{\mathbf{p}}';\mathbf{p},\mathbf{p}',t), \end{aligned}$$

and contains a direct ($\sim V_{ab}(|\mathbf{p}-\overline{\mathbf{p}}|)^2$) and an exchange scattering part (term with prefactor ∓ 1). The delta functions allow to perform one vector and one scalar integrations, so, in general, one is left with a five-fold integral. Further simplifications are possible only for special symmetries.

Isotropic case. Here, it is assumed that $f(\mathbf{p}) = f(p)$. Then, the direct integral transforms into [BSP+92, KBSK97][5]

$$
\begin{aligned}
I_{ab}^{ML}(p,t) &= m_b \frac{1}{(2\pi)^3} \int_0^\infty dq\, q \int_0^\infty dp'(p'+\overline{p}) \int_{-1}^{1} dz_1 V^2(q) \qquad \text{(F.10)} \\
&\times \; \Phi_{ab}\left(\sqrt{q^2+p^2+2qpz_1}, \sqrt{(p'+\overline{p})^2 - \left[\frac{m_b}{m_a}(q^2+2qpz_1)\right]}; p, p', t \right).
\end{aligned}
$$

where $\mathbf{q} = \mathbf{p} - \overline{\mathbf{p}}$ and $z_1 = \cos(\mathbf{q}, \mathbf{p})$.

Banyai's method for the direct integral.[6] In the isotropic case, one can remove an additional integration in the direct scattering integral if the potential can be integrated analytically, as in our case:

$$
\int \frac{dq}{(q^2+\kappa^2)^2} = \frac{1}{2\kappa^2}\left[\frac{1}{\kappa}\arctan\left(\frac{q}{\kappa}\right) + \frac{q}{q^2+\kappa^2}\right]. \qquad \text{(F.11)}
$$

This allows to derive

$$
\begin{aligned}
I_{ab}^{ML}(p,t) &= \frac{(4\pi e_a e_b)^2}{2\kappa^2}\frac{m_b}{p}\frac{1}{(2\pi)^3} \int_0^\infty dx\, x \int_0^\infty dy\, y \left[\frac{1}{\kappa}\arctan\left(\frac{q}{\kappa}\right) + \frac{q}{q^2+\kappa^2}\right]_{q_1}^{q_2} \\
&\times \; \Phi_{ab}(x, y, p, \sqrt{y^2 + \frac{m_b}{m_a}(x^2-p^2)}, t), \qquad \text{(F.12)} \\
q_1 &= \max\{|p-\sqrt{x}|, |p'-\sqrt{y}|\}; \quad q_2 = \min\{p+\sqrt{x}, p'+\sqrt{y}\}, \quad \text{(F.13)}
\end{aligned}
$$

where $p' = \sqrt{x^2+y^2-p^2}$ and $q_1 < q_2$. This expression is simpler than Eq. (F.10), however, it requires some care, because the integrand is varying rapidly as a function of x and y (for details, see [KBSK97]).

Markovian Boltzmann (T-matrix) collision integral

T–matrix scattering rates. The scattering rates in T-matrix approximation are derived from the Markov limit of Eq. (8.42):

$$
\begin{aligned}
\Sigma_a^{\gtrless}(\boldsymbol{p_a}, t) &= \frac{1}{i\mathcal{V}\hbar}\sum_b \int \frac{d\boldsymbol{p_b}}{(2\pi\hbar)^3}\frac{d\overline{\boldsymbol{p}}_{\boldsymbol{a}}}{(2\pi\hbar)^3}\frac{d\overline{\boldsymbol{p}}_{\boldsymbol{b}}}{(2\pi\hbar)^3}\, 2\pi\,\delta\left(E_a + E_b - \bar{E}_a - \bar{E}_b\right) \\
& \frac{1}{2!}\left|\langle\, \boldsymbol{p_a}\,\boldsymbol{p_b}\,|\mathbf{T}_{ab}(E_{ab}+i\epsilon)|\,\overline{\boldsymbol{p}}_{\boldsymbol{b}}\,\overline{\boldsymbol{p}}_{\boldsymbol{a}}\,\rangle^{\pm}\right|^2 \bar{f}_a^{\lessgtr}\, f_b^{\gtrless}\, \bar{f}_b^{\lessgtr}, \qquad \text{(F.14)}
\end{aligned}
$$

---

[5] The exchange integral is treated analogously and will be omitted.

[6] This method has been introduced independently by various authors, among them L. Banyai. Notice that it does not work for the exchange integral, see also ref. [KBSK97].

where $T_{ab}(E_{ab} + i\epsilon)$ is the retarded (anti-)symmetrized ($\pm$) on-shell T-matrix, which is related to the differential scattering cross section by [Joa79]

$$\frac{d\sigma_{ab}(p,\Omega)}{d\Omega} = (2\pi\hbar)^6 (2\pi)^4 \hbar^2 m_{ab}^2 \left| \langle \boldsymbol{p} \,|\mathbf{T}_{ab}|\, \overline{\boldsymbol{p}} \rangle^{\pm} \right|^2_{|\boldsymbol{p}|=|\overline{\boldsymbol{p}}|}. \tag{F.15}$$

Here, $\boldsymbol{p}$ is the momentum of relative motion and $m_{ab} = m_a m_b/(m_a + m_b)$ denotes the reduced mass.

For the derivation of explicit expressions for the scattering rates, we follow [GKSB98]. Considering a non-degenerate spatially homogeneous system and introducing relative and center of mass variables $\mathbf{p} = \mathbf{p}_a - \mathbf{p}_b$, $2\mathbf{P} = \mathbf{p}_a + \mathbf{p}_b$, we can write the T–matrix in the following form

$$\left| \langle \boldsymbol{p_a}\, \boldsymbol{p_b} \,|\mathbf{T}_{ab}|\, \overline{\boldsymbol{p_b}}\, \overline{\boldsymbol{p_a}} \rangle^{\pm} \right|^2 = (2\pi\hbar)^3 \delta(\boldsymbol{P} - \overline{\boldsymbol{P}}) \left| \langle \boldsymbol{p} \,|\mathbf{T}_{ab}|\, \overline{\boldsymbol{p}} \rangle \right|^2. \tag{F.16}$$

Furthermore, we introduce the angles $\angle(\boldsymbol{p}, \overline{\boldsymbol{p}}) = \vartheta$, $\angle(\boldsymbol{p}, \boldsymbol{p_1}) = \vartheta_1$, and $\angle(\overline{\boldsymbol{p}}, \boldsymbol{p_1}) = \vartheta_2$ with the abbreviations $\cos(\vartheta) = x$, $\cos(\vartheta_1) = x_1$, and $\cos(\vartheta_2) = x_2$ and make use of the well–known relation of spherical trigonometry $x_2 = x\, x_1 + \sin(\vartheta)\sin(\vartheta_1)\cos(\varphi_x)$. In the case of isotropic distribution functions, $f(\mathbf{p}) = f(p)$, part of the integration in Eq. (F.14) can be performed, and we get for the scattering rates

$$\begin{aligned} \Sigma_a^<(p_a,t) &= \frac{4\pi}{i(2\pi\hbar)^3} \sum_b \frac{m_b^3}{m_{ab}^4} \int_0^\infty dp \int_{-1}^1 dx_1 \int_{-1}^1 dx \int_0^{2\pi} d\varphi_x \, p^3 \, \frac{d\sigma(p,\Omega)}{d\Omega} \\ & \quad f_a\left(p_a^2 + 2p^2 - 2\, p_a\, p\, x_1 + 2\, p_a\, p\, x_2\right) \\ & \quad f_b\left(\gamma^2 \left[p_a^2 + p^2 - p_a\, p\, x\right] + 2\,\gamma \left[p^2\, x - p_a\, p\, x_2\right] + p^2\right), \end{aligned} \tag{F.17}$$

$$\begin{aligned} \Sigma_a^>(p_a,t) &= \frac{4\pi}{i(2\pi\hbar)^3} \sum_b \frac{m_b^3}{m_{ab}^4} \int_0^\infty dp \int_{-1}^1 dx_1 \; p^3\, \sigma^{tot}(p) \\ & \quad f_b\left(\gamma^2\, p_a^2 + (1+\gamma)^2\, p^2 - 2\gamma(1+\gamma)\, p_a\, p\, x_1\right). \end{aligned} \tag{F.18}$$

Here $\gamma = m_b/m_a$ is the mass ratio, $\sigma^{tot}(p,t)$ is the total cross section (angle integrated differential cross section) which depends on the macroscopic time via the screening parameter and the nonequilibrium distribution functions, and p denotes the modulus of the relative momentum. These expressions are straightforwardly generalized to include Pauli blocking, but this is only appropriate if these effects are consistently included in the cross sections (i.e. in the Schrödinger equation).

Equilibrium scattering rates. For non-degenerate charged particles in thermal equilibrium, $f_b(p^2) = f_b^0(p^2) = (n_b\Lambda_b^3)/(2s_b+1)\, \exp[-p^2/2m_b k_B T]$, where $\Lambda_b = (2\pi\hbar^2/m_b k_B T)^{1/2}$ is the thermal wave length. In this case, considerable simplification of the scattering rates are possible, and it follows

$$\begin{aligned}\Sigma_a^{>}(p_a) &= \frac{4\pi}{i(2\pi\hbar)^3}\frac{m_b^2 m_a}{m_{ab}^3}\frac{n_b\Lambda_b^3\, k_B T}{p_a}\int_0^\infty dp\, p^2\,\sigma^{tot}(p)\,\mathrm{e}^{m_b/2k_B T} \\ & \quad \left[\mathrm{e}^{-(p_a/m_a - p/m_{ab})^2} - \mathrm{e}^{-(p_a/m_a + p/m_{ab})^2}\right], \qquad \text{(F.19)}\end{aligned}$$

whereas $\Sigma_a^{<}(p_a)$ can be calculated from $\Sigma_a^{<}(p_a)$ using the condition of detailed balance, i.e. $\Sigma_a^{<}(p_a) = \Sigma_a^{>}(p_a)\, f_a^0(p_a)$.

Scattering cross section. For non-degenerate systems, the scattering cross sections are efficiently calculated from a phase shift analysis. Using a partial wave expansion of the radial Schrödinger equation for the two–particle scattering states, the differential cross section can be expressed in terms of the scattering phase shifts by [Tay72, GKSB98]

$$\begin{aligned}\frac{d\sigma}{d\Omega} &= \frac{\hbar^2}{p^2}\sum_{l,l'}^{\infty}(2l+1)(2l'+1)\,\sin\delta_l\sin\delta_{l'}\cos(\delta_l-\delta_{l'})\,P_l(\cos\vartheta)P_{l'}(\cos\vartheta) \\ &\times \left\{1+\delta_{a,b}\frac{1}{4}\left[A(l,l')+B(l,l')\right]+\frac{3}{4}\left[A(l,l')-B(l,l')\right]\right\}, \qquad \text{(F.20)}\end{aligned}$$

where $P_l(\cos\vartheta)$ are the Legendre polynomials, δ_l the scattering phase shifts, and l denotes the quantum number of angular momentum. Furthermore, we introduced the functions $A(l,l') \equiv (-1)^l(-1)^{l'}$ and $B(l,l') \equiv (-1)^l + (-1)^{l'}$. The second term in the paranthesis (proportional to $\delta_{a,b}$) is due to exchange effects in the case of identical particles.

The total cross section for scattering of different particles can be written as [Joa79]

$$\sigma_{ab}^{tot}(p) = \frac{4\pi\hbar^2}{p^2}\sum_{l=0}^{\infty}(2l+1)\,\sin^2\delta_l, \qquad (a\neq b), \qquad \text{(F.21)}$$

and for scattering of identical particles, the exchange contribution has to be included (second term)

$$\sigma_{aa}^{tot}(p) = \frac{2\pi\hbar^2}{p^2}\sum_{l=0,2,4}^{\infty}(2l+1)\,\sin^2\delta_l + \frac{6\pi\hbar^2}{p^2}\sum_{l=1,3,5}^{\infty}(2l+1)\,\sin^2\delta_l. \qquad \text{(F.22)}$$

The scattering phase shifts are obtained from solving the radial Schrödinger equation using a partial wave expansion [Tay72, GKSB98].

Evaluation of non-Markovian quantum collision integrals

Non-Markovian collision integrals contain the distributions at retarded times (cf. Ch. 6) and are of the form[7]

$$I_a(\mathbf{p},t) = \sum_b I_{ab}(\mathbf{p},t) = \int_0^{t-t_0} d\tau \Big\{ \Sigma_a^<(\mathbf{p},\tau,t-\tau) f_a^>(\mathbf{p},t-\tau) - \Sigma_a^>(\mathbf{p},\tau,t-\tau) f_a^<(\mathbf{p},t-\tau) \Big\}. \quad \text{(F.23)}$$

Time integral. Compared to the Markov case, there is an additional time integration to perform. It does not cause principal problems and can be computed as the momentum integrals (see above). Both, a constant time step or an adaptive (varying step) scheme are feasible. Care has to be taken for cases where $[E_{ab} - \overline{E}_{ab}]\tau/\hbar$ is large (especially with increasing time t) due to rapid oscillations of the integrand of the time integral. But typically, these values appear with a small weight. Moreover, usually a finite memory depth (determined by γ) reduces this problem.

Non-Markovian quantum Landau collision integral

$$I_{ab}^{L\pm}(\mathbf{p}t) = \frac{2}{\hbar^2} \int_0^{t-t_0} d\tau \int \frac{d\mathbf{p}'}{(2\pi)^3} \frac{d\overline{\mathbf{p}}}{(2\pi)^3} \frac{d\overline{\mathbf{p}}'}{(2\pi)^3} V_{ab}(|\mathbf{p}-\overline{\mathbf{p}}|) \times$$
$$[V_{ab}(|\mathbf{p}-\overline{\mathbf{p}}|) \mp V_{ab}(|\mathbf{p}-\overline{\mathbf{p}}'|)]\, e^{-(\gamma_{ab}+\overline{\gamma}_{ab})\tau/\hbar} \cos[(\epsilon_{ab}-\overline{\epsilon}_{ab})\tau/\hbar]$$
$$\times (2\pi)^3 \delta(\overline{\mathbf{p}}+\overline{\mathbf{p}}'-\mathbf{p}-\mathbf{p}')\Phi_{ab}(\overline{\mathbf{p}},\overline{\mathbf{p}}';\mathbf{p},\mathbf{p}',t-\tau), \quad \text{(F.24)}$$

where the renormalized energy and damping are given by $\epsilon_{ab} = E_a + E_b + \Delta_a + \Delta_b$ and $\gamma_{ab} = \gamma_a + \gamma_b$ and depend on momentum and on the actual time t. A manageable and consistent approximation is (cf. Ch. 7)

$$\gamma_a(\mathbf{p}t) = \sum_b \frac{2}{\hbar} \int_0^{t-t_0} d\tau \int \frac{d\mathbf{p}'}{(2\pi)^3} \frac{d\overline{\mathbf{p}}}{(2\pi)^3} \frac{d\overline{\mathbf{p}}'}{(2\pi)^3} V_{ab}(|\mathbf{p}-\overline{\mathbf{p}}|) \times$$
$$[V_{ab}(|\mathbf{p}-\overline{\mathbf{p}}|) \mp V_{ab}(|\mathbf{p}-\overline{\mathbf{p}}'|)]\, e^{-(\gamma_{ab}+\overline{\gamma}_{ab})\tau/\hbar} \cos[(\epsilon_{ab}-\overline{\epsilon}_{ab})\tau/\hbar]$$
$$\times (2\pi)^3 \delta(\overline{\mathbf{p}}+\overline{\mathbf{p}}'-\mathbf{p}-\mathbf{p}') R_{ab}(\overline{\mathbf{p}},\overline{\mathbf{p}}';\mathbf{p}',t-\tau), \quad \text{(F.25)}$$

[7] The additional integrals arising from initial correlations are simpler (they do not contain a time integral) and will not be discussed.

while Δ_a follows from the substitution cos $\rightarrow$ sin, and R_{ab} is given by

$$\begin{aligned} R_{ab}(\overline{p},\overline{p}';p',t) &= f_a(\overline{p},t)f_b(\overline{p}',t)\,[1-f_b(p',t)] \\ &+ f_b(p',t)\,[1-f_a(\overline{p},t)]\,[1-f_b(\overline{p}',t)]\,. \end{aligned} \tag{F.26}$$

Carrying out one momentum integration using the delta function, the direct term can be rewritten as (and, analogously for the exchange term)

$$\begin{aligned} I^L_{ab}(\mathbf{p}t) &= \frac{2}{\hbar^2}\int_0^{t-t_0} d\tau \int \frac{d\mathbf{p}'}{(2\pi)^3}\frac{d\mathbf{q}}{(2\pi)^3} V^2_{ab}(q) e^{-(\gamma_{ab}+\overline{\gamma}_{ab})\tau/\hbar} \\ &\times \cos\left[(\epsilon_{ab}-\overline{\epsilon}_{ab})\tau/\hbar\right] \Phi_{ab}(|\mathbf{p}+\mathbf{q}|,|\mathbf{p}'-\mathbf{q}|;\mathbf{p},\mathbf{p}',t-\tau). \end{aligned} \tag{F.27}$$

Isotropic case. If $f(\mathbf{p}) = f(p)$, one can introduce spherical coordinates with $z_{1,2}$ being the cosine of the angles between $\mathbf{p},\mathbf{q}$ and $\mathbf{p}',\mathbf{q}$, respectively, and perform the two polar angle integrations,

$$\begin{aligned} I^L_{ab}(\mathbf{p}t) &= \frac{2}{\hbar^2}\int_0^{t-t_0} d\tau \int_0^\infty \frac{dp'}{(2\pi)^2}\frac{dq}{(2\pi)^2} \int_{-1}^1 dz_1 dz_2 V^2_{ab}(q) e^{-(\gamma_{ab}+\overline{\gamma}_{ab})\tau/\hbar} \\ &\times \cos\left[(\epsilon_{ab}-\overline{\epsilon}_{ab})\tau/\hbar\right] \Phi_{ab}(|\mathbf{p}+\mathbf{q}|,|\mathbf{p}'-\mathbf{q}|;p,p',t-\tau), \end{aligned} \tag{F.28}$$

where $|\mathbf{p}+\mathbf{q}| = \sqrt{p^2+q^2+2z_1pq}$ and $|\mathbf{p}'-\mathbf{q}| = \sqrt{p^{2'}+q^2-2z_1p'q}$. This integral is feasible, but very time consuming. For the direct integral, essential simplification (reduction to three momentum integrations) is possible using, as above, *Banyai's method*, with the result

$$\begin{aligned} I^L_{ab}(\mathbf{p}t) &= \frac{2}{\hbar^2}\int_0^{t-t_0} d\tau \int_0^\infty \frac{dp'}{(2\pi)^2} \int_0^\infty dx\, x\, dy\, y \left[\frac{1}{\kappa}\arctan\left(\frac{q}{\kappa}\right) + \frac{q}{q^2+\kappa^2}\right]_{q_1}^{q_2} \\ &\times e^{-(\gamma_{ab}+\overline{\gamma}_{ab})\tau/\hbar} \cos\left[(\epsilon_{ab}-\overline{\epsilon}_{ab})\tau/\hbar\right] \Phi_{ab}(x,y;p,p',t-\tau), \end{aligned} \tag{F.29}$$

where $q_{1,2}$ are given by Eq. (F.13). The selfenergy shift and damping γ_a and Δ_a are treated analogously.

An even more efficient integration scheme is based on *Fast Fourier transforms*, which we explain for the selfenergies (see below),[8] and which allows to treat the full anisotropic problem.

[8]For its application one has to rewrite in Eq. (F.27) the cos factor as real part of a complex exponential, which renders the momentum integrals in the form of two successive convolutions.

Kadanoff-Baym equations

As an example, we consider the case of the interband Kadanoff-Baym equations, cf. Sec. 12.5[9], given by the pairwise adjoint equations

$$\begin{aligned}
\left[i\hbar\frac{\partial}{\partial t_1} - \epsilon_{\mu_1}(\mathbf{k})\right] g^{\gtrless}_{\mu_1\mu_2}(\mathbf{k}t_1t_2) &= \sum_{\bar{\mu}} \hbar\Omega_{\mu_1\bar{\mu}}(\mathbf{k}t_1) g^{\gtrless}_{\bar{\mu}\mu_2}(\mathbf{k}t_1t_2) + I^{\gtrless}_{\mu_1\mu_2}(\mathbf{k}t_1t_2), \\
\left[-i\hbar\frac{\partial}{\partial t_2} - \epsilon_{\mu_2}(\mathbf{k})\right] g^{\gtrless}_{\mu_1\mu_2}(\mathbf{k}t_1t_2) &= \sum_{\bar{\mu}} g^{\gtrless}_{\mu_1\bar{\mu}}(\mathbf{k}t_1t_2)\hbar\Omega_{\bar{\mu}\mu_2}(\mathbf{k}t_2) - I^{\gtrless *}_{\mu_2\mu_1}(\mathbf{k}t_2t_1),
\end{aligned}$$

where the effective Rabi energy contains the total electric field $\mathcal{E}$, following from Maxwell's equations, and the Hartree-Fock renormalization

$$\hbar\Omega_{\mu_1\mu_2}(\mathbf{k}t) = -d_{\mu_1\mu_2}\mathcal{E}(t)(1-\delta_{\mu_1\mu_2}) + i\hbar\sum_{\mathbf{k}'} g^{<}_{\mu_1\mu_2}(\mathbf{k}'tt)V(\mathbf{k}-\mathbf{k}') \ , \qquad \text{(F.30)}$$

and the collision integrals are given by (we drop the initial correlation term)

$$\begin{aligned}
I^{\gtrless}_{\mu_1\mu_2}(\mathbf{k}t_1t_2) &= \sum_{\bar{\mu}} \int_{t_0}^{t_1} d\bar{t}[\sigma^{>}_{\mu_1\bar{\mu}}(\mathbf{k}t_1\bar{t}) - \sigma^{<}_{\mu_1\bar{\mu}}(\mathbf{k}t_1\bar{t})] g^{\gtrless}_{\bar{\mu}\mu_2}(\mathbf{k}\bar{t}t_2) \qquad \text{(F.31)} \\
&- \sum_{\bar{\mu}} \int_{t_0}^{t_2} d\bar{t}\sigma^{\gtrless}_{\mu_1\bar{\mu}}(\mathbf{k}t_1\bar{t})[g^{>}_{\bar{\mu}\mu_2}(\mathbf{k}\bar{t}t_2) - g^{<}_{\bar{\mu}\mu_2}(\mathbf{k}\bar{t}t_2)] \ .
\end{aligned}$$

The *time integral* is treated exactly as discussed for the non-Markovian integrals above, while the idea of the time stepping was explained above, so we focus on the selfenergies.

Selfenergy in Born approximation

For the direct scattering term, we have

$$\begin{aligned}
\sigma^{\gtrless}_{\mu_1\mu_2}(\mathbf{k}t_1t_2) &= i\hbar\sum_{\mathbf{k}'} V(\mathbf{k}'-\mathbf{k},t_1)\, V(\mathbf{k}'-\mathbf{k},t_2)\, \pi^{\gtrless}(\mathbf{k}-\mathbf{k}',t_1t_2)\, g^{\gtrless}_{\mu_1\mu_2}(\mathbf{k}'t_1t_2), \\
\pi^{\gtrless}(\mathbf{q},t_1t_2) &= -i\hbar\sum_{\bar{\mathbf{k}}\lambda\mu} g^{\gtrless}_{\mu\lambda}(\bar{\mathbf{k}}+\mathbf{q},t_1t_2)\, g^{\lessgtr}_{\lambda\mu}(\bar{\mathbf{k}},t_2t_1), \qquad \text{(F.32)}
\end{aligned}$$

and similar for the exchange term. The structure of $\Sigma^{\gtrless}$ is analogous to the non-Markovian Landau collision integral, and all integration schemes discussed

[9] Other special cases differ mainly in the actual operator on the l.h.s.

there, apply here too. We, therefore, focus on the FFT method which is the most efficient one for the direct scattering term.

Fast Fourier transform method. Obviously, $\Sigma^{\gtrless}$ has the form of two successive convolutions (one in Π), which allows to apply the convolution theorem: Both convoluted factors are first Fourier transformed, multiplied in Fourier space and transformed back. Using three-dimensional FFT, this yields $\Sigma^{\gtrless}$ *at once for all momenta* **k**. There is a remarkable performance gain observed, which is achieved due to the existence of highly efficient FFT routines. We give a brief characterization of the capabilities of this approach [KBBK98][10]. In a naive estimate, the numerical effort needed to evaluate $\Sigma^{\gtrless}$ in the statically screened Born Approximation for one carrier component, would scale with the time duration of the run T (which may be replaced by the smaller memory depth) and the number of grid points per dimension N as T^2N^{3d} where d is the number of dimensions of the system. In an $M-$band system the number of collision terms, including polarization scattering, increases like M^4 (however, the number of operations increases only like M^2, cf. Eq. (F.32). This scaling looks quite prohibitive. However, (1) the FFT approach has a much more favorable scaling with the number of k-points like $T^2(N\log N)^d$. Furthermore, (2) it shows surprisingly high stability (very good conservation of carrier number and total energy) even with a large time step. (3) In contrast to other techniques, one always computes the full six-dimensional momentum integral, allowing for arbitrary anisotropic situations (e.g. warping, anisotropic distributions [BKB97]). (4) 2D calculations are becoming essentially simpler than 3D ones. (5) The scheme works for all non-Markovian scattering integrals, including those for one-time functions (Wigner distribution). (6) The Markov limit is readily obtained from the same code by modifying the time integration [Köh95].

Of course the applicability of this scheme is limited to scattering integrals of convolution type. It is not possible to treat directly exchange scattering integrals, strongly inhomogeneous systems (where momentum conservation does not hold) or strong scattering (beyond the Born approximation), For further details of the computational scheme, see Ref. [KKY].

[10]This technique was first used in solving KBE by Danielewicz [Dan84b] with a Fourier-Bessel Transform in cylindrical coordinates. Cartesian coordinates FFT was first used by Köhler [Köh95] and has the advantage of being able to treat arbitrary anisotropic situations.

Solution of Dyson-type integral equations

We briefly discuss how to solve integral equations of the Dyson type. As the simplest example, we consider the non-relativistic random phase approximation for the selfenergy, cf. Ch. 12,

$$
\begin{aligned}
\sigma_a^{\gtrless}(kt_1t_2) &= i\sum_{k'} V_s^{\gtrless}(k'-k,t_1t_2)\, g_a^{\gtrless}(k't_1t_2), && \text{(F.33)}\\
V_s^{\gtrless}(t_1t_2) &= \int_{t_0}^{t_1} d\bar{t}_1 \int_{t_0}^{t_2} d\bar{t}_2 V_s^{+}(t_1\bar{t}_1)\,\pi^{\gtrless}(\bar{t}_1\bar{t}_2)V_s^{-}(\bar{t}_2,t_2), && \text{(F.34)}\\
V_s^{\pm}(t_1t_2) &= V\delta(t_1-t_2)+V\int_{t_1}^{t_2} d\bar{t}\,\pi^{\pm}(t_1\bar{t})\,V_s^{\pm}(\bar{t}t_2), && \text{(F.35)}
\end{aligned}
$$

where Eqs. (F.34) and (F.35) are fulfilled for each value of the momentum, and V is the static potential. First, it is convenient to separate the singular part by defining $\tilde{V}_s^{\pm}(t_1t_2) = V_s^{\pm}(t_1t_2) - V\delta(t_1-t_2)$. Then Eqs. (F.34) and (F.35) transforms into

$$
\begin{aligned}
V_s^{\gtrless}(t_1t_2) &= V^2\pi^{\gtrless}(t_1t_2) + V\int_{t_0}^{t_2} d\bar{t}_2\pi^{\gtrless}(t_1\bar{t}_2)\tilde{V}_s^{-}(\bar{t}_2,t_2) && \text{(F.36)}\\
&+ V\int_{t_0}^{t_1} d\bar{t}_1\tilde{V}_s^{+}(t_1\bar{t}_1)\,\pi^{\gtrless}(\bar{t}_1t_2) + \int_{t_0}^{t_1} d\bar{t}_1\int_{t_0}^{t_2} d\bar{t}_2\tilde{V}_s^{+}(t_1\bar{t}_1)\,\pi^{\gtrless}(\bar{t}_1\bar{t}_2)\tilde{V}_s^{-}(\bar{t}_2t_2),\\
\tilde{V}_s^{\pm}(t_1t_2) &= V^2\pi^{\pm}(t_1t_2) + V\int_{t_1}^{t_2} d\bar{t}\,\pi^{\pm}(t_1\bar{t})\,\tilde{V}_s^{\pm}(\bar{t}t_2). && \text{(F.37)}
\end{aligned}
$$

These equations are readily solved by discretizing the two time arguments of all functions. The time integrals are treated as in the case of the non-Markovian collision integrals above. On the grid, one can even solve Eq. (F.37) analytically: one solves for $\tilde{V}_s^{\pm}(t_1t_2)$ (which appears under the time integral also with a certain integration weight) in terms of known values of $\tilde{V}_s^{\pm}$.[11] With $\tilde{V}_s^{\pm}$ found, Eq. (F.37) yields immediately $V_s^{\gtrless}$, and from Eq. (F.33) follow directly $\sigma^{\gtrless}$.

For more complicated cases, such as the ladder approximation where Eq. (F.35) is replaced by the Lippmann-Schwinger equation, the solution is similar, however, it is essentially more involved due to the complex dependence of all functions on the momentum arguments and a mixing of different momenta. There one has to use matrix inversion techniques.

---

[11] i.e. values known from previous time steps

Bibliography

[AB59] A.I. Akhiezer and V.B. Berestezki. *Quantum Electrodynamics.* Fizmatgiz, Moscow, 2nd edition, 1959. (russ.).

[ABR84] A.F. Aleksandrov, L.S. Bogdankievich, and A.A. Rukhadze. *Principles of Plasma Electrodynamics.* Springer-Verlag, 1984.

[AGD62] A.A. Abrikosov, L. Gor'kov, and I.E. Dzialoshinskii. *Methods of Quantum Field Theory in Statistical Physics.* Nauka, Moscow, 1962. (russ.), engl. transl.: Dover, New York, 1975.

[Aic91] J. Aichelen. "Quantum" Molecular Dynamics – a dynamical microscopic n–body approach to investigate fragment formation and the nuclear equation of state in heavy ion collisions. *Phys. Rep.*, **202**:233, 1991.

[AM98] V.M. Axt and Mukamel. Nonlinear optics of semiconductor and molecular nanostructures; a common perspective. *Rev. Mod. Phys.*, **70**:145, 1998.

[AP77] A.I. Akhiezer and S.V. Peletminskij. *Methods of Statistical Physics.* Nauka, Moscow, 1977. (russ.), Engl. Transl.: Pergamon Press 1980.

[AS94a] V.M. Axt and A. Stahl. *Z. Phys. B*, **93**:195, 1994.

[AS94b] V.M. Axt and A. Stahl. *Z. Phys. B*, **93**:205, 1994.

[AT87] M.P. Allen and D.J. Tildesley. *Computer Simulation of Liquids.* Clarendon Press, Oxford, 1987.

[Bak88] P. Bakshi. *J. Appl. Phys.*, **64**:2243, 1988.

[Bal60] R. Balescu. *Phys. Fluids*, **3**:52, 1960.

[Bal63] R. Balescu. *Statistical Mechanics of Charged Particles.* Interscience Publishers, 1963.

[Bär69] K. Bärwinkel. Die Vielteilchen–T–Matrix und ihre Anwendung in der Theorie realer Gase von mittlerer Dichte 1,2. *Z. Naturforschung*, **24a**:22 and 38, 1969.

[Bar71] A.A. Barker. *J. Chem. Phys.*, **55**:1751, 1971.

[Bar73] J.R. Barker. *J. Phys. C*, **6**:2663, 1973.

[BB57] S.T. Belyaev and G.I. Budker. *Soviet Phys. - Doklady*, **1**:218, 1957.

[BB66] G.A. Baraff and S.J. Buchsbaum. Surface-wave instability in helicon-wave propagation. *Phys. Rev.*, **144**:266, 1966.

[BBK93] M. Bonitz, R. Binder, and S.W. Koch. Carrier-acoustic plasmon instability in semiconductor quantum wires. *Phys. Rev. Lett.*, **70**:3788, 1993.

[BBK97] M. Bonitz, R. Binder, and H. S. Köhler. Quantum kinetic equations: Correlation dynamics and selfenergy. *Contr. Plasma Phys.*, **37**:101, 1997.

[BBS+93] M. Bonitz, R. Binder, D.C. Scott, S.W. Koch, and D. Kremp. Plasmons and instabilities in quantum plasmas. *Contrib. Plasma Phys.*, **33**:536, 1993.

[BBS+94] M. Bonitz, R. Binder, D.C. Scott, S.W. Koch, and D. Kremp. Theory of plasmons in quasi-one-dimensional degenerate plasmas. *Phys.Rev.E*, 1994.

[BCK90] P. Bakshi, J. Cen, and K. Kempa. Current driven plasma instability in quantum wires. *Solid State Commun.*, **76**:835, 1990.

[BD64] J.D. Bjorken and S. D. Drell. *Relativistic Quantum Mechanics.* McGraw-Hill Book Company, 1964.

[BD72] B. Bezzerides and D.F. DuBois. Quantum electrodynamics of nonthermal relativistic pasmas: Kinetic theory. *Ann. Phys. (N.Y.)*, **70**:10, 1972.

[BD79] D.B. Boercker and J.W. Dufty. Degenerate quantum gases in the binary collision approximation. *Annals of Phys. (N.Y.)*, **119**:43, 1979.

[BD81] D.B. Boercker and J.W. Dufty. Quantum kinetic theory of timecorrelation functions. *Phys. Rev. A*, **23**:1952, 1981.

[BDK98] M. Bonitz, J.W. Dufty, and C.S. Kim. Density operator approach to generalized non-Markovian semiconductor Bloch equations. *phys. stat. sol. (b)*, **206**:181, 1998.

[Be95] L. Banyai et al. Exciton-LO-phonon quantum kinetics: evidence of memory effects in bulk GaAs. *Phys. Rev. Lett.*, **75**:2188, 1995.

[Bea97] M. Borghesi et al. Relativistic channeling of a picosecond laser pulse in a near-critical preformed plasma. *Phys. Rev. Lett.*, **78**:879, 1997.

[Bec81] A. Bechler. Two-point Green's function in quantum electrodynamics at finite temperature and density. *Ann. Phys. (N.Y.)*, **135**:19, 1981.

[BG57] H.A. Bethe and J. Goldstone. *Proc. Roy. Soc. (London)*, **A 238**:551, 1957.

[BG58] K.A. Brueckner and J.H. Gammel. *Phys. Rev.*, **109**:1023, 1958.

[BG67] K. Bärwinkel and S. Grossmann. On the derivation of the Boltzmann–Landau equation from the quantum mechanical hierarchy. *Z. Phys.*, **198**:277, 1967.

[BG68] P. Bakshi and E.P. Gross. Kinetic theory of nonlinear electrical conductivity. *Ann. Phys.*, **49**:513, 1968.

[BGK57] I.B. Bernstein, J.M. Greene, and M.D. Kruskal. Exact nonlinear plasma oscillations. *Phys. Rev.*, **108**:546, 1957.

[BHG98] L. Banyai, H. Haug, and P. Gartner. Self-consistent RPA retarded polaron Green function for quantum kinetics. *Europ. Phys. J. B*, **1**:209, 1998.

[Bin79] K. Binder, editor. *Monte Carlo Methods in Statistical Physics*, Berlin, 1979. Springer.

[BK61] G. Baym and L.P. Kadanoff. Conservation laws and correlation functions. *Phys.Rev.*, **124**:287, 1961.

[BK95] R. Binder and S.W. Koch. Nonequilibrium semiconductor dynamics. *Progress in Quantum Electronics*, **19**:307–462, 1995.

[BK96] M. Bonitz and D. Kremp. Kinetic energy relaxation and correlation time of nonequilibrium many–particle systems. *Phys. Lett. A*, **212**:83, 1996.

[BKB97] R. Binder, S.H. Köhler, and M. Bonitz. Memory effects in the momentum orientation relaxation of electron hole plasmas in semiconductors. *Phys. Rev. B*, **55**:5110, 1997.

[BKDK] M. Bonitz, D. Kremp, J.W. Dufty, and C.S. Kim. Non-Markovian Lenard-Balescu equation. A density operator approach. *to be published.*

[BKK+] M. Bonitz, N. H. Kwong, D. Kremp, R. Binder, and H.S. Köhler. RPA–Kadanoff–Baym results for optically excited semiconductors. to be published.

[BKKS96] Th. Bornath, D. Kremp, W.D. Kraeft, and M. Schlanges. Kinetic equations for a nonideal quantum system. *Phys. Rev. E*, **54**:3274, 1996.

[BKS+] M. Bonitz, D. Kremp, D.C. Scott, R. Binder, W.D. Kraeft, and H.S. Köhler. Memory effects in two–particle collisions. page 185. of Ref. [KSHB95].

[BKS+96] M. Bonitz, D. Kremp, D.C. Scott, R. Binder, W.D. Kraeft, and H.S. Köhler. Numerical analysis of memory effects in the intraband relaxation in semiconductors. *Journal of Physics: Condensed Matter*, **8**:6057, 1996.

[BKSK] M. Bonitz, N. Kwong, D. Semkat, and D. Kremp. Ultrafast relaxation in strongly coupled coulomb systems. in Ref. [Kal98].

[BKW96] V.V. Belyi, Yu. A. Kukharenko, and J. Wallenborn. Pair correlation function and nonlinear kinetic equation for spatially uniform polarizable nonideal plasma. *Phys. Rev. Lett.*, **76**:3554, 1996.

[BM90] W. Botermans and R. Malfliet. Quantum transport theory of nuclear matter. *Phys. Reports*, **198**:115, 1990.

[BN63] J. Bok and P. Nozieres. Instabilities of transverse waves in a drifted plasma. *J. Phys. Chem. Solids*, **24**:709, 1963.

[BNNT93] I.M. Suarez Barnes, M. Nauenberg, M. Nockleby, and S. Tomosovic. *Phys. Rev. Lett.*, **71**:71, 1993.

[Bog46] N.N. Bogolyubov. *Problems of Dynamical Theory in Statistical Physics.* Gostekhisdat, 1946. (russ.).

[Bog61] N.N. Bogolyubov. In G. Uhlenbeck and J.deBoer, editors, *Studies in Statistical Mechanics*, volume **1**. North-Holland, Amsterdam, 1961.

[Bol72] L. Boltzmann. Weitere Studien über das Wärmegleichgewicht unter Gasmolekülen. *Wien. Akad. Sitzungsber.*, **66**:275–370, 1872.

[Bol96] L. Boltzmann. Entgegnung auf die Wärmetheoretischen Betrachtungen des Hrn. Zermelo. *Ann. Phys.*, **57**:773–784, 1896.

[Bon] M. Bonitz. Energy conservation in non-Markovian kinetic equations with selfenergy. *to be published.*

[Bon91] M. Bonitz. *Reaction Diffusion Processes in Nonideal Plasmas and Entropy for Structures in Nonequilibrium.* PhD thesis, Rostock University, Rostock, FRG, 1991. unpublished.

[Bon94] M. Bonitz. Impossibility of plasma instabilities in isotropic quantum plasmas. *Phys. Plasmas*, **1**:832, 1994.

[Bon95] M. Bonitz. Reply to comment on 'Impossibility of plasma instabilities in isotropic quantum plasmas'. *Phys. Plasmas*, **2**:1017, 1995.

[Bon96] M. Bonitz. Correlation time approximation in kinetic theory. *Phys. Lett. A*, **221**:85, 1996.

[BP53] D. Bohm and D. Pines. A collective description of electron interactions: III. Coulomb interactions in a degenerate electron gas. *Phys.Rev.*, **92**:609, 1953.

[BSBK94] M. Bonitz, D.C. Scott, R. Binder, and S.W. Koch. Nonlinear carrier-plasmon interaction in a one-dimensional quantum plasma. *Phys. Rev. B*, **50**:15095, 1994.

[BSHB] M. Bonitz, D. Semkat, H. Haug, and L. Banyai. Improved spectral function for Coulomb quantum kinetics. submitted for publication.

[BSK97] M. Bonitz, D. Semkat, and D. Kremp. Short-time dynamics of correlated many-particle systems: Molecular Dynamics vs. Quantum Kinetics. *Phys. Rev. E*, **56**:1246, 1997.

[BSP+92] R. Binder, D. Scott, A.E. Paul, M. Lindberg, K. Henneberger, and S.W. Koch. Carrier-carrier scattering and optical dephasing in highly excited semiconductors. *Phys. Rev. B*, **45**:1107, 1992.

[BU48] G.E. Beth and E. Uhlenbeck. *Physica*, **4**:916, 1948.

[BW34] G. Breit and J. Wheeler. *Phys. Rev.*, **46**:1087, 1934.

[CAH+96] F.X. Camescasse, A. Alexandrou, D. Hulin, L. Banyai, D.B. Tran Thoai, and H. Haug. Ultrafast electron redistribution through Coulomb scattering in undoped GaAs: Experiment and theory. *Phys. Rev. Lett.*, **77**:5429, 1996.

[Cas59] K.M. Case. Plasma oscillations. *Ann. Phys.*, **7**:349, 1959.

[Cha71] A.V. Chaplik. Possible crystallization of charge carriers in low-density inversion layers. *Zh. Eksp. Teor. Tiz.*, **62**:746, 1971. (Sov. Phys. JETP 35, 395 (1972)).

[Cha90] B. Chakraborty. *Principles of Plasma Mechanics (2nd Ed.)*. John Wiley & Sons, New York, 1990.

[Che87] L. Chen. *Waves and Instabilities in Plasmas.* World Scientific Publishing Co. Pte. Ltd., 1987.

[CKB88] J. Cen, K. Kempa, and P. Bakshi. *Phys. Rev.B*, **38**:10051, 1988.

[CKRU94] B.N. Chichkov, Y. Kato, H. Ruhl, and S. A. Uryupin. Electron distribution function in a thin plasma layer and possible x-ray laser emission due to a sharp temperature gradient. *Phys. Rev. A*, **50**:2691, 1994.

[Col93] J.H. Collet. Screening and exchange in the theory of the femtosecond kinetics of the electron-hole plasma. *Phys. Rev. B*, **47**:10279, 1993.

[CP85] R. Car and M. Parrinello. *Phys. Rev. Lett.*, **55**:3471, 1985.

[CPZ92] J. Clerouin, E.L. Pollock, and G. Zerah. *Phys. Rev. A*, **46**:5130, 1992.

[CSU92] B.N. Chichkov, S.A. Shumsky, and S. A. Uryupin. Nonstationary electron distribution functions in a laser field. *Phys. Rev. A*, **45**:7475, 1992.

[Dan84a] P. Danielewicz. Quantum theory of nonequilibrium processes, I. *Ann. Phys. (N.Y.)*, **152**:239, 1984.

[Dan84b] P. Danielewicz. Quantum theory of nonequilibrium processes. II. Application to nuclear collisions. *Ann. Phys. (N.Y.)*, **152**:305, 1984.

[Dan90] P. Danielewicz. Operator expectation values, self-energies, cutting rules, and higher-order processes in many-body theory. *Ann. Phys. (N.Y.)*, **197**:154, 1990.

[Dav89] R.C. Davidson. Kinetic waves and instabilities in a uniform plasma. In M.N. Rosenbluth and R.Z. Sagdeev, editors, *Basic Plasma Physics: Selected Chapters*. Elsevier Science Publishers B.V., 1989.

[Daw36] B. Dawydov. Über die Geschwindigkeitsverteilung der sich im elektrischen Felde bewegenden Elektronen. *Phys. Z. der Sowjetunion*, **9**:443, 1936.

[DB89] J.W. Dufty and D.B. Boercker. Classical and quantum kinetic equations with exact conservation laws. *J. Stat. Phys.*, **57**:827, 1989.

[DC67] J. Dorfman and W. Cohen. Difficulties in the kinetic theory of dense gases. *J. Math. Phys.*, **8**:282, 1967.

[DDR87] C. Dorso, S. Duarte, and J. Randrup. *Phys. Lett. B*, **188**:287, 1987.

[Den65] J. Denavit. *Phys. Fluids*, **8**:471, 1965.

[DF28] P. Debye and H. Falkenhagen. *Phys. Z.*, **29**:121, 1928.

[DJWC96] P.A. Deymier, G.E. Jabbour, J.D. Weinberg, and F.J. Cherne. Electronic and atomic structure of liquid potassium via path integral molecular dynamics with non–local quantum exchange. *Modelling and Simulation in Materials Science and Engineering*, **4**:137, 1996.

[DK68] I.E. Dzyaloshinskij and E.I. Kats. Superconductivity and quasi-one-dimensional (tread-like) structures. *Zh. Eksp. Teor. Fiz.*, **55**:338, 1968. (Sov. Phys. JETP **28**, 178, 1969).

[DKBB97] J.W. Dufty, Chang Sub Kim, M. Bonitz, and R. Binder. Density matrix methods for semiconductor Coulomb dynamics. *Int. J. Quantum Chemistry*, **56**(5), 1997.

[DL72] C. Deutsch and M. Lavaud. *Phys. Lett. A*, **39**:253, 1972.

[Don92] Dong. *Sol. State Comm.*, **84**:785, 1992.

[DP62] W.E. Drummond and D. Pines. *Nucl. Fusion, Suppl, Pt. 2*, page 1049, 1962.

[DR96] J.C. Diels and W. Rudolph. *Ultrashort Laser Pulse Phenomena.* Academic Press Inc., 1996.

[Dre59] H. Dreicer. Electron and ion runaway in a fully ionized gas. I. *Phys. Rev.*, **115**:238, 1959.

[Dre60a] H. Dreicer. Electron and ion runaway in a fully ionized gas. II. *Phys. Rev.*, **117**:329, 1960.

[Dre60b] H. Dreicer. Electron velocity distribution in a partially ionized gas. I. *Phys. Rev.*, **117**:343, 1960.

[Dru30] Druyvesteyn. Influence of energy loss by elastic collisions in the theory of electron diffusion. *Physica*, **10**:61, 1930.

[DuB67] D.F. DuBois. In W.E. Brittin, editor, *Nonequilibrium quantum statistical mechanics of plasmas and radiation*, volume IX C of *Lectures in Theoretical Physics.* Gordon and Breach, New York, 1967.

[Ebe76] W. Ebeling. Bound state effects in quantum transport theory. *Ann. Physik (Leipzig)*, **33**:350, 1976.

[EC59] H. Ehrenreich and M. Cohen. Consistent field approach to the many-electron problem. *Phys. Rev.*, **115**:786, 1959.

[Eck72] G. Ecker. *The Theory of Fully Ionized Plasmas.* Academic Press, New York and London, 1972.

[Eea84] W. Ebeling et al., editors. *Transport properties of dense plasmas*, Basel/Boston/Stuttgart, 1984. Birkhäuser Verlag.

[EHK67] W. Ebeling, H.J. Hoffmann, and G. Kelbg. *Beitr. Plasmaphys.*, 7:233, 1967.

[EKK76] W. Ebeling, W.D. Kraeft, and D. Kremp. *Theory of Bound States and Ionization Equilibrium in Plasmas and Solid.* Akademie-Verlag, Berlin, 1976.

[EKKR] W. Ebeling, Yu.L. Klimontovich, W.D. Kraeft, and G. Röpke. Kinetic equations and linear response theory for dense Coulomb fluids. in Ref. [Eea84].

[EM97] W. Ebeling and B. Militzer. *Phys. Lett. A*, **226**:298, 1997.

[Fai73] F.H.M. Faisal. Multiple absorption of laser photons by atoms. *J. Phys. B*, **6**:L86, 1973.

[Fal71] H. Falkenhagen. *Theorie der Elektrolyte.* S. Hirzel Verlag, Leipzig, 1971. unter Mitwirkung von W. Ebeling.

[FBS95] H. Feldmaier, K. Bieler, and J. Schnack. *Nucl. Phys. A*, **586**:493, 1995.

[Fel90] H. Feldmaier. *Nucl. Phys. A*, **515**:147, 1990.

[Fer58] R. A. Ferrell. Predicted radiation of plasma oscillations in metal films. *Phys. Rev.*, **111**:1214, 1958.

[Fet73] A.L. Fetter. Electrodynamics of a layered electron gas. I. Single layer. *Ann. Physics*, **81**:367, 1973.

[Fet74] A.L. Fetter. Electrodynamics of a classical electron surface layer. *Phys. Rev. B*, **10**:3739, 1974.

[FH65] R.P. Feynman and A.R. Hibbs. *Quantum Mechanics and Path Integrals.* McGraw-Hill, New York, 1965.

[Fil96] V.S. Filinov. Wigner approach to quantum statistical mechanics and quantum generalization molecular dynamics method. Parts 1,2. *J. Mol. Phys.*, **88**:1517,1529, 1996.

[FJ91] D.K. Ferry and C. Jacoboni, editors. *Quantum Transport in Semiconductors*, New York, 1991. Plenum Press.

[FMK95] V.S. Filinov, Yu.V. Medvedev, and V.L. Kaminskii. *J. Mol. Phys.*, **85**:711, 1995.

[FN90] H. Furukawa and K. Nishihara. *Phys. Rev. A*, **42**:3532, 1990.

[Fra57] W. Franz. *Z. Naturforsch.*, **139**:484, 1957.

[Fuj65] S. Fujita. Thermodynamic evolution equation for a quantum statistical gas. *J. Math. Phys.*, **6**:1877, 1965.

[FW71] A.L. Fetter and J.D. Walecka. *Quantum Theory of Many-Particle Systems*. McGraw-Hill, New York, 1971.

[GA76] C.C. Grimes and G. Adams. Observation of two-dimensional plasmons and electron-ripplon scattering in a sheet of electrons on liquid helium. *Phys. Rev. Lett.*, **36**:145, 1976.

[GD67] H.A. Gould and H.E. DeWitt. *Phys. Rev.*, **155**:68, 1967.

[Ge91] A.R. Goni etal. One-dimensional plasmon dispersion and dispersionless intersubband excitations in GaAs quantum wires. *Phys. Rev. Lett.*, **67**:3298, 1991.

[Gea88] E. Gornik et al. *Sol. State Electron.*, **31**:751, 1988.

[GF67] G. Goldman and E. Frieman. Logarithmic density behavior of a nonequilibrium Boltzmann gas. *J. Math. Phys.*, **8**:1410, 1967.

[GGE69] V.M. Galitski, S.P. Goreslavski, and V.F. Elesin. *Zh. Eksp. Teor. Fiz.*, **57**:207, 1969.

[GH98] H. Güldner and K. Henneberger. Photon kinetics on ultrashort time scales. a principal study. *phys. stat. sol. (b)*, **206**:413, 1998.

[GK92] K.I. Golden and G. Kalman. Phenomenological electrodynamics of two-dimensional Coulomb systems. *Phys. Rev. B*, **45**:5834, 1992.

[GKRT93] T. Gherega, R. Krieg, P.-G. Reinhard, and C. Toepffer. Dynamics of correlations in a solvable model. *Nucl. Phys. A*, **560**:166, 1993.

[GKSB] D. Gericke, S. Kosse, M. Schlanges, and M. Bonitz. T-matrix effects in nonequilibrium semiconductors. submitted for publication.

[GKSB98] D. Gericke, S. Kosse, M. Schlanges, and M. Bonitz. Strong coupling (t-matrix) effects in electron-hole plasmas in semiconductors. *phys. stat. sol. (b)*, **206**:257, 1998.

[Glu71] P. Gluck. Two-time Green's functions and collective effects in a Fermi system. *Nuovo Cimento*, **38**:67, 1971.

[Gol47] I.I. Goldman. *ZhETF*, **17**:681, 1947. (Sov. Phys. JETP).

[GP64] Yu.V. Gulyaev and V.I. Pustovoit. Amplification of surface waves in semiconductors. *Zh. Eksp. Teor. Tiz.*, **47**:2251, 1964. (Sov. Phys. JETP **20**, 1508 (1965)).

[Gre53] H. Green. *The molecular theory of fluids.* Oxford, 1953.

[GRH91] E.K.U. Gross, E. Runge, and O. Heinonen. *Many-Particle Theory.* Adam Hilger, 1991. German edition: B.G. Teubner, Stuttgart.

[GSK96] D. Gericke, M. Schlanges, and W.D. Kraeft. Stopping power of a quantum plasma - T-matrix approximation and dynamical screening. *Phys. Lett. A*, **222**:241, 1996.

[Gur60] A.V. Gurevich. Theory of the electron runaway effect. *ZhETF*, **39**:1296, 1960.

[GWR94] C. Greiner, K. Wagner, and P.-G. Reinhard. Memory effects in relativistic heavy ion collisions. *Phys. Rev. C*, **49**:1693, 1994.

[Hal75] A.G. Hall. Non-equilibrium Green functions: generalized Wicks's theorem and diagrammatic perturbation theory with initial correlations. *J. Phys. A: Math. Gen.*, **8**:214, 1975.

[Han73] J.P. Hansen. *Phys. Rev. A*, **8**:3096, 1973.

[Har28] D.R. Hartree. *Proc. Cambr. Soc.*, **24**:89, 1928.

[Har48] D.R. Hartree. *Repts. Progr. in Phys.*, **11**:113, 1948.

[Har62] M.J. Harrison. Collective excitation of degenerate plasmas in solids. *J. Phys. Chem. Solids*, **23**:1079, 1962.

[Has65] A. Hasegawa. Resistive instabilities in semiconductor plasmas. *J. Phys. Soc. Japan*, **20**:1072, 1965.

[Hay63] J. Hayes. *Il Nuovo Cimento*, **XXX**:1048, 1963.

[HB96] H. Haug and L. Banyai. Improved spectral functions for quantum kinetics. *Solid State Comm.*, **100**:303, 1996.

[HE92] H. Haug and C. Ell. Coulomb quantum kinetics in a dense electron gas. *Phys. Rev. B*, **46**:2126, 1992.

[Hen88] K. Henneberger. *Physica A*, **150**:419, 1988.

[HFPH95] R. Haberlandt, S. Fritzsch, G. Peinel, and K. Heinzinger. *Molekulardynamik. Grundlagen und Anwendungen.* Vieweg & Sohn, Braunschweig, 1995.

[HH88] K. Henneberger and H. Haug. Nonlinear optics and transport in laser-excited semiconductors. *Phys. Rev. B*, **38**:9759, 1988.

[HJ96] H. Haug and A.P. Jauho. *Quantum Kinetics in Transport and Optics of Semiconductors.* Springer-Verlag, Heidelberg, New York, 1996.

[HK93] H.Haug and S.W. Koch. *Quantum Theory of the Optical and Electronic Properties of Semiconductors.* World Scientific Publishing Co. Pte. Ltd., 2nd edition, 1993.

[HM78] J.P. Hansen and I.R. McDonald. *Phys. Rev. Lett.*, **41**:1379, 1978.

[HM81] J.P. Hansen and I.R. McDonald. *Phys. Rev. A*, **23**:2041, 1981.

[HM83a] W. Hänsch and G.D. Mahan. *Phys. Rev. B*, **28**:1886, 1983.

[HM83b] W. Hänsch and G.D. Mahan. *Phys. Rev. B*, **28**:1902, 1983.

[HMV79] J.P. Hansen, I.R. McDonald, and P. Vielillefosse. *Phys. Rev. A*, **20**:2590, 1979.

[HNCM93] D. Hohl, V. Natoli, D.M. Ceperley, and R.M. Martin. *Phys. Rev. Lett.*, **71**:541, 1993.

[HP97] U. Hohenester and W. Pötz. A density matrix approach to non–equilibrium free–carrier screening in semiconductors. *Phys. Rev. B*, **56**:13177, 1997.

[HQ86] P. Hawrylak and J.J. Quinn. Amplification of bulk and surface plasmons in semiconductor superlattices. *Appl. Phys. Lett.*, **49**:280, 1986.

[Iaf] G.J. Iafrate. Quantum transport in solids: The density matrix. page 53. of Ref. [FJ91].

[Ich73] S. Ichimaru. *Basic Principles of Plasma Physics.* Benjamin, London, 1973.

[IK88] G.J. Iafrate and J.B. Krieger. *Solid State Electronics*, **31**:517, 1988.

[Jac75] J.D. Jackson. *Classical Electrodynamics.* John Wiley & Sons, New York, 1975.

[Jah96] Frank Jahnke. *A Many-Body Theory for Laser Emission and Excitonic Effects in Semiconductor Microcavities.* Philipps-Universität Marburg, 1996. (Habilitationsschrift).

[Jau] A.-P. Jauho. Green's function methods: Nonequilibrium, high-field transport. page 141. of Ref. [FJ91].

[Jau83] A.P. Jauho. *J. Phys. F: Met. Phys.*, **13**:L203, 1983.

[JD94] G.E. Jabbour and P.A. Deymier. Discretized quantum path integral molecular dynamics with a non–local pseudopotential: simulation of the 3s and 3p states in the sodium atom. *Modelling and Simulation in Materials Science and Engineering*, **2**:1111, 1994.

[Joa79] C.J. Joachain. *Quantum Collision Theory.* North-Holland, Amsterdam, 1979.

[JW84] A.P. Jauho and J.W. Wilkins. *Phys. Rev. B*, **29**:1919, 1984.

[Kad68] B.B. Kadomtsev. *Plasma Turbulence.* Academic Press, New York, 1968.

[Kal98] G. Kalman, editor. *Physics of Strongly Coupled Coulomb Systems.* Plenum Press, 1998.

[Kam55] N.G. Van Kampen. On the theory of stationary waves in plasmas. *Physica*, **21**:949, 1955.

[Kam57] N.G. Van Kampen. *Physica*, **23**:641, 1957.

[KB60] M.D. Kruskal and I.B. Bernstein. Runaway electrons in an ideal Lorentz plasma. , 1960.

[KB89] L.P. Kadanoff and G. Baym. *Quantum Statistical Mechanics.* Addison-Wesley Publ. Co. Inc., 2nd edition, 1989.

[KBBK] N.H. Kwong, M. Bonitz, R. Binder, and H.S. Köhler. Two–time Kadanoff-Baym results for ultrafast carrier relaxation in semiconductor quantum wells. *Phys. Rev. B.* to be published.

[KBBK98] N.H. Kwong, M. Bonitz, R. Binder, and H.S. Köhler. Semiconductor Kadanoff-Baym equations results for optically excited electron-hole plasmas semiconductor quantum wells. *phys. stat. sol. (b)*, **206**:197, 1998.

[KBBS96] D. Kremp, Th. Bornath, M. Bonitz, and M. Schlanges. Quantum kinetic equations, memory effects, conservation laws. *Physica B*, **228**:72–77, 1996.

[KBCX91] K. Kempa, P. Bakshi, J. Cen, and H. Xie. Spontaneous generation of plasmons by ballistic electrons. *Phys. Rev. B*, **43**:9273, 1991.

[KBKS97] D. Kremp, M. Bonitz, W.D. Kraeft, and M. Schlanges. Non-Markovian Boltzmann equation. *Ann. of Phys. (NY)*, **258**:320, 1997.

[KBSK97] S. Kosse, M. Bonitz, M. Schlanges, and W.D. Kraeft. Evaluation of the quantum Landau collision integral. *Contr. Plasma Phys.*, **37**:499, 1997.

[KBX93] K. Kempa, P. Bakshi, and H. Xie. Current-driven plasma instabilities in solid-state layered systems with a grating. *Phys. Rev. B*, **47**:4532, 1993.

[KC78] M.V. Krasheninnikov and A.V. Chaplik. Plasma-acoustic waves on the surface of a piezoelectric crystal. *Zh. Eksp. Teor. Tiz.*, **75**:1907, 1978. (Sov. Phys. JETP 48, 960 (1978)).

[KE62] Yu.L. Klimontovich and W. Ebeling. Hydrodynamic description of the motion of charged particles in a weakly ionized plasma. *ZhETF*, **43**:145, 1962. [Soviet Physics JETP].

[KE72] Yu.L. Klimontovich and W. Ebeling. Quantum kinetic equations for a nonideal gas and a nonideal plasma. *ZhETF*, **63**:905, 1972. [Soviet Physics JETP **36**, 476 (1973)].

[KEKS83] D. Kremp, W. Ebeling, H. Krienke, and R. Sändig. HNC-type approximations for transport processes in electrolytic solutions. *J. Stat. Phys.*, **33**:99, 1983.

[Kel57] L.V. Keldysh. *Sov. Phys. JETP*, **7**:788, 1957.

[Kel63] G. Kelbg. *Ann. Physik (Leipzig)*, 12:219, 1963.

[Kel64a] L.V. Keldysh. Diagram technique for nonequilibrium processes. *ZhETF*, **47**:1515, 1964. [Soviet Phys. JETP, **20**, 1018 (1965)].

[Kel64b] L.V. Keldysh. Ionization in the field of a strong electromagnetic wave. *ZhETF*, **47**:1945, 1964.

[Kel95] L.V. Keldysh. Correlations in the coherent transient electron-hole system. *phys. stat. sol. (b)*, **188**:11, 1995.

[KF69] K.L. Kliewer and R. Fuchs. Lindhard dielectric function with a finite electron lifetime. *Phys. Rev.*, **181**:552, 1969.

[KG94] G. Kalman and K.I. Golden. *Correlated Dynamics of Layered Systems.* Nova Science Publishers, Inc., 1994.

[KH95] J. Kohanoff and J.P. Hansen. *Phys. Rev. Lett.*, **74**:626, 1995.

[KI86] J.B. Krieger and G.J. Iafrate. *Phys. Rev. B*, **33**:5494, 1986.

[KI87] J.B. Krieger and G.J. Iafrate. *Phys. Rev. B*, **35**:9644, 1987.

[KK68] G.A. Korn and T.M. Korn. *Mathematical Handbook for scientists and enginieers.* McGraw-Hill Book Company, 1968.

[KK74] Yu.L. Klimontovich and W.D. Kraeft. *Teplofiz. Vys. Temp. (russ.)*, **12**:239, 1974.

[KK81] Yu.L. Klimontovich and D. Kremp. *Physica A*, **109**:517, 1981.

[KK95] N.H. Kwong and H. S. Köhler. Separable *nn* potentials from inverse scattering for nuclear matter studies. *Phys. Rev. C*, **55**:1650, 1995.

[KKE71] D. Kremp, W.D. Kraeft, and W. Ebeling. *Physica A*, **51**:146, 1971.

[KKER86] W.-D. Kraeft, D. Kremp, W. Ebeling, and G. Röpke. *Quantum Statistics of Charged Particle Systems.* Akademie-Verlag, Berlin, 1986.

[KKK+84] D. Kremp, M.K. Kilimann, W.D. Kraeft, H. Stolz, and R. Zimmermann. *Physica A*, **127**:646, 1984.

[KKK87] Yu.L. Klimontovich, D. Kremp, and W.D. Kraeft. Kinetic theory for chemically reacting gases and partially ionized plasmas. *Advances of Chemical Physics*, **58**:175, 1987.

[KKS98] D. Kremp, W.D. Kraeft, and M. Schlanges. *Quantum Statistics of Strongly Coupled Plasmas.* Springer, 1998.

[KKY] H. S. Köhler, N.H. Kwong, and H. Yousif. in preparation.

[KL57] W. Kohn and J.M. Luttinger. *Phys. Rev.*, **108**:590, 1957.

[Kli57] Yu.L. Klimontovich. On the method of second quantization in phase space. *ZhETF*, **33**:982, 1957. [Soviet Physics JETP].

[Kli59] Yu. L. Klimontovich. *ZhETF*, **36**:1405, 1959. [Sov. Physics JETP **36**, 999 (1959)].

[Kli60a] Yu. L. Klimontovich. *Soviet Phys. - JETP*, **10**:524, 1960.

[Kli60b] Yu. L. Klimontovich. *Soviet Phys. - JETP*, **11**:876, 1960.

[Kli75] Yu.L. Klimontovich. *Kinetic Theory of Nonideal Gases and Nonideal Plasmas*. Nauka, Moscow, 1975. (russ.), Engl. transl.: Pergamon Press, Oxford 1982.

[KM93] H.S. Köhler and Rudi Malfliet. Extended quasiparticle approximation and Brueckner theory. *Phys. Rev. C*, **48**:1034, 1993.

[KMRT] D. Klakow, H. Matuszok, P.-G. Reinhard, and C. Toepffer. Wave packet molecular dynamics simulations of matter under extreme conditions. page 37. of Ref. [KSHB95].

[KMSR93] D. Kremp, K. Morawetz, M. Schlanges, and V. Rietz. Impact ionization in nonideal plasmas in a strong electric field. *Phys. Rev. E*, **47**:635, 1993.

[Kog62] Sh. M. Kogan. On the theory of hot electrons in semiconductors. *Fiz. Tverdovo Tela*, **4**:2474, 1962. [Soviet Physics - Solid State **4**, 1813, (1963)].

[Köh85] H.S. Köhler. Microscopic calculation of pre-equilibrium emission. *Nuclear Physics A*, **438**:564, 1985.

[Köh95] H.S. Köhler. Memory and correlation effects in nuclear collisions. *Phys. Rev. C*, **51**:3232, 1995.

[Köh96] H. S. Köhler. Memory and correlation effects in the quantum theory of thermalization. *Phys. Rev. E*, **53**:3145, 1996.

[Kor66] V. Korenman. *Ann. Phys. (N.Y.)*, **39**:72, 1966.

[KP74] Yu.L. Klimontovich and V.A. Puchkov. Influence of plasma polarization on kinetic properties in the presence of a strong electric field. *ZhETF*, **67**:556, 1974.

[KRG94] G. Kalman, Y. Ren, and K.I. Golden. Determination of the energy gap in the acoustic excitation of a superlattice. *Phys. Rev. B*, **50**:2031, 1994.

[KS52a] Yu. L. Klimontovich and V. P. Silin. *Dokl. Akad. Nauk*, **82**:361, 1952. [Sov. Physics Doklady].

[KS52b] Yu. L. Klimontovich and V. P. Silin. *ZhETF*, **23**:151, 1952. [Sov. Physics JETP].

[KS60] Yu. L. Klimontovich and V. P. Silin. *Uspekhi Fiz. Nauk*, **70**:247, 1960. [Sov. Phys. - Uspekhi **3**, 84 (1969)].

[KSB] D. Kremp, D. Semkat, and M. Bonitz. Initial correlations in generalized Kadanoff–Baym equations. to be published.

[KSB85] D. Kremp, M. Schlanges, and Th. Bornath. Nonequilibrium Real Time Green's Functions and the Condition of Weakening of Initial Correlation. *J. Stat. Phys.*, **41**:661, 1985.

[KSB86] D. Kremp, M. Schlanges, and Th. Bornath. The method of Green's functions in statistical mechanics of nonequilibrium systems. In H. Stolz, editor, *Proceedings of the ZIE-School on Kinetic Equations and Nonlinear Optics in Semiconductors*, number 86-3, page 33, Berlin, 1986.

[KSHB95] W.D. Kraeft, M. Schlanges, H. Haberland, and Th. Bornath, editors. *Physics of Strongly Coupled Plasmas*, Singapore, 1995. World Scientific Publ.

[KTR94] D. Klakow, C. Toepffer, and P.-G. Reinhard. *Phys. Lett. A*, **192**:55, 1994.

[Kub57] R. Kubo. *J. Phys. Soc. Japan*, **12**:570, 1957.

[Kuz91] A.V. Kuznetsov. *Phys. Rev. B*, **44**:8721, 1991.

[KW80] C.L. Kirschbaum and L. Wilets. *Phys. Rev. A*, **21**:834, 1980.

[KY75] E.A. Kaner and V.M. Yakovenko. Hydrodynamic instability in solid-state plasma. *Usp. Fiz. Nauk*, **115**:41, 1975. [Sov. Phys. Usp. **18**, 21 (1975)].

[Lan28] I. Langmuir. *Proc. Nat. Acad. Sci.*, **14**:627, 1928.

[Lan46] L.D. Landau. On the vibrations of the electron plasma. *J.Phys. (U.S.S.R.)*, **10**:25, 1946.

[LD92] C.Y. Lee and P.A. Deymier. Electron–phonon coupling in $Ba_{0.6}K_{0.4}Bi_3$ by discretized quantum path integral molecular dynamics. *Physica C*, **190**:299, 1992.

[Len60] A. Lenard. On Bogoluibov's kinetic equation for a spatially homogeneous plasma. *Ann. Phys.*, **10**:390, 1960.

[Lev69] I.B. Levinson. *Zh.Eksp.Teor.Fiz.*, **30**:660, 1969. [Sov. Phys. JETP **30**, 362 (1970)].

[LHBK94] M. Lindberg, Y.Z. Hu, R. Binder, and S.W. Koch. $\chi^{(3)}$ formalism in optically excited semiconductors and its applications in four-wave-mixing spectroscopy. *Phys. Rev. B*, **50**:18060, 1994.

[Lin54] J. Lindhard. *Kgl. Danske Videnskab. Selskab. Mat. Fys. Medd.*, **28**:8, 1954.

[LK88] M. Lindberg and S.W. Koch. Effective bloch equations for semiconductors. *Phys. Rev. B*, **38**:3342, 1988.

[LL62] L.D. Landau and E.M. Lifshitz. *Course of Theoretical Physics: Quantum Mechanics*, volume 3. Pergamon, Oxford, 1962.

[LL80a] L.D. Landau and E.M. Lifshitz. *Course of Theoretical Physics: Kinetics*, volume 10. Pergamon, Oxford, 1980.

[LL80b] L.D. Landau and E.M. Lifshitz. *Course of Theoretical Physics: Relativistic Quantum Mechanics*, volume 4. Akademie-Verlag, Berlin, 1980. German ed.

[LS91] Q.P. Li and S. Das Sarma. Elementary excitation spectrum of one-dimensional electron system in confined semiconductor structures: zero magnetic field. *Phys. Rev. B*, **43**:11768, 1991.

[LvV86] P. Lipavský, V. Špička, and B. Velický. Generalized Kadanoff-Baym ansatz for deriving quantum transport equations. *Phys. Rev. B*, **34**:6933, 1986.

[Mah81] G.D. Mahan. *Many Particle Physics.* Plenum Press, New York, 1981.

[Mat76] R.D. Mattuck. *A Guide to Feynman Diagrams in the Many-Body Problem.* McGraw-Hill, New York, 1976.

[McL89a] J.A. McLennan. Boltzmann equation for a dissociating gas. *J. Stat. Phys.*, **57**:887, 1989.

[McL89b] J.A. McLennan. *Introduction to Nonequilibrium Statistical Mechanics.* Prentice Hall, Eaglewood Cliffs, N.J., 1989.

[Mer70] N.D. Mermin. Lindhard dielectric function in the relaxation-time approximation. *Phys. Rev. B*, **1**:2362, 1970.

[MH83] G.D. Mahan and W. Hänsch. *J. Phys. F*, **13**:L47, 1983.

[MHH+95] G. Manzke, K. Henneberger, J. Heeg, K. El Sayed, S. Schuster, and H. Haug. Dynamics of screening and field fluctuations on ultrashort time scales. *phys. stat. sol. b*, **188**:395, 1995.

[Mik75] A.B. Mikhailovski. *Theory of Plasma Instabilities.* Atomizdat, Moscow, 1975.

[MKR+89] F. Morales, M.K. Kilimann, R. Redmer, M. Schlanges, and F. Bialas. Dynamical screening and the dc conductivity in a fully ionized plasma. *Cont. Plasma Phys.*, **29**:425, 1989.

[Mor95] K. Morawetz. The Landau equation including memory and energy conservation. *Phys. Lett. A*, **199**:241, 1995.

[MR94] K. Morawetz and G. Röpke. Memory effects and virial correction in nonequilibrium dense systems. *Phys. Rev. E*, **51**:4246, 1994.

[MRR+53] N. Metropolis, A.W. Rosenbluth, M.N. Rosenbluth, A.H. Teller, and E. Teller. *J. Chem. Phys.*, **21**:1087, 1953.

[MS59] P. Martin and J. Schwinger. Theory of many-particle systems. I. *Phys.Rev.*, **115**:1342, 1959.

[MSK93] K. Morawetz, M. Schlanges, and D. Kremp. Nonlinear conductivity and composition of partially ionized plasmas in a strong electric field. *Phys. Rev. E*, **48**:2980, 1993.

[MvL] K. Morawetz, V. Špička, and P. Lipavský. Formation of binary correlations in plasmas. submitted for publication.

[NBK+98] Ch. Nacke, W. Bathe, F. Kieseling, M. Seemann, H. Stolz, Ch. Heyn, S. Bargstädt-Franke, W. Hauser, and D. Heitmann. Phase sensitive resonant ultrafast reflection from GaAs quantum wells. *phys. stat. sol. (b)*, **206**:307, 1998.

[New82] R.G. Newton. *Scattering Theory of Waves and Particles.* Springer-Verlag, 1982. 2nd Ed.

[NS68] G.E. Norman and A.N. Starostin. *Teplofiz. Vys. Temp. (Soviet Phys. - High Temp.)*, 6:410, 1968.

[NV79] G.E. Norman and A.A. Valuev. *Plasma Phys.*, **21**:531, 1979.

[Nyq32] H. Nyquist. Regeneration theory. *Bell Systems Tech. J.*, **11**:126, 1932.

[OSE97] J. Ortner, F. Schautz, and W. Ebeling. Quasiclassical molecular-dynamics simulations of the electron gas: Dynamic properties. *Phys. Rev. E*, **56**(4), 1997.

[PBCM94] C. Pierleoni, B. Bernu, D.M. Ceperley, and W.R. Magro. Equation of state of the hydrogen plasma by path integral Monte Carlo simulations. *Phys. Rev. Lett.*, **73**:2145, 1994.

[PCZ] J.I. Penman, J. Clerouin, and G. Zerah. The equation of state of hydrogen plasma by orbital free molecular dynamics. page 78. of Ref. [KSHB95].

[Pei55] R.E. Peierls. *Quantum Theory of Solids.* Oxford University Press, 1955.

[Pen60] O. Penrose. Electrostatic instabilities of a uniform non-Maxwellian plasma. *Phys. Fluids*, **3**:258, 1960.

[PG96] S. Pfalzner and P. Gibbon. *Many-body Tree Methods in Physics.* Cambridge University Press, 1996.

[PKM93] N. Peyghambarian, S.W. Koch, and A. Myserowicz. *Introduction to semiconductor optics.* Prentice Hall Inc., Eaglewood Cliffs, NJ, USA, 1993.

[PM94] M.D. Perry and G. Mourou. Terawatt to Petawatt class subpicosecond laser. *Science*, **264**:917, 1994.

[Poz81] J. Pozhela. *Plasma and Current Instabilities in Semiconductors.* Pergamon Press, Oxford, 1981.

[Pri63] I. Prigogine. *Non-Equilibrium Statistical Mechanics.* Interscience Publ., New York, 1963.

[PS61] D. Pines and J.R. Schrieffer. Collective behavior in solid-state plasmas. *Phys.Rev.*, **124**:1387, 1961.

[PS62] D. Pines and J. R. Schrieffer. Approach to equilibrium of electrons, plasmons, and phonons in quantum and classical plasmas. *Physical Review*, **125**:804, 1962.

[PtV96] A. Pukhov and J. Meyer ter Vehn. Relativistic magnetic self-channeling of light in near-critical plasma: Three-dimensional particle-in-cell simulation. *Phys. Rev. Lett.*, **76**:3975, 1996.

[PTVF92] W.H. Press, S.A. Teukolsky, W.T. Vetterling, and B. P. Flannery. *Numerical Recipes.* Cambridge University Press, Cambridge, 2nd edition, 1992.

[Puc75] V.A. Puchkov. Polarization effects in the plasma kinetic equation under the influence of a high frequency electric field. *Vestnik MGU*, **16**:385, 1975.

[Puf61] R. D. Puff. *Ann. Phys.*, **13**:317, 1961.

[Pus69] V.I. Pustovoit. Interaction of electron streams with elastic lattice waves. *Usp. Fiz. Nauk*, **97**:257, 1969. [Sov. Phys. Uspekhi **12**, 105 (1969)].

[Rei80] H.R. Reiss. *Phys. Rev. A*, **22**:1786, 1980.

[Res65] P. Resibois. *Physica*, **31**:645, 1965.

[RHK94] F. Rossi, S. Haas, and T. Kuhn. Ultrafast relaxation of photoexcited carriers: The role of coherence in the generation process. *Phys. Rev. Lett.*, **72**:152, 1994.

[Rie97] V. Rietz. *Besetzungskinetik nichtidealer Wasserstoffplasmen im starken elektrischen Feld.* PhD thesis, Rostock University, Rostock, FRG, 1997. (unpublished).

[Rit57] R.H. Ritchie. Plasma losses by fast electrons in thin films. *Phys. Rev.*, **106**:874, 1957.

[RJPE93] D. Richards, B. Jusserand, H. Peric, and B. Etienne. Intrasubband excitations and spin-splitting anisotropy in GaAs modulation-doped quantum wells. *Phys. Rev. B*, **47**:16028, 1993.

[Röp87] G. Röpke. *Statistische Mechanik für das Nichtgleichgewicht.* VEB Deutscher Verlag der Wissenschaften, Berlin, 1987.

[RS67] B.B. Robinson and G.A. Schwartz. Two-stream instability in semiconductor plasmas. *J. Appl. Phys.*, **38**:2461, 1967.

[RT90] U. Reimann and C. Toepffer. *Laser and Particle Beams*, **8**:763 and 771, 1990.

[RT93] P.-G. Reinhard and C. Toepffer. Correlations in nuclei and nuclear dynamics. *Int. J. Mod. Phys. E*, **3**:435, 1993.

[RT94] P.-G. Reinhard and C. Toepffer. Correlations in nuclei and nuclear dynamics. *Int. J. Mod. Phys. E*, **3**:435, 1994.

[Sae65] A.W. Saenz. *J. Math. Phys.*, **6**:859, 1965.

[SATL77] Jr. S.J. Allen, D.C. Tsui, and R.A. Logan. Observation of the two-dimensional plasmon in Silicon inversion layers. *Phys. Rev. Lett.*, **38**:980, 1977.

[SB87] M. Schlanges and Th. Bornath. Quantum kinetic equations for systems with Coulomb interaction. Generalized Balescu-Lenard equation. *Wiss. Zeitschr. Univ. Rostock*, Heft 1:65, 1987.

[SBBK94] D.C. Scott, R. Binder, M. Bonitz, and S.W. Koch. Multiple undamped acoustic plasmons in three–dimensional two–component nonequilibrium plasmas. *Phys. Rev. B*, **49**:2174, 1994.

[SBH94] K. El Sayed, L. Banyai, and H. Haug. Coulomb quantum kinetics and optical dephasing on the femtosecond timescale. *Phys. Rev. B*, **50**:1541, 1994.

[SBK] D. Semkat, M. Bonitz, and D. Kremp. Ultrafast kinetic energy relaxation in dense plasmas. to be published.

[SBK88] M. Schlanges, Th. Bornath, and D. Kremp. *Phys. Rev. A*, **38**:2174, 1988.

[SBK92] D.C. Scott, R. Binder, and S.W. Koch. Ultrafast dephasing through acoustic plasmon undamping in nonequilibrium electron-hole plasmas. *Phys. Rev. Lett*, **69**:347, 1992.

[SBRK96] M. Schlanges, Th. Bornath, V. Rietz, and D. Kremp. Atomic level population for nonideal plasmas in strong electric fields. *Phys. Rev. E*, **53**:2751, 1996.

[SBT95] M. Schlanges, M. Bonitz, and A. Tschttschjan. Plasma phase transition in fluid hydrogen–helium mixtures. *Contrib. Plasma Phys.*, **35**:109, 1995.

[Sch61] J. Schwinger. *J. Math. Phys.*, **2**:407, 1961.

[Sch96] W. Schäfer. Influence of electron-electron scattering on femtosecond four-wave-mixing in semiconductors. *JOSA B*, **13**:1291, 1996.

[Sco93] D.C. Scott. *Carrier Relaxation and Collective Phenomena in Nonequilibrium Semiconductor Electron-Hole Plasmas.* PhD thesis, University of Arizona, 1993. unpublished.

[Sem96] D. Semkat. Anfangskorrelationen und Memoryeffekte in der kinetischen Theorie. Master's thesis, Rostock University, Rostock, FRG, 1996. (Diplom) unpublished.

[Sha89] J. Shah. *Solid State Electronics*, **32**:1051, 1989.

[Sil52] V.P. Silin. *ZhETF*, **23**:641, 1952. [Sov. Phys. JETP].

[Sil62] V.P. Silin. *Soviet Phys. - JETP*, **13**:1244, 1962.

[Sil67] V.P. Silin. *Kinetic Equations for a gas of charged particles.* MIR, Moscow, 1967. Supplement to the russian ed. of [Bal63], in Russ.

[Sil73] V.P. Silin. *Parametric effect of high-power radiation on plasmas.* Nauka, Moscow, 1973. (in Russian).

[SKM94] J. Schilp, T. Kuhn, and G. Mahler. Electron-phonon quantum kinetics in pulse-excited semiconductors: memory and renormalization effects. *Phys.Rev. B*, **50**:5435, 1994.

[SKM95] J. Schilp, T. Kuhn, and G. Mahler. Quantum kinetics of the coupled carrier-phonon system in photoexcited semiconductors. *phys. stat. sol. (b)*, **188**:417, 1995.

[SM85] D. Strickland and G. Mourou. *Opt. Commun*, **56**:219, 1985.

[SR69] G.A. Swartz and B.B. Robinson. Coherent microwave instabilities in a thin-layer solid-state plasma. *J. Appl. Phys.*, **40**:4598, 1969.

[SRT90] K. Schmitt, P.-G. Reinhard, and C. Toepffer. Truncation of time-dependent many-body theories. *Z. Phys. A*, **336**:123, 1990.

[SSH+94] K. El Sayed, S. Schuster, H. Haug, F. Herzel, and K. Henneberger. Subpicosecond plasmon response: Buildup of screening. *Phys. Rev. B*, **49**:7337, 1994.

[SSTH98] R. Schepe, T. Schmielau, D. Tamme, and K. Henneberger. Damping and t-matrix in dense e-h plasmas. *phys. stat. sol. b*, **206**:273, 1998.

[ST86] W. Schäfer and J. Treusch. An approach to the nonequilibrium theory of highly excited semiconductors. *Z. Phys. B*, **63**:407, 1986.

[Ste67] F. Stern. Polarizability of a two-dimensional electron gas. *Phys. Rev. Lett.*, **18**:546, 1967.

[Sto74] H. Stolz. *Einführung in die Vielelektronentheorie der Kristalle.* Akademieverlag, Berlin, 1974.

[Tay72] J.R. Taylor. *The Quantum Theory of Nonrelativistic Collisions.* John Wiley & Sons, 2nd edition, 1972.

[Tea80] D.C. Tsui et al. Far infrared emission from plasma oscillations of Si inversion layers. *Sol. State Comm.*, **35**:875, 1980.

[Tea94] M. Tabak et al. Ignition and high gain with ultrapowerful lasers. *Phys. Plasmas*, **1**:1626, 1994.

[TH93] D.B. Tran Thoai and H. Haug. Coulomb quantum kinetics in pulse-excited semiconductors. *Z. Phys. B*, **91**:199, 1993.

[THWS96] W. Theobald, R. Hässner, C. Wülker, and R. Sauerbrey. Temporally resolved measurement of electron densities ($> 10^{23} cm^{-3}$) with high harmonics. *Phys. Rev. Lett.*, **77**:298, 1996.

[TK93] R. Trebino and D.J. Kane. Single-shot measurement of the intensity and phqase of an arbitrary ultrashort pulse by using frequency resolved optical gating. *J. Opt. Soc. Am.*, **A 10**:1101, 1993.

[TKS78] T.N. Theis, J.P. Kotthaus, and P.J. Stiles. Wavevector dependence of the two-dimensional plasmon dispersion relationship in the (100) silicon inversion layer. *Sol. State Comm.*, **26**:603, 1978.

[TL29] L. Tonks and I. Langmuir. Oscillations in ionized gases. *Phys. Rev.*, **33**:195, 1929.

[Tom50] S. Tomonaga. *Prog. Theor. Phys.*, **5**:544, 1950.

[vdL94] D. von der Linde. Materie in extrem intensiven Laserfeldern. *Naturwissenschaften*, **81**:329, 1994. in German.

[vL94] V. Špička and P. Lipavský. Quasiparticle Boltzmann equation in semiconductors. *Phys. Rev. Lett.*, **73**:3439, 1994.

[Vla38] A.A. Vlasov. The vibrational properties of an electron gas. *Zh. Eksp. Teor. Fiz.*, **8**:291, 1938.

[Vla45] A.A. Vlasov. *J. Phys. (U.S.S.R.)*, **9**:25, 1945.

[vN55] J. v. Neumann. *Mathematical Foundations of Quantum Mechanics.* Princeton University Press, Princeton, N.J., 1955.

[Vol35] D.M. Volkov. *Z. Physik*, **94**:125, 1935.

[VVS62] A. Vedenov, E. Velikhov, and R. Sagdeev. *Nucl. Fusion, Suppl., Pt. 2*, page 465, 1962.

[WB74] P.F. Williams and Aaron N. Bloch. Selfconsistent dielectric response of a quasi-one-dimensional metal at high frequencies. *Phys. Rev. B*, **10**:1097, 1974.

[WC85] S.-J. Wang and W. Cassing. Explicit treatment of N-body correlations within a density matrix formalism. *Ann. Phys. (N.Y.)*, **159**:328, 1985.

[WD69] R.H. Williams and H.E. DeWitt. *Phys. Fluids*, **12**:2326, 1969.

[Wea93] C. Wirner et al. Direct observation of hot electron distribution function in GaAs/GaAlAs heterostructures. *Phys. Rev. Lett.*, **70**:2609, 1993.

[Wig32] E. Wigner. On the quantum correction for thermodynamic equilibrium. *Phys. Rev.*, **40**:749, 1932.

[XKB92] H. Xie, K. Kempa, and P. Bakshi. Growth rates of current-excited plasma waves in semiconductor layered systems. *J. Appl. Phys.*, **72**:4767, 1992.

[ZCP92] G. Zerah, J. Clerouin, and E.L. Pollock. *Phys. Rev. Lett.*, **69**:446, 1992.

[Zer96] E. Zermelo. Über einen Satz der Dynamik und die mechanische Wärmetheorie. *Ann. Phys.*, **57**:485–494, 1896.

[Zim87] R. Zimmermann. *Many-Particle Theory of Highly Excited Semiconductors.* Teubner, Leipzig, 1987.

[Zim92] R. Zimmermann. Carrier kinetics for ultrafast optical pulses. *J. Lumin.*, **53**:187, 1992.

[ZMR96] D.N. Zubarev, V. Morozov, and G. Röpke. *Statistical Mechanics of Nonequilibrium Processes.* Akademieverlag, Berlin, 1996.

[ZTR] G. Zwicknagel, C. Toepffer, and P.-G. Reinhard. Molecular dynamics simulations of strongly coupled plasmas. page 45. of Ref. [KSHB95].

[Zub71] D.N. Zubarev. *Nonequilibrium Statistical Thermodynamics.* Nauka, Moscow, 1971. ((russ.), Engl. transl.: Plenum Press 1974).

[Zwa63] R. Zwanzig. *Phys. Rev.*, **129**:486, 1963.

[Zwi94] G. Zwicknagel. *Energieverlust schwerer Ionen in stark gekoppelten Plasmen.* PhD thesis, University Erlangen, 1994. unpublished.

Index